EYE MOVEMENT RESEARCH
Mechanisms, Processes and Applications

STUDIES IN VISUAL INFORMATION PROCESSING 6

Series Editors:

Rudolf Groner
Institute for Psychology
University of Bern
Switzerland

Géry d'Ydewalle
Department of Psychology
University of Leuven
Belgium

ELSEVIER
AMSTERDAM • LAUSANNE • NEW YORK • OXFORD • SHANNON • TOKYO

EYE MOVEMENT RESEARCH
Mechanisms, Processes and Applications

Edited by

John M. FINDLAY
Robin WALKER
Robert W. KENTRIDGE
Department of Psychology
University of Durham
England

1995

ELSEVIER
AMSTERDAM • LAUSANNE • NEW YORK • OXFORD • SHANNON • TOKYO

NORTH-HOLLAND
ELSEVIER SCIENCE B.V.
Sara Burgerhartstraat 25
P.O. Box 211, 1000 AE Amsterdam, The Netherlands

ISBN: 0 444 81473 6

PRINTED IN THE NETHERLANDS

INTRODUCTION

What might jumping spiders, Japanese readers and basketball players have in common ? All feature amongst the contents of this book because all have been the subject of investigation by oculomotor scientists. The 1990s have been designated as the 'Decade of the Brain' and the brain, amongst many other accomplishments, controls the movements of the eyes. Explaining the nature of this control is the quest of the eye movement researcher. As with other areas of brain function, answers can be given at a series of hierarchical levels. Much of the excitement of the area comes about because of the expectation that the overlap of these levels will bridge the gulf between brain workers and students of mental processes. The chapters in this book are all derived from papers given at the Seventh European Conference on Eye Movements, held in Durham, England in September 1993. A feature of the ECEM conferences has always been to provide a forum for participants from all levels of the hierarchy.

The chapters in the book include the four invited lectures of the conference and selected contributions from over one hundred papers on various topic areas which were presented. Selection was based on recommendations by the scientific committee (R Abadi, G Barnes, G d'Ydewalle, J Findlay, A Gale, R Groner, K O'Regan, G Lüer and D Zambarbieri). It is deeply regretted that many excellent pieces of work could not be included.

One of the drawbacks of proceedings based on a conference is the uneven distribution of topics. Work on oculomotor physiology is under-represented because few papers were submitted to the meeting. Work on binocular eye movements is almost absent for the same reason. In contrast, topics relating to saccade control and attention, saccade control in reading are particularly well represented.

We are most grateful to the authors for the quality of their papers, for meeting the deadlines set and for attempting to follow the labyrinthine sets of instructions for preparing camera-ready copy of a uniform format. A number of individuals assisted in various ways in the preparation of the material and we would like to acknowledge the assistance of Paula Beuster, Val Brown, Clare and Catherine Findlay in particular.

Durham, September 1994

John Findlay
Robin Walker
Bob Kentridge

CONTENTS

DISPLAYS AND APPLICATIONS

CONTRIBUTORS

ABADI, R V — Department of Optometry and Vision Sciences, UMIST, P.O. Box 88, Manchester, M60 1QD, England

ACCARDO, A P — Dipartimento di Elettrotecnica e Informatica, University of Trieste, Via Valerio 10, Trieste I-34100, Italy
Email: *accardo@gnbts.univ.trieste.it*

BARNES, G R — MRC Human Movement and Balance Unit, The Institute of Neurology, Queen Square, London WC1N 3BG, England
Email: *skgtgrb@ucl.ac.uk*

BEAUVILLAIN, C — Laboratoire de Psychologie Expérimentale, Université René Descartes, CNRS URA 316, 28 rue Serpente, Paris 75006, France
Email: *labexp@frmop22.cnusc.fr*

BECKER, W — Sektion Neurophysiologie, Universität Ulm, Albert-Einstein Allee 11, D-89081 Ulm, Germany
Email: *wolfgang.becker@medizin.uni-ulm.de*

BISCALDI, M — Department of Neurophysiology, University of Freiburg, Hansastrasse 9, D-79104 Freiburg, Germany
Email: *biscaldi@sun1.ruf.uni-freiburg.de*

BOTTI, F — Neurologische-Universitätsklinik, Albert-Ludwigs Universität, Hansastrasse 9, D 79104 Freiburg, Germany

BOUQUEREL, A — Laboratoire de Neurobiologie Humaine, Université de Provence, U R A CNRS 372, Marseille, France

BOUR, L J — Department of Clinical Neurophysiology, Academic Medical Centre H2-214, Graduate School Neurosciences, Meibergdreef 9, 1105 AZ Amsterdam, The Netherlands
Email: *bour@amc.uva.nl*

BRANS, J — Department of Clinical Neurophysiology, Academic Medical Centre H2-214, Graduate School Neurosciences, Meibergdreef 9, 1105 AZ Amsterdam, The Netherlands

BREMMER, F — Department of Zoology, Ruhr University Bochum, D–44780 Bochum , Germany
Email: *bremmer@neuroinformatik.ruhr-uni-bochum.de*

BUCCI, M P — Laboratoire de Physiologie de la Perception et de l'Action, 15, rue de L'École de Médecine, 75270 Paris Cedex 06, France

CRAWFORD, T J — Academic Unit of Neuroscience, Charing Cross and Westminster Medical School, Fulham Palace Road, London W6 8RF, England

DA POZZO, S — Department of Ophthalmology, Children's Hospital of Trieste, Trieste, Italy

DE GRAEF, P — Laboratory of Experimental Psychology, Katholieke Universiteit Leuven, Tiensestraat 102, B-3000 Leuven , Belgium

DE TROY, A — Laboratory of Experimental Psychology, Katholieke Universiteit Leuven, Tiensestraat 102, B-3000 Leuven , Belgium
Email: *andreas%psl%psy@cc3.kuleuven.ac.be*

DEFFNER, G — Digital Imaging Venture Projects, Texas Instruments, P.O. Box 655474, MS 429, Dallas, Texas 75265, U.S.A.

DEMARIA, J L — Laboratoire de Neurobiologie Humaine, Université de Provence, U R A CNRS 372, Marseille, France

DEUBEL, H — Max-Planck Institut für Psychologische Forschung, Leopoldstrasse 24, D-80802 München, Germany
Email: *deubel@mpipf-muenchen.mpg.d400.de*

DORÉ, K — Laboratoire de Psychologie Expérimentale, Université René Descartes, CNRS URA 316, 28 rue Serpente, Paris 75006, France
Email: *labexp@frmop22.cnusc.fr*

D'YDEWALLE, G
Laboratory of Experimental Psychology, Katholieke
Universiteit Leuven, Tiensestraat 102, B-3000 Leuven , Belgium
Email: gery=dYdewalle%psl%psy@cc3.kuleuven.ac.be

EGGERT, T
Laboratoire de Physiologie de la Perception et de L'Action,
15, rue de L'École de Médecine, 75270 Paris Cedex 06, France

ERKELENS, C J
Utrecht Biophysics Research Unit, University of Utrecht,
Princetonplein 5, 3584 CC Utrecht, The Netherlands
Email: erkelens@fys.ruu.nl

FINDLAY, J M
Department of Psychology, University of Durham, South Road,
Durham, DH1 3LE, England
Email: j.m.findlay@durham.ac.uk

FISCHER, B
Department of Neurophysiology, University of Freiburg,
Hansastrasse 9, D-79104 Freiburg, Germany
Email: aiple@sun1.ruf.uni-freiburg.dbp.de

FRENCK-MESTRE, C
CREPCO, Université de Provence - CNRS, 29 Avenue
Robert Schuman, 13621 Aix-en-Provence Cedex, France
Email: crepco@frmop11.cnusc.fr

GAMLIN, P D R
Department of Physiological Optics, University of Alabama at
Birmingham, Birmingham, Alabama 35294, U.S.A.

GAUTHIER, G M
Laboratoire de Contrôles Sensorimoteurs, Université de Provence,
Centre Scientifique de St-Jérôme, avenue Escadrille Normandie-
Niemen, F-13397 Marseille Cedex 20, France
Email: labocsm@frmrs11.bitnet

GAYMARD, B
Laboratoire INSERM 289, Hôpital de la Salpêtrière, 47 Boulevard
de l'Hôpital, 75653 Paris cédex 13, France

GOLDBERG, J H
Department of Industrial Engineering, Pennsylvania State
University, 207 Hammond Building, University Park,
PA 16802-1401, U.S.A.
Email: jhgie@engr.psu.edu

GREALY, M A
MRC Human Movement and Balance Unit, The Institute of
Neurology, Queen Square, London WC1N 3BG, England

HAEGAR, B
Academic Unit of Neuroscience, Charing Cross and Westminster
Medical School, Fulham Palace Road, London W6 8RF, England

HALLETT, P E
Department of Physiology, University of Toronto, Toronto,
Ontario, Canada M5S 1A8
Email: peter@biovision.utoronto.ca

HARRIS, C M
Department of Ophthalmology, The Hospitals for Sick
Children, Great Ormond Street, London WC1N 3JH, England

HELLER, D
Institut für Psychologie, Technical University of
Aachen, Jägerstrasse 17/19, D-52056 Aachen , Germany
Email: ds0701u@dacth11

HEMFORTH, B
Center for Cognitive Science, Institute for Computer Science and
Social Research, Albert-Ludwigs University, Friedrichstrasse 50,
D-79098 Freiburg, Germany

HENDERSON, L
Academic Unit of Neuroscience, Charing Cross and Westminster
Medical School, Fulham Palace Road, London W6 8RF, England

HODGSON, T L
Department of Psychology, Birkbeck College, University of
London, Malet Street, London WC1E 7HX, England
Email: ubjtp78@ccs.bbk.uk.ac

HOFFMANN, K-P
Department of Zoology, Ruhr University Bochum,
D–44780 Bochum , Germany
Email: hoffman@rjba.rz.ruhr-uni-bochum.de

HOFMEISTER, J
Institut für Psychologie, Technical University of
Aachen, Jägerstrasse 17/19, D-52056 Aachen , Germany

HONDA, H
Department of Psychology, Niigata University, Ikarashi,
Niigata 950-21, Japan
Email: psyhonda@cc.niigata-u.ac.jp

HYÖNÄ, J   Department of Psychology, University of Turku, SF-20500 Turku, Finland
Email: hyona@sara.utu.fi

INCHINGOLO, P   Dipartimento di Elettrotecnica e Informatica, University of Trieste, Via Valerio 10, Trieste I-34100, Italy
Email: inchin@gnbts.univ.trieste.it

KALESNYKAS, R   Department of Physiology, University of Toronto, Toronto, Ontario, Canada M5S 1A8

KAPOULA, Z   Laboratoire de Physiologie de la Perception et de l'Action, 15, rue de L'École de Médecine, 75270 Paris Cedex 06, France
Email: zk@ccr.jussieu.fr

KENNARD, C   Academic Unit of Neuroscience, Charing Cross and Westminster Medical School, Fulham Palace Road, London W6 8RF, England

KENTRIDGE, R W   Department of Psychology, University of Durham, South Road, Durham, DH1 3LE, England
Email: Robert.Kentridge@durham.ac.uk

KONIECZNY, L   Center for Cognitive Science, Institute for Computer Science and Social Research, Albert-Ludwigs University, Friedrichstrasse 50, D-79098 Freiburg, Germany
Email: lars@cognition.iig.uni-freiburg.de

KRAPPMANN, P   Brain Research Institute NW2, University of Bremen, P O B 330440, D-28334 Bremen, Germany

KRUMMENACHER, J   Institut für Psychologie, Technical University of Aachen, Jägerstrasse 17/19, D-52056 Aachen, Germany
Email: jkrummen@roger.psycho.rwth-aachen.de

LAINE, M   Department of Psychology, University of Turku, SF-20500 Turku, Finland
Email: matlaine@sara.cc.utu.fi

LAND, M F   Sussex Centre for Neuroscience, School of Biological Sciences, University of Sussex, Brighton BN1 9QG, England
Email: bafj9@sussex.ac.uk

LAWDEN, M C   Academic Unit of Neuroscience, Charing Cross and Westminster Medical School, Fulham Palace Road, London W6 8RF, England

LEKWUWA, G U   MRC Human Movement and Balance Unit, The Institute of Neurology, Queen Square, London WC1N 3BG, England
Email: skgtgrb@ucl.ac.uk

MANNAN, S   Department of Physics (Biophysics), Blackett Laboratory, Imperial College, Prince Consort Road, London SW7 2BZ, England

MAYS, L E   Department of Physiological Optics, University of Alabama at Birmingham, Birmingham, Alabama 35294, U.S.A.
Email: optp014@earn.uabdpo

MERGNER, Th   Neurologische-Universitätsklinik, Albert-Ludwigs Universität, Hansastrasse 9, D 79104 Freiburg, Germany
Email: mergner@sun1.ruf.uni-freiburg.de

MILES, F   Laboratory of Sensorimotor Research, National Eye Institute, Bethesda, MD 20892, U.S.A.
Email: fam@lsr.nei.nih.gov

MÜLLER, H J   Department of Psychology, Birkbeck College, University of London, Malet Street, London WC1E 7HX, England
Email: ubjta52@uk.ac.bbk.cu

MUNRO, N A R   Department of Neurology, The Middlesex Hospital, Mortimer Street, London W1N 8AA, England
Email: regtnam@ucl.ac.uk

NIEMI, J   Department of General Linguistics, University of Joensuu, Finland

ONKEN, R — Fakultät für Luft- und Raumfahrttechnik, Universität der Bundeswehr München, Werner-Heisenberg-Weg 39, D-85377 Neubiberg, Germany

O'REGAN, J K — Groupe Regard, Laboratoire de Psychologie Expérimentale, 28 rue Serpente, F-75006 Paris, France
Email: *oregan@moka.ccr.jussieu.fr*

OSAKA, M — Faculty of Letters, Department of Psychology, Kyoto University, Kyoto 606, Japan
Email: *b52046@tansei.cc.u-tokyo.ac.jp*

OSAKA, N — Department of Psychology, Osaka University of Foreign Studies, Mino, Osaka 562, Japan

PASCAL, E — Department of Vision Sciences, Glasgow Caledonian University, Glasgow, G 4 0BA, Scotland

PENSIERO, S — Department of Ophthalmology, Children's Hospital of Trieste, Trieste, Italy

PERISUTTI, P — Department of Ophthalmology, Children's Hospital of Trieste, Trieste, Italy
Email: *perisutti@gnbts.univ.trieste.it*

PIERROT-DESEILLIGNY, Ch — Laboratoire INSERM 289, Hôpital de la Salpêtrière, 47 Boulevard de l'Hôpital, 75653 Paris cédex 13, France

PYNTE, J — CREPCO, Université de Provence - CNRS, 29 Avenue Robert Schuman, 13621 Aix-en-Provence Cedex, France
Email: *crepco@frmop11.cnusc.fr*

QUACCIA, D — Laboratoire de Contrôles Sensorimoteurs, Université de Provence, Centre Scientifique de St-Jérôme, avenue Escadrille Normandie-Niemen, F-13397 Marseille Cedex 20, France
Email: *labocsm@frmrs11.bitnet*

RADACH, R — Institut für Psychologie, Technical University of Aachen, Jägerstrasse 17/19, D-52056 Aachen, Germany
Email: *raradach@roger.psycho.rwth-aachen.de*

RAYNER, K — Department of Psychology, University of Massachusetts, Tobin Hall, Amherst, MA 01003, U.S.A.
Email: *rayner@psych.umass.edu*

REINGOLD, E M — Department of Psychology, Erindale Campus, University of Toronto, 3359 Mississauga Road, Mississauga, Ontario, Canada L5L 1C6
Email: *reingold@psych.toronto.edu*

RIVAUD, S — Laboratoire INSERM 289, Hôpital de la Salpêtrière, 47 Boulevard de l'Hôpital, 75653 Paris cédex 13, France

ROLL, J P — Laboratoire de Neurobiologie Humaine, Université de Provence, U R A CNRS 372, Marseille, France

ROLL, R — Laboratoire de Neurobiologie Humaine, Université de Provence, U R A CNRS 372, Marseille, France

RUDDOCK, K H — Department of Physics (Biophysics), Blackett Laboratory, Imperial College, Prince Consort Road, London SW7 2BZ, England

SCHEEPERS, C — Center for Cognitive Science, Institute for Computer Science and Social Research, Albert-Ludwigs University, Friedrichstrasse 50, D-79098 Freiburg, Germany

SCHNEIDER, W X — Max-Planck Institut für Psycholgische Forschung, Leopoldstrasse 24, D-80802 München, Germany
Email: *wxs@mpipf-muenchen.mpg.dbp.de*

SCHRYVER, J C — Cognitive Systems and Human Factors Group, Engineering Physics and Mathematics Division, Oak Ridge National Laboratory, Oak Ridge, TN 37831-6360, U.S.A.

SCHULTE, A — Fakultät für Luft- und Raumfahrttechnik, Universität der Bundeswehr München, Werner-Heisenberg-Weg 39, D-85377 Neubiberg, Germany
Email: *ld2brbeh@rz.unibw-muenchen.de*

SCHWEIGART, G   Neurologische-Universitätsklinik, Albert-Ludwigs
Universität, Hansastrasse 9, D 79104 Freiburg, Germany

SHEINBERG, D   Department of Cognitive and Linguistic Sciences, Box 1978,
Brown University, Providence, Rhode Island 02912, U.S.A.

STAMPE, D M   Department of Psychology, Erindale Campus, University of Toronto,
3359 Mississauga Road, Mississauga, Ontario, Canada L5L 1C6
Email: dstampe@psych.toronto.edu

STEIN, J F   University Laboratory of Physiology, Parks Road, Oxford, England

STRUBE, G   Center for Cognitive Science, Institute for Computer Science and
Social Research, Albert-Ludwigs University, Friedrichstrasse 50,
D-79098 Freiburg, Germany

STUHR, V   Department of Neurophysiology, University of Freiburg,
Hansastrasse 9, D-79104 Freiburg, Germany
Email: aiple@sun1.ruf.uni-freiburg.dbp.de

TSUJI, H   Educational Center for Information Processing, Kyoto University,
Kyoto 606, Japan

VAN DIEPEN, P M J   Laboratory of Experimental Psychology, Katholieke
Universiteit Leuven, Tiensestraat 102, B-3000 Leuven , Belgium
Email: paul%psl%psy@cc3.kuleuven.ac.be

VAN RENSBERGEN, J   Laboratory of Experimental Psychology, Katholieke
Universiteit Leuven, Tiensestraat 102, B-3000 Leuven , Belgium
Email: johan%sys%psy@cc3.kuleuven.ac.be

VAN'T ENT, D   Department of Clinical Neurophysiology, Academic Medical
Centre H2-214, Graduate School Neurosciences, Meibergdreef 9,
1105 AZ Amsterdam, The Netherlands

VELAY, J L   Laboratoire de Neurobiologie Humaine, Université de
Provence, U R A CNRS 372, Marseille, France
Email: lnh@frmop22.bitnet

VERCHER, J-L   Laboratoire de Contrôles Sensorimoteurs, Université de Provence,
Centre Scientifique de St-Jérôme, avenue Escadrille Normandie-
Niemen, F-13397 Marseille Cedex 20, France
Email: labocsm@frmrs11.bitnet

VERFAILLIE, K   Laboratory of Experimental Psychology, University of
Leuven, Tiensestraat 102, B-3000 Leuven, Belgium.
Email: Karl.Verfaillie@ psy.kuleuven.be

VICKERS, J N   Neuro-Motor Psychology Laboratory, The University of
Calgary, Calgary, Alberta, Canada T2N 1N4
Email: vickers@acs.ucalgary.ca

VITU, F   Groupe Regard, Laboratoire de Psychologie Expérimentale, 28 rue
Serpente, F-75006 Paris, France
Email: vitu@frmop22.cnusc.fr

VOGELS, I M L C   Utrecht Biophysics Research Unit, University of Utrecht,
Princetonplein 5, 3584 CC Utrecht, The Netherlands

WALKER, R   Department of Psychology, University of Durham, South Road,
Durham, DH1 3LE, England
Email: Robin.Walker@durham.ac.uk

WEBER, H   Department of Neurophysiology, University of Freiburg,
Hansastrasse 9, D-79104, Freiburg, Germany
Email: aiple@sun1.ruf.uni-freiburg.dbp.de

WRIGHT, J R   Department of Physics (Biophysics), Blackett Laboratory,
Imperial College, Prince Consort Road, London SW7 2BZ, England

ZAMBARBIERI, D   Dipartimento di Informatica e Sistemistica, University of Pavia,
Via Abbiategrasso 209, I-27100 Pavia, Italy
Email: daniela@bioing2.univpv.it

ZELINKSY, G J   Department of Cognitive and Linguistic Sciences, Box 1978,
Brown University, Providence, Rhode Island 02912, U.S.A.
Email: gjz@clouseau.cog.brown.edu

INVITED LECTURES

EYE MOVEMENTS AND COGNITIVE PROCESSES IN READING, VISUAL SEARCH, AND SCENE PERCEPTION

Keith Rayner

Department of Psychology, University of Massachusetts, Amherst, MA 01003 USA

Abstract

Recent research dealing with five issues concerning eye movements in reading is discussed. A summary is provided of research dealing with (1) the span of effective vision, (2) integration of information across eye movements, (3) where to fixate next, (4) when to move the eyes, and (5) models of eye movement control. It is argued that since word frequency effects are found indpendent of landing position in a word that a model which allows lexical variables to influence when to move the eyes should be favored over one which does not. Finally, research on eye movements during scene perception and visual search is also reviewed. It is argued that the basic mechanisms of eye movement control are similar across tasks (reading, scene perception, and visual search), but that the trigger to move the eyes differs as a function of the specific task.

Keywords

eye movement control, reading, perceptual span, integration of information across saccades, visual search, scene perception, fixation duration, saccade length

Introduction

During the past twenty years, a dichotomy has developed with respect to research on eye movements and reading. Namely, two different types of research have been undertaken. One type of research has focused on the characteristics of eye movements per se during reading, with the primary topic of interest being eye movement control. The second type of research has utilized eye movement data to infer something about the process of reading. The characteristic of this latter type of work is that the researcher seems to be only incidentally interested in eye movements per se, and primarily interested in what eye movements can tell us about reading. One interesting aspect to this dichotomy is that researchers in each of these groups don't often seem to have much tolerance for the other group. As I just noted, those interested primarily in using eye movements to study reading seem to have little interest in the details of eye movements (such as where the eye lands, where it came from, saccade latency, etc), whereas those interested in eye movements per se seem to be constantly worried that effects that those interested in language processing obtain are somehow artifacts associated with eye movements. My own belief, and part of what I want to argue in this chapter is that advances are being made by both types of research group and that workers in each group should pay more attention to the other group.

To a large extent, the preceding paragraph is a caricature, and like most caricatures, it undoubtedly oversimplifies the true state of affairs. There are clearly a number of

Eye Movement Research/J.M. Findlay et al. (Editors)

people interested in both aspects of the dichotomy I proposed. Indeed, in my own laboratory, we have engaged in both types of research. One interesting sociological phenomenon that I have sensed is that those at the extreme ends of the dichotomy typically perceive our work as being part of the other group. Thus, even though we have done a considerable amount of research on where the eyes land in words, saccade latency, and eye movement control, some researchers primarily interested in the oculomotor aspects of reading typically seem to think of us as working primarily on language processing, while those interested primarily in language processing often seem to think that we work primarily on oculomotor aspects of reading. As I hope to document in this chapter, considerable advances have been made on both fronts. My goal is to review some recent findings regarding eye movements in reading. I will primarily describe findings where there seems to be some agreement, but will not invariably do this. I will also note some instances where controversy exists. Near the end of the chapter, I will come back to the issue I began with (the dichotomy of research interests) and present some relevant data to discriminate between different types of models of eye movement control in reading. Finally, although the title of the chapter lists scene perception and visual search, I will focus for the most part on research on reading, but discuss those topics at the end of the chapter.

I will begin by discussing 5 issues: (1) the span of effective vision, (2) integration of information across eye movements, (3) eye movement control: where to fixate, (4) eye movement control: when to move, and (5) models of eye movement control. For the first four issues, I will primarily review prior research and make what seem to me to be appropriate conclusions. Since, to my mind, the most controversy exists concerning the last two issues, I will discuss some recent preliminary results that we have obtained which bear on the controversy. Prior to discussing the five issues, I will first provide a brief overview of some basic facts about eye movements and reading (see Rayner & Pollatsek, 1987 for more details).

Basic Facts about Eye Movements and Reading

During reading, we make a series of eye movements (referred to as *saccades*) separated by periods of time when the eyes are relatively still (referred to as *fixations*). The typical saccade is about eight to nine letter spaces; this value is not affected by the size of the print as long as it is not too small or too large (Morrison & Rayner, 1981). The appropriate metric to use when discussing eye movements therefore is letter spaces, and not visual angle (generally, 3-4 letter spaces is equivalent to 1 degree of visual angle). Because of the high velocity of the saccade, no useful information is acquired while the eyes are moving; readers only acquire information from the text during fixations (Wolverton & Zola, 1983). The average fixation duration is 200-250 ms. The other primary characteristic of eye movements is that about 10-15% of the time readers move their eyes back in the text (referred to as *regressions*) to look at material that has already been read.

Eye movements during reading are necessary because of the acuity limitations in the visual system. A line of text extending around the fixation point can be divided into three regions: foveal, parafoveal, and peripheral. In the foveal region (extending 1 degree of visual angle to the left and right of fixation), acuity is sharpest and the letters can be easily resolved. In the parafoveal region (extending to 5 degrees of visual angle on either side of fixation) and the peripheral region (everything on the line beyond the parafoveal region), acuity drops off markedly so that our ability to identify letters is not very good even in the near parafovea. The purpose of eye movements in reading is therefore to place the foveal region on that part of the text to be processed next.

It is important to note that as text difficulty increases, fixation durations increase, saccade lengths decrease, and regression frequency increases. More importantly, the values presented above for fixation duration, saccade length, and regression frequency are averages and there is considerable variability in all of the measures. Thus, although the average fixation duration might be 250 ms and the average saccade length might be 8 letter spaces for a given reader, for others these values might be somewhat higher or lower. This between reader variability (which also exists for regression frequency) is perhaps not as important as the fact that there is considerable within reader variability. In other words, although a reader's average fixation duration is 250 ms, the range can be from under 100 ms to over 500 ms within a passage of text. Likewise, the variability in saccade length can range from 1 letter space to over 15 letter spaces (though such long saccades typically follow regressions). The reasons for this variability will hopefully become clearer later in this chapter.

The Span of Effective Vision

How much information does a reader acquire on each fixation?

This basic question has inspired a great deal of research in my laboratory and other labs. In order to investigate this question, George McConkie and I developed what has become known as the *eye-contingent display change* paradigm. Around the same time that we developed the technique, Steve Reder and Kevin O'Regan were also working on developing the technique. In this paradigm, a reader's eye movements are monitored (generally every millisecond) by a highly accurate eye-tracking system. The eyetracker is interfaced with a computer which controls the display monitor from which the subject reads. The monitor has a rapidly decaying phosphor and changes in the text are made contingent on the location of the reader's eyes. Generally, the display changes are made during saccades and the reader is not consciously aware of the changes.

There are three primary types of eye-contingent paradigms: *the moving window, foveal mask,* and *boundary* techniques. With the moving window technique (McConkie & Rayner, 1975), on each fixation a portion of the text around the reader's fixation is available to the reader. However, outside of this window area, the text is replaced by other letters, or by *x*s (see Figure 1). When the reader moves his or her eyes, the window moves with the eyes. Thus, wherever the reader looks, there is readable text within the window and altered text outside the window. The rationale of the technique is that when the window is as large as the region from which a reader can normally obtain information, reading will not differ from when there is no window present. The foveal mask technique (Rayner & Bertera, 1979) is very similar to the moving window paradigm except that the text and replaced letters are reversed. Thus, wherever the reader looks, the letters around the fixation are replaced by *x*s while outside of the mask area the text remains normal (see Figure 1). Finally, in the boundary technique (Rayner, 1975), an invisible boundary location is specified in the text and when the reader's eye movement crosses the boundary, an originally displayed word or letter string is replaced by a target word (see Figure 1). The amount of time that the reader looks at the target word is computed both as a function of the relationship between the initially displayed stimulus and the target word and as a function of the distance that the reader was from the target word prior to launching a saccade that crossed the boundary.

The major findings from experiments dealing with the size of the span of effective vision (also referred to as the *perceptual span*) are as follows:

```
      move  smoothly  across  the  page  of  text          Normal
                      *
      ────────────────────────────────────────
      XXXX  XXXothly  across  the  XXXX  XX  XXXX
                      *                                Moving  Window

      XXXX  XXXXXXXX  XXXXss  the  page  of  teXX
                                      *
      ────────────────────────────────────────
      move  smoothlXXXXXXXs  the  page  of  text
                      *                                Foveal  Mask

      move  smoothly  acrosXXXXXXXage  of  text
                              *
      ────────────────────────────────────────
      move  smoothly  across  the  date  of  text
                *               *                      Boundary

      move  smoothly  across  the  page  of  text
                                      *
```

Figure 1. Examples of the moving window, foveal mask, and boundary paradigms. The first line shows a normal line of text with the fixation location marked by an asterisk. The next two lines show an example of two successive fixations with a window of 17 letter spaces and the other letters replaced with Xs (and spaces between words preserved). The next two lines show an example of two successive fixations with a 7 letter foveal mask. The bottom two lines show an example of the boundary paradigm. The first line shows a line of text prior to a display change with fixation locations marked by asterisks. When the reader's eye movement crosses an invisible boundary (the letter e in the), an initially displayed word (date) is replaced by the target word (page). The change occurs during the saccade so that the reader does not see the change.

1. The span extends 14-15 character spaces to the right of fixation (DenBuurman, Boersema, & Gerrisen, 1981; McConkie & Rayner, 1975; Rayner, 1986; Rayner & Bertera, 1979; Rayner, Inhoff, Morrison, Slowiaczek, & Bertera, 1981; Rayner, Well, Pollatsek, & Bertera, 1982).
2. The span is asymmetric - to the left of fixation, it extends to the beginning of the currently fixated word, or 3-4 letter spaces (McConkie & Rayner, 1976; Rayner, Well, & Pollatsek, 1980). For readers of languages printed from right-to-left (such as Hebrew), the span is asymmetric but in the opposite direction from English so that it is larger left of fixation than right (Pollatsek, Bolozky, Well, & Rayner, 1981).
3. No useful information is acquired below the line of text (Inhoff & Briihl, 1991; Inhoff & Topolski, 1992; Pollatsek, Raney, LaGasse, & Rayner, 1993).
4. The *word identification span* (or area from which words can be identified on a given fixation) is smaller than the total span of effective vision (Rayner et al, 1982; Underwood & McConkie, 1985). The word identification span generally does not exceed 7-8 letter spaces to the right of fixation.
5. The size of the span of effective vision and the word identification span is not fixed, but can be modulated by word length. For example, if three short words occur in succession, the reader may be able to identify all of them. If the upcoming word is constrained by the context, readers acquire more information from that word (Balota, Pollatsek, & Rayner, 1985) and if the fixated word is difficult to process, readers obtain less information from the upcoming word (Henderson & Ferreira, 1990; Inhoff, Pollatsek, Posner, & Rayner, 1989; Rayner, 1986).
6. Orthography influences the size of the span. Specifically, experiments with Hebrew readers (Pollatsek et al., 1982) suggest that their span is smaller than that of English readers and experiments with Japanese readers (Ikeda & Saida, 1978; Osaka, 1992) suggests that their span is even smaller. Hebrew is a more densely packed language than English and Japanese is more densely packed than Hebrew; densely packed refers to the fact that it takes more characters per sentence in English than Hebrew, for example.
7. Reading skill influences the size of the span. Beginning readers (at the end of second grade) have a smaller span than skilled readers (Rayner, 1986) and adult dyslexic readers have smaller spans than skilled readers (Rayner, Murphy, Henderson, & Pollatsek, 1989). On the other hand, Underwood and Zola (1986) found no differences between good and poor fifth grade readers. One reason for the difference may be that there was less of a gap in ability levels in the Underwood and Zola study than in the other studies.

Integration of Information Across Saccades

What kind of information is integrated across saccades in reading? Experiments using both the moving window and the boundary technique have demonstrated a *preview* benefit from the word to the right of fixation; information obtained about the parafoveal word on fixation *n* is combined with information on fixation *n+1* to speed identification of the word when it is subsequently fixated (Blanchard, Pollatsek, & Rayner, 1989).

A number of experiments using the boundary paradigm have varied the orthographic, phonological, morphological, and semantic similarity between an initially displayed stimulus and a target word in attempts to determine the basis of the preview effect. The major findings from these studies are as follows:

1. There is facilitation due to orthographic similarity (Balota et al., 1985; Rayner, 1975; Rayner, Ehrlich, & McConkie, 1978; Rayner, Zola, & McConkie, 1980; Rayner et al., 1982) so that *chest* facilitates the processing of *chart*. However, the facilitation is not strictly due to visual similarity because changing the case of letters from fixation to fixation (so that *ChArT* becomes *cHaRt* on the next) has little effect on reading behavior (O'Regan & Lévy -Schoen, 1983; McConkie & Zola, 1979; Rayner et al., 1980).

2. The facilitation is in part due to abstract letter codes associated with the first few letters of an unidentified parafoveal word (Rayner et al., 1980, 1982), though there may be some facilitation from other parts of the word to the right of fixation besides the beginning letters (see Inhoff & Tousman, 1990). However, the bulk of the preview effect is due to the beginning letters. Inhoff's research shows that the effect is not simply due to spatial proximity because there is facilitation from the beginning letters of words when readers are asked to read sentences from right to left, but with letters within words printed from left to right (Inhoff et al., 1989).

3. There is facilitation due to phonological similarity, so that *beech* facilitates *beach* and *shoot* facilitates *chute*, with less facilitation in the latter than in the former case (Pollatsek, Lesch, Morris, & Rayner, 1992).

4. Although morphological factors can influence fixation time on a word, they are not the source of the preview benefit (Inhoff, 1987, 1989; Lima, 1987; Lima & Inhoff, 1985).

5. There is no facilitation due to semantic similarity: *song* as the initial stimulus does not facilitate the processing of *tune*, even though such words yield semantic priming effects under typical priming conditions (Rayner, Balota, & Pollatsek, 1986).

Eye Movement Control: Where to Fixate Next

There are two components to the issue of how eye movements are controlled during reading: (1) where to fixate next and (2) when to move the eyes. It appears that there are separate mechanisms involved in these decisions (Rayner & McConkie, 1976; Rayner & Pollatsek, 1981), and they will accordingly be discussed separately. The primary findings concerning where to fixate next are as follows:

1. Word length seems to be the primary determinant of where to fixate next. When word length information about the upcoming word is not available, readers move their eyes a much shorter distance than when such information is available

(McConkie & Rayner, 1975; Morris, Rayner, & Pollatsek, 1990; Pollatsek & Rayner, 1982; Rayner & Bertera, 1979). Also, the length of the word to the right of fixation strongly influences the size of the saccade (O'Regan, 1979, 1980; Rayner, 1979).

2. There is a *landing position* effect such that readers tend to fixate about halfway between the beginning and the middle of words (Dunn-Rankin, 1978; McConkie, Kerr, Reddix, & Zola, 1988; O'Regan, 1981; O'Regan, Lévy -Schoen, Pynte, & Brugaillere, 1984; Rayner, 1979). Extensive research efforts on this effect have examined the consequences of being fixated at locations other than this optimal viewing location (McConkie, Kerr, Reddix, Zola, & Jacobs, 1989; Vitu, 1991; Vitu, O'Regan, & Mittau, 1990) and it has been found that the consequences are more serious when words are presented in isolation than when they are in text. This result suggests either that contextual information overrides low-level visual-processing constraints or that readers are somewhat flexible about where they can acquire information around fixation.

3. There is a **launch site** effect such that where readers land in a word is strongly influenced by where the saccade came from (McConkie et al., 1988). Thus, whereas the most frequent landing position may be near the middle of the word, if the prior saccade was launched some distance (8-10 letters) from the target word then the landing position will be shifted to the left of center. Likewise, if the prior saccade was launched close (2-3 characters) to the beginning of the target word, the landing position will be shifted to the right of center.

4. Given the two preceding findings, the optimal strategy would be to fixate near the middle of each successive word. However, since short words can often be identified when they are to the right of the currently fixated word, they are often skipped. Factors such as this result in the landing position distribution being spread somewhat due to the launch site effect.

5. Although it has been suggested that semantic information within an as yet-unfixated parafoveal word can influence the landing position in that word (Hyönä, Niemi, & Underwood, 1989; Underwood, Clews, & Everatt, 1990), Rayner and Morris (1992) were unable to replicate the effect (see also Hyönä, 1993). At this point, it seems that there is no semantic preprocessing effect.

Eye Movement Control: When to Move

What determines when we move our eyes? This issue is at the heart of the controversy I described at the beginning of this chapter. Language researchers are convinced that the amount of time a reader fixates on a word or segment of text reveals something about the cognitive processes associated with comprehending that word or segment. The oculomotor researchers don't seem so sure about this. Some relevant findings are as follows:

1. During reading, information gets into the processing system very early in a fixation (thus leaving a lot of time for processes associated with word recognition and

other necessary processes). Experiments using the foveal mask paradigm in which the onset of the mask is delayed following a saccade have demonstrated that if the reader has 50 ms to process the text prior to the onset of the mask then reading proceeds quite normally (Ishida & Ikeda, 1989; Rayner, Inhoff, Morrison, Slowiaczek, & Bertera, 1981). If the mask occurs earlier, reading is disrupted. Although readers may typically acquire the visual information needed for reading during the first 50 ms of a fixation, they can extract information at other times during a fixation as needed (Blanchard, McConkie, Zola, & Wolverton, 1984).

2. There have been many demonstrations that various lexical, syntactic, and discourse factors influence fixation time on a word. In particular, there are demonstrations that the following variables influence fixation time: (1) word frequency (Just & Carpenter, 1980; Henderson & Ferreira, 1993; Inhoff & Rayner, 1986; Raney & Rayner, 1994; Rayner, 1977; Rayner & Duffy, 1986; Rayner, Sereno, Morris, Schmauder, & Clifton, 1989; Schmauder, 1991); (2) contextual constraint (Balota et al., 1985; Ehrlich & Rayner, 1981; Schustack, Ehrlich, & Rayner, 1987; Zola, 1984); (3) semantic relationships between words in a sentence (Carroll & Slowiazek, 1986; Morris, 1994; Sereno & Rayner, 1992; (4) anaphora and co-reference (Duffy & Rayner, 1990; Ehrlich & Rayner, 1983; Garrod, O'Brien, Morris, & Rayner, 1990; O'Brien, Shank, Myers, & Rayner, 1988; (5) lexical ambiguity (Dopkins, Morris, & Rayner, 1992; Duffy, Morris, & Rayner, 1988; Rayner & Duffy, 1986; Rayner & Frazier, 1989; Rayner, Pacht, & Duffy, 1994; Sereno, Pacht, & Rayner, 1992); and (6) syntactic disambiguation (Clifton, 1993; Ferreira & Clifton, 1986; Ferreira & Henderson, 1990; Frazier & Rayner, 1982; Rayner, Carlson, & Frazier, 1983; Rayner & Frazier, 1989; Rayner, Garrod, & Perfetti, 1992).

The above types of effects are now legion. One would therefore think that this should be enough to convince those on the oculomotor edge of the continuum that I outlined at the beginning of this chapter that eye fixations can be used to study moment-to-moment comprehension and language processing effects. But, apparently not since these researchers continue to argue that such effects may really be due to oculomotor factors rather than to underlying language processing effects. What exactly is their argument? To deal with this question, let's move to the next topic.

Models of Eye Movement Control

There are now quite a few proposals for how eye movements are controlled in reading. For the sake of simplicity, I'd like to lump the various proposals into two general categories: (1) those that allow for lexical variables, or something like lexical variables, where on-going comprehension processes are controlling the movement of the eyes versus (2) those that maintain that the movement of the eyes is not directly related to on-going language processing but is primarily due to oculomotor factors. In the first category would be proposals such as Morrison's (1984) model and the various modifications of it (Henderson & Ferreira, 1990, Pollatsek & Rayner, 1990), as well as models such as that of Just and Carpenter (1980). In the second category would be proposals such as O'Regan's (1990, 1992; O'Regan & Lévy -Schoen, 1987) strategy-tactics model, as well as proposals of Kowler and Anton (1987) and McConkie et al. (1988, 1989). In order to better clarify each type of model, let me provide an overview of the Morrison and the O'Regan model.

Morrison's model. According to Morrison's model, each fixation begins with visual attention focused on the word currently centered at the fovea. After processing of the foveal word has reached a criterion level (maybe lexical access), attention shifts to the parafoveal word to the right of the foveal word. The shift of attention gates processing of the word at the newly attended location and signals the eye movement system to prepare a program to move the eyes. The motor program is executed once it is completed, and the eyes then follow attention to the new word. Because there is a lag between the shift of attention and the movement of the eyes due to programming latency, information is acquired from the parafoveal word before it is fixated. Attention will sometimes shift again to the word beyond the parafoveal word if the parafoveal word is relatively easy to identify. In these cases, the eye movement program will be changed to send the eyes two words to the right and the parafoveal word will be skipped. Usually there is some cost to do this and the duration of fixation prior to a skip is inflated (see Hogaboam, 1983, and Pollatsek, Rayner, & Balota, 1986, for evidence that the fixation prior to a skip is inflated). The cost is because the reader had to cancel the motor program for the saccade that was started to go to the parafoveal word. If the reader is too far into the motor program to cancel the saccade, there will either be (1) a short fixation on word n+1 followed by a saccade to word n+2 or (2) a saccade which lands at an intermediate position. The model can explain a number of facts about eye movement behavior in reading, including some that puzzled reading researchers for a long time (such as (1) why there are very short fixations in text given that saccade latencies in simple oculomotor tasks are typically on the order of 175-200 ms and (2) why the eyes sometimes land in unusual places).

O'Regan's model. According to O'Regan's model, eye movements during reading are determined by a pre-established program based on the eye's initial landing position in a word. If the eye lands in a region that is generally optimal (near the word's middle), the eye will directly leave the word. However, if the eye lands at a non-optimal position, it will refixate the word before leaving it. Individual eye fixations depend on the tactics adopted on the word and also on the eye's first position in the word. Fixation durations according to this scheme are mainly determined by oculomotor constraints, with a tradeoff between the first and second fixation durations when two are made in the word. Linguistic factors influence only the second of two fixations in a word, and the duration of single fixations when these are long. According to this position, refixation probability is primarily determined not by linguistic processing, but by factors available early in the fixation - that is, low-level visual factors. Gaze durations, however, can depend on linguistic factors such as word frequency, particularly when two fixations occur or when a single fixation is long.

Morrison's model is not really a prototypical model of those doing language research - they probably don't have a model of eye movement control in mind. In fact, they probably feel that it's not necessary to specify a model dealing with where the eye lands. All they care about is that gaze durations are variable as a function of various linguistic variables. Thus, Morrison's model falls on the oculomotor end of the continuum. But, it does specifically indicate that lexical processing influences when the eyes move. O'Regan's model is the champion these days of those who believe that low-level visual processes and oculomotor constraints are most important for understanding eye movements in reading. The model allows linguistic processes to have an influence, but any such influence is relatively weak - on the second fixation when two are made or if a single fixation is long.

What I want to do now is focus on the distinction that is inherent in these two models concerning the extent to which linguistic variables can influence fixation duration. Let me be clear that the results I'm going to discuss next are not a test of O'Regan's model. Rather they are intended to illuminate the extent to which linguistic variables influence fixations.

I'd also like to make another point. Often many researchers who are on the oculomotor end of the continuum have qualms about using gaze duration as a dependent variable. The argument is that it is clear when a fixation begins and ends, but it is not as clear what gaze duration measures. These researchers therefore prefer cases in which only single fixations occur on a word. Given this preference, Gary Raney and I examined some data on the word frequency effect in which we relied on cases where readers made only a single fixation on the target word. But, for comparison the means for the single fixation cases are compared with first fixation duration and gaze duration means in Table 1. As is clear in the table, single and first fixation data yield very similar values and the difference between high and low frequency words is very similar in the two cases.

	Single	First	Gaze
Low F.	264	256	301
High F.	242	238	258

Table 1. A comparison of single fixation duration, first fixation duration, and gaze duration (all in milliseconds) for low and high frequency words.

Back to the issue of lexical involvement in deciding when to move the eyes - what Raney and I did is examine fixation time on high and low frequency words as a function of where the eye lands in the word when only a single fixation is made. It seems to me that if oculomotor factors are primarily responsible for when we move our eyes, we would not expect frequency differences when the eyes land on the ends of words away from the preferred location. However, contrary to this expectation (which would be consistent with the oculomotor view) we found a frequency effect independent of landing position. That is, if the reader landed on the first or last letter of a seven letter word, they still looked longer at a low frequency word than a high frequency word. This frequency effect occurred for all letter positions in words which we examined that were five to nine letters long.

Let me be a bit more specific about what we did. We examined a large corpus of data in which subjects were reading passages of text. Within the passages of text, for words that were 5, 6, 7, 8, or 9 letters in length, we identified ten high and ten low frequency words for each word length. We made certain that all of the target words occurred in the middle of a line of text and that they did not begin or end a sentence. We then computed the landing position and the fixation duration for each word in the sample.

Another issue often discussed by researchers favoring oculomotor control of fixation times has to do with the frequency of effects analysis (McConkie, Zola, & Wolverton, 1985; McConkie, Reddix, & Zola, 1992). In particular, the issue is: are the differences observed in means due to a relatively small proportion of the data (means at the ends of the distribution) or to a shift in the entire distribution? Figure 2 shows the frequency distribution for the data. It is obvious that the effect is not due just to some long fixations at the end of the distribution, but there is a shift of the low frequency distribution relative to the high frequency distribution.

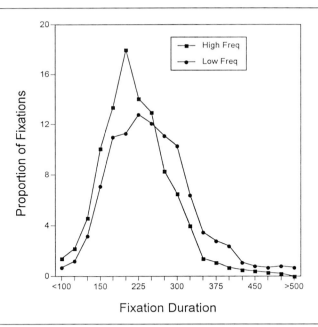

Figure 2. Frequency distributions for high- and low- frequency words. Duration shown in 25ms bins.

It may be the case that those advocating the oculomotor control position could find a way around our finding that the word frequency effect emerges independent of the landing position, but the finding seems to me to pretty strongly implicate lexical involvement in the decision to move the eyes. At this point, it seems safe to conclude that fixation times on target words reflect something about the processes involved in comprehending the word.

Eye Movements in Scene Perception and Visual Search

To what extent are eye movements in reading characteristic of or similar to eye movements in scene perception and visual search? On the one hand, it seems pretty clear that one doesn't want to generalize from research on one of these tasks to another since what the subject has to do differs considerably depending on whether he/she is reading, looking at a scene, or engaged in a visual search task (Loftus, 1983; Rayner, 1984). For example, there are differences in eye movement parameters when one compares reading to search using the same materials (Spragins, Lefton, & Fisher, 1976). And, there are gross differences in fixation duration and saccade length depending on whether one is reading, looking at a scene, or searching (Rayner, 1984; Rayner & Pollatsek, 1992).

On the other hand, it may be that the basic mechanisms involved in eye movement control in the three tasks might be rather similar. Rayner and Pollatsek (1992) proposed as a working hypothesis the idea that the basic mechanism of eye movement control is the same across tasks, even though the resulting pattern may differ. In particular, they suggested that the same basic mechanisms and sequence of operations that Morrison proposed for reading might operate in scene perception and visual search: namely, the subject engages in a cycle of processing the fixated material, shifting attention to some other part of the scene, with the actual saccade following at a later point.

There are two critical issues with respect to how this might work (just as there are in reading): a model of eye movement control in non-reading tasks should be able to account for both when an eye movement occurs and where the eyes go to. As noted earlier, the *where* decision in a model like Morrison s is determined by low-level cues (word length) and the reader typically sends attention to the next word. This kind of model would be incomplete for the free viewing situation that occurs in scene perception because there are no assumptions about where the next fixation should be when the task does not explicitly prescribe it. However, a considerable amount of work on eye movements in scene perception (see Rayner & Pollatsek, 1992 for a summary) has indicated that the pattern of eye movements is not random and that people tend to fixate areas of the scene that are judged as "informative" (either by the experimenter or by other subjects). One problem with the experiments is that "informative" is not very well-defined. In particular, "informative" areas tend to be both semantically important (e.g., contain objects whose identity is important to decode to understand the scene) and visually striking (e.g., contain lots of brightness changes, definite contours, etc.). But, at this point it seems quite reasonable to assume that low-level information is involved in the *where* decision in scenes just as it is in reading.

Rayner and Pollatsek (1992) suggested that the following expansion of Morrison's model might account for eye movements in free viewing. The subject inspects an object or some significant area of a display and when this object or area is identified, a command is sent out to move the eyes. For the most part, the location of where to move is guided by low-level information such as the nearest large discontinuity in the brightness pattern or a near significant contour. This is selected as the next area to be processed in the scene. Thus, the usual pattern will be to move from one object to the nearest object (Engel, 1977) with the decision being strongly influenced by low-level factors (Findlay, 1982). Since there is an "inhibition of return" mechanism (Posner & Cohen, 1984) whereby the location just processed is inhibited by the attentional system, the subject will not oscillate between two nearby objects. It thus seems plausible that such a "dumb" mechanism - fixate the nearest significant region but avoid the one just fixated - could insure that most of the significant areas of a scene would be inspected in a reasonably small number of fixations. If subjects are searching a scene for a particular target object (and so it is important to inspect all of the objects efficiently), a different control mechanism might be involved (where the eyes are programmed to go left-to-right, top-to-bottom, or something of this nature).

How about the *when* decision? One possibility is that, as noted above, something akin to identification of a fixated object serves as the trigger for an eye movement in scene perception (see Henderson, 1993), just as something like lexical access is the trigger in reading. When the scene viewing task involves object identification, this seems quite reasonable. However, if the task is to find a specific target object (visual search) then identification of all objects would not be necessary. Here, the trigger to move the eye may be something like "is the fixated object the same as the target" - if the answer is "no" the eyes move to another object.

Let me conclude with three other points that are relevant to the issue I just discussed. First, in a task in which subjects had to explore scenes and count the number of non-objects (objects that resemble real world objects in terms of size range and the presence of a clear and closed-part structure, but which are completely meaningless otherwise), DeGraef, Christiaens, and d'Ydewalle (1990) found that early fixations on objects which violated various aspects of a scene (e.g., in a position violation, a dump truck would be on top of a building and in a support violation a car would be suspended in mid-air with no visible means of support) did not differ from fixations on the same objects in a control (no-violation) condition; it was only fixations late in viewing that revealed fixation time differences. Apparently, subjects initially scanned the scene in a visual search mode and then examined it more carefully (during which time they detected the violations).

Second, Gary Raney and I compared performance with respect to the word frequency effect when subjects read passages versus when different subjects searched the same passages for the presence of a target word. Not surprisingly, in the search task subjects average fixations were shorter and their saccades were longer. Furthermore, when we examined fixation times on high and low frequency words, we found the normal frequency effect during reading. However, when the task was to search for the presence of a target word, we found no difference in either first fixation or gaze duration between high and low frequency words (none of which were the target word, though they were the same word length as the target). This seems to suggest that when comprehending the meaning of the text is not relevant, the trigger to move the eyes becomes something very different than in reading. In the present case, it would seem like a simple decision as to whether or not the fixated word was the target would suffice.

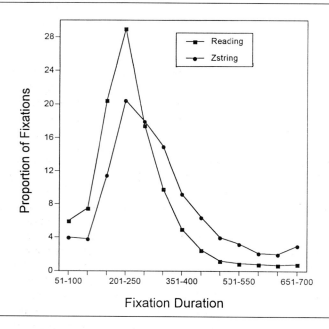

Figure 3. Frequency distributions for reading and z-strings fixation

Third, Vitu, O'Regan, Inhoff, and Topolski (1994) have recently compared normal reading to what they refer to as "mindless reading". In the mindless reading situation, subjects are presented strings comprised of the letter z and asked to move their eyes across the line as if they were reading. The letters in the normal text were replaced by zs to create the z-string condition. They found that there was some similarity in the distributions of saccade lengths, fixation durations, and landing positions between normal reading and "reading" the z-strings. They argued that the resemblance observed between the global and local characteristics of eye movements during normal text reading and during the scanning of meaningless letter strings might suggest that predetermined oculomotor strategies are an important part of the determinants of the oculomotor behavior during normal text reading. From my point of view, while the distributions are similar, it's clear that the mean values for fixation duration and saccade length are very different. In Figure 3, I have replotted their data so that the two distributions are in the same graph. It appears that the difference between the distributions is like the difference between the distributions the distributions of high and low frequency words shown in Figure 2. Given the task, maybe it's not surprising that a lifetime of reading habits would yield some resemblance to reading. Moreover, if the basic mechanisms of eye movement control are similar across different tasks as Rayner and Pollatsek (1992) argued, but the trigger to move the eye differs as a function of the specific task, then the results are not too surprising. Basically, we would expect data that mimic normal reading, but differ in some fundamental ways.

Summary

In this chapter, I summarized a number of recent findings relevant to eye movement research. I also made a distinction between researchers primarily interested in the oculomotor aspects of reading and those interested in using eye movement data to infer something about the reading process. Hopefully, it is clear that my view is that both types of researchers have provided valuable information. I also argued that there is very good reason to believe that a model of eye movement control that allows for lexical processing to be involved in the decision about when to move the eyes is to be preferred over models that do not. I also argued that in tasks like scene perception and visual search, while the basic properties of eye movement control are similar to those in reading, the trigger mechanism involved in making the decision about when to move varies as a function of the specific task. Finally, I would like to make it clear that by arguing for models of eye movement control that involve lexical processes in the decision to move the eyes, I do not mean to demean results that have emerged from researchers interested mostly in the oculomotor aspects of reading. The results that they have obtained, such as demonstrations of the landing position effect and the launch site effect, are certainly very important to a full understanding of reading and other tasks. The issue is really one of emphasis: as should be clear from this chapter, I'm persuaded that lexical and other cognitive factors are much more centrally involved in eye movement control in reading than models stressing oculomotor factors would concede.

Acknowledgment

Preparation of this chapter was supported by Grant HD26765 from the National Institute of Health and by Grant SBR-9121375 from the National Science Foundation. Chuck Clifton and Sandy Pollatsek provided helpful comments on the chapter.

References

Balota, D.A., Pollatsek, A., & Rayner, K. (1985). The interaction of contextual constraints and parafoveal visual information in reading. Cog. Psychol. 17, 364-390.

Blanchard, H.E., McConkie, G.W., Zola, D., & Wolverton, G.S. (1984). Time course of visual information utilization during fixations in reading. J. Exp. Psychol. Human Per. Performan. 10, 75-89.

Blanchard, H.E., Pollatsek, A., & Rayner, K. (1989). Parafoveal processing during eye fixations in reading. Perception and Psychophysics 46, 85-94.

Carroll, P., & Slowiaczek, M.L. (1986). Constraints on semantic priming in reading: A fixation time analysis. Memory & Cognition 14, 509-522.

Clifton, C. (1993). The role of thematic roles in sentence parsing. Can. J. Exp. Psychol. 47, 222-246.

DeGraef, P., Christiaens, D., & d'Ydewalle, G. (1990). Perceptual effects of scene context on object identification. Psychol. Res. 52, 317-329.

DenBuurman, R. Boersema, T., & Gerrisen, J.F. (1981). Eye movements and the perceptual span in reading. Reading Res. Quart. 16, 227-235.

Dopkins, S., Morris, R.K., & Rayner, K. (1992). Lexical ambiguity and eye fixations in reading: A test of competing models of lexical ambiguity resolution. J. Memory Language 31, 461-476.

Duffy, S.A., Morris, R.K., & Rayner, K. (1988). Lexical ambiguity and fixation times in reading. J. Memory Language 27, 429-446.

Duffy, S.A., & Rayner, K. (1990). Eye movements and anaphor resolution: Effects of antecedent typicality and distance. Language and Speech 33, 103-119.

Dunn-Rankin, P. (1978). The visual characteristics of words. Scientific American 238, 122-130.

Ehrlich, K., & Rayner, K. (1983). Pronoun assignment and semantic integration during reading: Eye movements and immediacy of processing. J. Verbal Learning Verbal Beh. 22, 75-87.

Ehrlich, S.F., & Rayner, K. (1981). Contextual effects on word perception and eye movements during reading. J. Verbal Learning Verbal Beh. 20, 641-655.

Engel, F.L. (1977). Visual conspicuity, directed attention and retinal locus. Vision Res. 17, 95-108.

Ferreira, F., & Clifton, C. (1986). The independence of syntactic processing. J. Memory Language 25, 348-368.

Ferreira, F., & Henderson, J.M. (1990). The use of verb information in syntactic parsing: Evidence from eye movements and word-by-word self-paced reading. J. Exp. Psychol. Learning Memory Cognition 16, 555-568.

Findlay, J. (1982). Global processing for saccadic eye movements. Vision Res. 22, 1033-1045.

Frazier, L., & Rayner, K. (1982). Making and correcting errors during sentence comprehension: Eye movements in the analysis of structurally ambiguous sentences. Cog. Psychol.14, 178-210.

Garrod, S., O'Brien, E.J., Morris, R.K., & Rayner, K. (1990). Elaborative inferencing as an active or passive process. J. Exp. Psychol. Learning Memory Cognition 16, 250-257.

Henderson, J.M. (1993). Eye movement control during visual object processing: Effects of initial fixation position and semantic constraint. Can. J. Exp. Psychol. 47, 79-98.

Henderson, J.M., & Ferreira, F. (1990). Effects of foveal processing difficulty on the perceptual span in reading: Implications for eye movement control. J. Exp. Psychol. Learning Memory Cognition 16, 417-429.

Henderson, J.M., & Ferreira, F. (1993). Eye movement control during reading: Fixation measures reflect foveal but not parafoveal processing difficulty. Can. J. Exp. Psychol. 47, 201-221.

Hogaboam, T. (1983). Reading patterns in eye movement data. In Eye Movements in Reading: Perceptual and Language Processes, K. Rayner Ed, Academic Press, New York, pp 309-332.

Hyönä, J. (1993). Do irregular letter combinations attract readers' attention? Evidence from fixation locations in words. Submitted for publication.

Hyönä, J., Niemi, P., & Underwood, G. (1989). Reading long words embedded in sentences: Informativeness of word parts affects eye movements. J. Exp. Psychol. Human Per. Performan. 15, 142-152.

Ikeda, M., & Saida, S. (1978). Span of recognition in reading. Vision Res. 18, 83-88.

Inhoff, A.W. (1987). Lexical access during eye fixations in sentence reading: Effects of word structure. In Attention and Performance 12, M. Coltheart Ed., Erlbaum, Hillsdale, NJ, pp 403-418.

Inhoff, A.W. (1989). Lexical access during eye fixations in reading: Are word codes used to integrate lexical information across interword fixations? J. Memory Lang. 28, 444-461.

Inhoff, A.W., & Briihl, D. (1991). Semantic processing of unattended text during selective reading: How the eyes see it. Perception and Psychophysics 49, 289-294.

Inhoff, A.W., Pollatsek, A., Posner, M.I., & Rayner, K. (1989). Covert attention and eye movements during reading. Quart. J. Exp. Psychol. 41A, 63-89.

Inhoff, A.W., & Rayner, K. (1986). Parafoveal word processing during eye fixations in reading: Effects of word frequency. Perception & Psychophysics 40, 431-439.

Inhoff, A.W., & Topolski, R. (1992). Lack of semantic activation from unattended text during passage reading. Bull. Psychonomic Soc. 30, 365-366.

Inhoff, A.W., Tousman, S. (1990). Lexical priming from partial-word previews. J. Exp. Psychol. Human Per. Performan. 16, 825-836.

Ishida, T., & Ikeda, M. (1989). Temporal properties of information extraction in reading by a text-mask replacement technique. J. Opt. Soc. America A6, 1624-1632.

Just, M.A., & Carpenter, P.A. (1980). A theory of reading: From eye fixations to comprehension. Psychol. Rev. 87, 329-354.

Kowler, E., & Anton, S. (1987). Reading twisted text: Implications for the role of saccades. Vision Res. 27, 45-60.

Lima, S.D. (1987). Morphological analysis in sentence reading. J. Mem. Lang. 26, 84-99.

Lima, S.D., & Inhoff, A.W. (1985). Lexical access during eye fixations in reading: Effects of word-initial letter sequence. J. Exp. Psychol. Human Per. Performan. 11, 272-285.

Loftus, G. R. (1983). Eye fixations on text and scenes. In Eye Movements in Reading: Perceptual and Language Processes, K. Rayner Ed, Academic Press, New York, pp 359-376.

McConkie, G.W., Kerr, P.W., Reddix, M.D., & Zola, D. (1988). Eye movement control during reading: I. The location of initial fixations on words. Vision Res. 28, 1107-1118.

McConkie, G.W., Kerr, P.W., Reddix, M.D., Zola, D., & Jacobs, A.M. (1989). Eye movement control during reading: II. Frequency of refixating a word. Perception & Psychophysics 46, 245-253.

McConkie, G.W., & Rayner, K. (1975). The span of the effective stimulus during a fixation in reading. Perception & Psychophysics 17, 578-586.

McConkie, G.W., & Rayner, K. (1976). Asymmetry of the perceptual span in reading. Bull. Psychonomic Soc. 8, 365-368.

McConkie, G.W., Reddix, M.D., & Zola, D. (1992). Perception and cognition in reading: Where is the meeting point? In Eye Movements and Visual Cognition: Scene Perception and Reading, K. Rayner Ed, Springer-Verlag, New York, pp 293-303.

McConkie, G.W., & Zola, D. (1979). Is visual information integrated across successive fixations in reading? Perception & Psychophysics 25, 221-224.

McConkie, G.W., Zola, D., & Wolverton, G.S. (1985). Estimating frequency and size of effects due to experimental manipulations in eye movement research. In Eye Movements and Human Information Processing, R. Groner, G.W. McConkie, and C. Menz Eds, North Holland, Amsterdam, pp 137-148.

Morris, R.K. (1994). Lexical and message-level sentence context effects on fixation times in reading. J. Exp. Psychol. Learning Memory Cognition, in press.

Morris, R.K., Rayner, K., & Pollatsek, A. (1990). Eye movement guidance in reading: The role of parafoveal letter and space information. J.Exp.Psychol. Human Per. Performan.16, 268-281.

Morrison, R.E. (1984). Manipulation of stimulus onset delay in reading: Evidence for parallel programming of saccades. J. Exp. Psychol. Human Per. Performan. 10, 667-682.

Morrison, R.E., & Rayner, K. (1981). Saccade size in reading depends upon character spaces and not visual angle. Perception & Psychophysics 30, 395-396.

O'Brien, E.J., Shank, D.M., Myers, J.L., & Rayner, K. (1988). Elaborative inferences during reading: Do they occur on-line? J. Exp. Psychol. Learning Memory Cognition 12, 346-352.

O'Regan, J.K. (1979). Eye guidance in reading: Evidence for the linguistic control hypothesis. Perception & Psychophysics 25, 501-509.

O'Regan, J.K. (1980). The control of saccade size and fixation duration in reading: The limits of linguistic control. Perception & Psychophysics 28, 112-117.

O'Regan, J.F. (1981). The convenient viewing hypothesis. In Eye Movements: Cognition and Visual Perception, D.F. Fisher, R.A. Monty, and J.W.Senders Eds, Erlbaum, Hillsdale, NJ,pp 289-298.

O'Regan, J.K. (1990). Eye movements and reading. In Eye Movements and Their Role in Visual and Cognitive Processes, E. Kowler Ed, Elsevier, Amsterdam, pp 395-453.

O'Regan, J.K. (1992). Optimal viewing position in words and the strategy-tactics theory of eye movements in reading. In Eye Movements and Visual Cognition: Scene Perception and Reading, K. Rayner Ed, Springer-Verlag, New York, pp 333-354.

O'Regan, J.K., & Lévy-Schoen, A. (1983). Integrating visual information from successive fixations: Does transaccadic fusion exist? Vision Res. 23, 765-768.

O'Regan, J.K., & Lévy-Schoen, A. (1987). Eye movement strategy and tactics in word recognition and reading. In Attention and Performance 12, M. Coltheart Ed, Erlbaum, Hillsdale, NJ, pp 363-383.

O'Regan, J.K., Lévy -Schoen, A., Pynte, J., & Brugaillere, B. (1984). Convenient fixation location within isolated words of different length and structure. J. Exp. Psychol. Human Per. Performan. 10, 250-257.

Osaka, N. (1992). Size of saccade and fixation duration of eye movements during reading: Psychophysics of Japanese text processing. J. Opt. Soc. America A9, 5-13.

Pollatsek, A., Bolozky, S., Well, A.D., & Rayner, K. (1981). Asymmetries in the perceptual span for Israeli readers. Brain and Lang. 14, 174-180.

Pollatsek, A., Lesch, M., Morris, R.K., & Rayner, K. (1992). Phonological codes are used in integrating information across saccades in word identification and reading. J. Exp. Psychol. Human Per. Performan. 18, 148-162.

Pollatsek, A., Raney, G.E., LaGasse, L., & Rayner, K. (1993). The use of information below fixation in reading and in visual search. Can. J. Exp. Psychol. 47, 179-200.

Pollatsek, A., & Rayner, K. (1982). Eye movement control in reading: The role of word boundaries. J. Exp. Psychol. Human Per. Performan. 8, 817-833.

Pollatsek, A., & Rayner, K. (1990). Eye movements and lexical access in reading. In Comprehension Processes in Reading, D.A. Balota, G.B. Flores d'Arcais, and K. Rayner Eds, Erlbaum, Hillsdale, NJ, pp 143-163.

Pollatsek, A., Rayner, K, & Balota, D.A. (1986). Inferences about eye movement control from the perceptual span in reading. Perception & Psychophysics 40, 123-130.

Posner, M., & Cohen, Y. (1984). Components of visual orienting. In Attention and Performance 10, H. Bouma and D.G. Bouwhuis, Ed, Erlbaum, Hillsdale, NJ. pp 531-556.

Raney, G.E., & Rayner, K. (1994). Word frequency effects and eye movements during two readings of a text. Can. J. Exp. Psychol., in press.

Rayner, K. (1975). The perceptual span and peripheral cues in reading. Cog. Psychol. 7, 65-81.

Rayner, K. (1977). Visual attention in reading: Eye movements reflect cognitive processes. Memory & Cognition 4, 443-448.

Rayner, K. (1979). Eye guidance in reading: Fixation locations in words. Perception 8, 21-30.

Rayner, K. (1984). Visual attention in reading, picture perception, and visual search: A tutorial review. In Attention and Performance 12, H. Bouma and D. Bouwhuis Eds, Erlbaum, Hillsdale, NJ, pp 67-96.

Rayner, K. (1986). Eye movements and the perceptual span in beginning and skilled readers. J. Exp. Child Psychol. 41, 211-236.

Rayner, K., Balota, D.A., & Pollatsek, A. (1986). Against parafoveal semantic preprocessing during eye fixations in reading. Can. J. Psychol. 40, 473-483.

Rayner, K., & Bertera, J.H. (1979). Reading without a fovea. Science 206, 468-469.

Rayner, K., Carlson, M., & Frazier, L. (1983). The interaction of syntax and semantics during sentence processing: Eye movements in the analysis of semantically biased sentences. J. Verbal Learning and Verbal Beh. 22, 358-374.

Rayner, K., & Duffy, S.A. (1986). Lexical complexity and fixation times in reading: Effects of word frequency, verb complexity, and lexical ambiguity. Memory & Cognition 14, 191-201.

Rayner, K., & Frazier, L. (1987). Parsing temporarily ambiguous complements. Quart. J. Exp. Psychol. 39A, 657-673.

Rayner, K., & Frazier, L. (1989). Selection mechanisms in reading lexically ambiguous words. J. Exp. Psychol. Learning Memory Cognition 15, 779-790.

Rayner, K., Garrod, S., & Perfetti, C.A. (1992). Discourse influences during parsing are delayed. Cognition 45, 109-139.

Rayner, K., Inhoff, A.W., Morrison, R.E., Slowiaczek, M.L., & Bertera, J.H. (1981). Masking of foveal and parafoveal vision during eye fixations in reading. J. Exp. Psychol. Human Per. Performan. 7, 167-179.

Rayner, K., & McConkie, G.W. (1976). What guide's a reader's eye movements? Vision Res. 16, 829-837.

Rayner, K., McConkie, G.W., & Ehrlich, S.F. (1978). Eye movements and integrating information across fixations. J. Exp. Psychol. Human Per. Performan. 4, 529-544.

Rayner, K., McConkie, G.W., & Zola, D. (1980). Integrating information across eye movements. Cog. Psychol. 12, 206-226.

Rayner, K., & Morris, R.K. (1992). Eye movement control in reading: Evidence against semantic preprocessing. J. Exp. Psychol. Human Per. Performan. 18, 163-172.

Rayner, K., Murphy, L.A., Henderson, J.M., & Pollatsek, A. (1989). Selective attentional dyslexia. Cog. Neuropsychol. 6, 357-378.

Rayner, K., Pacht, J.M., & Duffy, S.A. (1994). Effects of prior encounter and global discourse bias on the processing of lexically ambiguous words: Evidence from eye fixations. J. Memory Language, in press.

Rayner, K., & Pollatsek, A. (1981). Eye movement control during reading: Evidence for direct control. Quart. J. Exp. Psychol. 33A, 351-373.

Rayner, K., & Pollatsek, A. (1987). Eye movements in reading: A tutorial review. In Attention and Performance 12, M. Coltheart Ed, Erlbaum, Hillsdale, NJ, pp 327-362.

Rayner, K., & Pollatsek, A. (1992). Eye movements and scene perception. Can. J. Psychol. 46, 342-376.

Rayner, K., Sereno, S.C., Morris, R.K., Schmauder, A.R., & Clifton, C. (1989). Eye movements and on-line language comprehension processes. Lang. Cog. Processes 4 (Special Issue), 21-50.

Rayner, K., Well, A.D., & Pollatsek, A. (1980). Asymmetry of the effective visual field in reading. Perception & Psychophysics 27, 537-544.

Rayner, K., Well, A.D., Pollatsek, A., & Bertera, J.H. (1982). The availability of useful information to the right of fixation in reading. Perception & Psychophysics 31, 537-550.

Schmauder, A.R. (1991). Argument structure frames: A lexical complexity metric? J. Exp. Psychol. Learning Memory Cognition 17, 49-65.

Schustack, M.W., Ehrlich, S.F., & Rayner, K. (1987). The complexity of contextual facilitation in reading: Local and global influences. J. Memory Language 26, 322-340.

Sereno, S.C., Pacht, J.M., & Rayner, K. (1992). The effect of meaning frequency on processing lexically ambiguous words: Evidence from eye fixations. Psychol. Science 3, 296-300.

Sereno, S.C., & Rayner, K. (1992). Fast priming during eye fixations in reading. J. Exp. Psychol. Human Per. Performan. 18, 173-184.

Spragins, A.B., Lefton, L.A., & Fisher, D.F. (1976). Eye movements while reading and searching spatially transformed text: A developmental examination. Memory & Cognition 4, 36-42.

Underwood, G., Clews, S., & Everatt, J. (1990). How do readers know where to look next? Local information distributions influence eye fixations. Quart. J. Exp. Psychol. 42A, 39-65.

Underwood, N.R., & McConkie, G.W. (1985). Perceptual span for letter distinctions during reading. Reading Res. Quart. 20, 153-162.

Underwood, N.R., & Zola, D. (1986). The span of letter identification for good and poor readers. Reading Res. Quart. 21, 6-19.

Vitu, F. (1991). The influence of parafoveal processing and linguistic context on the optimal landing position effect. Perception and Psychophysics 50, 58-75.

Vitu, F., O'Regan, J.K., & Mittau, M. (1990). Optimal landing position in reading isolated words and continuous text. Perception and Psychophysics 47, 583-600.

Vitu, F., O'Regan, J.K., Inhoff, A.W., & Topolski, R. (1993). Mindless reading: Eye movement characteristics are similar in scanning strings and reading text. Manuscript submitted for publication.

Wolverton, G.S., & Zola, D.A. (1983). The temporal characteristics of visual information extraction during reading. In Eye Movements in Reading: Perceptual and Language Processes, K. Rayner Ed., Academic Press, New York, pp 41-51.

Zola, D. (1984). Redundancy and word perception during reading. Perception & Psychophysics 36, 277-284.

MODELS OF OCULOMOTOR FUNCTION: AN APPRAISAL OF THE ENGINEER'S INTRUSION INTO OCULOMOTOR PHYSIOLOGY

Wolfgang Becker

Sektion Neurophysiologie, Universität Ulm, 89 081 Ulm, Germany

Abstract

Models have become a major tool in oculomotor physiology as a means of interpreting experimental results, formulating hypotheses, and understanding system function. Basically, a model is set of mathematical equations describing the theoretical or empirical relationship between the various physical and/or conceptual variables characterizing a system. In oculomotor physiology these equations are generally represented in the form of signal flow diagrams which are manipulated and analyzed using the inventory of systems control theory. We here distinguish between *conceptual* models sketching the direction and qualitative character of interactions between the constituents of a system, *descriptive* models summarizing empirical observations, *inferential* models inferring a system's internal structure from input-output measurements, and *homeomorphic* models reflecting all of a system's essential components and their actual arrangement. Early models of the saccadic and the smooth pursuit systems, and their subsequent evolution and contribution to current views, are considered in some detail. A basic problem of these systems (and of any other visuomotor system) is the appreciable transport delay enclosed by the retinal feedback loop which can lead to instability (oscillations). Whereas it is well established that the saccadic system obviates this problem in most situations by its discontinuous operation, the principle adopted by the pursuit system is still debated.

Introduction

In the wake of Norbert Wiener's influential work on Cybernetics (Wiener, 1949) physiologists in the 1950s became increasingly aware of the omnipresence of feedback control in biological systems and of the possibility of understanding function in terms of signal processing. The oculomotor system proved to be particularly attractive for such an approach for several reasons, for example:
- Its task is clearly defined: To bring visual objects of interest into, and stabilize them on, the foveal area of high visual resolution.
- The occurrence of negative feedback is quite obvious: Because of the rigid coupling of the sensory to the motor organ, retinal eccentricity (or slip) quantifies the difference between desired and actual eye position for velocity) and, being the primary determinant of the eye's corrective movements, always acts to reduce itself.
- The kinematics and dynamics of eye movements are relatively simple: There are only three degrees of freedom (of which only two are accessible to higher control mechanisms since torsion is automatically adjusted in compliance with Listing's law), and the mechanical load is constant, with inertial forces playing only a minor role

Early on, therefore, people with an interest in systems theory - many of them with an engineering background - ventured attempts to explain oculomotor control by cybernetical models. The present contribution reviews some of these early models, analyzes the underlying principles, and offers a categorization according to their speculative or factual nature; it discusses their subsequent evolution and tries to appraise their contribution to current views and to the development of the field in general.

To some degree, this review is a delayed answer to the reservations expressed by physiologists at the time when people with systems engineering views first intruded into their field. Physiologists deplored with some justification the engineers' ignorance of the intricacies of the nervous system and their tendency to construct models which merely demonstrate how the model's author would have realized what he thinks is the function of a system - if only the Lord had ceded that task to him. However, the physiologists also often failed to appreciate the merits of the engineer's approach which involves abstracting functions from their substrates to ease understanding or to "skip" gaps in detailed knowledge. Moreover, they often tended to equate the engineer's view of brain function with his tool, then often an awkward box containing an analog computer, instead of with the principles of signal processing demonstrated by means of that tool.

Most of these objections have vanished nowadays and the cybernetical view of oculomotor function has permeated the field to the point where the term "cybernetics" itself seems to be disappearing, simply because cybernetical reasoning has become an integral part of systems physiology and is no longer perceived as a separate science.

The Concept of "Model"

Much of the systems engineer's efforts are aimed at describing the system he wants to understand in terms a model. In a strict sense, a model is something very abstract, namely a set of mathematical equations describing the relationship between the input, the output and the internal variables of a system. As an example consider the following set of equations (see also Fig. 1)

$$F_C = c \cdot \Theta_C; \quad F_D = D \cdot \dot{\Theta}_C; \quad F_E = I \cdot \ddot{\Theta}_E; \quad \Theta_H = \Theta_E + \Theta_C; \quad F_E = F_C + F_D$$

which is known under the name of Steinhausen's torsion pendulum model (Steinhausen, 1933); it describes the forces arising during an acceleration of the head and the resulting displacements of the endolymph and the cupula in the vestibular organ (cf. Fig. 1A). We mention the Steinhausen model for several reasons here:

- It reminds us that modelling in biology has a far older tradition than the biocybernetical approach of the 1950s and following decades; Steinhausen's work was published in the early 1930s, and many other models can be traced back to the last century.
- It is the prototype of an essentially *homeomorphic, analytical* model; analytical, because it could be derived by rigorous analysis of the system components using well established physical laws; homeomorphic because it provides a 1:1 representation of all the functionally relevant elements of the system under consideration. Yet, the homeomorphism has not been pushed to the degree where one can't see the wood for the trees; there is for example no submodel of the cupula to explain its elasticity in terms of its biomechanical ultrastructure - such an addition would be of little help in grasping what the semicircular canals are good for.

- It can be used to illustrate the cybernetical approach of representing models in the form of flow diagrams by converting physical and conceptual entities into signals, and mathematical relationships into processing stages (Fig. 1B). At the left of Fig. 1B, for example, the accelerations of the head, the endolymph and the cupula are treated as signals, and their relationship is represented by a summing junction, that is, by a processing stage. Flow diagrams are by far the most common form of model representation in the oculomotor literature; for the experienced viewer they constitute visual metaphors of the system under consideration which he can manipulate in his mind to make qualitative predictions (e.g. what happens if the viscous friction D increases?)
- Finally, the Steinhausen model has been very successful - successful to the degree that people often are no longer interested in an analysis of its details but collapse these to obtain a *descriptive* model in the form of a single handy box which then can be used as an elementary component of more complex systems. According to this box, within a certain frequency band, cupula excursion is proportional to head velocity (description in the frequency domain; Fig. 1C) or, what is an equivalent statement, in response to a velocity step, the cupula is rapidly deflected and then slowly returns to its resting position (time domain; Fig. 1D).

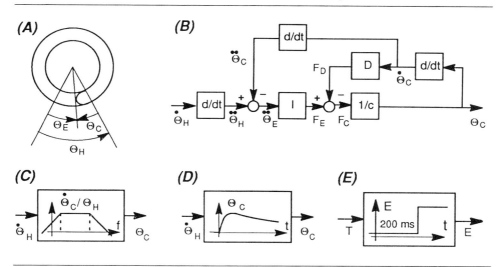

Fig. 1 - A-D, Steinhausen model of cupula deflection. A, physical situation. Θ_H, Θ_E, Θ_C, angular displacements of head, endolymph and cupula. B, flow diagram; F_E, inertial force acting on endolymph; F_D, frictional force between endolymph and canal wall; F_C, force acting on cupula. I, inertial moment of endolymph; D, coefficient of friction; c, stiffness of cupula. Dots and double dots indicate first and second derivatives, respectively. C, summarizing description in the frequency domain (Bode diagram). D, summarizing description in the time domain (cupula deflection after a step in angular velocity). E, empirical and grossly simplified black box model of saccadic system responding to single small target steps. T, target position; E, eye position.

For pragmatic reasons, the concept of "model" has been extended also to purely empirical descriptions of systems which, unlike the syntheses in Fig. 1C and D do not reflect an analytic insight. As an example, consider Fig. 1E which depicts an extreme case of a *black box* model; the box in Fig. 1E is a very simplified scheme of the saccadic system which reflects the *empirical* fact that the system responds to single, small, steps of a target with steps of similar size delayed by 200 ms, or so.

Black box models have been bitterly denounced on the grounds that they merely formalize our ignorance of the systems being represented. Indeed, the "delay line model" of the saccadic system in Fig. 1E tells us nothing about how targets are recognized, how decisions are taken and how the saccadic motor command is generated. Yet this purely descriptive model can be perfectly appropriate if we want to summarize on a very coarse scale the operation of the saccadic system. Conceive of a human operator who is part of a man-machine system and has to look at various instruments in order to make appropriate reactions. If we want to estimate the performance of this man-machine system, we will not be interested in details of saccade generation. Instead, the essential thing to notice will be the delay added by the saccadic system, and this delay is perfectly modelled by our box.

In summary, there is no reason to condemn descriptive black box models altogether. They can be very convenient tools for summarizing empirical knowledge and can help estimate a subsystem's function within a more complex structure on the same terms as those of the collapsed single box descriptions of analytical models. As a general rule, models should always be judged in the light of the purpose they are meant to fulfil.

Early Models of the Oculomotor system

The earliest flow diagram of oculomotor function I am aware of was published by Ludvigh (1952; p 446) who was concerned with the role of the then recently discovered spindle organs in the extraocular muscles. His speculative scheme addresses a basic issue of motor control in general, namely: How do motor systems achieve their accuracy given that their effector organs, the muscles, are likely to vary in efficiency? Ludvigh suggested a solution principle, borrowed from control theory, which, by now, has become almost commonplace: Parametric adjustment, or adaptation.

The idea is straightforward: If the muscles become weak, magnify the neural command sent to the muscles. But, how does the oculomotor system know that the muscles are weakening? Ludvigh's scheme conceived of two *a posteriori* mechanisms: (1) An visual comparison of intended and achieved eye position (or velocity) and (2) a non-visual comparison of the centrally commanded eye position to the position signalled by proprioceptive feedback.

Whereas parametric adjustment through visual signals is well documented now, a corresponding role for the proprioceptive afferents has not been established although they exhibit many ingredients expected of a system involved in adaptive processes, such as mossy and climbing fibre projections to the cerebellar cortex.

Ludvigh's flow diagram constitutes a *conceptual* model. Conceptual models suggest the direction and the qualitative character of interactions among identified or hypothesized constituents of a system without making quantitative assumptions; they are helpful in organizing and communicating ideas and often have been the precursors of more quantitative suggestions.

In 1960 and 61 Sünderhauf and Vossius both published flow diagrams (cf. Fig. 3) which - although still very qualitative - created an awareness for a very basic quantitative problem in the control of goal directed movements (Sünderhauf, 1960; Vossius, 1961). These authors were inspired by work of Küpfmüller (a professor of engineering and cofounder of the journal *Kybernetik* - later *Biological Cybernetics*) who had investigated the stability and optimum adjustment of servomechanisms with inherent delay times. Studying rapid goal-directed hand movements, Küpfmüller had suggested that these movements could be controlled by a non-visual, "local" feedback loop (*innerer Regelkreis*) with visual feedback intervening only discontinuously, at the completion of a rapid movement, to decide whether a further rapid movement should be generated to reach the target (Küpfmüller & Poklekowski, 1956). By applying these ideas to the oculomotor system, Sünderhauf and Vossius arrived at conceptual precursors of the more widely known sampled data model advanced by Young in 1962.

The work of the above authors illustrates one of the fundamental contributions of control theory, which is calculating the consequences of closing a feed back loop; this is an ever recurring problem in the oculomotor system, and often not a trivial one because of the unavoidable presence of delay times associated with visual processing. For a better appreciation of this point, it may be helpful to introduce a few elementary notions of feedback control; we shall do so by considering an idealized, highly simplified oculomotor system (Fig. 2), before returning in more detail to the Vossius and Young models:

Excursus: Of Proportional and Integrating Controllers, and Transport Delays

Our fictitious system consists of
- The retina which signals the position error e, this being the difference between eye position E and target position T.
- A delay time, D, which represents the lumped transport and processing delays.
- A controller, C, which transforms the delayed retinal error signal e' into a motor command, c; taken together, C and D represent the central stages of the oculomotor system (the sensori-motor interface in its broadest sense).
- The plant, P, which is the system that is being controlled (lumped properties of eye muscles, conjunctivae, eye ball); for the sake of simplicity, P is assumed to have no dynamics so that E equals c at all times.

The most simple form of a controller is a *proportional* one; in response to a given magnitude of the delayed error, e', it generates a motor command which is strictly proportional to e': $c = g_p \cdot e'$ (g_p, gain of controller). An implication of this law is that E will never exactly match T (except for T = 0) since, in order to generate an eye position E, the controller must be "fed" with an error of magnitude $e = E/g_p$. Moreover, whenever vision is obscured, eye position E would drop to zero, for no retinal error signal would drive the controller. Clearly, the real oculomotor system does not behave like a system with proportional control because, within a large range of eccentricities, targets can be exactly fixated upon and eye positions can be maintained in the absence of vision; these properties suggest that, on a global scale, the central stages of the oculomotor system be equated to an *integrating* controller.

With an integrating controller, the *rate of change* of the motor command c is proportional to the existing error: $dc/dt = g_i \cdot e'$; therefore, the motor command itself, and hence also eye position, reflect the time integral, that is, the past history of the error: $c = g_i \cdot \int e(t)$. An integrating controller continues to adjust (increase or decrease) its motor command until there is no more error. It can also be considered a memory; when the error signal is disconnected (e.g., by obscuring vision), it maintains its command at the level reached by the time of disconnection.

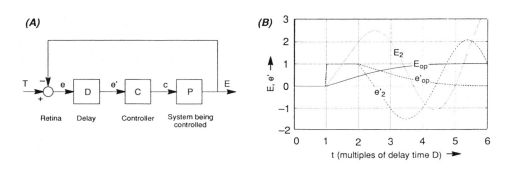

Fig. 2 - Fictitious oculomotor system. **A**, *flow diagram.* T, *target position;* e, *retinal error;* e', *retinal error after time delay D;* c, *motor command issued by controller;* E, *eye position.* **B**, *system response to a step of magnitude 1 occurring at time 0 ("step response") given an integrating controller* ($c = g_i \cdot \int e'(t) dt$). E_2, e'_2, *unstable behaviour with* $g_i = 2/D$. E_{op}, e'_{op}, *optimal step response (1% overshoot) with* $g_i = 0.44$.

To be useful, our fictitious oculomotor system should not only be able to fixate without error, it should also be fast, that is, after a change in target position, it should be able to readjust eye position in reasonable time. At first glance this seems to be fairly easy to achieve; by increasing the gain, g_i, we can accelerate the rate at which the motor command, and hence eye position, change. However, at this point we can no longer ignore the inherent delay time (D). Suppose the target makes a step of magnitude 1. After one delay time, the controller will begin to react and change eye position at a speed of g_i (see traces E_{op} and E_2 in Fig. 2B), and the error will begin to decrease at the same rate: $e = 1 - g_i(t-2D)$ (t, time; note that the fixation error caused by the step is initially also of magnitude 1). However, in spite of the decreasing error, the controller continues to command the same rate of change until a further delay time has passed, for only then does it "see" the decrease in error resulting from its own action (cf. traces e'_{op} and e'_2). It now becomes obvious that we have to be careful with the gain; for example, with a gain of $g_i = 2/D$ (=20/sec if we assume D = 100 msec for simplicity), the eye position would reach twice the desired value by the time the controller first notices a decrease of the error (cf. traces E_2 and e'_2 in Fig. 2B). Even then the controller would continue to increase eye position, albeit at a lower rate; it would take a further time lapse of $1/g_i$ (= D/2 in our example) before it would notice the eye has already gone too far and that it must reverse its action. It is fairly easy, also without formal calculus, to continue the above semiquantitative reasoning and to convince oneself that, with the chosen value of g_i, our system will undergo ever increasing oscillations.

Formal calculation indicates that, in order to avoid such instability, g_i must stay below a value of $\pi/2D$, and that only if g_i is less than $1/(D \cdot e)$ ($e \approx 2.72$, base of natural logarithms), will the change of eye position follow an asymptotic course without initial overshoot and "ringing". In the latter case it takes the eye more than 6 delay times to approach the target to within 1%. To accelerate the response of our system, we could, as a compromise, increase g_i to a value of about 0.44 and accept an overshoot of 1% (which is barely noticeable, cf. trace E_{op} in Fig. 2B), whereby we would reduce the settling time to a little less than 4D. The latter figure represents the main conclusion of our technical excursus: A feedback controlled eye positioning system with integrating controller and inherent transport delay requires at least four delay periods to align the eye with a new target.

Inferential models of the saccadic system:
Vossius' model and Young's sampled data model.

With the above conclusion we can return back to Vossius and the early 1960s. Vossius felt that saccades represented the step response of a servo involving the combination of a proportional and an integrating controller, and a plant for which he tried to make realistic assumptions. However, when he considered prototypical saccades of 70 ms duration, he realized that the delay time in the servo loop would have to be less than 20 ms; in fact, in order to fit the step response of the hypothesized servo to the time course of experimentally recorded human saccades, the transport delay had often to be adjusted to values below 10 msec (cf. Vossius, 1960, his Fig. 4a).

Therefore, the hypothesized servo could not be based on visual feedback which, even under the most favourable conditions, has a delay of no less than 80 ms. To account for this problem, Vossius suggested two principles that are still considered valid today, although in somewhat changed form, namely discrete error sampling and local feedback control of saccade execution (Fig. 3). Specifically, he postulated that the saccadic system responded only to discrete samples of the visually signalled error. These samples would be stored in a memory ("sample and hold" circuit) and forwarded to a local feedback loop as a command of desired eye position. Sampling would be such that the next sample of the visual error signal is delayed until a time which will allow for the effect of the saccade that was elicited by the preceding sample (i.e. one delay time after the end of the saccade at the earliest) thereby avoiding the "confusion" arising from the continuous usage of a delayed, and therefore outdated, error signal which, as shown in Fig. 2B, can cause instability. (Basically, this view is still correct; however later work has shown that the error signal can also be updated by an extraretinal, predictive feedback signal and thereby allow shorter sampling intervals, cf. Becker & Jürgens, 1979).

As its name indicated, the "local" feedback loop included only a local section of the total system, comprising the motor nuclei, the plant, and the muscle spindles which were thought to provide the feedback signal; accordingly, the transport delays around the loop would be small, of the order of 10 ms or so. Much of the details of Vossius' assumptions concerning this local loop are no longer tenable. For the present

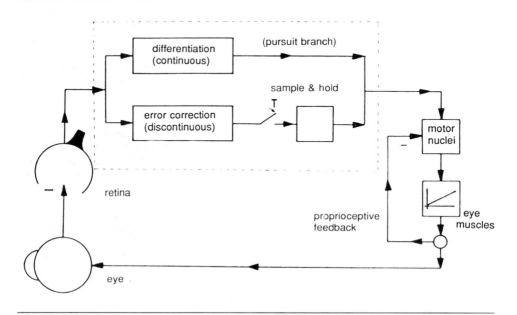

Fig. 3 - Vossius model of oculomotor system. The eye muscles are thought to behave as an integrator (upon receiving a step of innervation they would contract at a constant rate).

scope, it is sufficient to recognize that its hypothesized behaviour is close to that of an optimally adjusted servo with integrating controller and a delay time of about 10 ms, and that its response to the step-wise changing commands of the sample and hold mechanism would determine the time course of the saccade.

The idea of discrete sampling was further elaborated by Young (1962, 1963); his sampled data model was the first model to yield to a comprehensive computer simulation of oculomotor behaviour in response to a variety of target movement patterns. The model extended the idea of sampling also to the pursuit system, but abandoned the local feedback proposal which, in its early form advocated by Vossius, was incompatible with physiology as mentioned above. Instead, Young adopted the earlier view of Westheimer (1954a) according to which saccades represented the response of the plant to a step of innervation, the plant (extraocular muscles, connective tissue, eye ball) being a heavily damped spring-mass system.

The Vossius and Young models are examples of *inferential* models. Like Ludvigh, their authors had very little experimental data on the inner workings of the oculomotor system. However, they went one step beyond the merely conceptual discussion of its possible complexion; they inferred the system's inner structure, in terms of specific operations (sampling, integration), from observations of its input-output behaviour. What do we gain in formulating such models? On an extreme note, one could argue that we have merely found a special form to represent an observed behaviour and commonsense reasoning, without any cognitive value in the sense of explaining hitherto unexplicable phenomena. Yet, even if that were true, deriving these models would nonetheless be useful because the mere task of translating commonsense ideas into the rigorous rules of a testable signal flow diagram provides for a very effective control of consistency and helps to recognize basic principles and to ignore minor points.

However, the impact of good inferential models such as those of Vossius or Young has certainly gone much deeper; they have pinpointed specific operations that *must* take place. For example, both models *prove* that the saccadic branch of the oculomotor system must contain an integrator which acts as a memory of desired eye position and the content of which is changed in discrete steps (note that in Young's diagrams the integrator is easily overlooked because it is lumped with the transport delay of the visual pathway and labelled "computing delay"; e.g., Young, 1962, his Fig. 2). These models thus pose the challenge of identifying the corresponding neural substrate (concerning the nature of which - circumscribed structure or distributed property of the neural pathways as a whole? - they can obviously give no answer at all). In summary, good inferential models provide a framework that helps organize the questions one may wish to ask when exploring a system's internal structure.

Analysis of the final common pathway: A first step toward homeomorphism.

The first systematic steps into the inner life of the oculomotor system made under the guidance of model considerations are due to D.A. Robinson who worked his way backward from the mechanics of the plant to the signal processing in the motor nuclei and pre-motor structures. Robinson showed that the dynamics of the plant are mainly determined by its elastic and viscous components whereas inertia plays only a minor role (Robinson, 1964). He demonstrated that, accordingly, saccades result from a pulse-step pattern of innervation with the pulse overcoming the viscoelasticity of the plant and rotating the eye at high velocity, and the step holding the eye at the position reached by the end of the pulse - a pattern that was later clearly confirmed by recordings from the oculomotor nuclei (Robinson, 1970, Fuchs & Luschei, 1970). Ultimately, these observations led Robinson to the suggestion that the step component might be derived from a neural precursor of the pulse component through a "neural integrator" (cf. below).

Frequency domain descriptions, and the tools used for reasoning within them (e.g., Laplace transform, Nyquist diagram, Bode diagram), constitute another fundamental contribution of systems theory to oculomotor physiology. The fertility of this approach is well illustrated by Robinson's work on the vestibulo-ocular reflex (VOR) which we can afford to consider only briefly here, because its exemplary character has been authoritatively detailed, from a similar point of view as ours, by the author himself (e.g. Robinson, 1986; Robinson, 1987). This work proved that the VOR-pathway must also include an integrator, and emphasized that the classical three neuron arc is not the sole backbone of this reflex, but mainly a source of signals proportional to head velocity which compensate for the viscous damping of the plant, whereas the compensatory *displacement* of the eyes depends critically on an integrating pathway in parallel to it.

In retrospect, given that (1) the cupula system signals only head *velocity*, whereas the eye has to counteract variations of head *position*, and that (2) the eye, due to its mechanical damping, develops an increasing lag at frequencies above 0.7 Hz, or so, with respect to the innervation driving it, these results may appear quite straightforward and trivial. However, this unwarranted impression of triviality is a result of our increasing education by the engineering way of thinking which enables us to cognitively manipulate complicated relationships using a powerful, high level language. One can only speculate whether research into the VOR today would still be puzzling about unclear polysynaptic effects in dealing with the integrating pathway of the VOR, had this intrusion of systems theory not taken place.

Fig. 4 gives a highly simplified synopsis of Robinson's models of the oculomotor system's final common pathway (motor nuclei, plant) and its immediate prenuclear organization. On a prenuclear level the VOR slow phase and the saccadic commands are specified in terms of desired eye velocity, $\dot{E}$, and the same probably holds also for the pursuit command. These velocity commands must be converted by a *neural integrator* into the nuclear position commands driving the extraocular muscles. In the absence of evidence to the contrary, it is a logical step to assume that the neural structure subserving this function is common to all oculomotor subsystems. With regard to the saccadic system, the neural integrator converts the prenuclear, pulse-like velocity command into the step signal which causes the permanent displacement of the eye; therefore, the saccade amplitude is determined by the area under the pulse (product of intensity and duration of activity). By the same token, the neural integrator functions as the saccadic system's sample and hold stage which stores the agonist-antagonist imbalance needed to compensate the elastic forces in the position to be maintained after the saccade. The contributions of the direct (velocity) pathways are given a weight of magnitude T (where T is the time constant of the plant) and therefore compensate for the sluggishness of the plant. Initially a teleological postulate (Skavenski & Robinson, 1973), this compensation has been verified by showing that the ratio of velocity to position sensitivity of an average motoneuron is of the order of 0.25 sec (Robinson, 1981) and therefore an approximate match of the 0.15 - 0.2 sec time constant obtained with mechanical manipulations of the plant (Robinson, 1964). For many purposes therefore, the section of the oculomotor system enclosed by the heavily outlined box in Fig. 4 can be collapsed into a single function: Integration of the velocity commands received from the various oculomotor subsystems.

Fig. 4 is an intermediate step on the way from an inferential to a homeomorphic model; by and large, the final common pathway is rendered with reasonable accuracy, although it is true that there are many phenomena which require a more detailed modelling (non-linearities, push-pull arrangement of muscles, higher order damping, etc). The integrator, on the other hand, is homeomorphic only in that it is not a vague speculation but a definitely existing function; however, it is obviously far from being a homeomorphic representation of the actual neural wiring subserving this function.

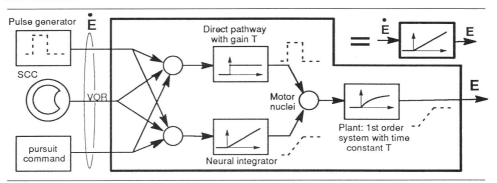

Fig. 4 - Final common pathway of oculomotor system. Input signals are velocity commands (Ė) from various subsystems, output is eye position; overall behaviour therefore is that of an integrator. Components are characterised by their step responses (sketched inside boxes); dashed profiles illustrate processing of saccadic pulse. SCC, semicircular canals. Global behaviour of stages inside heavily outlined box can be equated to that of an integrator.

Although the scheme sketched in Fig. 4 has been widely adopted, it has not gone completely unchallenged. Galiana, for example, has suggested that neural integration takes place only "outside" of saccades, but is suspended during saccades (Munoz & Galiana, 1990). Her arguments have been developed in the context of eye-head coordination and VOR-saccade interaction and are beyond the scope of our discussion.

The Local Feedback Hypothesis

The discovery of saccade related burst neurons in the pons (Luschei and Fuchs, 1972) has beautifully confirmed the notion of the prenuclear pulse as the primordial activity eliciting saccades. The number of discharges emitted by these neurons during a burst is closely related to the amplitude of the corresponding saccade and the discharge frequency correlates with saccade velocity (Keller, 1974; Scudder *et al.*, 1988) - thus, there is a close correspondence with the theory depicted in Fig. 4, if we equate the burst frequency with the pulse amplitude and the number of discharges with the pulse area. How does the oculomotor circuitry of the brain stem create this activity? This question has been repeatedly tackled during the last 20 years and has spawned what I consider one of the most fascinating examples of interaction between engineering and physiology. First, to stay in touch with a general view of the saccadic system, let us recall that the burst we are talking about represents the discrete sample of current error - or of desired eye displacement - postulated by the sampled data model. The burst must code this displacement in terms of its area so that, when summed up in the neural integrator, eye position is indeed displaced by the desired amount.

Originally it was felt that burst intensity was fairly constant so that only its duration would determine saccade amplitude - a hypothesis that was founded on the very tight and linear relation between saccade amplitude and duration. In view of the coding of desired displacement by the locus of activity on a collicular map, burst generation was termed the "spatial-to-temporal translation". An early, speculative model of this process invoked delay lines under the form of cerebellar parallel fibres (Kornhuber, 1971). Large saccades would be generated by selecting long or very thin fibres that sequentially activate a large number of converging neurons making for a long lasting burst activity, while small saccades would be obtained by activating short or thick fibres.

However, it soon became clear that burst generation is certainly not based on such a hardwired program. Instead, all experimental evidence indicates that it is controlled by a local servomechanism which adjusts burst duration as a function of burst intensity such that burst area equals desired displacement. To recall only a few, (1) the amplitude of goal-directed saccades is not affected by variations of their peak velocity (and hence of the saccadic pulse signal's amplitude), be these variations spontaneous or artificially induced by drugs (Becker et al., 1981); (2) when goal-directed saccades of monkeys are interrupted in "mid-flight" by electrical stimulation of the so called omnipause neurons, the movement is continued immediately after the end of stimulation and without intervention of visual mechanisms, putting the eye on its target with about normal accuracy; (3) similarly, if saccades of human subjects are transiently decelerated or arrested by sensory stimuli, a compensatory resurge of saccadic velocity is subsequently observed which leads the eye to its goal (Becker, 1994).

As mentioned above, the idea of a local feedback loop controlling saccade execution was first advanced by Vossius, but was abandoned because of the lack of evidence for the physiological substrate assumed by Vossius. The idea of local feedback then was reintroduced, as a means of controlling burst area, by Robinson in 1975 at a time when single unit studies just had begun to reveal a panoply of saccade related neurons, such as burst neurons, tonic neurons, pauser neurons, and so on. Fig. 5 depicts the model both in terms of its neural wiring (5A) and under the form of a signal flow diagram (5B). The model is fascinating because of its close relation to the then known neuroanatomy and the very plausible role assigned to the various types of neurons, and for some time it appeared even to be on the verge of allowing a transition from inferential reasoning to explaining oculomotor function on a neuron-by-neuron basis.

Yet, if we read the model from left to right, it begins on a very speculative note: it postulates that an efference copy of current eye-in-head position be added to the retinal afferents to reconstruct a representation of target position in *space* - a principle also invoked by models of the smooth pursuit system which we shall consider in more detail in that context (cf.below). The local feedback circuit proper calculates the difference between this reconstructed target-in-space position and current eye position - eye position being obtained again from an efference copy. This difference, the dynamic motor error, drives excitatory burst neurons (EBNs) which - technically speaking - act as non-linear, high-gain amplifiers. Burst activity, in turn, is fed into the neural integrator to which the tonic neurons (TNs) are thought to belong. Their output, then, determines eye position, and a collateral of it is used as the efference copy required for calculating target-in-space and dynamic motor error. Burst activity continues until the dynamic motor error approaches zero and can no longer drive the burst neurons.

There is one obvious problem, however: The local feedback loop inevitably involves time delays. Combined with the high gain of the burst neurons, these delays would cause the loop to oscillate. Robinson suggested that such oscillations be prevented by omnipauser neurons (OPNs); these would act as a latch inhibiting EBNs and thereby preventing burst activity unless explicitly told by a decision to unleash them. Once triggered by such a decision, the excitatory burst neurons would disinhibit themselves by activating inhibitory burst neurons (IBNs) which, in turn, would inhibit the pause neurons. The EBN-IBN-OPN latch loop would thus keep the EBNs clear of inhibition until the dynamic error, and hence the burst activity, begins to dwindle, at which moment pauser activity, driven by a tonic background, would pop in again to inactivate the local servo.

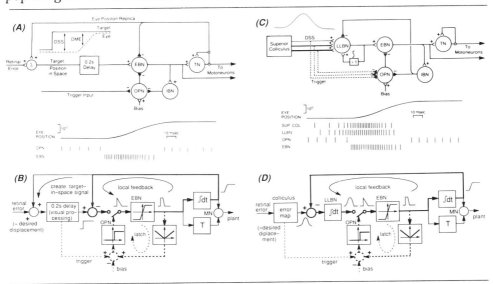

*Fig. 5 - Local feedback models of pulse generation. **A** and **B**, Robinson model, neural wiring diagram (A) and signal flow scheme (B). **C** and **D**, Scudder model, neural wiring (C) and signal flow scheme (D). DSS, desired saccade size; DME, dynamic motor error; EBN, excitatory burst neuron; IBN, inhibitory burst neuron; LLBN, long lead burst neuron; MN, motoneuron; OPN, omnipause neuron; TN, tonic neuron.*

The local feedback hypothesis has justly been heeded as a paradigmatic example of a model that thoroughly takes into account known physiology, suggests explanations for a wide range of phenomena, and inspires specific experiments to prove or falsify it. Above, we have already mentioned experiments which interfere with the normal functioning of the burst neurons and which confirm the automatic adjustment of burst duration. However, one can also manipulate the normal function of the latch circuit, for example by eliminating the pauser neurons. According to the model, this should lead to permanent back-and-forth saccades, and it has indeed been speculated that the saccadic oscillations seen in certain patients might be caused by a damage to the omnipauser system (Zee *et al.*, 1979). However, experimental lesions of the pauser area in monkey do not confirm this conjecture; instead of oscillations they produce slow saccades (Kaneko & Fuchs, 1991). On the other hand, such a slowing with pauser lesions is in keeping with a later model suggested by Scudder. The Scudder model is shown in Fig. 5 CD in a similar format to the Robinson model (Fig. 5AB) to demonstrate how the same fundamental concept, feedback control of burst generation, can assume different topologies in the light of new facts and under the guidance of a different minded intuition; we mention intuition here to underline the fact that integrating a growing body of physiological facts into a model is a creative act like the development of any other scientific theory, and should not be viewed as a routine application of engineering handbook rules.

The Scudder model avoids the problematic reconstruction of the target-in-space position and adopts the view of Jürgens *et al.* (1981) who suggested that it is the desired saccadic eye *displacement* rather than the desired postsaccadic eye *position* which is being controlled by local feedback. The model starts from a representation of desired eye displacement by the locus of activity on a collicular map. The colliculus is assumed to transmit a bell shaped burst containing an always fixed number of spikes to a pool of long lead burst neurons; the desired saccade amplitude would be indicated by the weight of the collicular projection. The long-lead burst neurons act as an integrator which can be viewed as an up- and down counter; the count goes up with the number of spikes received from the colliculus and goes down with the number of spikes emitted by the excitatory burst neurons. Initially, the burst neurons are inhibited by the pausers and cannot fire, so that the counter goes strictly up. Only after the collicular burst has reached sufficient intensity are the omnipause neurons shut down via an additional "trigger projection" from the colliculus; the counts accumulated by the long lead bursters then flood unimpeded to the excitatory burst neurons and initiate the burst activity, which, much as in the Robinson model drives integrating tonic neurons, keeps the excitatory bursters clear of pauser inhibition and, via the local feedback pathway, starts the down counting of the long lead bursters. Down counting goes on until the number of spikes sent to the neural integrator matches the received collicular activity, at which point the saccadic burst automatically terminates because the integrator is empty. Like the Robinson model, the Scudder model can explain how natural and artificially inflicted variations of burst intensity are automatically compensated by a reciprocal variation of burst duration.

In addition, it accounts also for the slowing of saccades after lesions to the omnipauser system: Suppose the pauser neurons are out of service. As soon as the first collicular spikes trickle into the long lead burst neurons, they are promptly handed down to the excitatory burst neurons in the absence of inhibition from pauser neurons. The excitatory neurons therefore immediately start firing, although at a very low frequency, and reduce the count of the integrator, so that there will never build up a significant drive. Hence, the saccade will be slow, but will nonetheless reach its target since accurate spike counting in the long lead bursters is not affected. If this is too much abstract reasoning, consider a mechanical metaphor of the Scudder model: Imagine a water tank into which we slowly pour a fixed amount of water. The tank corresponds to our integrating long lead burst neurons. At its bottom is a stopper which represents the inhibition by the pauser neurons. If we remove the stopper only after the tank is already halfway filled, we get a strong flush of water, that is, a vigorous saccadic burst. However, if the stopper is defective, the water will flow out of the tank without any particular thrust, that is, we get a slow saccade.

In spite of its partial improvement on the original Robinson model the Scudder model is still far from being the final truth. One conceptual flaw is particularly obvious. The model actually leaves the crucial task of generating the appropriate burst area to the superior colliculus - it is the colliculus which must provide the correct number of spikes - and the Scudder circuitry merely shapes the diluted collicular burst into a more brisk event. So, at present we must acknowledge that despite of many years of modelling, of recording in animals and observing the effects of lesions in humans we are still not certain of the neural circuitry of burst generation. About the only fact we know for sure at present is that burst generation profits from a local, or extraretinal, feedback control. However, we are still unable to answer even such basic questions as to whether the superior colliculus is situated ahead of the local feedback circuit or is a part of it. Whereas both Robinson and Scudder place the colliculus ahead, more recent hyptheses include it in the loop and actually view it as the summing junction where current eye displacement is subtracted from desired eye displacement (Waitzman *et al.*, 1988; Guitton, 1992).

Models of the pursuit system

In contrast to the saccadic system and to the VOR, the smooth pursuit branch of the oculomotor system has proven to be much more elusive in face of attempts to understand it by modelling. Smooth pursuit reactions act to reduce, and ideally to zero, the velocity ("slip") of a moving object's image on the retina. Thus, in contrast to the saccadic system, the variable that is being controlled is not position but velocity; however, since this control is based on visual feedback, the pursuit subsystem is "plagued" by the same problem as the saccadic system, and any other visuomotor system, namely an appreciable inherent transport delay.

Early observation showed that the human smooth pursuit mechanism will outsmart any attempt to treat it as a conventional, linear, time-invariant control system; already Westheimer (1954b) noted a phase lead developing after a few cycles of sinusoidal tracking. All later experiments confirmed that there must be a very efficient mechanism of pattern analysis and, based thereupon, pattern prediction, which permanently tries to over-ride the system's inherent transport delay and to synchronize the eye with the target.

Of course these observations spurred the ambition of modellers who came up with a variety of inventions to explain these characteristics. However, on a closer look, many of these suggestions reduce to the equivalent of a homunculus - in the form of very intelligent controller boxes which do all the job and explain almost nothing (e.g. "target selective adaptive controller", Bahill & McDonald, 1983). Yet, as unlikely as the details of the various technical solutions which have been they proposed may be, models of the predictive component of the smooth pursuit system have one basic insight in commont: We first have to have a notion of how a target moves before we can attempt to imitate and predict its movement with our eyes. To acquire this notion we must reconstruct the target's motion in space from its movement on the retina, that is, we have to add a signal of eye-in-head movement - a postulate familiar from theories of motion perception.

Reconstruction of target behaviour in space may play a role also during the pursuit of *random* movements. Indeed, the pursuit system's inherent delay time and its built-in negative visual feedback limit the possibility of improving performance by simply having the motor output react more vigorously to a given velocity error - as discussed above, we would risk oscillations. Therefore, in the first revision of the sampled data model (Young et al., 1968; Young, 1971), it was already implicitly assumed that the pursuit system uses reconstructed target velocity *permanently* and not only for the purpose of predicting. Later on, Yasui and Young (1975) elaborated this proposition in more detail and suggested that an extraretinal signal representing current eye velocity (variously termed "efference copy", "corollary discharge", "effort of will") is added to the retinal afferents (Fig. 6A). In terms of signal flow this is tantamount to a virtual inactivation of the visual feedback pathway and the construction of an internal signal, $\hat{T}'$, representing the target's movement in space; this signal would determine both the pursuit response and movement perception ("perceptual feedback hypothesis"). Because visual feedback and efference copy cancel each other, the overall behaviour of the hypothesized system can be described by an equivalent structure of pure feedforward character shown in the heavy outline of Fig.6A.

There is an obvious objection against the scheme in Fig. 6A: It ignores the physiological fact that much of the delay times occur in the visual leg of the pursuit system, ahead of any conceivable site for the addition of an efference copy, rather than in the "oculomotor command generator". Therefore, efference copy and visual feedback would not be in temporal register and their straightforward addition would create a chaotic picture of target movement. A formal remedy to this objection would be what is called model feedback: The idea is to feed the efference copy through a model, or replica, of the visual system and thereby subject it to the same delay before adding it. Robinson and colleagues (Robinson et al. 1986) have proposed a model of pursuit based on just that strategy. A simplified version of their scheme is shown in Fig. 6B. Da, Dc, and De stand for the delays of the afferent leg, of the central controller and of the efferent leg of the pursuit system. The primed functions in the model feedback loop stand for replicas of the delays encountered in the "outer loop"

(De→∫dt→P→d/dt→Da; the operations ∫dt and d/dt need not be replicated because they cancel each other). After passing these replicas, the delayed efference copy (c') is now in synchrony, at the site of addition, with the modulation of the visual afferents caused by the eye movement and can therefore eliminate it. The resulting signal (T') therefore represents a delayed version of the target's movement in space and the system's overall behaviour becomes indistinguishable from what the heavily outlined forward path alone would produce. Accordingly, the damped oscillations which frequently are observed in the pursuit system's step response would arise from within the central controller C.

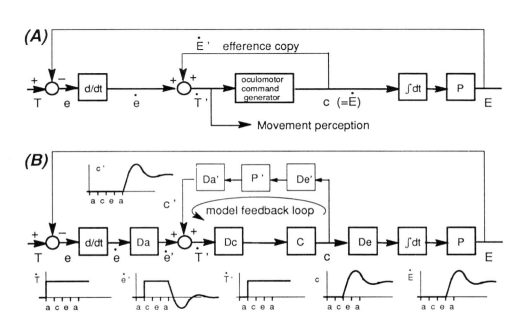

*Fig. 6 - Models of the pursuit system. **A**, perceptual feedback hypothesis (Yasui & Young, 1975). T, target position; E, eye position; e, retinal position error; ė, retinal slip velocity; Ṫ', reconstructed target velocity; c motor command (= commanded eye velocity, E) issued by controller (= oculomotor command generator), Ė, eye velocity; Ė', copy of motor command; P, plant. **B**, model feedback hypothesis (Robinson et al., 1986)). Da, Dc, De, delay times associated with afferent processes, controller operation and efferent processes, respectively. Da', P', De', neural replicas of Da, P, and De. ė', central representation of slip velocity; Ṫ, target velocity; Ṫ', central reconstruction of target velocity; c,' delayed copy of motor command. Other symbols as in A. Insets show signal waveforms during step response; a, c, e on time scale denote delays corresponding to Da, Dc, and De.*

Intelligent and plausible as this may be from the viewpoint of an engineer who tries to synthesize a control mechanism under the presently known constraints of the pursuit system, for the physiologist the scheme in Fig. 6B is mostly speculation unless any of the postulated processing stages can be identified. It is true that there are suggestions of cells in the cerebellar vermis (Suzuki & Keller, 1988) and the medial superior temporal cortex (Newsome *et al.*, 1988) which could reflect reconstructed target velocity. However, one should be aware that there are other topologies that achieve the same overall behaviour.

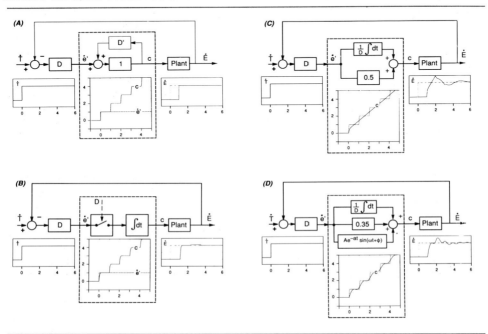

*Fig. 7 - Relationship between perceptual feedback hypothesis (**A**) and other theories of smooth pursuit control (**B-D**). In each panel, the dashed box encloses the controller. Insets inside dashed boxes characterize the controllers by their step responses. Time axes scaled in multiples of delay time D. Staircase curves in C and D repeat step response of A for comparison. Insets on right side of each panel show the model's response (in terms of eye velocity, Ė; continuous profile) to a step of target velocity, Ṫ (the dashed profile is a copy of the left inset profile showing target velocity). D', replica of D; other symbols as in Fig. 6. See text for more details.*

Interestingly these alternatives establish a relation between seemingly very different hypotheses of pursuit control. To appreciate this point, consider a simplified version of the perceptual feedback model (Fig. 7A), reduced to a visual delay D, an oculomotor command generator with a gain of one (motor command identical to perceived target velocity), an ideal plant, and an efference copy via a replica, D', of the visual delay (positive "model feedback") which cancels the effect of the negative visual feedback; moreover, to simplify matters further, the model argues in terms of velocity only, omitting the conversions from position to velocity and back from velocity to position in the afferent and efferent legs, respectively. The behaviour of this fictitious system is virtually identical to that of its heavily outlined forward pathway: Given our idealized assumptions, the pursuit movement would be a delayed but otherwise faithful image of the target motion.

However, the topology of Fig. 7A can also be looked at in a different way: The oculomotor command generator and the positive feedback around it can be conflated (dashed box) and viewed as the controller of a "conventional" negative feedback system. To describe the properties of this controller, we can specify its response to a step function (the step response provides a complete description for any linear system or system component). It is easy to convince oneself that a step of e' at its input will cause the controller to output a staircase with steps separated by intervals of duration D (see middle inset in Fig. 7A; note that this is a "thought experiment"; within the real setting of the pursuit system the controller is unlikely to receive a sustained step input unless the retinal error signal is artificially "clamped" at some constant value). Interestingly, on a coarse time scale this staircase is an approximation of the step response an integrating controller would produce (which is a ramp instead of a staircase).

A similar staircase would result, if we sampled the retinal error velocity at intervals D and transmit these samples to an integrator - this is exactly the mechanism that was suggested by Young in his very first version of the sampled data model. Given this identical behaviour of the sampling controller, the idealized sampled data system of Fig. 7B exhibits indeed the same response to the sudden onset of a constant velocity target movement as the perceptual feedback model of Fig. 7A (however, we must point out that, because of the discrete sampling, the similarity between the systems in Fig. 7A and B deteriorates if the target velocity undergoes variations that are fast in comparison to the sampling period D).

It is also conceivable to approximate the staircase response of the controller in Fig. 7A by a continuous processing of the error. As an example, Fig. 7C shows a coarse approximation by the parallel feed forward of the position and velocity errors with gains 1/D and 0.5, respectively - a structure which is, at least remotely, reminiscent of the pursuit models advocated by Lisberger and collaborators (e.g. Goldreich *et al.*, 1992). Because the approximation is rather crude, with such a controller the pursuit response to the sudden onset of target motion will exhibit dampened oscillations, like the real system. The period of these oscillations is proportional to the pursuit system's delay time.

Goldreich *et al.* have exploited this fact to discriminate between their own and Robinson's pursuit models. When they manipulated the pursuit delay time of monkeys (by either varying target salience or electronically delaying the visual feedback) the animals' oscillation period indeed exhibited a linear dependence on the delay time. In contrast, as already mentioned, under Robinson's model feedback hypothesis (Fig. 6B) the oscillations are an intrinsic property of the oculomotor command generator and unrelated to the delay time; however, when the visual delay time is increased without a corresponding increase of the model delay, additional oscillations will arise whose frequency is related to the resulting mismatch, i.e. to D-D', a phenomenon that was clearly not observed in the experiments of Goldreich et al. We have mentioned this point in some detail here because it constitutes a beautiful example how control systems reasoning, and purely external measurement and manipulations, can direct the choice between basic alternatives regarding a system's inner life - a choice where intuitive reasoning would probably be at lost.

Finally, one could improve the approximation of the desired staircase by adding an element oscillating at a frequency of 1/D as shown in Fig. 7D; this again yields a pure forward arrangement of processing stages, seemingly without need for the internal, positive feedback required in A. With this type of controller, the response of our fictitious pursuit system to the onset of target motion would exhibit clearly smaller oscillations than with the crude approximation in Fig. 7C, and by adding higher order harmonics to the controller's characteristic one could even more approach the "ideal" behaviour of the perceptual feedback topology in Fig. 7A. However, one should not be misled by the apparent simplicity of the scheme in Fig. 7D. To set up an oscillation requires at least local feed back through a pool of neurons, and tuning the oscillation frequency to the visual delay may be as difficult as delaying the efference copy by an appropriate amount of time.

There are several lessons to be drawn from the above metamorphoses: (1) Our transformation of a model which seems to be the most parsimonious expression of a basic idea, perceptual feed back in our case, into other models with different, but also parsimonious internal structure and similar behaviour, unravels a corresponding transformation of the initial idea into equivalent ones. For example, instead of constructing target velocity by adding a delayed efference copy to the current error, we could also compute it on the basis of the current error and its past history, as in scheme D, where the history is stored under the form of oscillations. We can do so because we can infer, from our knowledge of the system, the motor commands engendered by the previous error signals and the eye movements caused by these commands. (2) In interpreting neural data, the mere similarity of a recorded discharge pattern to a waveform occurring within a model is intellectually stimulating but far from conclusive. For example, the occurrence of patterns resembling the velocity of a target with respect to space does not yet prove the existence of perceptual feedback. The scheme in D shows that it could have been generated in a different way. On the other hand, as pointed out by Deno et al. (1989), discharge patterns with no readily interpretable relationship to eye, error or target velocity nevertheless might contribute to the reconstruction of target velocity from its slip velocity on the retina, if we consider the possibility of distributed processing in D. It is probably fair to say that notwithstanding the encouraging results of Goldreich et al. (cf. above), so far, none of the principles suggested for the pursuit system has been able to gain the status of an "inferential reality" comparable to that of the local feedback hypothesis for the saccadic system.

The future of modelling: Take-over by neural network models?

In the last decade researchers in the field of oculomotor physiology have increasingly adopted learning neural networks as a modelling tool. There are, in fact, a number of features that make computational neural networks (CNNs) a much closer analogue of biological neural networks (BNNs) than the "classical" cybernetic description can be - no wonder, given that the neural networks of informatics draw on speculations about the operation of biological neurons. Particularly attractive is the principle of step wise self-optimization by trial and error which also underlies the evolution of biological systems during phylogenesis and their fine tuning during ontogenesis. Moreover, with regard to the oculomotor system, the layered structure of CNNs is reminiscent of the sequence of visual and visuomotor maps which make up most of the system's afferent leg.

CNNs have been trained to mimic such diverse functions as the neural integrator of the oculomotor system (Arnold & Robinson, 1991), the translation of vestibular stimuli about arbitrary axes into the extraocular muscle activations subserving the corresponding vestibulo-ocular reflex (Anastasio & Robinson, 1990), or the presumed shift of a visual object's representation on a visual map by an efference copy of eye position (Zipser & Andersen, 1988). It is revealing that in all of the above examples the function to be imitated has first been defined in terms borrowed from, or at least inspired by, systems theory. This underlines the fact that CCNs are not an alternative *replacing* the classical modelling approach but a highly welcome *complement*: CNNs help to demonstrates that, in principle, a hypothesized function can be realized by a pool of neurons. To the degree that their hitherto often arbitrary topology can be adjusted to reflect the known neuroanatomy and electrophysiology of a function's hypothesized biological substrate, they will finally also help to prove (or disprove) hypotheses. However, most present day CNN models still try to transplant the life of a highly structured, but largely unknown topology of specialized neurons into a random, unstructured matrix of stereotyped neurons using a few trial and error cycles, whereas the real system has evolved over million of years and has passed a plethora of tests most of which we are probably unaware of, as yet.

CNNs probably can learn almost any set of input-output relations. In doing so they do not need to have an "insight" into the structure of the underlying processing in terms of abstract concepts - such as "integration", or "threshold", or "local feedback". Important as such concepts are for the human in his search for a continuous chain of intelligible causes and effects and for arriving at a holistic view of problems, they do not exist as separable entities in a general purpose CNN. From the viewpoint of a practical engineer it is this ability of CNNs to synthesize a set of input-output transformations which neither he nor the network understand that makes these networks particularly advantageous. For the physiologist who tries to analyze the mechanism of existing transformations this advantage can turn into a clear handicap: Having taught a CNN to mimic the behaviour of a biological system, he will face a similar problem as with the original BNN, namely to infer from the activity and from the connections of a multitude of single neurons how the behaviour has been realized and what it might mean.

In summary, classical systems considerations (including non-linear and stochastic methods) are likely to play a twofold role in the context of CNN modelling: (1) Identify and isolate the major processing steps of the system under consideration in order to segment it into stages that are amenable to the attempt at understanding their function in terms of a neural network, preferentially one which reflects the neuroanatomy and physiology of the candidate substrate of that function. For example, having recognized that a comparison between desired and actual eye displacement must occur somewhere ahead of the saccadic system's burst neurons and that the superior colliculus is a structure where this operation might take place, it is sensible to investigate neural network models of the colliculus in order to understand how such a comparison could be achieved by this structure. (2) Analyze how a successful CNN produces the behaviour it has learned, in much the same way as one would try to analyze the behaviour of the biological system. Suppose, for example, the controller, C, of the pursuit system in Fig. 6B is a CNN which we have trained to produce step responses of the type observed in human or simian subjects. The challenge then is to understand how the CNN controller realizes this behaviour.

On a longer perspective, it can be anticipated that the painstaking investigations of oculomotor structures in animals that are being carried out worldwide will eventually, although not tomorrow, provide all the essential information that is needed to understand the motor part of at least the saccadic system in terms of homeomorphic neural network models. Even then, however, it will often be preferable to base functional considerations on a macroscopic signal flow diagram as a high level language which obviates the pain of reading the machine code consisting of individual neurons and their interconnections. However, those of us who do not dispose of a laboratory equipped for such experiments may need to recognize that the number of significant discoveries that can be obtained by transforming some students into black boxes and inferential modelling of their input-output relations is likely to decline.

References

Anastasio, T.J. & Robinson, D.A. (1990) Distributed parallel processing in the vertical vestibulo-ocular reflex: Learning networks compared to tensor theory. Biol. Cybern., 63: 161-167.

Arnold, D.B. & Robinson, D.A. (1991) A learning network model of the neural integrator of the oculomotor system. Biol. Cybern., 64: 447-454.

Bahill, A.T, McDonald, J.D. (1983) Model emulates human smooth pursuit system producing zero-latency target tracking. Biol.Cybern., 48: 213-222

Becker, W. Jurgens, R. (1979) An analysis of the saccadic system by double step stimuli. Vision Res., 19:967-983

Becker, W., King, W.M., Fuchs,A.F., Jürgens, R., Johanson, G. and Kornhuber, H.H. (1981) Accuracy of goal directed saccades and mechanisms of error correction. In: Fuchs, A.F. and Becker, W. (eds) Progress in Oculomotor Research,Elsevier/North-Holland, New York, pp 29-37.

Becker, W. (1994) Banging the bang-bang circuit: Experimental tests of Robinson's local feedback theory of saccade generation. In: Fuchs, A.F., Brandt, Th., Büttner, U. & Zee, D.S. (eds) Contemporary Ocular Motor and Vestibular Research: A Tribute to David A. Robinson, Thieme Verlag, Stuttgart (in press).

Deno, D.C., Keller, E.L. & Crandall, W.F. (1989) Dynamical Neural Network Organization of the Visual Pursuit System, IEEE. Trans. Biomed. Eng., 36: 85-92.

Fuchs, A.F, Luschei, E.S. (1970) Firing patterns of abducens neurons of alert monkeys in relationship to horizontal eye movement. J.Neurophysiol., 33: 382-392 .

Goldreich, D., Krauzlis, R.J., Lisberger, S.G. (1992) Effect of changing feedback delay on spontaneous oscillations in smooth pursuit eye movements of monkeys. J. Neurophysiol., 67: 625-638.

Guitton, D. (1992) Control of eye-head coordination during orienting gaze shifts. TINS, 15:174-179

Jürgens, R., Becker, W. & Kornhuber HH (1981) Natural and drug-induced variations of velocity and duration of human saccadic eye movements: Evidence for a control of the neural pulse generator by local feedback. Biol. Cybern., 39: 87-96

Kaneko, C.R.S. & Fuchs, A.F. (1991) Saccadic eye movement deficits following ibotenic acid lesions of the nuclei raphe interpositus and prepositus hypoglossi in monkey. Acta Otolaryngol., Suppl.481: 213-215.

Keller, E.L. (1974) Participation of medial pontine reticular formation in eye movement generation in monkey. J. Neurophysiol., 37: 316-332.

Kornhuber, H.H. (1971) Motor functions of cerebellum and basal ganglia. Kybernetik 8:157-162.

Küpfmüller, K, Poklekowski, G. (1993) Der Regelmechanismus willkürlicher Bewegungen. Z.Naturforsch., 11b: 1-7.

Ludvigh, E. (1952) Possible role of proprioception in the extraocular muscles. A.M.A.Arch.Ophthalmol., 160: 436-441

Luschei, E.S. & Fuchs, A.F. (1972) Activity of brain stem neurons during eye movements of alert monkey.].Neurophysiol. 35:445-461.

Munoz, D.P. & Galiana, H.L. (1990) Gaze control in the cat: Studies and modelling of the coupling between orienting eye and head movements in different behavioral tasks. J. Neurophysiol., 64: 509-531.

Newsome, W.T., Wurtz, R.H. & Komatsu, H. (1988) Relation of cortical areas MT and MST to pursuit eye movements. II. Differentiation of retinal from extraretinal inputs. J. Neurophysiol., 60: 604-620.

Robinson, D.A. (1964) The mechanics of human saccadic eye movement. J. Physiol.(Lond.), 174: 245-264

Robinson, D.A. (1970) Oculomotor unit behavior in the monkey. J. Neurophysiol., 33: 393-404

Robinson, D.A. (1975) Oculomotor control signals. In: Lennestrand G & Bach-y-Rita P (eds) Mechanisms of Ocular Motility and their Clinical Implications. Pergamon, Oxford, pp 337-374

Robinson, D.A. (1981) Control of eye movements. In: Brookhart, J.M., Mountcastle, V.B., Brooks, V.B. & Geiger S.R. (eds) Handbook of Physiology - The Nervous System II, Am. Physiol. Soc., Bethesda, Maryland.

Robinson, D.A. (1986) The systems approach to the oculomotor system. Vision Res. 26:91-99

Robinson, D.A. (1987) The windfalls of technology in the oculomotor system. Invest. Opthalmol. Visual Science 28:1912-1924

Robinson, D.A., Gordon, J.L. and Gordon, S.E. (1986) A model of the smooth pursuit eye movement system. Biol. Cybern., 55: 43-57.

Scudder, C.A., Fuchs, A.F. & Langer, T.P. (1988) Characteristics and functional identification of saccadic inhibitory burst neurons in the alert monkey. J. Neurophysiol., 59: 1430-1454.

Skavenski, A.A.., Robinson, D.A. (1973) Role of abducens neurons in vestibuloocular reflex. J.Neurophysiol., 36: 724-739

Steinhausen, W. (1933) Über die Beobachtung der Cupula in den Bogengangsampullen des Labyrinths des lebenden Hechtes. Pflügers Arch.Ges.Physiol., 232: 500-512.

Sünderhauf A (1960) Untersuchungen über die Regelung der Augenbewegungen. Klin.Monatsbl.Augenheilk., 136: 837-852

Suzuki, D.A. & Keller, E.L. (1988) The role of the posterior vermis of monkey cerebellum in smooth-pursuit eye movement control. II. Target velocity-related purkinje cell activity. J. Neurophysiol., 59: 19-39.

Vossius G (1960) Das System der Augenbewegung (I). Z.Biol., 112: 27-57

Vossius G (1961) Die Regelbewegung des Auges. In: NTG im VDE (ed) Aufnahme und Verarbeitung von Nachrichten durch Organismen. Hirzel-Verlag, Stuttgart, pp 149-156.

Waitzman, T.P.Ma., Optican, L.M. & Wurtz, R.H. (1988) Superior colliculus neurons provide the saccadic motor error signal. Exp. Brain Res., 72: 649-652.

Westheimer, G. (1954a) Mechanism of saccadic eye movements. A.M.A.Arch. Ophthalmol., 52: 710-724.

Westheimer, G. (1954b) Eye movement responses to a horizontally moving visual stimulus. A.M.A.Arch.Ophthalmol., 52: 932-941.

Wiener, N. (1949) Cybernetics. Wiley, New York.

Yasui, S. & Young, L.R. (1975) Perceived visual motion as effective stimulus to pursuit eye movement system. Science 190: 906-908.

Young, L.R. (1962) A sampled data model for eye tracking movements. Sc.D.Thesis. Massachusetts Institute of Technology

Young, L.R. (1963) A sampled data model for eye tracking movements. In: Broid, V. (ed) Automatic and Remote Control. Proc. 2nd Internat. Congr. IFAC, Basel, 1963. Butterworth, London, pp 454-463

Young, L.R., Forster, J.D., Van Houtte, N. (1968) A revised stochastic sampled data model for eye tracking movements. 4th Annual NASA-University Conference on Manual Control, University of Michigan, Ann Arbor, Michigan.

Young, L.R., Forster, J.D. & Van Houtte, N. (1968) A revised stochastic sampled data model for eye tracking movements. Fourth Annual NASA-University Conference on Manual Control, U. of Michigan, Ann Arbor, Michigan.

Young, L.R. (1971) Pursuit eye tracking movements. In: Bach-y-Rita P, Collins CC, Hyde JE (eds) The Control of Eye Movements. Academic Press, New York, pp 429-443

Zee, D., Robinson, D.A. & Eng, D. (1979) A hypothetical explanation of saccadic oscillations. Ann. Neurol., 5: 405-414.

Zipser, D. & Andersen, R.A. (1988) A back-propagation programmed network that simulates response properties of a subset of posterior parietal neurons. Nature, 331: 679-684.

THE SENSING OF OPTIC FLOW BY THE PRIMATE OPTOKINETIC SYSTEM

F. A. Miles

Laboratory of Sensorimotor Research, National Eye Institute, Bethesda, MD 20892, USA.

Abstract

Primates have several reflexes that generate eye movements to compensate for bodily movements that would otherwise disturb their gaze and undermine their ability to process visual information. Two vestibulo-ocular reflexes compensate selectively for rotational (RVOR) and translational (TVOR) disturbances of the head, receiving their inputs from the semicircular canals and otolith organs, respectively. Two independent visual tracking systems deal with any residual disturbances of gaze (global optic flow) and are manifest in the two components of the optokinetic response: the early, or direct, component (OKNe) with brisk dynamics and the delayed, or indirect, component (OKNd) with sluggish dynamics. I hypothesize that OKNd - like the RVOR - is phylogenetically old, being found in all animals with mobile eyes, and that it evolved as a backup to the RVOR to compensate for residual rotational disturbances of gaze. In contrast, OKNe seems to have evolved much more recently in animals with significant binocular vision and, I suggest, acts as a backup to the TVOR (also recently evolved?) to deal primarily with translational disturbances of gaze. I also suggest that highly complex optic flow patterns (such as those experienced by the moving observer who looks a little off to one side of his direction of heading) are dealt with by a third visual tracking mechanism, the smooth pursuit system, which is the most recently evolved of all and spatially filters visual motion inputs so as to exclude all but the motion of the object of interest (local optic flow).

Keywords

optic flow, optokinetic response, vestibulo-ocular reflex, translation, rotation.

Introduction

Eye movements exist to facilitate vision, often by preventing movement of the image(s) on the retina which, if excessive, can impair visual acuity. Motion of the observer poses a serious challenge to the stability of retinal images and is dealt with chiefly by labyrinthine reflexes that generate compensatory eye movements. The primate labyrinth has two types of receptor organ which provide the input for two vestibulo-ocular reflexes that compensate selectively for rotational and translational disturbances of the head. These labyrinthine reflexes, which I shall refer to as the rotational and translational vestibulo-ocular reflexes (RVOR and TVOR), operate open-loop insofar as they produce an output, eye movement, that does not influence their input, head movement. One serious consequence of this is that if these open-loop reflexes fail to compensate completely, which is not uncommon, then the eye

Eye Movement Research/J.M. Findlay et al. (Editors)
1995 Elsevier Science B.V.

movements will not completely offset the head movements and the image of the world on the retina will tend to drift. However, such retinal image slip activates visual tracking mechanisms that operate as closed-loop negative feedback systems to rotate the eyes so as to reduce the slip. In studies of these visual backup systems it has been usual to examine the visual compensatory mechanisms by rotating the visual surroundings around the stationary subject, the ensuing ocular following being termed *optokinetic nystagmus* (OKN). Only recently have translational visual stimuli been introduced, and this has led to a reinterpretation of some of the responses to conventional optokinetic stimuli. Thus, the primate optokinetic response has two components and my colleagues and I have recently suggested that, in the real world, one component is mostly concerned with *rotational* disturbances - as generally supposed - but the other is mostly concerned with *translational* problems.

Two Vestibulo-ocular Reflexes

The RVOR, which has been studied extensively in a wide variety of animals, senses angular accelerations of the head through the semicircular canals and generates compensatory eye movements that offset head turns. The canals are largely insensitive to translation (Goldberg & Fernandez, 1975) and, in monkeys, the gain of the RVOR - the ratio of the output (eye rotation) and the input (head rotation) - is close to unity so that the retinal image of the world is reasonably stable during pure head turns. For the moment, I shall ignore the fact that the eyes lie some distance in front of the axis of rotation of the head and so undergo some translation during normal head turns.

The TVOR, which has attracted relatively little attention until very recently, senses linear accelerations of the head through the otolith organs and generates compensatory eye movements to offset head translations. The otolith organs are largely insensitive to rotations in the horizontal plane (Goldberg & Fernandez, 1975). [Head rotations in the vertical plane activate the otolith organs because they alter the orientation of the head with respect to the earth's gravity. This can give rise to ocular counter-rolling, an otolith-ocular reflex that acts to maintain the *orientation* of the eyes. This reflex is very weak in monkeys (Krejcova *et al.*, 1971) and I shall not consider it here.] To be optimally effective, the output of the TVOR should accord with the proximity of the object of interest, nearby objects necessitating much greater compensatory eye movements than distant ones in order for their retinal images to be stabilized by the passing observer. Recent experiments on monkeys indicate that TVOR responses to lateral translation in this species are linearly related to the inverse of the viewing distance even when precautions are taken to exclude any contribution from visual tracking (Schwarz *et al.*, 1989; Schwarz & Miles, 1991). Interestingly, the compensatory eye movements generated by the TVOR during forward motion (in the dark, to exclude visual tracking) are gaze dependent, operating to increase any eccentricity of the eyes with respect to the direction of heading, exactly as would be required by the optic flow pattern in this situation during normal viewing: if the observer's gaze is directed downwards during the forward motion then his/her compensatory eye movements are downward, while if gaze is directed to the right of the direction of heading then the compensatory eye movements are rightward, and so forth (Paige *et al.*, 1988).

Two Visual Stabilization Mechanisms

In the traditional optokinetic test situation the stationary subject is usually seated inside a cylindrical enclosure with vertical patterned walls that can be rotated about him. Usually, the experiment starts with the subject in the dark to allow the cylinder time to reach the desired constant speed, at which point the lights are turned on for a period generally lasting many seconds. When the lights come on the subject tracks the continuously moving walls of the cylinder with his eyes (slow phases), necessitating regular saccades (quick phases) to recenter the eyes: OKN. If the cylinder rotates rapidly (>60°/sec), the slow phases can take considerable time to approach the speed of the cylinder and the development of the response shows two distinct episodes that are generally thought to reflect two distinct mechanisms: Firstly, there is an initial rapid rise in slow-phase eye speed during the first few hundred milliseconds that generally leaves the eyes somewhat short of the cylinder speed; this has been termed the "direct" component of OKN by Cohen *et al* (1977) but I shall refer to it simply as the *early* component (OKNe). Secondly, there is subsequently a gradual increase in slow-phase eye speed extending over a period of perhaps half a minute during which the eyes reach an asymptotic speed more nearly approaching that of the cylinder; this has been termed the "indirect" component by Cohen *et al* (1977) but I shall refer to it as the *delayed* component (OKNd); Cohen and his coworkers attributed OKNd to the gradual charging up of a central velocity storage integrator (Cohen *et al.*, 1977; Matsuo & Cohen, 1984; Raphan *et al.*, 1977; Raphan *et al.*, 1979). If the lights are then extinguished, the sequence of events is reversed: there is an immediate drop in eye speed reflecting the rapid loss of OKNe, followed by a roughly exponential drop extending over a period of many seconds reflecting the more gradual loss of OKNd, perhaps due to discharge of the proposed velocity storage integrator (optokinetic afternystagmus, OKAN).

My colleagues and I have recently suggested that only OKNd evolved as a true visual backup to the RVOR to help compensate for rotational disturbances of gaze, and that OKNe evolved as a backup to the TVOR to help compensate for translational problems (Schwarz *et al.*, 1989; Busettini *et al.*, 1991; Miles & Busettini, 1992; Miles *et al.*, 1992a,b; Miles, 1993). The major distinguishing features of these two proposed visual tracking mechanisms are listed in Table 1. Whereas the decomposition of head movements into rotational and translational components by the labyrinth is complete and unequivocal in primates, the decomposition of optic flow into rotational and translational components by the optokinetic system is incomplete and has only indirect supporting evidence. Before considering this evidence it is necessary to discuss some general aspects of the optic flow experienced by the moving observer.

Table 1. Synosis of the hypothesis that two visual tracking systems have evolved to deal with rotational and translational disturbances of gaze (after Miles *et al*, 1992b).

THE ROTATIONAL MECHANISM	THE TRANSLATIONAL MECHANISM
Delayed component of OKN: • Long time-constant.[a] • Strong after-nystagmus.[a] • Sensitive to low speed.[ab] • Centripetal/fugal asymmetry.[d]	*Early* component of OKN: • Short time-constant.[a] • Weak after-nystagmus.[a] • Sensitive to high speed.[ac] • No centripetal/fugal asymmetry?
Backup to canal-ocular reflex (RVOR)? • Sensitive to gain of RVOR[e] • Insensitive to gain of TVOR?	Backup to otolith-ocular reflex (TVOR)? • Insensitive to gain of RVOR[e] • Sensitive to gain of TVOR.[f]
Organized in canal planes?	Organized in otolith planes?
Helps stabilize gaze against *en masse* global disturbances? • Dumped by motion parallax? • Insensitive to disparity? • No fig/grd discrimination? • Input from entire visual field?	Helps stabilize gaze on *local depth plane of interest*? • Can utilize motion parallax.[c] • Sensitive to disparity.[g] • Primitive fig/grd discrimination.[c] • Input from binocular field only.[c]
Pretectum/Accessory Optic System.[h]	Direct Cortico-Pontine-Cerebellar System.[j]
In all animals with mobile eyes?	Only animals with good binocular vision?

[a] Cohen *et al* (1977). [b] Zee *et al* (1987). [c] Miles *et al* (1986). [d] Van Die & Collewijn (1982); Naegele & Held (1982); Westall & Schor (1985); Ohmi *et al* (1986). [e] Lisberger *et al* (1981). [f] Busettini *et al* (1991). [g] Howard & Gonzalez (1987); Kawano *et al* (1994). [h] Schiff *et al* (1990). [j] Kawano *et al* (1990; 1992a; 1992b); Shidara & Kawano (1993).

Optic Flow

Pure rotation of the observer results in *en masse* motion of the retinal image if the observer is passive (that is, makes no compensatory eye movements) and if the second-order translational effects due to the eccentricity of the eyes with respect to the axis of head rotation are ignored. In this situation, both the direction and the speed of the optic flow at all points are dictated entirely by the observer's motion - the 3-D structure of the visual scene is irrelevant. Provided that any compensatory eye movements are in the correct direction, only the *speed* of flow will be altered - the *global flow pattern* will remain the same. (Even with *over*compensation, when the direction of flow is reversed, the general pattern is still otherwise preserved.) In principle, appropriate compensatory eye movements could completely offset the visual effects due to rotational disturbances so that the entire scene would be stabilized on the retina.

If the passive observer undergoes pure translation, the optic flow consists of streams of images emerging from a point straight ahead and disappearing into another point behind. One is generally not very aware of this except when travelling at high speed, as in a car on the highway: ahead the world seems to be expanding while behind it seems to be contracting. As with rotational disturbances, the *direction* of flow at any given point here depends solely on the motion of the observer, but in contrast, the *speed* of the flow at any given point now also depends on the *viewing distance* at that location. As a consequence, the visual motion experienced during translation includes complex image shear as the nearby objects move across the field of view much more rapidly than the more distant ones: *motion parallax* (Gibson, 1950; Gibson, 1966). The sensation here is of the visual world pivoting around the far distance, an effect most readily appreciated during lateral translation, as when looking out from a fast moving train: see Fig. 1A.

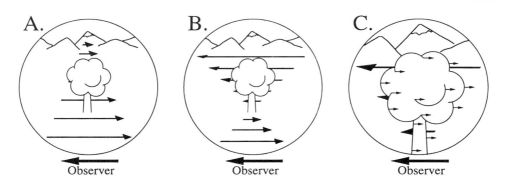

Fig. 1. *Optic flow associated with lateral translation of the observer, as when looking from a train, for example. (A) The observer makes no compensatory eye movements so that the moving scene appears to pivot about the distant mountains (effective infinity). (B) The observer attempts to stabilize the retinal image of a particular object in the middle ground (tree), presumably using the TVOR and OKNe, the result being that the scene now appears to pivot about the tree. (C) Same situation as in (B), except that the central field is enlarged and compensation is assumed to be less than adequate so that the scene actually pivots about a point just beyond the tree; note the opposing flow in the central and peripheral regions of the field. The length of the arrows denotes the speed of flow in particular parts of the field. (From Miles et al, 1992b.)*

This dependence on proximity also means that ocular compensation for linear disturbances of the observer can never stabilize the entire retinal image if the visual scene has 3-D structure. Furthermore, the compensatory eye movements associated with translational disturbances can have a profound effect on both the speed and the pattern of the optic flow. For example, with a pure lateral translation of the observer, "compensatory" rotations of the eyes would at best stabilize only images of objects at one particular viewing distance: in the case where the observer moves to the left, objects nearer than this "plane of stabilization" would appear to move rightwards while more distant objects would appear to move leftwards (motion parallax). The sensation here would be of the visual world pivoting around the stabilized object: see Fig. 1B. There is evidence that the visual stabilization mechanism in monkeys has special features to deal with this motion parallax.

OKNd: Provides Backup to the RVOR?

Evidence that OKNd is linked to the RVOR comes from the observation that changes in the gain of the RVOR, which can be induced with magnifying or minifying spectacles (Miles & Fuller, 1974; Miles & Eighmy, 1980), result in parallel changes in the amplitude of OKNd but *not* of OKNe (Lisberger *et al.*, 1981). This is consistent with the idea that OKNd shares some central pathway(s) with the RVOR - indeed, that portion of the pathway(s) containing the variable gain element(s) responsible for adaptive gain control of the RVOR - and supports the notion that these two systems are truly synergistic, combining to compensate selectively for rotational disturbances of the observer: see Fig. 2A The presumption here is that shared properties reflect shared anatomy and shared function.

Canal-ocular reflexes with brisk dynamics (like the RVOR) complemented by optokinetic reflexes with sluggish dynamics (like OKNd) appear to be ubiquitous among contemporary vertebrates with mobile eyes, suggesting that these compensatory mechanisms evolved early in a common ancestor. My colleagues and I have suggested that these primordial visuo-vestibular mechanisms originated in a lateral-eyed progenitor and compensated almost exclusively for rotational disturbances of the observer, the visual backup having special features that rendered it largely blind to the commonest translational disturbances of gaze - those associated with forward locomotion - and, further, that the primordial visual system actually relied on the motion parallax associated with locomotion to decode the 3-D structure of the environment.

This view of the primordial optokinetic system comes in part from consideration of a contemporary lateral-eyed mammal, the rabbit, whose oculomotor system is largely insensitive to both the visual and the vestibular consequences of forward locomotion (Baarsma & Collewijn, 1975; Collewijn & Noorduin, 1972). Presumably, this feature is desirable in a lateral-eyed animal since temporalward rotation of the two eyes (divergence) would disrupt the images of the scene ahead - the very region of the field where potentially interesting new images are being actively sought (Howard, 1982). Recordings from single neurons assumed to mediate OKN in the rabbit suggest that in this species OKN has the same frame of reference as the RVOR: individual neurons have very large (binocular) visual receptive fields and respond best to global rotations of the entire visual scene about axes that coincide with the rotational axes that are optimal for activating particular semicircular canals (Graf *et al.*, 1988; Leonard *et al.*, 1988; Simpson *et al.*, 1988; Soodak & Simpson, 1988). Presumably, this arrangement facilitates the orderly summation of rotational information from the retina and the canals.

However, the mere fact that the visual receptive fields of the neurons mediating OKN are organized to respond optimally to rotations does not necessarily render those neurons totally insensitive to translation. Consider, for example, the directionally selective neurons in the rabbit's optokinetic system that provide the backup for the vestibulo-ocular reflexes emanating from the horizontal canals and are activated by temporalward motion in one eye and nasalward motion in the other. During forward locomotion, when both eyes sees temporalward motion, the visual

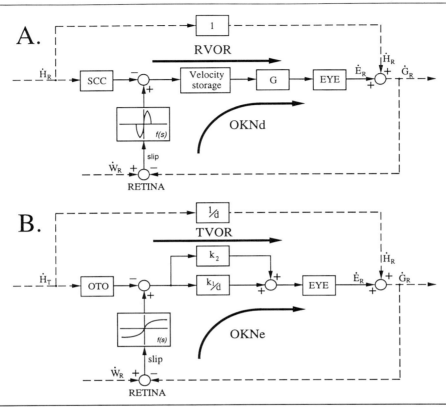

Fig. 2. Block diagrams showing the proposed linkages between the visual and vestibular reflexes operating to stabilize gaze. (A) The open-loop RVOR and the closed-loop OKNd generate eye movements, $\dot{E}_R$, that compensate for rotational disturbances of the head, $\dot{H}_R$. These reflexes share (a) a velocity storage element, which is responsible for the slow build-up in OKN and the gradual decay in RVOR with sustained rotational stimuli, and (b) a variable gain element, G, which mediates long-term regulation of RVOR gain. SCC, semicircular canals. The element, f(s), indicates that the visual input is sensitive to low slip speeds only. (From Miles et al, 1992a.) (B) The open-loop TVOR and the closed-loop OKNe generate eye movements that compensate for translational disturbances of the head, $\dot{H}_T$, which affect gaze in inverse proportion to the viewing distance, d. These reflexes share (a) a variable gain element, k_1/d, which gives them their dependence on proximity, and (b) a fixed gain element, k_2, which generates a response irrespective of proximity. OTO, otolith organs. (From Schwarz et al, 1989). Dashed lines represent physical links: $\dot{H}_T$, head velocity in linear coordinates; $\dot{H}_R$, $\dot{E}_R$, $\dot{G}_R$ and $\dot{W}_R$, velocity of head, eyes (in head), gaze and visual surroundings, respectively, in angular coordinates.

input through one eye will tend to raise the activity of such neurons while the input through the other will tend to depress it, i.e., the inputs from the two eyes will tend to cancel. Thus, if the activation from one eye is balanced by the suppression from the other then there will be no net eye rotation. However, differences in the 3-D arrangement of objects on either side of the animal would cause asymmetrical optic flow speeds at the two eyes so that the motion inputs to the eye seeing the nearer objects would be more potent than those to the other eye. Differences in the number and saliency of the objects to either side also would result in marked asymmetries in the potency of the motion stimulus at each eye. Clearly, it would be rare for the motion inputs from the two eyes to exactly balance, and if the two eyes are yoked the net result would be neural activity that would tend to rotate the eyes towards the side with the nearer and/or more salient objects. I assume that such arbitrary behavior would be undesirable. Regardless, it is nonetheless apparent that visual receptive fields that are organized to respond optimally to rotations are not necessarily insensitive to translation. This raises two obvious questions: "How does the rabbit's optokinetic system avoid responding to translational optic flow?" "How does the rabbit's visual system process such translational flow signals?"

Selective Insensitivity to Translational Optic Flow.

The rabbit's optokinetic system has at least three special features that operate specifically to reduce its sensitivity to the temporalward motion created by the animal's forward movement. Firstly, temporalward motion generally has only a suppressive effect on the activity of the neurons in the optokinetic pathway so that the major drive comes from the withdrawal of the resting maintained discharge, which is often low (Collewijn, 1975; Simpson *et al.*, 1988; Soodak & Simpson, 1988). Secondly, these neurons are insensitive to motion in the lower visual field (Collewijn, 1975; Simpson *et al.*, 1981; Simpson *et al.*, 1979), thereby excluding a major potential source of translational contamination since objects in the lower visual field are likely to be the ones most near and hence their retinal images most sensitive to translation. Thirdly, these neurons are sensitive only to low speeds which is appropriate for a visual backup to the RVOR since the latter functions sufficiently well that any retinal slip due to rotational disturbances of gaze will generally be minor and hence within the operating range; as Dr.J.I. Simpson pointed out to me, substantial disturbances of gaze must therefore generally emanate from translational disturbances of gaze and will be ignored because they exceed the system's speed range. The net result is that in normal everyday conditions the rabbit's optokinetic system responds selectively to the rotational component of optic flow, being driven by the nasalward motion seen by the leading eye.

Considerations such as the above reinforce the common assumption that the classical optokinetic system - manifest in primates as OKNd - originally evolved to deal selectively with rotational disturbances of gaze. The critical question now is, "To what extent is the rabbit's optokinetic system a good model for the primate OKNd?". Curiously, the data for primates are in some respects less complete. For example, there have been no published attempts to show that OKNd in primates shares the coördinate framework of the semicircular canals. During forward locomotion, the pole of the optic flow field is located in the far periphery of the retina in lateral-eyed animals like the rabbit with their panoramic field of view, whereas for frontal-eyed animals such as ourselves and monkeys with a comparatively narrow field of view the pole is much more central and we see mostly an expanding flow field (centrifugal optic flow). If gaze is aimed directly at the pole of the flow field, pure foveofugal retinal image motion will result and no compensatory eye movements are called for to stabilize the foveal image. (The more common situation in which gaze is off to one side of the direction of heading will be considered later.) Once more, however, differences in the saliency and 3-D arrangement of the objects to either side could cause asymmetric optic flow that might tend to deviate the eyes. As in the rabbit, the system seems to deal with this potential problem by means of selective directional asymmetries: when tested monocularly with stimuli confined to one hemifield, the human optokinetic response is weaker when the motion is foveofugal than when it is foveopetal (Naegele & Held, 1982; Ohmi *et al.*, 1986; Van Die & Collewijn, 1982; Westall & Schor, 1985). Unfortunately, it is not clear whether the asymmetry is in OKNd and/or OKNe because the recordings were done during maintained stimulation when both components are present. There is evidence that the primate OKNd shares the rabbit optokinetic system's preference for low slip speeds (Zee *et al.*, 1987).

Translational Flow as a Source of 3-D Information.

One consequence of allowing the eyes to be dragged passively through the environment during locomotion - as the rabbit seems to do - is that the velocity gradients in the retinal image (motion parallax) accurately reflect the 3-D layout of objects in the scene, with high slip speeds indicating that objects are nearby and low slip speeds that they are distant. My colleagues and I have suggested that the primordial visual system may actually have depended on the optic flow associated with forward motion to sense the 3-D layout of the scene to either side, perhaps using relative-motion detectors such as those described in pigeon and primate tecta (Frost & Nakayama, 1983; Frost *et al.*, 1981; Davidson & Bender, 1991). Of course, this veridical relationship between the 2-D velocity gradients in the retinal image and the 3-D structure of the scene would be lost if the animal were to compensate for the translation.

OKNe: Provides Backup to the TVOR?

That there might be a visual complement specific to the TVOR reflex was first suggested by Nakayama (1985) and Koenderink (1986). There are now several lines of evidence which are consistent with the idea that OKNe acts synergistically with the TVOR to compensate selectively for translational disturbances of the head. We have seen that viewing distance is a critical parameter governing the optic flow associated with translations - but not rotations - and recent findings which indicate that the OKNe is sensitive to absolute and relative depth cues support the idea that this component of the optokinetic response is concerned primarily with translational optic flow.

Sensitivity to Absolute Viewing Distance.

Recent experiments on monkeys indicate that the initial ocular following responses to sudden movements of the visual scene share the TVOR's dependence on proximity, i.e., the amplitude of these visual tracking responses is inversely proportional to viewing distance, even when measures are taken to ensure that the retinal stimulus remains the same at all distances (Schwarz *et al.*, 1989; Busettini *et al.*, 1991). These data led to the suggestion that OKNe and the TVOR share a central pathway whose efficacy is modulated by absolute distance cues. This shared arrangement has a strong formal resemblance to that proposed earlier for the OKNd and the RVOR: compare A and B in Fig 2.

Sensitivity to Relative Depth Cues: Motion Parallax.

If OKNe is to combine successfully with the TVOR to compensate for translational disturbances of the observer it must be able to deal with the associated motion parallax, whereby retinal images of objects nearer than the "plane of stabilization" move in one direction while those beyond this plane have the reverse motion: Fig. 1B. If, as seems likely, the observer fails to compensate fully for his own motion then the scene will pivot around a point slightly beyond the object of regard so that the retinal image of the latter will slowly drift back across his central retina while the image of the distant background will be swept forwards across his peripheral retina (antiphase motion): Fig. 1C. Interestingly, there is evidence from studies of both monkeys and humans that the optokinetic system can utilize motion parallax cues such as these to improve its performance. Thus, when the scene is partitioned into separate central and peripheral zones, concurrent motion in the surround that is opposite in direction to that at the center actually improves the tracking of motion at the center - antiphase enhancement; conversely, surround motion that is in the same direction as that at the center degrades tracking performance - inphase suppression (Guedry Jr *et al.*, 1981; Hood, 1975; Miles *et al.*, 1986; Ter Braak, 1957; Ter Braak, 1962). Significantly, these effects due to concurrent motion in the peripheral retina are evident at short latency (<100 ms) and hence must be a characteristic of OKNe (Miles *et al.*, 1986). It is apparent from this that *en masse* motion is not the optimal stimulus for OKNe, which can use the motion parallax associated with translation to help stabilize the images of objects off to one side. My colleagues and I have proposed a preliminary model of OKNe with provision for antiphase enhancement and inphase suppression. Our basic scheme consists of a negative feedback tracking system that is *driven* by foreground images moving in the central retina and *modulated* by background images moving in the peripheral retina (Miles *et al.*, 1986; Miles *et al.*, 1992b).

The early component of OKN is completely eliminated by occipital lobectomy (Zee *et al.*, 1987) and compromised by lesions restricted to the medial superior temporal (MST) area of cortex (Dürsteler & Wurtz, 1988) or one of its relays through the ventral paraflocculus (Miles *et al.*, 1986). Interestingly, the antiphase pattern of retinal image motion that is optimal for driving OKNe in the monkey is also a very effective stimulus for some neurons in the middle temporal area (MT) of the monkey's cortex (Allman *et al.*, 1985; Tanaka *et al.*, 1986), a region that is known to project to MST (Maunsell & Van Essen, 1983; Ungerleider & Desimone, 1986).

Sensitivity to Relative Depth Cues: Disparity.

It has been shown recently that human OKN is much better when the moving visual scene is binocularly fused than when disparate (Howard & Gonzalez, 1987). These effects of disparity occurred quite rapidly, consistent with mediation by OKNe, and this study supports the idea that the ocular stabilization mechanism can respond selectively to the motion of objects in the plane of fixation and ignore the motion of objects that are nearer or further. This is consistent with a scheme in which the visual stabilization mechanism receives its main drive from neurons that are tuned for disparity and driven only by stimuli in and around the plane of fixation (Miles *et al.*, 1992b). However, it is also possible that these effects of disparity on OKN are secondary to shifts in attention, single fused images attracting attention much more successfully than disparate ones. Kawano *et al* (1994) have shown that when the visual scene is subdivided into central and peripheral regions, the tracking of motion at the center is enhanced when the images in the periphery have uncrossed binocular disparities (as though more distant) and tracking is diminished when those images have crossed disparities (as though nearer). However, this could arise indirectly from effects on the perceived distance to the moving display and need not necessarily indicate that the neurons decoding the motion have surrounds that are disparity selective.

The Smooth Pursuit System: A Spatial Filter?

There are many everyday situations in which it makes no sense for the oculomotor system to attempt a global analysis of the optic flow. As discussed earlier, frontal-eyed animals such as ourselves and monkeys mainly see an expanding flow field during forward movement (centrifugal flow), and at any given moment eye movements can compensate for the flow only in a particular region of the field. If the selected region lies to the right of the direction of heading, for example, then the observer's compensatory eye movements should be rightward, while if the selected region lies to the left of the direction of heading then the eye movements should be leftward and so forth. I have already mentioned that this gaze dependence is found in the TVOR (Paige & Tomko, 1991) and the only option for a visual stabilization mechanism here is to abandon the global analysis of the flow field and to concentrate solely on the local flow in the region of particular interest. Thus, appropriate ("compensatory") tracking in this situation first requires a decision as to which region should take precedence, presumably based on some assessment of the potential significance of the various features present, and then some spatial filtering to eliminate the visual inputs coming from other regions.

It is known that the primate smooth pursuit system - traditionally viewed as functioning to track small moving objects - can initiate ocular tracking with brisk dynamics in response to the motion of eccentric retinal images when the subject's attention is directed towards such targets (Lisberger & Westbrook, 1985; Lisberger & Pavelko, 1989; Newsome *et al.*, 1985; Rashbass, 1961; Tychsen & Lisberger, 1986).

Indeed, I suggest that it is through this mechanism that human subjects can arrange for their optokinetic responses to accord with the flow in a restricted region of the visual field, regardless of whether that region is foveal or extra-foveal (Cheng & Outerbridge, 1975; Van Den Berg & Collewijn, 1987; Dubois & Collewijn, 1979; Howard *et al.*, 1989; Murasugi *et al.*, 1986). I assume that monkeys too are capable of this and further suggest that the need to concentrate from time to time upon selected elements of the shifting scene on the retina provided the major pressure to evolve the pursuit system. Thus, I question the general supposition that the pursuit system evolved to track small moving objects, useful though this ability is. In the scheme that my colleagues and I have proposed, the pursuit system evolved as part of an attentional focussing mechanism that spatially filters the visual motion inputs driving the oculomotor system. It is not difficult to imagine how, once evolved, such a mechanism could also be deployed to track small moving targets, even across textured backgrounds - the response property that has been regarded as the hallmark of the pursuit system. I suggest that the pursuit system represents yet a third visual tracking mechanism that is substantially different from OKNd and OKNe and has evolved to stabilize the eyes on *local features* of interest in the often busy, swirling scene confronting the moving observer. Of course, pursuit is not a single entity and is known to have several components - the initiation has at least two phases which are in turn quite distinct from the maintenance phase (Lisberger & Pavelko, 1989; Lisberger & Westbrook, 1985; Morris & Lisberger, 1987).

Bibliography

Allman, J., Miezin, F. & McGuinness, E. (1985). Stimulus specific responses from beyond the classical receptive field: Neurophysiological mechanisms for local--global comparisons in visual neurons. Ann. Rev. Neurosci., *8*, 407-430.

Baarsma, E. A. & Collewijn, H. (1975). Eye movements due to linear accelerations in the rabbit. J. Physiol., *245*, 227-247.

Busettini, C., Miles, F. A. & Schwarz, U. (1991). Ocular responses to translation and their dependence on viewing distance. II. Motion of the scene. J. Neurophysiol., *66*, 865-878.

Cheng, M. & Outerbridge, J. S. (1975). Optokinetic nystagmus during selective retinal stimulation. Exp. Brain Res., *23*, 129-139.

Cohen, B., Matsuo, V. & Raphan, T. (1977). Quantitative analysis of the velocity characteristics of optokinetic nystagmus and optokinetic after-nystagmus. J. Physiol. (Lond.), *270*, 321-344.

Collewijn, H. (1975). Direction-selective units in the rabbit's nucleus of the optic tract. Brain Res., *100*, 489-508.

Collewijn, H. & Noorduin, H. (1972). Conjugate and disjunctive optokinetic eye movements in the rabbit, evoked by rotatory and translatory motion. Pflügers Arch., *335*, 173-185.

Davidson, R. M. & Bender, D. B. (1991). Selectivity for relative motion in the monkey superior colliculus. J. Neurophysiol., *65*, 1115-1133.

Dubois, M. F. W. & Collewijn, H. (1979). Optokinetic reactions in man elicited by localized retinal stimuli. Vision Res., *19*, 1105-1115.

Dürsteler, M. R. & Wurtz, R. H. (1988). Pursuit and optokinetic deficits following chemical lesions of cortical areas MT and MST. J. Neurophysiol., *60*, 940-965.

Frost, B. J. & Nakayama, K. (1983). Single visual neurons code opposing motion independent of direction. Science, *220*, 744-745.

Frost, B. J., Scilley, P. L. & Wong, S. C. P. (1981). Moving background patterns reveal double-opponency of directionally specific pigeon tectal neurons. Exp. Brain Res., *43*, 173-185.

Gibson, J. J. (1950). In *The Perception of the Visual World*, , Houghton Mifflin, Boston.

Gibson, J. J. (1966). In *The Senses Considered as Perceptual Systems*, , Houghton Mifflin, Boston.

Goldberg, J. M. & Fernandez, C. (1975). Responses of peripheral vestibular neurons to angular and linear accelerations in the squirrel monkey. Acta Otolaryngol., *80*, 101-110.

Graf, W., Simpson, J. I. & Leonard, C. S. (1988). Spatial organization of visual messages of the rabbit's cerebellar flocculus. II. Complex and simple spike responses of Purkinje cells. J. Neurophysiol., *60*, 2091-2121.

Guedry Jr, F. E., Lentz, J. M., Jell, R. M. & Norman, J. W. (1981). Visual-vestibular interactions: The directional component of visual background movement. Aviat. Space Environ. Med., *52*, 304-309.

Hood, J. D. (1975). Observations upon the role of the peripheral retina in the execution of eye movements. J. Otorhinolaryngol., *37*, 65-73.

Howard, I. (1982). In *Human Visual Orientation*, , Wiley, London.

Howard, I. P., Giaschi, D. & Murasugi, C. M. (1989). Suppression of OKN and VOR by afterimages and imaginary objects. Exp. Brain Res., *75*, 139-145.

Howard, I. P. & Gonzalez, E. G. (1987). Human optokinetic nystagmus in response to moving binocularly disparate stimuli. Vision Res., *27*, 1807-1816.

Kawano, K., Inoue, Y., Takemura, A. & Miles, F. A. (1994). Effect of disparity in the peripheral field on short-latency ocuar following responses. Vis. Neurosci., In press.

Kawano, K., Shidara, M., Watanabe, Y. & Yamane, S. (1992a). Short-latency responses of neurons in dorsolateral pontine nucleus and cortical area MST of alert monkey to movement of large-field visual stimulus. In *Vestibular and Brain Stem Control of Eye, Head and Body Movements*, H. Shimazu, Y. Shinoda, ed., pp. 397-404, Japanese Scientific Societies Press.

Kawano, K., Shidara, M. & Yamane, S. (1992b). Neural activity in dorsolateral pontine nucleus of alert monkey during ocular following responses. J. Neurophysiol., *67*, 680-703.

Kawano, K., Watanabe, Y., Kaji, S. & Yamane, S. (1990). Neuronal activity in the posterior parietal cortex and pontine nucleus of alert monkey during ocular following responses. In *Vision, Memory and the Temporal Lobe*, E. Iwai, M. Mishkin, ed., pp. 311-315, Elsevier, New York.

Koenderink, J. J. (1986). Optic flow. Vision Res., *26*, 161-179.

Krejcova, H., Highstein, S. & Cohen, B. (1971). Labyrinthine and extra-labyrinthine effects on ocular counter-rolling. Acta Otolaryng., *72*, 165-171.

Leonard, C. S., Simpson, J. I. & Graf, W. (1988). Spatial organization of visual messages of the rabbit's cerebellar flocculus. I. Typology of inferior olive neurons of the dorsal cap of Kooy. J. Neurophysiol., *60*, 2073-2090.

Lisberger, S. G., Miles, F. A., Optican, L. M. & Eighmy, B. B. (1981). Optokinetic response in monkey: Underlying mechanisms and their sensitivity to long-term adaptive changes in vestibuloocular reflex. J. Neurophysiol., *45*, 869-890.

Lisberger, S. G. & Pavelko, T. A. (1989). Topographic and directional organization of visual motion inputs for the initiation of horizontal and vertical smooth-pursuit eye movements in monkeys. J. Neurophysiol., *61*, 173-185.

Lisberger, S. G. & Westbrook, L. E. (1985). Properties of visual inputs that initiate horizontal smooth pursuit eye movements in monkeys. J. Neurosci., *5*, 1662-1673.

Matsuo, V. & Cohen, B. (1984). Vertical optokinetic nystagmus and vestibular nystagmus in the monkey: Up-down asymmetry and effects of gravity. Exp. Brain Res., *53*, 197-216.

Maunsell, J. H. R. & Van Essen, D. C. (1983). The connections of the middle temporal visual area (MT) and their relationship to a cortical hierarchy in the macaque monkey. J. Neurosci., *3*, 2563-2586.

Miles, F. A. (1993). The sensing of rotational and translational optic flow by the primate optokinetic system. In *Visual Motion and its Role in the Stabilization of Gaze*, F. A. Miles, J. Wallman, ed., pp. 393-403, Elsevier, Amsterdam.

Miles, F. A. & Busettini, C. (1992). Ocular compensation for self motion: visual mechanisms. In *Sensing and Controlling Motion: Vestibular and Sensorimotor Function*, B. Cohen, D. L. Tomko, F. Guedry, ed., pp. 220-232, Ann. NY Acad. Sci., New York.

Miles, F. A., Busettini, C. & Schwarz, U. (1992a). Ocular responses to linear motion. In *Vestibular and Brain Stem Control of Eye, Head and Body Movements*, H. Shimazu, Y. Shinoda, eds., pp. 379-395, Springer-Verlag/Japan Scientific Societies Press, Tokyo .

Miles, F. A. & Eighmy, B. B. (1980). Long-term adaptive changes in primate vestibuloocular reflex. I. Behavioral observations. J. Neurophysiol., *43*, 1406-1425.

Miles, F. A. & Fuller, J. H. (1974). Adaptive plasticity in the vestibulo-ocular responses of the rhesus monkey. Brain Res., *80*, 512-516.

Miles, F. A., Kawano, K. & Optican, L. M. (1986). Short-latency ocular following responses of monkey. I. Dependence on temporospatial properties of the visual input. J. Neurophysiol., *56*, 1321-1354.

Miles, F. A., Schwarz, U. & Busettini, C. (1992b). The decoding of optic flow by the primate optokinetic system. In *The Head-Neck Sensory-Motor System*, A. Berthoz, W. Graf, P. P. Vidal, ed., pp. 471-478, Oxford University Press, New York.

Morris, E. J. & Lisberger, S. G. (1987). Different responses to small visual errors during initiation and maintenance of smooth-pursuit eye movements in monkeys. J. Neurophysiol., *58*, 1351-1369.

Murasugi, C. M., Howard, I. P. & Ohmi, M. (1986). Optokinetic nystagmus: the effects of stationary edges, alone and in combination with central occlusion. Vision Res., *26*, 1155-1162.

Naegele, J. R. & Held, R. (1982). The postnatal development of monocular optokinetic nystagmus in infants. Vision Res., *22*, 341-346.

Nakayama, K. (1985). Biological image motion processing: a review. Vision Res., *25*, 625-660.

Newsome, W. T., Wurtz, R. H., Dürsteler, M. R. & Mikami, A. (1985). Deficits in visual motion processing following ibotenic acid lesions of the middle temporal visual area of the macaque monkey. J. Neurosci., *5*, 825-840.

Ohmi, M., Howard, I. P. & Eveleigh, B. (1986). Directional preponderance in human optokinetic nystagmus. Exp. Brain Res., *63*, 387-394.

Paige, G. D. & Tomko, D. L. (1991). Eye movement responses to linear head motion in the squirrel monkey. II. Visual-vestibular interactions and kinematic considerations. J. Neurophysiol., *65*, 1183-1196.

Paige, G. D., Tomko, D. L. & Gordon, D. B. (1988). Visual-vestibular interactions in the linear vestibulo-ocular reflex (VOR). Invest. Ophthalmol. Vis. Sci. (Suppl.), *29*, 342.

Raphan, T., Cohen, B. & Matsuo, V. (1977). A velocity-storage mechanism responsible for optokinetic nystagmus (OKN), optokinetic after-nystagmus (OKAN) and vestibular nystagmus. In *Developments in Neuroscience, Volume 1*, R. Baker, A. Berthoz, ed., pp. 37-47, Elsevier/North-Holland, Amsterdam.

Raphan, T., Matsuo, V. & Cohen, B. (1979). Velocity storage in the vestibulo-ocular reflex arc (VOR). Exp. Brain Res., *35*, 229-248.

Rashbass, C. (1961). The relationship between saccadic and smooth tracking eye movements. J. Physiol. (Lond.), *159*, 326-338.

Schiff, D., Cohen, B., Büttner-Ennever, J. & Matsuo, V. (1990). Effects of lesions of the nucleus of the optic tract on optokinetic nystagmus and after-nystagmus in the monkey. Exp. Brain Res., *79*, 225-239.

Schwarz, U., Busettini, C. & Miles, F. A. (1989). Ocular responses to linear motion are inversely proportional to viewing distance. Science, *245*, 1394-1396.

Schwarz, U. & Miles, F. A. (1991). Ocular responses to translation and their dependence on viewing distance. I. Motion of the observer. J. Neurophysiol., *66*, 851-864.

Shidara, M. & Kawano, K. (1993). Role of Purkinje cells in the ventral paraflocculus in short-latency ocular following responses. Exp. Brain Res., *93*, 185-195.

Simpson, J. I., Graf, W. & Leonard, C. (1981). The coordinate system of visual climbing fibers to the flocculus. In *Progress in Oculomotor Research*, A. Fuchs, W. Becker, ed., pp. 475-484, Elsevier North Holland, Amsterdam.

Simpson, J. I., Leonard, C. S. & Soodak, R. E. (1988). The accessory optic system of the rabbit. II. Spatial organization of direction selectivity. J. Neurophysiol.,   *60*, 2055-2072.

Simpson, J. I., Soodak, R. E. & Hess, R. (1979). The accessory optic system and its relation to the vestibulocerebellum. In *Reflex Control of Posture and Movement*, R. Granit, O. Pompeiano, ed., pp. 715-724, Elsevier, Amsterdam.

Soodak, R. E. & Simpson, J. I. (1988). The accessory optic system of the rabbit. I. Basic visual response properties. J. Neurophysiol., *60*, 2037-2054.

Tanaka, K., Hikosaka, K., Saito, H.-A., Yukie, M., Fukada, Y. & Iwai, E. (1986). Analysis of local and wide-field movements in the superior temporal visual areas of the macaque monkey. J. Neurosci. , *6*, 134-144.

Ter Braak, J. W. G. (1957). \"Ambivalent" optokinetic stimulation. Folia. Psychiat. Neurol. Neerl., *60*, 131-135.

Ter Braak, J. W. G. (1962). Optokinetic control of eye movements, in particular optokinetic nystagmus. Proc. 22th Int. Congr. physiol. Sci,. Leiden, *1*, 502-505.

Tychsen, L. & Lisberger, S. G. (1986). Visual motion processing for the initiation of smooth-pursuit eye movements in humans. J. Neurophysiol., *56*, 953-968.

Ungerleider, L. G. & Desimone, R. (1986). Cortical connections of visual area MT in the macaque. J. Comp. Neurol., *248*, 190-222.

Van Den Berg, A. V. & Collewijn, H. (1987). Voluntary smooth eye movements with foveally stabilized targets. Exp. Brain Res., *68*, 195-204.

Van Die, G. & Collewijn, H. (1982). Optokinetic nystagmus in man. Human Neurobiol., *1*, 111-119.

Westall, C. A. & Schor, C. M. (1985). Asymmetries of optokinetic nystagmus in amblyopia: the effect of selected retinal stimulation. Vision Res., *25*, 1431-1438.

Zee, D. S., Tusa, R. J., Herdman, S. J., Butler, P. H. & Güçer, G. (1987). Effects of occipital lobectomy upon eye movements in primate. J. Neurophysiol., *58*, 883-907.

THE FUNCTIONS OF EYE MOVEMENTS IN ANIMALS REMOTE FROM MAN

Michael F Land

Sussex Centre for Neuroscience, School of Biological Sciences,
University of Sussex, Brighton BN1 9QG, U.K.

Abstract

All vertebrates share a characteristic pattern of eye-movements which consists of periods of stationary fixation, separated by fast gaze-relocating saccades. The underlying reason for this strategy is the need to keep the retinal image almost stationary, to avoid blur. Primates, and a few other vertebrates, have an additional system for tracking small targets. If vision has the same basic requirements in all sighted animals, then evolutionarily unrelated creatures should share this pattern. Cuttlefish, crabs and many insects all show this pattern of fixations and saccades, with reflex compensation for body rotation. In some flying insects the same eye movements occur, but - unencumbered by contact with the ground - it is now the whole body that makes the saccades and fixations, or in some cases tracks a target.

There are, however, a few animals which employ a quite different strategy, taking in information when the eye is moving (scanning). Some sea-snails have a narrow retina which scans perpendicular to its long dimension in order to detect prey in the surrounding ocean. Mantis shrimps have a strip of retina across their compound eyes which contains their colour-vision system, and they move this so as to "colour in" the monochrome image in the rest of the eye. Jumping spiders have both conventional and scanning eyes viewing the same fields; the former detect motion and the latter elucidate pattern, distinguishing potential mates from prey. In all these cases the scanning movements are unlike saccades in being sufficiently slow for the receptors to generate fully modulated responses.

Keywords

Eye-movements, scanning, invertebrates, flow fields, motion blur, resolution

Introduction

For all their apparent variety, human eye movements are controlled by a small number of well-defined mechanisms. Gaze changes are made by the fast (saccadic) system, and the eye is held almost still during the intervening fixations by two powerful reflexes - the vestibulo-ocular reflex (VOR) and optokinetic nystagmus (OKN). This "saccade and fixate" system is supplemented in primates by vergence and smooth pursuit, the former concerned with keeping the two eyes in register for

Eye Movement Research/J.M. Findlay et al. (Editors)

objects at different distances, and the latter ensuring that small moving objects are kept on or near the fovea. Vergence and pursuit, although not confined to primates, are fairly uncommon amongst other vertebrates, as they evolved to deal with the special visual needs of front-eyed, foveate animals such as ourselves. The saccade and fixate strategy, however, seems to be nearly universal amongst vertebrates (Walls, 1962, Carpenter, 1988).

Why is this pattern so important, and how universal is it? Is it confined to the vertebrates, linked, perhaps, to our kind of camera-like eye? Or is it found in other phyletic groups with other kinds of eye? Do the cephalopod molluscs show it - animals with eyes like ours but of quite different evolutionary origins? Do insects and crustaceans with compound eyes share this strategy? The answers should tell us something important about the role of eye movements in vision. If they really *are* phylogenetically universal, then this argues very strongly that they are fundamental to the process of vision, and not just an interesting set of habits retained from our particular ancestors.

Powerful arguments have been advanced for thinking that our eye movement strategy is concerned above all with keeping retinal image velocity within a range that the retina can cope with: a few degrees per second in humans (Carpenter, 1988, 1992). The basis of the argument is that receptors with finite response times will not give fully modulated signals if structures in the image move too fast across them - in just the same way that slow shutter speeds blur photographs. If this is true then we would expect all animals with good eyesight to adopt some measures for keeping the image still, and, since animals move in the world, for shifting it from time to time as well.

I hope to show that by and large this is true, and that most animals with reasonable eyesight use a saccade and fixate strategy not very different from ours - even though, in flying insects for example, it may take unexpected forms. However, there are some remarkable exceptions, three of which I shall describe briefly later. These are animals whose eyes really do "pan" across the scene, taking in information as they do so. It would seem at first sight that their existence contradicts the "shutter time" idea just commended, but in fact it doesn't. The scanning rates seem to be nicely judged to be just below the speed at which image quality would suffer, and thus they strengthen rather than weaken the basic argument.

Turning a corner.

During locomotion that involves turning, all animals must change the direction of their gaze from time to time. They could, of course, just let the direction of the eyes follow that of the head or body, but if avoiding image blur is important we would expect to see a saccade and fixate strategy instead. Fig. 1 shows records of the eye movements made during turns by animals from three different phyla - the chordata, arthropoda and mollusca - in which eyes evolved independently (see Land & Fernald, 1992). The records all show the same features, namely that during turns the eyes make fast movements into the turn, followed by periods in which the eye counter-rotates relative to the head or body, ensuring that gaze direction (eye + head) stays more or less constant in the intervals between the fast saccadic movements.

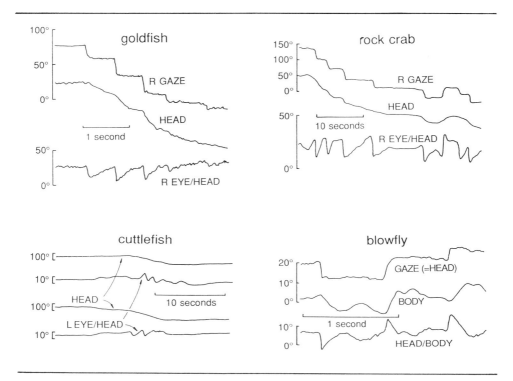

Fig. 1. Four examples of eye and body movements made by animals with different phylogenetic origins, during locomotion involving rotation. In each case the eye movements (lower records) have the double function of changing gaze direction with fast saccades, and stabilizing gaze between saccades by moving the eyes in the opposite direction to the head or body movements. This results in fast gaze changes separated by almost stationary fixations (upper records, not shown for the cuttlefish). Compiled and modified from Easter et al (1974), goldfish; Paul et al (1990), rock crab; Collewijn (1970), crayfish; Land (1973), blowfly.

Fig. 2 shows an insect example - a stalk-eyed fly- in more detail. The fly's body turns smoothly through 90°, but the eyes, built into the ends of the stalks attached rigidly to the head, make two 45° fast saccades. The head counter-rotates relative to the body in the intervening intervals, again keeping gaze direction impressively still. (Sadly, the role of the impressive eye-stalks in these animals has more to do with aggressive display than with vision). The examples given demonstrate clearly that a saccade and fixate strategy, involving a stabilising system for counter-rotating the eye, has evolved a number of times in evolution: we may tentatively assume for the same reasons.

Some arthropods track small moving objects, as primates do, but as in the vertebrates this ability is uncommon. A particularly good example is the praying mantis, whose capture technique involves tracking a moving prey with the head (and

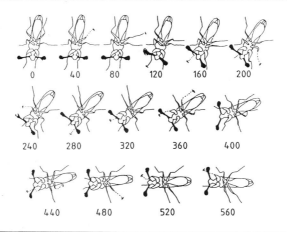

Fig. 2. A stalk-eyed fly, turning through 90°, filmed from below through a glass plate. Time in ms. The head (and eyes) make two saccades (after ca 120 and 380 ms) and remain stationary in the intervals between as the body rotates continuously. Data from a film by W Wickler and U Seibt.

hence the eyes) prior to making a lunge to catch it with the forelegs as it comes into range. Rossel (1980) found that mantids can track targets accurately and smoothly at slow speeds, but as the target movement speeds up, so the pursuit becomes increasingly saccadic in nature (as does human pursuit). Of particular interest in Rossel's study was the finding that pursuit becomes more saccadic as the contrast of the background is increased (Fig. 3). A problem for any pursuit system is that it has to overcome the ubiquitous optokinetic response - a visual feedback loop (OKN in humans) whose function is precisely to keep the image of the background stationary on the retina. One way round the problem is to move the eyes at speeds beyond the range of the optokinetic system, and by switching to saccades, this is what the mantis seems to be doing. These issues are further discussed by Land (1992) and Collett et al (1993).

Flies as disembodied eye movements

Our eyes are attached indirectly to the ground via a body with substantial inertia, and so to shift gaze fast we must make eye movements. For a small flying insect this need not be the case. With low mass and high manoeuvrability, eye movement can be achieved by body movement. Although most insects are able to make limited head (eye) movements around all three axes, they do not always choose to do so, with the result that flight behaviour and eye movements become the same thing. An excellent example of this is the small hoverfly *Syritta pipiens*. Female flies hover around flowers, feeding on nectar, whilst the males spend much of their time in stealthy

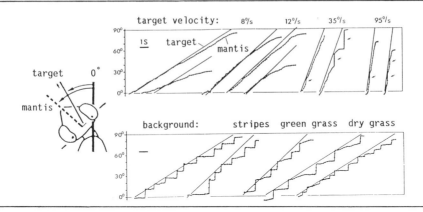

Fig. 3. Pursuit eye movements in the praying mantis. Upper records show smooth pursuit against a plain background. Note the occurrence of saccades (arrows) at the highest velocities. Lower records show that pursuit becomes saccadic when the background is strongly textured. With the relative weak contrast of dry grass, pursuit is mixed. Modified from Rossel (1980).

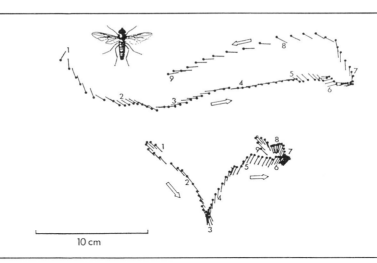

Fig. 4. Insect flight manoeuvres as eye movements. The hoverfly Syritta pipiens filmed from above, showing the flight path of a female (above) being tracked by a male (below). Notice that the female's flight has a "saccade and fixate" pattern, with very little rotation between saccades, although there is no restriction on translation. The male, however, tracks the female smoothly, keeping her within 5° of his midline. Here there is a region of high acuity, absent in the female. Also notice that he maintains a constant distance from the female. Corresponding times are numbered every 400ms. From Collett & Land (1975)

pursuit of the females (Collett & Land, 1975). The males have an advantage in that they have an "acute zone" in the front-facing part of the compound eye, where the resolution is about 3 times better than anywhere in the female eye. Thus the males can shadow the females around until they land, whilst remaining effectively out of sight. Fig. 4 shows an example of this. It is clear that the flight behaviour of the female (above) and male (below) are not the same. Although the female's flight is continuous, her turning is not. She makes rotational saccades from time to time (e.g. just before 3, just after 5) and between these the body does not rotate, even though translational flight may occur in any direction. The flight of non-tracking males is similar. As soon as they begin to track, however, the pattern changes dramatically. Throughout the 3.6s period shown in Fig. 4 the male points directly towards the female, tracking smoothly, and keeping her within the ± 5° forward sector containing his acute zone. Notice too that he maintains a roughly constant distance of about 10cm, which is important if he is to remain undetected. Interestingly, if the female moves fast he switches to a saccadic mode of tracking, just as we do. Unlike mantids, *Syritta* is able to track smoothly against a textured background. The responses of the optokinetic and tracking control systems simply add together, with the result that a male tracking a female in a rotating environment can do so, but with a small position error (Collett, 1980). From the point of view of visuo-motor coordination, it is not far-fetched to think of male *Syritta* flight manoeuvres as analogous to primate eye movements, and those of the females to non-primate (rabbit, say) eye movements.

How still does the eye need to be?

Having made the case for thinking that all well-sighted animals avoid a moving retinal image where they can, and shift gaze as fast as possible when motion is unavoidable, it is useful to look in more detail at the at the factors that determine how much image slip can be tolerated. We will consider later the related question: What kinds of slip are actually desirable?

If we ask how fast the image can move before blur becomes a problem, it turns out that the answer depends not only on the response time of the receptors, but also on the the fineness with which the image is sampled by the receptor mosaic. Consider a receptor whose field of view (or acceptance angle) is, say, 1°. Images of objects 1° and larger will, if stationary, fully stimulate the receptor, but smaller ones will do so only partially. Thus *spatial* degradation begins when image detail is smaller than a receptor's acceptance angle (this assumes that the eye's optics resolve adequately). A parallel argument applies to *temporal* degradation, which will start to occur when the receptor has inadequate time to respond. Suppose the receptor takes 20ms to respond fully to a small light flash. This "flash response time" sets the minimum time required for the cell to produce a response to any type of stimulus (Howard et al, 1984, Land et al, 1990). Thus if an object takes less than 20ms to pass through the receptor's field of view the response will be only partial, but if it takes longer the response will be complete. Returning to the 1° object that was just fully resolved spatially, it is now clear that to elicit a full response from the receptor it must take at least 20ms to pass through its field of view, which means that it can move across the retina at a maximum velocity of $(1/0.02) = 50°.s^{-1}$. We can generalise this result to say that *the maximum tolerable velocity across the retina, without loss of usable contrast, is given by the receptor acceptance angle divided by the response time*. Interestingly, this relation predicts that the maximum acceptable velocity should increase as the spatial resolution of the eye decreases, which means that the relatively coarse (1°) mosaic of insect eyes should be more tolerant to image slip than the 0.5' foveal mosaic of humans by about two orders of magnitude, if the response times are similar. Conversely, excellent resolution like ours requires particularly good image stabilization. Applying the "one acceptance angle per response time" rule to humans would give a value of rather less than $1°.s^{-1}$ as the maximum speed that will not degrade the finest resolvable grating image. This turns out to be a little pessimistic; Westheimer and McKee (1975) estimate that measurable contrast loss begins at 2-$3°.s^{-1}$.

This argument leaves no doubt about the need to stabilize the eye, if its resolving power is not to be compromised, and the better the eye the truer this is. However, it does also suggest an alternative way of acquiring visual information, particularly in eyes where the receptor mosaic is relatively coarse. Provided the speed given by the rule above is not exceeded, an eye may make movements that scan the retina across the image, without loss of spatial resolution. In conventional eyes with 2-dimensional retinae such a strategy might merely produce confusion. However, there are a few eyes with narrow, almost linear retinae, which do indeed move in a way that shows that they really are scanning the image. Four of these unusual eyes are considered in the next section.

Scanning vision: sea snails, water fleas, mantis shrimps and jumping spiders.

The most straight-forward scanning eye I know of is in the carnivorous planktonic sea-snail *Oxygyrus* (Fig. 5). It has been known for a century that this group of gastropod molluscs, the heteropods, have very narrow retinae (Hesse, 1900), but the reason for this has only recently become apparent. *Oxygyrus* has a lens eye not unlike a fish eye, except that the retina is only 3 receptors wide by about 410 receptors long, and covers a field of about 3° by 180°. The 1-dimensional structure of this retina would make very little sense unless it moved in some way, and indeed the eyes do scan (Land, 1982). The eyes move so that the retina sweeps through a 90° arc at right angles to its long dimension. The scanning pattern is a sawtooth, and the slower upward component has a velocity of $80°.s^{-1}$. The eye scans through the dark field below the animal, and the suggestion is that it is searching for food particles glinting against the dark of the abyss.

The oceanic copepod *Labidocera* exhibits a similarly straight-forward scanning pattern (Land, 1988). The animal has a pair of eye-cups directed dorsally (Fig. 6). The combined retina has a set of 10 slab-like receptor structures - 5 per eye - arranged as a line. These are pulled backwards and forwards by a combination of a pair of small muscles behind, and elastic ligaments in front, so that the linear retina scans the water above as shown. The muscle-powered movement is the slower one, but even this is fast, more than $200°.s^{-1}$. Interestingly, only the males have these specialized eyes, and we must assume that this scanning arrangement is part of the way they

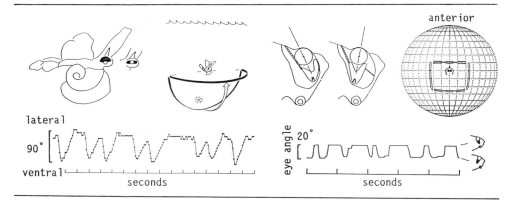

Fig. 5. The sea-snail Oxygyrus (left), with one eye pointing downwards. The inset shows the appearance of the eye when directed laterally. Diagram on the right shows the visual field of the eye during a scanning movement, and its probable role in detecting plankton. The time course of 8 scans is given below. Mainly from Land (1962).

Fig. 6. Head of the copepod Labidocera from the side, showing the eyecup at the extreme positions of a scan (left). The plot on the right shows the upward-pointing field of view of the line of rhabdoms projected onto a hemisphere above the animal. The time course of a number of scans is shown below. Mainly from Land (1988).

find females. These do have rather dark elongated bodies, so a scanning linear array might well be an appropriate detector.

The third example is more complicated. The mantis shrimps are quite large crustaceans, very distantly related to the more familiar decapod shrimps. Like their insect namesakes they are ambush predators, with a legendary ability to destroy their prey with smashing or spearing appendages (Caldwell & Dingle, 1976). Their eyes are basically compound eyes of the ordinary apposition type, and these provide an erect 2-dimensional image. However, stretching more or less horizontally across each eye is a band of enlarged facets, 6 rows wide (Fig. 7). This mid-band, which has a field of view only a few degrees wide, contains the animals' extraordinary colour vision system (Cronin & Marshall, 1989, Marshall et al, 1991). This consists of 4 of the mid-band rows (the other 2 subserve polarization vision) and in each row the receptors are in two tiers. Each of these 8 tiers contains a different visual pigment, giving the animal *octo-chromatic* colour vision. In adopting this impressive system, however, the mantis shrimps have set their eye movement system a daunting task. The outer parts of the eye operate as normal compound eyes - and are subject to the kinds of image stability considerations discussed earlier. The mid-band, however, has to move or it will not be able to register the colour of objects in the environment outside a very narrow strip. The result of this visual schizophrenia is a repertoire of eye movements which is quite unlike anything else in the animal kingdom (Land et al, 1990). In addition to the "normal" eye movements - fast saccades, tracking and

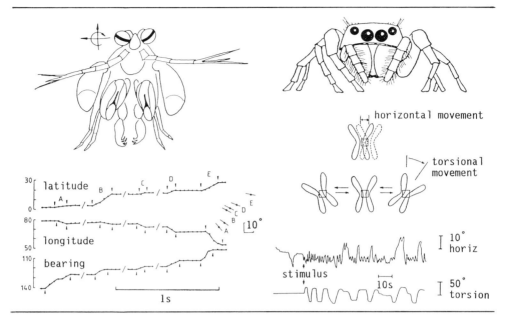

Fig. 7. The mantis shrimp Odontodactylus from the front, showing the large compound eyes and the 6-row mid-bands indicated in black. Arrows indicate the three rotational axes of the eyes. Below is a record of 5 scanning movements, showing their small amplitude and low velocity, and the independence of movement around each axis. The insert (right) shows the angular trajectory the eye's centre and mid-band, projected onto a sphere. From Land et al (1990).

Fig. 8. The jumping spider Phidippus showing the large movable principal eyes, and smaller fixed antero-lateral eyes. Below is a diagram and record of the movements of the boomerang-shaped retinae of the two principal eyes while scanning a novel target. These movements are conjugate, and consist of a stereotyped pattern of fast horizontal oscillations and slower torsional rotations. From Land (1969).

optokinetic stabilizing movements - there is a special class of frequent, small (ca 10°) and relatively slow (40°.s^{-1}) movements, which give the animal a strange inquisitive appearance, perhaps because they resemble human saccades in their frequency of occurrence. They are, however, not saccades, which are much faster. These movements, illustrated in Fig. 7, are typically at right angles to the band, and the only plausible explanation is that they are the scanning movements the band uses to "colour in" the monochrome picture provided by the rest of the eye.

Jumping spiders stalk insect prey rather as cats stalk birds. They have eight simple (camera-type) eyes, although two are usually rudimentary (Fig. 8). Of the remaining six, four are fixed to the carapace and act only as motion detectors. If something moves in the surroundings these eyes initiate a turn, which results in the target being acquired by the larger, forward-facing pair of "principal" eyes (Homann, 1928). These eyes have narrow retinae shaped like boomerangs, subtending about 20° vertically by 1° horizontally in the central region, which is only about 6 receptor rows wide (Land, 1985, Blest, 1985). The resolution is very high, with receptor spacings of 10' fairly typical, and as low as 2.5' in one genus (*Portia*). The principal retinae can move, horizontally and vertically by as much as 50°, and they can also rotate about the optic axis (torsion) by a similar amount (Land, 1969). When presented with a novel target, the eyes scan it in a stereotyped way, moving slowly from side-to-side at speeds between 3 and $10°.s^{-1}$, and rotating through ±25° as they do so. We actually know what they are looking for: legs! Drees (1952) showed that jumping spiders are relatively indifferent to the appearance of potential prey, so long as it moves, but males are quite particular in what they regard as potential mates. Drawings consisting of a central dot with leg-like markings on the sides, however, will elicit courtship displays. Whatever its other functions may be, scanning in these spiders really seems to be concerned with feature extraction, the procedure itself apparently designed to detect the presence and orientation of linear structures in the target. The dual system of fixed and moveable eyes of jumping spiders, with one set acting as target finder and the other as analyser, does seem to have much to commend it, compared with the cumbersome time-sharing arrangement in mantis shrimps where the two functions are combined in the same eye.

Table 1   **Scanning eyes: inverse relation of speed and resolution.**

Animal	Scan rate (s) $°s^{-1}$	Receptor subtense (r) °	"Dwell time" (t=r/s) msec
Labidocera (Copepod)	219	3.5	16
Oxygyrus (Mollusc)	80	1.1	15
Odontodactylus (Stomatopod)	40	1.0	25
Metaphidippus (Spider)	6.2	0.15	24

The four examples of scanning given in this section represent a range of different functions, from simple detection, to colour and feature extraction. Nevertheless, they should all be expected to obey the rule given earlier, that the scanning speed should not exceed the receptor acceptance angle divided by the response time. We would not expect the speed to be much slower than this, however, as that would merely waste time. Clearly there is an optimum. Table I gives the scanning speeds and acceptance angles for the four animals discussed. In the Table it is clear that there is an inverse relationship between resolution and scanning speed, as indeed there should be; the high resolution jumping spider is slowest, and the copepod the fastest. The response times of the receptors, estimated as the time it takes a receptor to move through its own acceptance angle, are also shown, and they all fall nicely into the range of 15 to 25ms. Although we do not know the true response times for these animals, these values are well within the range of insect flash response times, for which data are available (Howard et al, 1984).

Conclusions: good and bad retinal motion.

Whilst most animals' visual systems go to considerable lengths to protect the stability of gaze from the vagaries of body movement, a few, as we have seen, actually exploit the tolerance of the receptor response to modest velocities to scan the image with systematic eye movements. The question arises: Do we all do this, in one way or another? To answer this we need to look at the sources of image motion our eyes are subject to. Gibson (1950) pointed out that whenever we move there is a pattern of image movement across the retina - the "flow field" - that is the inevitable consequence of our locomotion. Generally speaking the flow field has two components, one due to translational (linear) motion, and one to rotation. The translational flow field is a pattern of velocity vectors expanding from a central stationary "pole" that corresponds to the direction of motion. Rotation, by contrast, causes the image to move in the same direction everywhere, and with the same angular velocity. The only information in the rotational pattern concerns the speed of rotation itself, which is unlikely to be of great interest. On the other hand the translational flow field is rich and valuable, as it contains information about the distances of objects (nearer objects move past faster) and also the animal's current heading, as shown by the location of the pole.

The track of the female hoverfly in Fig. 4 illustrates one way of coping with the mixture of flow-field components, and it may be typical of animals generally. Rotation is strenuously prevented, being allowed only as brief saccades during which - because of the speeds involved - the fly must have very reduced acuity. Translation, however, is not obviously impaired, and one has no sense that the fly is being held back from moving in any direction it chooses, provided this does not involve rotation. The situation in man is similar; VOR prevents eye rotation, but does not interfere with our locomotion. The inference is that in insects and in man the oculomotor system leaves the translational flow field intact, by getting rid of rotational flow before it even happens. The reason for doing this probably lies in the difficulty of extracting the useful translational flow from the combined flow-field (see for example Buchner, 1984, Fig.1). And even though there is evidence that humans can do this under some circumstances (Warren & Hannon, 1990), that capability is probably only the second line of defence.

The retinal velocities involved in translational flow are not great (Kowler, 1991), especially around the direction of motion where they fall well within the range that the receptors can deal with. Implausibly, we still lack direct proof that flow-field information usefully influences human behaviour (although driving and playing tennis seem inconceivable without it). There is, however, convincing evidence that bees can learn their distances from objects by velocity information alone (Lehrer et al., 1988), and there is strong circumstantial evidence that translational flow is used by other animals (Davies & Green, 1990, Lee & Reddish, 1981). Some animals actually generate translational flow in order to measure distance. Collett (1978) showed that locusts judge their jump distance by the rate of image motion across the retina as they make stereotyped lateral "peering" movements. Because translational flow contains information about an animal's progress through the environment immediately ahead of it, it would be astonishing if it were not properly exploited for the control of locomotion.

To sum up. In nearly all animals with good eyesight the main function of the oculomotor system is to prevent rotational slip of the image. The overriding reason for this is the need to prevent loss of acuity resulting from the blur caused by the finite response time of the photoreceptors. However, in a few animals rotational motion is actually used to scan the image, but when this does occur the velocities involved do not exceed a "no blur" value, given by the acceptance angle of a receptor divided by its response time. These exceptions aside, animals who have stabilised their eyes against rotation are generally free to contemplate and exploit the remaining translational image motion that results from locomotion. This is usually slow enough to avoid blur, and contains much useful information about the structure of the world ahead of the animal.

Acknowledgements

The work on which this review is based was supported in large part by the SERC, UK.

References

Blest AD (1985) The fine structure of spider photoreceptors in relation to function. In: Neurobiology of Arachnids (ed Barth FG) pp 79-102, Berlin: Springer

Buchner E (1984) Behavioural analysis of spatial vision in insects. In: Photoreception and Vision in Invertebrates. (ed. Ali MA) pp 561-621. New York: Plenum

Caldwell RL, Dingle H (1976) Stomatopods. Sci Amer 234(1): 80-89

Carpenter RHS (1988) Movements of the Eyes. 2nd ed. London: Pion

Carpenter RHS (1991) The visual origins of ocular motility. In: Vision and Visual Dysfunction. vol 8. (ed. Carpenter RHS) pp 1-10. Basingstoke: Macmillan.

Collewijn H (1970) Oculomotor reactions in the cuttlefish, *Sepia officinalis*. J Exp Biol 52: 369-384

Collett TS (1978) Peering - a locust behaviour pattern for obtaining motion parallax information. J Exp Biol 76: 237-241

Collett TS (1980) Angular tracking and the optomotor response. An analysis of visual reflex interaction in a hoverfly. J Comp Physiol 140: 145-158

Collett TS, Land MF (1975) Visual control of flight behaviour in the hoverfly, *Syritta pipiens* L. J Comp Physiol 99: 1-66

Collett T, Nalbach H-O, Wagner H (1993) Visual stabilization in arthropods. In: Visual Motion and its Role in the Stabilization of Gaze (eds Miles FA, Wallman, J) pp 239-263. Elsevier

Cronin TW, Marshall NJ (1989) A retina with at least ten spectral types of photoreceptors in a mantis shrimp. Nature 339: 137-140

Davies MNO, Green PR (1990) Optic flow-field variables trigger landing in hawk but not in pigeons. Naturwissenschaften 77: 142-144

Drees O (1952) Untersuchungen über die angeborenen Verhaltensweisen bei Springspinnen (Salticidae). Z Tierpsychol 9: 169-207.

Easter SS, Johns PR, Heckenlively D (1974) Horizontal compensatory eye movements in goldfish (*Carrassius auratus*). I. The normal animal. J Comp Physiol 92: 23-35

Gibson JJ (1950) The Perception of the Visual World. Boston: Houghton Mifflen

Hesse R (1900) Untersuchungen über die Organe der Lichtempfindung bei neideren Thieren. VI. Die Augen einiger Mollusken. Z Wiss Zool 68: 379-477

Homann H (1928) Beiträge zur Physiologie der Spinnenaugen. I Untersuchungsmethoden. II Das Sehvermögen der Salticiden. Z Vergl Physiol 7: 201-269

Howard J, Dubs A, Payne R (1984) The dynamics of photo-transduction in insects. A comparative study. J Comp Physiol A:154: 707-718

Kowler E (1991) The stability of gaze and its implications for vision. In: Vision and Visual Dysfunction. Vol 8. (ed. Carpenter RHS) pp 71-92. Basingstoke: Macmillan

Land MF (1973) Head movements of flies during visually guided flight. Nature 243: 299-300

Land MF (1969) Movements of the retinae of jumping spiders (Salticidae: Dendryphantinae) in response to visual stimuli. J Exp Biol 51: 471-493

Land MF (1982) Scanning eye movements in a heteropod mollusc. J Exp Biol 96: 427-430

Land MF (1985) The morphology and optics of spider eyes. In: Neurobiology of Arachnids (ed Barth FG) pp 53-78, Berlin: Springer

Land MF (1988) The functions of eye and body movements in *Labidocera* and other copepods. J Exp Biol 140: 381-391

Land MF (1992) Visual tracking and pursuit: humans and arthropods compared. J Insect Physiol 38: 939-951

Land MF, Fernald RD (1992) The evolution of eyes. Annu Rev Neurosci 15: 1-29

Land MF, Marshall JN, Brownless D, Cronin TW (1990) The eye-movements of the mantis shrimp *Odontodactylus scyllarus* (Crustacea: Stomatopoda) J Comp Physiol A 167: 155-166

Lee DN, Reddish PE (1981) Plummeting gannets: a paradigm of ecological optics. Nature 293: 293-294

Lehrer M, Srinivasan MV, Zhang SW, Horridge GA (1988) Motion cues provide the bee's visual world with a third dimension. Nature 332: 356-357

Marshall NJ, Land MF, King CA, Cronin TW (1991) The compound eyes of mantis shrimps (Crustacea, Hoplocarida, Stomatopoda). Phil Trans R Soc B 334: 33-84

Paul H, Nalbach H-O, Varjú D (1990) Eye movements in the rock crab *Pachygrapsus marmoratus* walking along straight and curved paths. J Exp Biol 154: 81-97

Rossel S (1980) Foveal fixation and tracking in the praying mantis. J Comp Physiol A 139: 307-331

Walls GL (1962) The evolutionary history of eye movements. Vision Res 2: 69-80

Warren WH, Hannon DJ (1990) Eye movements and optical flow. J Opt Soc Amer A 7: 160-169.

Westheimer GA, McKee S (1975) Visual acuity in the presence of retinal image motion. J Opt Soc Am 65: 847-850

PURSUIT AND CO-ORDINATION

HUMAN EYE MUSCLE PROPRIOCEPTIVE FEEDBACK IS INVOLVED IN TARGET VELOCITY PERCEPTION DURING SMOOTH PURSUIT

J.L. Velay, R. Roll, J.L. Demaria, A. Bouquerel & J.P. Roll

Lab. Neurobiologie Humaine, Univ. de Provence, U.R.A. CNRS 372, Marseilles, France.

Abstract

We previously established that applying mechanical vibration to the subject's inferior rectus muscle results in the illusion that a luminous spot fixated in darkness is moving upward. This finding confirmed that eye muscle proprioception participates in the egocentric localization of a motionless visual target. The present experiment was designed to study whether ocular proprioception is involved in speed perception during smooth pursuit. The target was made to move 5 degrees upwards or downwards from a central position, at a velocity of 1 to 5 deg.sec^{-1} in total darkness. The subjects matched the target velocity manually by moving a stylus on a digitizing table, both with and without vibration applied to the inferior rectus of the right seeing eye. In the upward smooth pursuit situation, vibrating the lengthened inferior rectus increased the perceived target velocity with respect to the control situation. Conversely, in downward pursuit, vibrating the shortened inferior rectus decreased the perceived target speed. These results suggest that during smooth visual pursuit, the eye muscle proprioceptive feedback includes a velocity signal about the visual target which probably results from the combined processing of proprioceptive messages originating in the two antagonistic vertical muscles.

Keywords

eye proprioception, smooth pursuit, velocity perception, vibration, Human

Introduction

Efficient oculomotor performance and accurate visual motion perception involve a monitoring of the eye movements in orbital coordinates, and this extraretinal signal is processed at both the sensorimotor and perceptual levels. Whether the extraretinal signal is of central (corollary discharge) or peripheral (eye proprioception) origin is still a controversial issue (Steinbach, 1987). Although proprioceptive projections to numerous structures involved in oculomotricity have been found to exist, only a few experimental data are available that clearly demonstrate that proprioception is involved in visuomotor control (Donaldson & Knox, 1993), while other data argue against its participation (Guthrie et al, 1983; Keller and Robinson, 1971; Robinson, 1981).

In humans, it has recently been established that proprioception contributes to visual localisation (Campos et al, 1986; Gauthier et al, 1990; Roll and Roll, 1987; Roll et al, 1991; Steinbach and Smith, 1981; Velay et al, 1994).

A powerful means of specifically activating somatic muscle spindles in humans consists of applying tendon mechanical vibration; this provides a useful tool for inducing sensory illusions of limb movement. Vibration has also been applied to one eye of human subjects while they were fixating a luminous target in an otherwise totally darkened room (Roll and Roll, 1987; Roll et al, 1991; Velay et al, 1994). This gave rise to an illusory target movement, the direction of which depended on which muscle was vibrated: vibration of the lateral rectus of the right eye induced a leftward illusory motion of the luminous spot, for instance, whereas vibration of the inferior rectus generated an upward shift. Like the illusory perception of limb movement induced by somatic muscle vibration, these visual illusions were found to be frequency dependent, and this was consistent with the idea that they are of proprioceptive origin. This body of data confirmed that eye muscle proprioception participates in the egocentric localisation of a motionless visual target.

If proprioception participates in stability perception during fixation of a stationary point, might it not also be involved in motion perception during slow eye movements? Since the smooth eye movements that occur during fixation of a stationary target can be regarded as being essentially the same as those made when tracking moving targets, the processes underlying these two types of slow control might be partly the same (Kowler, 1990). On the other hand, the illusion of movement evoked when the eyeball is vibrated during fixation of a stable point is quite similar to the motion perception resulting from the real pursuit of a target moving at a very slow velocity. The present experiment was therefore designed to study whether human ocular proprioception might be involved in speed perception during smooth pursuit.

Methods

The subjects were required to monocularly pursue a red spot which was made to move 5 degrees upwards or downwards from a central position, at a velocity of 1 to 5 deg.s^{-1} ln total darkness. At the same time, they had to match the target velocity manually by moving a stylus on a digitizing table, both with and without vibration applied to the inferior rectus (IR) of the right seeing eye.

The experimental set-up used in this visuo-manual tracking task is shown in figure 1. The subject was seated in front of a vertical screen which was 114 cm away from his seeing eye. The visual target was a red spot (3 mm in diameter) generated by a laser beam. The spot could move upwards and downwards at a constant speed. The subject's head was immobilized by means of a head restraint and a bite board. Mechanical vibrations with a 0.2 to 0.4 mm peak to peak amplitude (rectangular pulse: 3 ms) were applied to the right eye by means of an electromagnetic vibrator (L.D.S. type 101) to which a small probe was attached. Its contacting surface was polished and had a concave shape which made it easily adaptable to the eyeball. The vibrator was mounted on a micro manipulator so that very fine movements were possible in all three spatial planes and the probe could be accurately positioned underneath the eyeball. The micro manipulator was in turn attached to a two dimensional column which was vertically and laterally adjustable to allow for larger displacements. The column was fixed to a firm support on the ground, which was separate from the head rest so that no vibration could be transmitted. Vibratory pulses were generated by a neurostimulator and transmitted to the vibrator via a power amplifier. In their right hand, the subjects held a stylus that they moved on the surface of a vertical digitizer (summagraphics) placed in front of them. The subject was seated with his head in the device. The vibrator probe was carefully adjusted underneath the right eyeball.

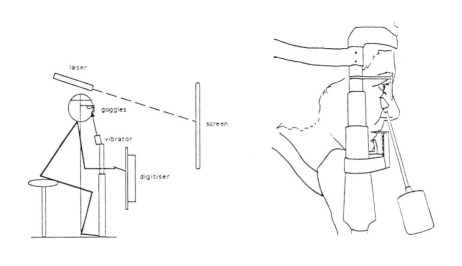

Figure 1. Schematic view of the experimental set-up.

A constant vibration frequency (80 Hz) was used. In one subject, the eye movements were recorded with an infrared-system, consisting of light emitting and sensing elements positioned in goggles (Ober2 system, Permobil Inc.). Both horizontal and vertical deviations could be measured with this system.

After occluding the left eye, the subject was left in total darkness, with his right hand holding the stylus on the digitizer. When the spot appeared, he had to gaze at it for 2 seconds in the central position and to pursue it as soon as it began to move vertically. The subject was instructed to move the stylus at the same speed as the spot he was visually pursuing. In the situations where vibration was applied under the eye, it began when the target spot appeared and stopped 1 sec after the target had reached the final position. Before the experiment, a practice session was run in order to make the subjects familiar with the tracking task requirements and to check whether they differentiated between and could copy the five target speeds correctly.

Five volunteers participated in this experiment. There were 3 experimental factors: 1: presence or not of the vibration, 2: direction of the target movement (upwards or downwards), and 3: speed of the target (1 to 5 deg/s). The various combinations of the 3 experimental factors resulted in 20 experimental situations. The subjects were also tested in a control situation in which the eye was vibrated while the target remained stable. A set of 5 measurements was taken in each experimental situation. The hand velocity was computed in all the situations with all the subjects. A three-way analysis of variance was performed on the data.

Results

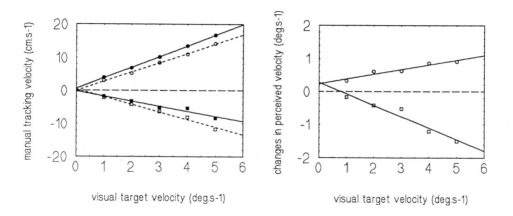

Figure 2A. Speed of the manual tracking as a function of the speed of the visual target. Circles: upward movements. Squares: downward movements. Full and clear symbols denote the vibrated and non-vibrated pursuit conditions, respectively. When no vibration was delivered, the subjects were able to unambiguously reproduce the five target speeds with their hand. In the case of both upward and downward movements, the speed of the hand movement was linearly correlated with the target speed.

Figure 2B. From the slope of the regression lines of the data recorded without applying vibration, it was possible to calculate the relationships between the visual target velocity and the manual tracking velocity. By subtracting the values recorded without vibration from those with vibration, the changes in the perceived target velocity could then be computed at each speed.

The mean manual speeds obtained across subjects in each experimental situation are given in figure 2. The negative sign denotes downward movements.

Overall, eye vibration was found to have affected the velocity of the manual tracking movement ($F(1,4)= 8.69$, $p<.05$). The changes differed qualitatively however depending on the direction of the target movement, that is on the direction of the pursuit. In upward pursuit, vibrating the lengthened IR muscle increased the speed of the manual tracking, indicating that the perceived target speed had increased; and conversely, in downward pursuit, vibrating the shortened IR decreased the perceived target speed (fig 2A). These direction-dependent differences in the perceived target speed induced by the ocular vibration were roughly proportional to the target velocity (figure 2B), particularly in the case of downward movements. The mean variation in the perceived speed was around 25% of the real target speed.

In the control situation, when the target remained immobile, applying vibration to the IR muscle induced an illusory visual target motion in the upward direction. The estimated speed of the illusory motion was 0.6 deg.s^{-1}, which is compatible with that of around 1 deg.s^{-1} obtained in previous studies with a different experimental procedure (Roll et al, 1991; Velay et al, 1994).

In the subject in whom the eye movements were monitored, the gain in the vertical pursuit was computed both with and without vibration. No significant changes in eye movements were found to occur when vibration was applied.

Discussion

Recalling what happens when limb muscles are vibrated can help to understand the effects of eye vibration. Somatic muscle spindles are known to be closely involved in both stabilized position and movement perception. Because it simulates the elongation of a limb muscle, vibration has been extensively used to induce illusory movement sensations of proprioceptive origin. Vibratory illusions are probably due to an artificial disequilibrium in the balance between the proprioceptive afferent messages originating from agonist and antagonist muscles (Gilhodes et al, 1986) and, at least with slow movements, velocity perception seems to involve a comparison between afferent information of antagonistic origins. During a slow, active or passive movement of the limb, when vibration is applied to the muscle which is being lengthened, the velocity of the movement is felt as being higher, and this increase in the perceived velocity is proportional to the actual speed of the movement; whereas during a passive movement, when vibration is applied to the shortened muscle, the velocity of the movement is perceived as being lower. The present data resulting from eye vibration are directly comparable with these perceptual consequences of limb vibration. Applying vibration to the IR muscle when it is stretched, i.e. during an upward eye movement, elicited the illusory impression that the pursued target was moving faster, and the same vibration applied to the same muscle during its shortening induced the opposite perception. Consequently, during smooth vertical pursuit, the eye muscle proprioceptive feedback might include a velocity signal about the visual target, which probably results from the combined processing of proprioceptive messages originating in both antagonistic vertical muscles, whether they play an agonist or antagonist role in the movement being performed.

At the present stage, these changes in the perceived speed do not seem to have resulted from any actual changes in the eye movement itself. This point might be important because it suggests that perceived motion and ocular pursuit may be independent in this experimental situation, and this is in disagreement with several results which seem to indicate that perceived motion is the effective stimulus in smooth pursuit (Steinbach, 1976; Waespe and Schwartz, 1987; Yasui and Young, 1975). Other data have suggested however, that perceived motion and pursuit are not so strictly correlated since neither motion after-effect nor induced motion seem to be accompanied by eye pursuit (Mack et al, 1979; Mack et al, 1982).

The preliminary results presented here require confirmation by extensive repetition before we can attempt to answer to the question as to know whether the same motion analysers subserve both perception and pursuit (Kowler, 1990). Whatever the underlying process may be, these data support the idea that eye muscle proprioception may be involved in target velocity perception during smooth pursuit.

References

Campos E.C., Chiesi C. & Bolzani R. (1986) Abnormal spatial localization in patients with herpes zoster ophthalmicus. Arch Ophthal 104:1176-1177

Donaldson, I.M.L. and Knox, P.C. (1993). Evidence for corrective effects of afferent signals from the extraocular muscles on single units in the pigeon vestibulo-oculomotor system. Exp. Brain Res., 95, 240-250.

Gauthier, G.M., Nommay, & D. Vercher, J.L. (1990). The role of ocular muscle proprioception in visual localization of targets. Science, 249, 58-61

Gilhodes J.C., Roll J.P. & Tardy-Gervet M.F. (1986) Perceptual and motor effects of agonist-antogonist muscle vibration in man, Exp. Brain Res., 61:395-402.

Guthrie, B.L., Porter, J.D., Sparks, D.L. (1983). Corollary discharge provides accurate eye position information to the oculomotor system. Science, 221, 1193-1195

Keller E.L. & Robinson D.A. (1971) Absence of a stretch reflex in extraocular muscles in the monkey, J. Neurophysiol 34:908-919.

Kowler E (1990) The role of visual and cognitive processes in the control of eye movement. in "Eye movements and their role in visual and cognitive processes", Reviews of Oculomotor Research, vol 4, Kowler E. (Ed), PP 1-70.

Mack A., Fendrich R. & Pleune J. (1979) Smooth pursuit eye movements: is perceived motion necessary? Science 203: 1361-1363.

Mack A., Fendrich R. Wong E. (1982) Is perceived motion a stimulus for smooth pursuit? Vision Res. 22: 77-88.

Robinson D.A. (1981) Control of eye movements, in Brookhart JM & Mountcastle VB (Eds), Handbook of physiology, sect. 1, the nervous system, vol 2, American Physiological Society, Bethesda, Md., 1275-1320

Roll, J.P. & Roll, R. (1987). Kinaesthetic and motor effects of extraocular muscle vibration in man. In: O'Regan, K. Lévy-Schoen, A. (Eds) Eye movements: from Physiology to Cognition. Elsevier, North-Holland, 57-68

Roll, R., Velay, J.L. & Roll, J.P. (1991). Eye and neck proprioceptive messages contribute to the spatial coding of retinal input in visually oriented activity. Exp. Brain Res., 85, 423-431

Steinbach, M.J. (1976) Pursuing the perceptual rather than the retinal stimulus. Vision Res., 16: 1371-1376.

Steinbach, M.J. (1987). Proprioceptive knowledge of eye position. Vision Res., 27, 1737-1744

Steinbach, M.J. & Smith, D.R. (1981). Spatial localization after strabismus surgery: evidence for inflow. Science, 213, 1407-1409

Velay J.L., Roll R., Lennerstrand G. & Roll J.P. (1994) Eye proprioception and visual localization in humans: influence of ocular dominance and visual context, Vision Res., in press.

Waespe W. & Schwartz U. (1987) Slow eye movements induced by apparent target motion in monkey, Exp. Brain Res., 67:433-435.

Yasui S. & Young L.R. (1975) Perceived visual motion as effective stimulus to pursuit eye movement system., Science, 190: 906-908.

VARIABILITY OF SINUSOIDAL TRACKING CHARACTERISTICS IN CHILDREN

Accardo A.P. [a], *Pensiero S.* [b], *Da Pozzo S.* [b], *Inchingolo P.* [a] *and Perisutti P* [b]

[a] Dipartimento di Electrotecnica, Elettronica e Informatica, University of Trieste
[b] Department of Ophthalmology, Children's Hospital of Trieste, Italy

Abstract

The variability throughout the test of the smooth pursuit (SP) response to a sinusoidal stimulation was studied in 10 normal children and and in 10 adults. This phenomenon is not described in the literature and its influence on values of the SP characteristics is unknown : similarly, the values of the SP characteristics in school ages have not yet been subject of experimental study. The parameters which describe the response to a sinusoidal stimulation studied here are SP phase and gain (peak eye velocity /peak target velocity) and overall phase and gain (peak eye position/ peak target position). It is shown that records obtained from children present a marked fluctuation, depending on less constant attention level. Nevertheless, these fluctuations are not responsible for the different mean values of the SP characteristics found in adults and children. These differences are only marginally due (i.e. at the highest stimulation frequency) to fatigue, while they can be mainly ascribed to an incomplete SP maturation in children.

This study concerned also the possible influence of the duration of each test on the mean values of the SP parameters. Our data demonstrate that a correct evaluation of each parameter is obtained only from records of longer than 6 cycles. Moreover, for a valid comparison among experimental groups, the same criterion of characteristics evaluation is needed: therefore all authors should explicitly mention the method used in order to allow comparisons among literature data.

Keywords

eye movements, smooth pursuit, normal children, variability, attention, fatigue, parameter evaluation.

Introduction

In newborns and infants, since a sinusoidal stimulation is not able to attract the attention of the child (Aslin, 1981), visual tracking has been studied in response to a constant velocity stimulation (Krementizer et al, 1979; Shea and Aslin, 1990). In the opinion of the above cited authors, smooth tracking can be already generated during the first months of life. Moreover Shea and Aslin (1990) reported that pursuit gain increases till the age of 8 months; the smooth pursuit system (SPS) maturation process after this age is still unknown.

Conventionally, pursuit in adults is measured during tracking of predictable, sinusoidal target motion (Leigh and Zee, 1991). The parameters which describe the response to this kind of stimulation are the SP phase and gain (velocity characteristics) and the overall phase and gain (position characteristics). When a stimulation below 0.8 Hz is used, the SP gain is close to 1 and a very low phase lag

Eye Movement Research/J.M. Findlay et al. (Editors)

is present because of the large target predictability and repeatability. Above this frequency, or more correctly, as the target acceleration increases, the velocity gain progressively decreases, with a corresponding increment of the phase lag. For acceleration smaller than 1000 deg/s/s (Lisberger et al., 1981) or 400 deg/s/s (Baloh et al., 1988) the velocity gains are normally larger than 0.7.

The values of the SP characteristics in school ages, the variability of the eye response throughout the test in adults and children, and the influence of this variability on the mean values of the SP characteristics are not known.

This work, starting from the study of variability, tries to answer these questions.

Materials and Methods

The variability of the SPS response has been studied in 10 normal children of age ranging from 8 to 12 years. All the investigated parameters have been compared to the corresponding ones evaluated in a group of 10 adults who underwent the same test. Informed consent was obtained from subjects or, in the case of the children, from their parents before any testing began.

The horizontal movements of both eyes in binocular vision were recorded by the limbus-tracking technique using EIREMA1 (Accardo et al., 1989), a device which detects the eye position with a resolution of about 5'. All subjects, with their head fixed on a chin rest, were asked to pursue a target moving cosinusoidally at frequencies of 0.2, 0.4, 0.8, 1.0 and 1.2 Hz on the horizontal plane, with a ± 8 deg amplitude, at the distance of 1 meter. In order to analyze the variability of the SPS response throughout the test, the target was presented for 5 to 15 cycles at increasing stimulation frequencies. Just before the test, a precise calibration of each eye position was obtained by asking the subject to track two series of 9 different target positions in a range of ± 16 deg. The duration of the full session was about 5 minutes during which the attention level was maintained high by means of continuous incitement.

For each test, the eye position signals were filtered using a third-order Butterworth low-pass analogue filter with 50 Hz cut-off frequency and sampled at 250 Hz.

In the first off-line procedure, the sampled signals were linearized by using for each subject his own non-linear calibration characteristic. After digital compensation of the phase lag introduced by the analogue filter, the eye velocity was evaluated by digitally differentiating the eye position signal; then, it was filtered by a zero-phase, third-order Butterworth low-pass digital filter with 11 Hz cut-off frequency, reducing the noise to less than ± 1deg/s.

From the smooth component of the velocity traces we evaluated the SPS gain (Velocity Gain, VG) and phase (Velocity Phase, VP). The overall gain (Position Gain, PG) and phase (Position Phase, PP) were evaluated from the position traces, including the contribution of the SPS and the Saccadic System (SS) working together. The values of these parameters were calculated separately for each half-cycle of stimulation using the amplitudes and the phases of two trigonometrical curves fitting the velocity and position responses.

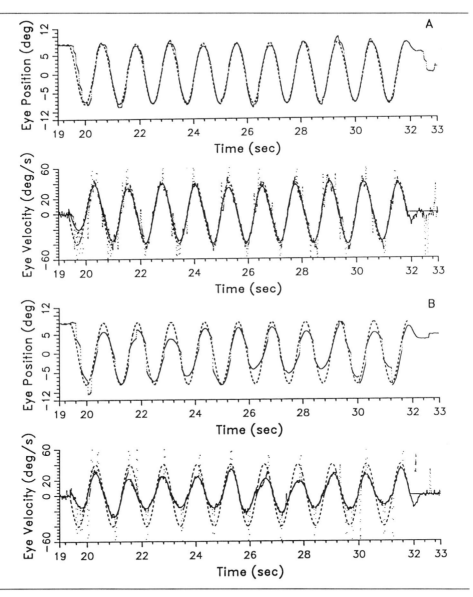

Figure 1 Examples of smooth tracking position (top) and velocity response (bottom), at a 0.8 Hz frequency stimulation, performed by an adult (A) and a child (B). Dashed line: target, dotted line: eye response, continuous line (bottom): best-fit sinusoidal curve used to evaluate the SP parameters (VG and VP).

A.P. Accardo et al.

Results

Figure 1 shows two examples (in A from an adult, in B from a child) of smooth tracking at a 0.8 Hz stimulation frequency. The wide variability present

Frequency Hz		Position Gain		Velocity Gain		Position Phase deg		Velocity Phase deg	
		rightw.	leftw.	rightw.	leftw.	rightw.	leftw.	rightw.	leftw.
Children									
0.2	mean	0.98	1.00	0.86	0.86	-2.70	-3.91	-1.47	-0.71
	±1SD	0.09	0.09	0.09	0.16	3.65	4.78	3.53	3.48
0.4	mean	1.03	1.04	0.85	0.81	-0.21	-2.42	-0.87	-2.25
	±1SD	0.11	0.12	0.16	0.12	6.22	7.54	3.73	4.29
0.8	mean	0.90	0.90	0.57	0.54	0.05	0.20	2.91	2.01
	±1SD	0.09	0.11	0.20	0.21	8.14	5.48	3.79	5.17
1.0	mean	0.84	0.81	0.46	0.42	13.97	13.06	7.31	3.60
	±1SD	0.08	0.11	0.12	0.09	8.46	9.50	7.29	9.54
1.2	mean	0.67	0.66	0.31	0.28	29.39	27.10	12.01	8.15
	±1SD	0.09	0.08	0.14	0.12	7.88	9.36	4.80	3.66
Adults									
0.2	mean	0.98	0.99	0.95	0.96	-0.67	-3.15	-0.28	-2.64
	±1SD	0.03	0.03	0.07	0.07	2.72	3.09	3.09	2.49
0.4	mean	0.99	0.99	0.93	0.95	-2.07	-1.58	-1.73	-0.57
	±1SD	0.05	0.04	0.10	0.08	3.25	4.00	2.55	3.35
0.8	mean	0.98	0.98	0.80	0.82	-3.02	2.74	-2.08	0.39
	±1SD	0.04	0.06	0.16	0.13	8.00	4.76	4.57	2.26
1.0	mean	0.97	0.98	0.72	0.76	-2.42	3.28	0.42	1.65
	±1SD	0.08	0.07	0.17	0.11	9.99	7.98	4.84	4.36
1.2	mean	0.86	0.88	0.59	0.62	5.15	9.12	3.87	4.12
	±1SD	0.21	0.19	0.24	0.21	9.13	10.86	4.58	4.67

Table 1. Rightward and leftward mean values of position and velocity gains and phases (±1SD) at the five stimulation frequencies in children and adults.

in B was detected in many child responses also at the other stimulation frequencies, both for the PG and VG values.

In all the tests the mean values of the VG were smaller in children than in adults, while the PG were smaller only at higher frequencies, with a marked intersubject variability in both groups (table 1). In the same table the mean values

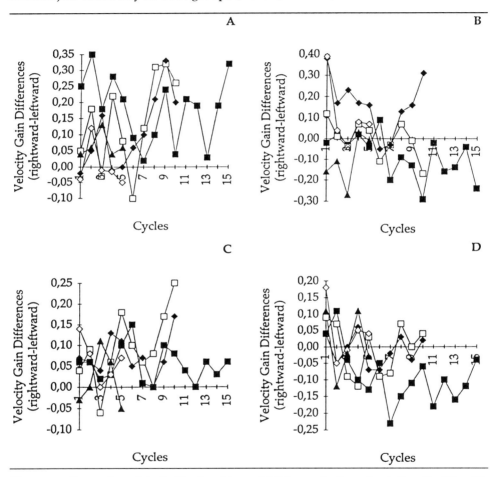

Figure 2. Examples of individual subjects (children: A, B and adults: C, D) with (A, C) or without (B, D) some kind of direction predominance. The diagrams show the differences between rightward and leftward velocity gains during each cycle of stimulation. In both A and C the points almost all fall above the zero line, showing consistent superiority of rightward performance for these two subjects. Subjects B and D show few systematic differences between rightward and leftward performance. Solid triangle: 0.2 Hz, open diamond: 0.4 Hz, solid diamond: 0.8 Hz, open square: 1.0 Hz, solid square: 1.2 Hz.

of the PP and VP are shown for both children and adults; they exhibited the same trend, going from a phase lead at 0.2 Hz frequency to a growing phase lag for increasing frequencies. The VP and PP in children exhibited larger absolute values than in adults, at almost all the considered frequencies, while the range in children was larger than in adults.

No significant difference (p > 0.05 at the Student's *t*-test) was found between rightward and leftward tracking mean parameters in both adults and children (figures 3 and 4), though certain subjects exhibited an idiosyncratic predominance

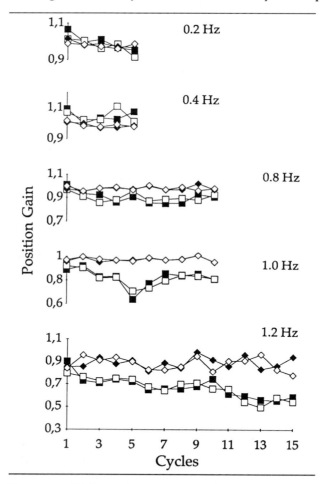

Figure 3. Rightward (open symbols) and leftward (solid symbols) intersubject mean values of position gain through the stimulation cycles in children (square) and adults (diamond) at the five stimulation frequencies.

for one direction at some stimulation frequencies (figure 2).

Therefore, in order to analyze the variability of the SP response through each test separately in adults and in children, in figures 3 and 4 the PG and the VG parameters, evaluated for each stimulation frequency, are shown. The PG and VG remained stable on average through the whole single test, though children

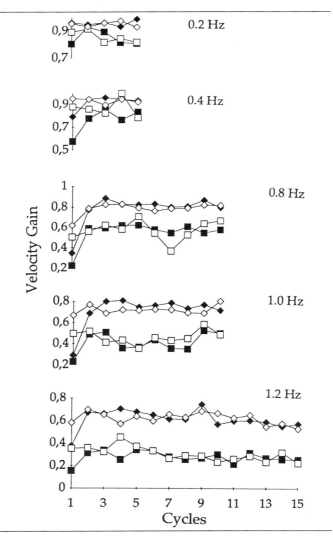

Figure 4. Rightward (open symbols) and leftward (solid symbols) intersubject mean values of velocity gain through the stimulation cycles in children (square) and adults (diamond) at the five stimulation frequencies.

exhibited constantly smaller absolute values and a more marked fluctuation than adults specially at lower frequencies. In children, at the 1.2 Hz frequency, a slight decrement of the VG and an important reduction of the PG was observed. As concerns the phases, they did not show significant variations throughout the test, remaining nearly constant around their mean value (table 1). The phases in children showed a slightly greater lag than in adults: the maximum delay difference between the groups was about 15 ms.

Discussion

The data obtained in this study show, for frequencies above 0.8 Hz, significant differences in gain and phase mean values between adults and children; as far as concerns the VG, the differences are significant also at lower frequencies.

The fluctuations shown in fig.3 and fig.4 might depend mainly on the attention level, which appears to be less constant in children than in adults, rather than on the fatigue. Indeed fatigue would determine an uniform decrease of gains along the test, while in our registrations the gains, also after an important fall, can become high again. Only at the 1.2 Hz stimulation frequency, the monotonic gain decrement throughout the test could be justified by fatigue.

Could the fluctuations be responsible for the gain mean values differences found between adults and children?

To reduce the influence of such fluctuations on the mean values of the parameters, it is proper to choose only the parts of the test in which a good tracking is performed, using, for example, criteria based on mean values and standard deviation (SD). As done in the literature, our parameters evaluation was made by averaging the values obtained from the best recorded cycles. In particular we considered, for each subject, all the half-cycle values within the range of ± 1.5 SD with respect to the mean of each registration. Using this method, the differences between adults and children reported in table 1 were found. We noted however that, rejecting the parts which are characterized by a bad tracking, the mean values for each subject show only small variations; on the contrary, as obvious, variance values decrease remarkably.

In order to eliminate any kind of influence of attention levels on smooth tracking, we took into account, for each subject, only the single half-cycle in which the best smooth response had been obtained; the parameters (which will be termed the maximum values) were then calculated on this single half-cycle, rather than on the average of all the good half-cycles of the same test. In table 2 the VG values obtained with these two different methods (mean and maximum) are shown. The differences between adults and children are small at 0.2 and 0.4 Hz because both adults and children can track with a VG close to 1, while they are relevant (>20%) and similar for the two methods above such frequencies. Indeed, differences between adults and children evaluated with both the maximum and the mean criteria ranged from 0.26 at 0.4 Hz to 0.32 at 1.2 Hz. Therefore, it is quite hard to affirm that differences between adults and children can be ascribed only to different attention levels: different maturation levels of the SPS must also be present.

Frequency Hz	Children				Adults			
	mean values rigthw. leftw.		max values rightw. leftw.		mean values rightw. leftw.		max values rightw. leftw.	
0.2	0.86	0.86	0.99	0.97	0.95	0.96	1.00	1.02
0.4	0.85	0.81	0.92	0.98	0.93	0.95	1.01	1.00
0.8	0.57	0.54	0.68	0.66	0.80	0.82	0.94	0.94
1.0	0.46	0.42	0.60	0.56	0.72	0.76	0.88	0.86
1.2	0.31	0.28	0.42	0.44	0.59	0.62	0.76	0.78

Table 2. Rightward and leftward mean and maximum values of velocity gains at the five stimulation frequencies in children and adults.

Is it possible that the duration of each test could influence the mean values of the studied parameters ?

To find out the influence of the number of utilized half-cycles on each mean value evaluation, the VG (figure 5) and the PG as the mean of successive half-cycles, excluding the first two, were calculated. After some cycles, depending on the stimulation frequency, the gain values become stable. In any case, for all frequencies, our results show that after 6 whole cycles from the beginning of the test the parameters are stable (figure 5). Therefore, if a sinusoidal stimulation is used, a correct evaluation of the parameters needs a test including at least 6 cycles being the mean calculated on five cycles, rejecting the first in which the VG is, as obvious, remarkably low.

Conclusions

The correct comparison of two experimental groups needs the same criterion of characteristics evaluation: only maximum values, mean values among a fixed number of cycles, or other criteria should be used. All authors should explicitly mention the used method to allow comparisons among literature data. Our study demonstrates that a correct parameter mean is obtained only from records larger than 6 cycles.

The present study was carried out using the same processing method in adults and children; hence the differences found between the two groups are real.

Records obtained from adults and children present different variability during the test: in particular, the traces of children present a marked fluctuation, depending on less constant attention levels. However, our data show that the dif-

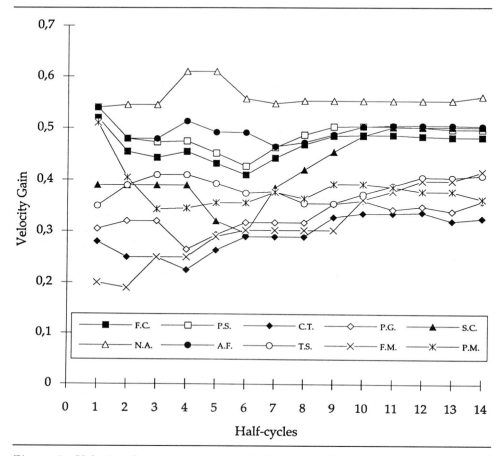

Figure 5. Velocity Gain vs. successive half-cycles utilized to evaluate it in each subject. Example in case of children in the 1.0 Hz test. The first cycle is not reported because it was rejected.

ferences between the mean (or maximum) values of parameters in adults and in children are not dependent on the attention levels, and only marginally (higher stimulation frequencies) on fatigue. They are mainly due to incomplete SPS maturation in children, perhaps including also a defective prediction mechanism. The latter hypothesis could be rejected because the very small values (tab.1) of the velocity phases (lower than 10 deg) produce a time delay (lower than about 23 ms) which could support the idea of an almost complete predictor mechanism in children.

Acknowledgments: work supported by MURST (60% and 40%).

References

Accardo A., Busettini C., Inchingolo P., dell'Aquila T., Pensiero S. and Perissutti P. (1989). EIREMA1: A device for the measurement of eye movements in strabismic children. In: Schmidt R. and Zambarbieri D. (eds). Proceed. of 5th European Conference on Eye Movements, ECEM5, Pavia (Italy), 235-237.

Aslin, R.N. (1981). Development of smooth pursuit in human infants. In Fisher, D.F., Monty R.A. & Senders, J.W. (eds), Eye movements: Cognition and visual perception. Hillsdale, NJ: Erlbaum, pp.31-51.

Baloh, R.W., Yee, R.D., Honrubia, V. & Jacobson, K. (1988). A comparison of the dynamics of horizontal and vertical smooth pursuit in normal human subjects. Aviat. Space Environ. Med. 59, 121-124.

Krementizer, J.P., Vaughan, H.G. Kurtzberg, D. & Dowling, K. (1979). Smooth pursuit eye movements in the newborn infant. Child Devl. 50, 442-448.

Leigh R.J. & Zee D.S. (1991). The neurology of eye movements (2nd Ed.). Philadelphia, PA: F.A. Davis Co.

Lisberger, S.G., Evinger, C., Johanson, G.W. & Fuchs, A.F. (1981). Relationship between eye acceleration and retinal image velocity during foveal smooth pursuit eye movements in man and monkey. J. Neurophysiol. 46, 229-249.

Shea, S.L. & Aslin, R.N. (1990). Oculomotor responses to step-ramp targets by young human infants. Vis. Res., 30, 1077-1092.

OCULAR TRACKING OF SELF-MOVED TARGETS: ROLE OF VISUAL AND NON-VISUAL INFORMATION IN VISUO-OCULO-MANUAL COORDINATION

J.-L. Vercher, D. Quaccia and G.-M. Gauthier

Laboratoire de Controles Sensorimoteurs, Universite de Provence, Avenue Escadrille Normandie Niemen, F-13397 Marseille Cedex 20, France
electronic mail: labocsm@frmrs11.bitnet

Abstract

Smooth pursuit (SP) of a self-moved target shows particular characteristics: SP delay is shorter and maximal velocity is higher than in eye-alone tracking. In the present study, we showed that if visual target movement is randomly inverted relative to arm movement at the onset of arm movement (same amplitude, same velocity but reversed direction of movement), SP shows a close spatio-temporal coupling with arm movement. The results show that during the first 200 ms, the eyes move in the same direction as the arm, while the visual target moves if the opposite direction. After 200 ms, the eyes inverse their course through a combined SP and saccadic motion. Subsequently, appropriate tracking of the visual target is observed: arm is moving in one direction, and SP in the opposite direction. These results confirm that non-visual information produced by the arm movement is able to trigger and control SP. Preliminary results with passive arm movements indicates that active control of the arm is necessary to trigger SP in the direction opposite to visual the visual target direction. The respective roles of visual, proprioceptive and efferent signals in controlling SP of self-moved target are discussed.

Keywords

smooth pursuit, Oculo-manual tracking, Coordination control, Sensory-motor control

Introduction

When a Subject tracks with his/her eyes a visual target attached to his/her own arm, previous studies have shown an improved performance of the smooth pursuit (SP) system. In particular, tracking accuracy increases (Steinbach & Held, 1968), the maximum velocity increases (from $40°/s$ up to $100°/s$) and eye-to-arm latency decreases from 100-120 ms to zero (Gauthier et al., 1988). This increase of performance has been described both in human beings and primates (Gauthier & Mussa-Ivaldi, 1988; Vercher & Gauthier, 1988). Performance can be further increased by the use of other sensory inputs related to the arm motion (tactile, auditory or proprioceptive inputs (Mather & Lackner, 1981). Gauthier et al. (1988)

Eye Movement Research/J.M. Findlay et al. (Editors)
1995 Elsevier Science B.V.

proposed a model of the coordination control exerted between the arm motor system and the oculomotor system. According to this model, during self-moved target tracking, non-visual information, generated during arm movement, is transmitted to the oculomotor system. Gauthier et al. (1988) suspected that efference copy was responsible for the synchrony between arm motion and eye motion, and that proprioceptive inflow played a role in the cross-calibration of both systems. Arm proprioception suppression in human (by ischemic block: Gauthier & Hofferer, 1976) as in primates (by section of the dorsal roots: Gauthier & Mussa-Ivaldi, 1988) removed the coordination between arm and eyes.

The aim of this study was to investigate further the role of visual versus non-visual information generated during arm movements, and used to increase the performance of the SP system during tracking of a self-moved target. In previous studies (Vercher et al., 1991; Vercher & Gauthier, 1992), we have shown by artificially delaying the visual feedback of the arm motion, that SP of a self-moved target exhibits particular characteristics such as a SP latency shorter and a maximal velocity higher than those of eye-alone tracking. In fact, if a Subject tracks a visual target controlled by his own arm, eye movement and arm movement are closely synchronised, even if target motion onset is delayed with respect to the arm motion by up to 450 ms. After a few seconds of tracking, SP becomes synchronised with visual target movement, except during arm movement velocity changes where SP correlates to arm movement.

In order to dissociate the two kinds of signals controlling arm motion (vision and non-visual information) we used a situation in which the motion of the self-moved target was reversed relative to arm motion i.e. when the Subject moved his arm toward the right, the target moved toward the left, at the same velocity and with the same amplitude. In order to eliminate the influence of the SP prediction operator (Bahill & McDonald, 1983), we concentrated our analysis on the few hundred ms after the onset of the movement. The respective roles of visual and non-visual information in oculo-manual coordination has been determined by comparing the tracking performance under different conditions, namely eye-alone tracking of a visual target, and eye tracking of self-moved target, with and without inversion of the visual feedback of the arm.

The findings were as follows. If visual target movement was randomly reversed relative to arm movement at the onset of arm movement (same amplitude, same velocity but reversed direction of movement), SP exhibited a close spatio-temporal coupling with arm movement. Data showed that, even while the visual target moved in the opposite direction, during the first 200 ms, the eyes nonetheless moved in the same direction as the arm. After 200 ms, eye motion direction changed through a combination of SP and saccadic motion. Subsequently, appropriate tracking of the visual target was observed: the arm moved in one direction, and SP in the opposite direction. These results confirm that non-visual information produced by the arm movement is able to trigger and control SP.

Methods

Subjects

Nine Subjects ranging in age between 19 and 32 participated to this study. They were undergraduate students or staff from our department. They were instructed and trained with the apparatus in a preliminary session and they all gave their informed consent for their participation in the study. They were all exempt from known visual or oculomotor problems, and used their preferred arm during the experiment (they were all right-handed).

Experimental setup

The Subject was seated at 171 cm in front of a projection screen (Fig. 1), with the head immobilised by a dental bite-bar. The arm of the Subject rested on a mobile horizontal semi-cylindrical gutter, the arm pointing in the direction of the screen. The Subject has no direct vision of his forearm, but had a visual cue of the ongoing movement through a target projected on the screen (*self-moved target*). Another moving dot could be used as an *external target*. The self-moved target motion on the screen was adjusted to provide a one-to-one relationship (judged by the Subject) between rotation of the handle and self-moved target movement.

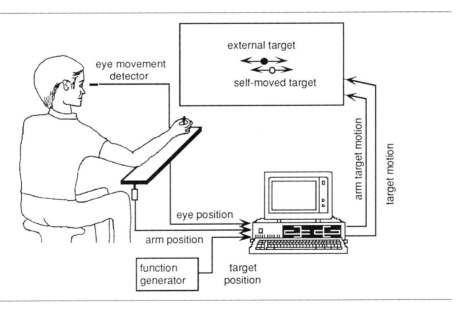

Fig. 1 - Experimental setup.

Horizontal eye movements were recorded by means of an infrared corneal reflection device (*IRIS Skalar, inc. Delft, The Netherlands*), with a resolution of 2 min of arc and a bandwidth ranging from DC to 100 Hz at -3 dB. The arm motion was recorded by means of a precision potentiometer placed at elbow level. Two visual targets could be projected on the screen. Their respective motions were controlled through galvanometer-mounted mirrors (*General Scanning, inc., Watertown, MA, USA*). The motion of the *external target*, controlled by a function generator, was sinusoidal, the frequency was 0.3-0.4 Hz, with an amplitude of ±5° as seen from the Subject's eyes. The maximum velocity of the target was around 12°/s. The *self-moved target* could be substituted for the first one. The movement of the *self-moved target* was controlled by the signal issuing from the potentiometer. Both targets were simultaneously presented on the screen only during calibration. The signals from both the potentiometer and the function generator driving the external target, were amplified to a level equivalent to the output of the eye movement monitor and filtered (low-pass filter, 100 Hz at -3 dB). Target position, eye position and arm position signals were digitised at 500 samples/s by a 12-bit A/D converter, displayed in real-time on a graphic screen, and sections of 5 s were recorded on a disk for off-line analysis, starting 200 ms before the onset of the target/arm movement.

Tracking conditions

• *Eye-alone tracking of an external target.* The target was presented on one side of the screen and started to move horizontally toward the other side of the screen. The Subject was instructed to track the visual target with his eyes, as accurately as possible. This condition was used as a reference.

• *Ocular tracking of a self-moved target.* Prior to each trial, the Subject placed the self-moved target at the centre of the screen. The instruction was to sinusoidally move the arm at learned frequency and amplitude and to track the self-moved target with the eyes. A GO signal was provided by the experimenter to the Subject, to start moving the arm.

Under the second condition, the relationship between the arm and the target could be reversed: when the Subject started to move his arm toward the right, the target moved toward the left. When both movements shared the same direction, coupling was stated as *direct*. When the arm and the target moved in opposite direction, coupling was stated as *reversed*. This alteration of the self-moved target relationship could be applied systematically or randomly during a same session.

Data analysis

Subjects were tested at least 3 times each on different days. Off-line signal analysis started with digital low-pass filtering (a running average of 11 coefficients providing a cut-off frequency of 20 Hz). The analysis determined the latency between target motion and eye motion and the direction of the initial eye movement relative to target motion, using interactive graphic software. Position signals were digitally differentiated, and velocity signals were displayed on the screen. A graphic cursor was moved to locate the beginning of each movement. Data were analysed by ANOVA. For statistical comparisons, a significant difference was taken as $p < 0.05$, and a highly significant difference as $p < 0.01$.

Results

Eye tracking of a external/self-moved target. Reference data

Fig. 2 shows typical records obtained from a Subject in eye-alone tracking (Fig. 2A) and self-moved target tracking (Fig. 2B) conditions. The target waveform used in eye-alone tracking has been chosen in order to imitate a typical arm movement (low acceleration at the beginning of motion and at turn-around). The average latencies were different in the two conditions: 152 ± 24.12 ms in eye-alone tracking and 8 ± 36.61 ms in self-moved target tracking (highly significant, $p<0.01$). Sometimes in this latter condition (as in Fig. 2B), the eyes clearly started to move *before* the arm. As opposed to eye-alone tracking (Fig. 2A) a saccade never occurred at the beginning of self-moved target tracking.

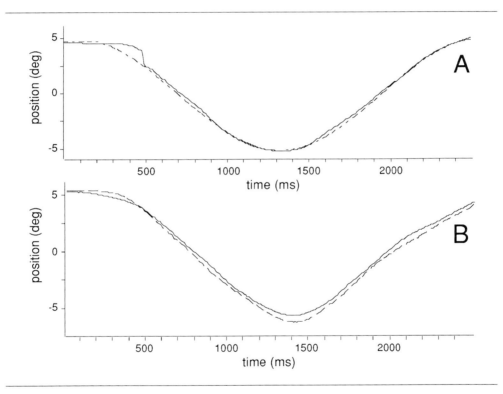

Fig. 2 - A: eye-alone tracking. Subjects were instructed to fixate the target and to track it when it started to move. Solid line represents eye movement and dashed line target movement. B: self-moved target tracking. Subjects were instructed to start moving the arm and to track with the eyes the self-moved target. Smooth pursuit started almost at the same time as the arm, and eventually, as here, before self-moved target motion, but always in the same direction as the arm.

Random inversion of arm to target coupling

After collecting reference data (previous section), the Subjects were exposed, in a different session, to a condition in which the motion of the self-moved target could be randomly reversed. On the average, one *reversed* trial occurred for every four *normal* trials. Under this condition, the Subject could not determine before moving his arm which type of coupling would be used. Figure 3 shows a typical response with reversed visual feedback. The eyes start moving at the same time, and in the same direction as the arm, in spite of the visual self-moved target motion occurring in the opposite direction. The average latency with reversed coupling was 13.2 ± 38.18 ms and the average latency measured in non-reversed trials was 10.1 ± 37.26 ms. The difference in latency between reversed and non reversed trials was not significant ($p > 0.5$).

After 139.0 ± 53.18 ms on average, the eye motion slowed down, the direction of motion changed, a saccade was triggered toward the target, and then the SP continued on the basis of visual information. Eye direction and velocity were then correlated to target direction and furthermore, eye motion was correlated to target motion in terms of both direction and velocity. The time of reversal of SP was not significantly different from the eye-to-target latency observed in eye-alone tracking ($p > 0.5$). The initial SP gain (measured during the first 100 ms) was 0.69 ± 0.36, eye velocity was poorly correlated to arm velocity ($R^2 = 0.42$).

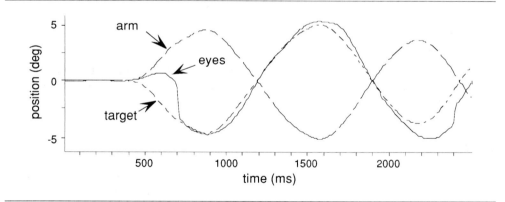

Fig. 3 - Eye tracking of self-moved target with random inversion of target motion relative to arm motion. In most trials, the eyes started to move at the same time and in the same direction as the arm, but with a low gain. After 100-150 ms, SP direction changed, 100 ms later a saccade was triggered allowing the eye to reach the visual target. From this moment on, the eyes tracked the visual target, with appropriate gain and low phase. Solid line denotes eye motion, and the dashed lines denote arm motion and target motion (same as arm motion, but reversed).

Smooth pursuit latencies

Figure 4 summarises eye latency data in the three conditions (eye-alone tracking, self-moved target tracking, and self-moved target tracking with reversed feedback) and shows that in the two latter conditions SP was triggered early, and occurred systematically in the direction of the actual arm movement but not in the direction of the corresponding visual target motion.

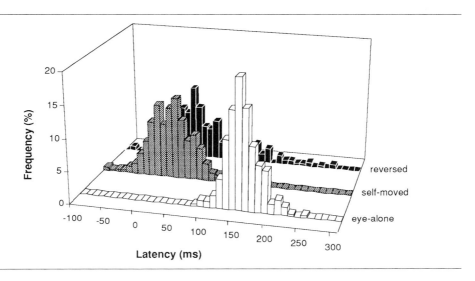

Fig. 4 - Latency histograms, in eye-alone tracking (white blocks) as compared to directly coupled self-moved target tracking (grey blocks) and reversed self-moved target tracking (black blocks).

Discussion

In this study as in a previous work (Vercher & Gauthier, 1992) we investigated the role played by non-visual information in the synchronisation of arm movement and ocular tracking of self-moved target. We randomly uncoupled the motion of the target from the motion of the arm, by reversing the direction of target movement. This paradigm allowed us to separate visual information coding target motion from non-visual information coding arm motion. The analysis was essentially based on measurements of ocular latency.

In eye-alone tracking, the average latency was markedly higher than values previously obtained and classically considered as the intrinsic SP delay: 90 ms (Carl & Gellman, 1987) to 130 ms depending of target motion path (Collewijn & Tamminga, 1984). This is certainly due to the kind of target motion we used. In fact, this motion was designed to imitate a typical movement of the forearm, as measured in the self-moved target tracking condition. Because smooth latency depends on the visual target motion, comparison of latencies obtained under different conditions can be done only if similar target motions are used. When the Subjects tracked the self-moved target with their eyes, the decrease of target-to-eye latency was similar to that obtained in previous studies (Steinbach & Held, 1968; Gauthier & Hofferer, 1976; Gauthier et al., 1988; Vercher & Gauthier, 1992) and clearly showed that SP latency can be shortened if the observer directly drives the target.

The aim of this study was to demonstrate that this short tracking latency was not due to prediction of visual origin. Indeed, when in our experiment the visual input was opposite to the driving movement, this short latency SP was always triggered in the direction of the *arm* rather than in the direction of the visual signal.

Obviously, the non-visual signals issued from the arm motor system cannot control SP as well as the visual inputs. Smooth pursuit of imaginary targets is poor and combined with saccadic pursuit (Gauthier & Hofferer, 1976). In the experiment reported here, when the target motion was reversed, the initial gain was lower than when the target was directly coupled to the arm motion (Fig. 2B, Fig. 3). After a few hundred ms of tracking, SP was essentially under visual control (Fig. 3). It must be noted than the time from arm motion onset and the decrease of eye velocity preceding the time when SP changed in direction was similar to the average latency measured in eye-alone tracking. This shows that in the early period of self-moved target tracking, SP was under direct non-visual control. Subsequently, that is around 100 ms later, the visual information became predominant.

Preliminary results with eye tracking of a *passively* moved arm-attached target seem to indicated that the synchronisation is mainly due to efferent signals rather than proprioceptive information generated during the arm motion. The afferent signal would play a major role in calibration and gain control of SP, essentially at the beginning (first second) of tracking. After the first second, in case of conflict between arm motion and vision, SP seems to be mostly under visual control.

Conclusion

This study showed that non-visual information generated by the arm motion is definitely responsible for eye-arm synchronisation, rather than the predictor operator proper to the SP system. In case of conflict between visual information coding the target and arm motion, SP is always triggered with short latency in the direction of the arm motion. In the early stages of eye motion (during the first one hundred ms) the predictor is not yet activated but when it becomes operative, SP is mostly due to visual signals.

We also showed that in spite of the predominant role played in oculo-manual coordination by visual information during SP of self-moved targets, signals issued by the arm motor system can trigger SP more precociously than vision. This synchronisation between arm motion and eye motion compensates for the "slowness" of visual information and allows a spatio-temporal consistency during manipulation under visual control. The role played by afferent and/or efferent information in manuo-ocular information is still open.

When conflicts occur between visual motion and kinesthetic information from the arm motion, the central control system first attempts to turn off or compensate the *hardwired* activation of eye in the direction of arm movement then slowly switches to a strategy which consists at favouring visual information as a driving input to the SP system.

We conclude that two mechanisms cooperate in ocular tracking of self-moved target: a short-latency activation mechanism with poor accuracy, mostly relaying non-visual signals, combined with a longer latency and higher accuracy mechanism, processing visual information.

References

Bahill A.T. & McDonald J.D. (1983). Smooth pursuit eye movements in response to predictable target motions. Vision Res., 23, 1573-1583.

Carl J.R. & Gellman R.S. (1987). Human smooth pursuit: stimulus-dependent responses. J. Neurophysiol., 57, 1446-1463.

Collewijn H. & Tamminga E.P. (1984). Human smooth and saccadic eye movements during voluntary pursuit of different target motions on different backgrounds. J. Physiol. (London), 351, 217-250.

Gauthier G.M. & Hofferer J.M. (1976). Eye tracking of self-moved targets in the absence of vision. Exp. Brain Res., 26, 121-139.

Gauthier G.M. & Mussa-Ivaldi F. (1988). Oculo-manual tracking of visual targets in monkeys: role of the arm afferent information in the control of arm and eye movements. Exp. Brain Res., 73, 138-154.

Gauthier G.M., Vercher J.-L., Mussa Ivaldi F. & Marchetti E. (1988). Oculo-manual tracking of visual targets: control learning, coordination control and coordination model. Exp. Brain Res., 73, 127-137.

Mather J.A. & Lackner J.R. (1981). The influence of efferent, proprioceptive, and timing factors on the accuracy of eye-hand tracking. Exp. Brain Res., 43, 406-412.

Steinbach M.J. & Held R. (1968). Eye tracking of observer-generated target movements. Science, 161, 187-188.

Vercher J.-L. & Gauthier G.M. (1988). Cerebellar involvement in the coordination control of the oculo-manual tracking system: effects of cerebellar dentate nucleus lesion. Exp. Brain Res., 73, 155-166.

Vercher J.-L. & Gauthier G.M. (1992). Oculo-manual coordination control: ocular and manual tracking of visual targets with delayed visual feedback of the hand motion. Exp. Brain Res, 90, 599-609.

Vercher J.-L., Zuber B.L. & Gauthier G.M. (1991). Ocular and manual tracking of visual targets with delayed visual feedback. Sixth European Conference on Eye Movements, Leuven, Belgium, September 15-18 1991.

EYE MOVEMENTS EVOKED BY LEG-PROPRIOCEPTIVE AND VESTIBULAR STIMULATION

Fabio M. Botti, Georg Schweigart and Thomas Mergner

Neurological University Clinic, Hansastrasse 9, D-79104 Freiburg, Germany

Abstract

Leg-proprioceptive stimulation in intact humans during rotation of the feet under the stationary body induces nystagmus. The slow component of this leg-eye response reaches a considerable magnitude only at low stimulus frequencies/velocities. It appears to sum linearly with the vestibulo-ocular reflex (VOR) and to prevent VOR gain attenuation at low frequencies, if the body is rotated on the stationary feet. Its normal function could be to aid eye stabilization during slow body sways.

Keywords

Vestibulo-proprioceptive interaction, eye movements, vestibulo-ocular reflex (VOR), arthrokinetic nystagmus

Introduction

The importance of vestibular and visual afferents for eye movement control is well established. In contrast, little is known to date about the role of somatosensory motion cues for this function. In particular, it has often been speculated whether somatosensory input may help to stabilize gaze. Stabilization of the eyes on the visual "scene" is required for accurate vision. During head movements, the vestibulo-ocular reflex (VOR) makes a major contribution to the stabilization of the eyes on the stationary visual scene. VOR gain in the standard testing condition (sinusoidal rotations in the dark), however, typically is below unity (of the order of 0.6). Somatosensory input arising from head movements during head-on-trunk or trunk-on-foot rotations could, indeed, be considered useful to aid the VOR in gaze (eye-in-space) stabilization.

A stereotype neck-eye response (cervico-ocular reflex, COR), has been described in several animal species and in new born human babies (see Jürgens and Mergner, 1989). It can be demonstrated selectively during passive horizontal plane trunk rotation under the stationary head in the dark, for instance. The response is oriented towards the trunk (i.e., towards the platform that carries the head). Thus, it is compensatory in the sense that it would help the VOR in gaze stabilization during a head rotation on the stationary trunk. A COR with considerable gain and compensatory direction also has been found in

Eye Movement Research/J.M. Findlay et al. (Editors)

patients with chronic bilateral loss of vestibular function (Bles et al., 1984; Bronstein and Hood, 1986; Huygen et al., 1991). In fact, Dichgans et al. (1973) showed that vestibular loss in monkey is followed by a gain increase of the COR, which then contributes to gaze stabilization following saccadic eye-plus-head shifts. Preliminary data suggest a similar mechanism in adult patients (Kasai and Zee, 1978). In intact human adults, by contrast, the COR is very small and its direction (phase) is very variable and mostly anti-compensatory (Jürgens and Mergner, 1989; Bronstein and Hood, 1986). There is a gain enhancement of the VOR during combined vestibular and neck stimulation, but this effect is, to a large degree, not direction specific since it occurs independently of how the two stimuli are combined (Jürgens and Mergner, 1989). Thus, it appears unlikely that the COR makes a major direct contribution to gaze stabilization in intact human adults.

Less is known about a gaze stabilizing role of leg proprioceptive input that arises during trunk rotation on the stationary feet. Stepping movements on a rotating platform, such that the body remains essentially stationary in space, may evoke a nystagmus (Bles, 1981) as does a pedaling of a pivotable platform by a stationary sitting subject (Lackner and DiZio, 1984). But it is not clear from these experiments whether the response stems from an efference copy-like signal or the proprioceptive input. More clear are the findings of Brandt et al. (1977). These authors rotated a big drum with constant velocity about a stationary subject in darkness. When the subject extended one of his arms and made contact with the rotating drum, a passive movement of the arm in the shoulder joint resulted. This stimulus evoked an illusory perception of self-motion in space in the direction opposite to the drum rotation. Furthermore, nystagmus-like eye movements were evoked ('arthrokinetic nystagmus'), with the slow phase being opposite with respect to the perceived body rotation (compensatory). The authors concluded that somatosensory afferents play an important role in the registration of self-motion in space and in the control of eye movements. There are, in fact, several experimental studies in animals that provide a neurophysiological basis for this notion, showing that rotation of the limbs about the stationary trunk evokes responses in vestibular neurons at several levels of the central nervous system (see Mergner et al., 1983).

The aim of the present study was to evaluate in some detail the leg proprioceptive eye response in intact human adults and its interaction with the VOR during a variety of vestibular-proprioceptive stimulus combinations. We wanted to compare this leg-eye response to (i) the neck-eye response, and (ii) to the leg and neck proprioceptively evoked human self-motion perception. The latter point was motivated by the fact that both leg and neck proprioceptive inputs exert analogous effects on human self-motion perception in space; the two inputs interact linearly with the vestibular input, thereby improving the perception of self-motion, particularly at low stimulus frequencies (Mergner et al., 1991, 1993). This led us to record, in addition to their eye movements, the subjects' self-motion perception.

Methods

Seven healthy subjects (age: 24-45 yrs.) participated in the study. Fig. 1A shows the experimental setup in a schematic form. Subjects were seated on a Barany chair for horizontal body rotation. Their feet rested on a foot-sled that could be rotated independently of the chair. Sled rotation led to a rotation of the subjects' legs on a circumferential path about the same rotation axis as the chair and, if the chair was stationary, it led to a leg-to-trunk excursion mainly in the hip joints. This represented the leg proprioceptive stimulus. Chair rotation together with the sled, such that no relative motion between legs and trunk occurred, represented the vestibular stimulus. Different combinations of leg and trunk rotations were used to generate various vestibular-proprioceptive combinations. The stimuli had a sinusoidal wave form. In some parts of the experiments stimulus frequency was varied (f= 0.025, 0.05, 0.1, 0.2, and 0.4 Hz) while peak angular displacement (dmax) was kept constant (at ±8°; note that peak angular velocity covaried with frequency in this "frequency series": v_{max}= 1.25, 2.5, 5, 10 and 20°/s, respectively). In other parts, frequency was kept constant at 0.05 Hz, while peak displacement was varied (dmax= ±1, 2, 4, and 8°; "amplitude series"), and thus peak velocity varied correspondingly (v_{max}= 0.3, 0.6, 1.25, 2.5°/s).

Subjects' heads, supported by a chair-fixed chin rest, were always held in a fixed alignment with their trunks. The experiments were performed in the dark, and auditory cues were reduced by plugging the subjects' ears with the help of rubber plugs.

Subjects' horizontal and vertical eye movements were recorded with the help of an infrared technique (IRIS, Skalar). In addition, we measured subjects' perception of trunk motion in space using a concurrent pointing procedure (Mergner et al., 1991, 1993). The position signals of chair, sled, pointer, horizontal and vertical eye movements, and the velocity signal of horizontal eye movements were fed into a computer at a sampling rate of 100 Hz. The analysis was performed off-line. Slow and fast components of the horizontal eye movements were separated using an interactive computer program. The slow component of the horizontal eye response and the pointer reading were analyzed in terms of gain and phase using a fundamental wave analysis (FFT).

For the frequency series, we present median values and their 95% confidence ranges of the individual subjects' mean values (n=2-6, depending on frequency). For the amplitude series, in contrast, only mean values across all subjects will be presented. The mean values were obtained by averaging six original slow responses for each subject and stimulus condition, before performing the fundamental wave analysis. This procedure was chosen in order to reduce noise in the eye responses as these became very small with decreasing stimulus amplitude.

F.M. Botti, G. Schweigart & T. Mergner

Results

Leg proprioceptive stimulation during foot-sled rotation about the stationary chair evoked a small, but clearly measurable eye response in all subjects. Fig. 1 B gives an example of such a response and its separation into slow and fast components. Note that slow and fast components were essentially opposite in direction to each other; a slow eye movement to the right, for instance, was interrupted by saccades to the left.

The magnitude of the slow component clearly depended on stimulus frequency. Its median gain was low at high frequencies (e.g., 0.05 at 0.4 Hz) and increased with decreasing frequency, reaching 0.25 at 0.025 Hz (Fig. 2A). Median phase lagged the foot rotation by approximately 20° across all frequencies tested. In Fig. 2A, the phase is plotted in relation to the relative trunk-to-foot excursion (thus, shifted by 180°), in order to make clear that it would be approximately compensatory during a trunk rotation on the stationary feet.

Fig. 2A shows, for comparison, the median gain and phase values of the subjects' VOR as a function of stimulus frequency. VOR gain was below unity at high frequencies (0.7 at 0.4 Hz) and further attenuated with decreasing frequency, reaching 0.3 at 0.025 Hz. VOR phase was essentially compensatory

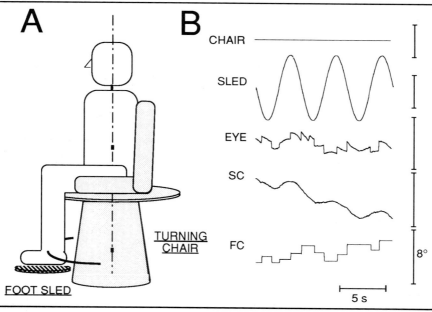

Figure 1A,B. A Schematic drawing of the stimulus setup. B Example of leg proprioceptive eye response. Horizontal positions of turning chair, foot sled, eye, slow component (SC) and fast component (FC) of the eye movement (rightward displacements, up).

with respect to the trunk-in-space rotation at high to mid-frequencies, but developed some lead at low frequency (20° at 0.025 Hz). This frequency dependent behaviour of the VOR is typical for intact adults during sinusoidal rotations in the dark.

Note that the gain increase observed for the leg proprioceptive response with decreasing frequency could equally well reflect a dependency on peak angular velocity, since peak velocity covaried with frequency (see Methods). In an additional experiment, we therefore varied stimulus amplitude (peak displacement and velocity) of the proprioceptive and vestibular stimuli while keeping frequency constant (0.05 Hz). As shown in Fig. 2B, the proprioceptive response to the 2.5°/s stimulus had a gain of 0.12, similar as in the above experiment, and also the phase was roughly similar. The gain rose with decreasing peak velocity, reaching 0.35 with the 0.3°/s stimulus. The phase, in contrast, remained essentially constant. Thus, the proprioceptive response behaved like being sensitive to very low angular velocities and saturating rapidly with increasing velocity.

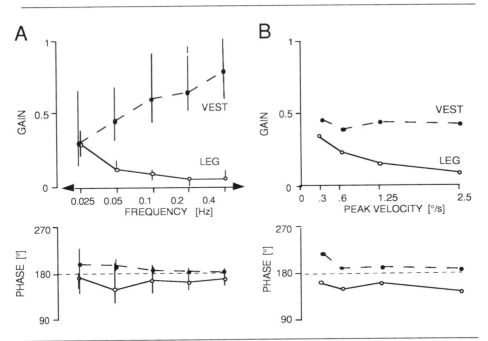

Figure 2 A,B. Gain and phase of the slow eye response during leg proprioceptive stimulation (LEG, open circles) and during vestibular stimulation (VEST, filled circles) plotted as a function of stimulus frequency (A) and as a function of stimulus velocity (B; 0.05 Hz).

VOR gain and phase in the amplitude series remained essentially constant (approximately 0.4 and 15°, respectively). Thus, there was no indication of a displacement or velocity non-linearity of the VOR in our data. This finding is at variance with the vestibular evoked self-motion perception which shows a clear velocity threshold (Mergner et al., 1991).

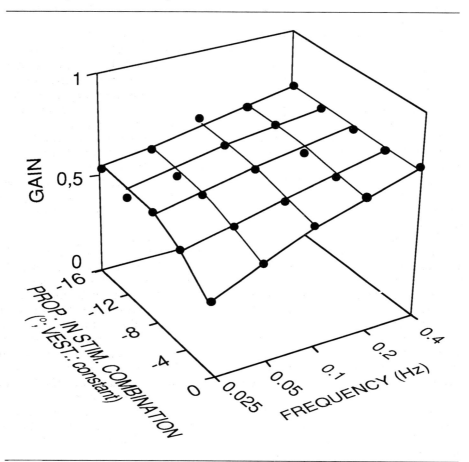

Figure 3. Median gain (dots) of the slow eye response as a function of both, stimulus frequency and of the amplitude of the proprioceptive stimulus (PROP.) in the stimulus combination (vestibular stimulus, VEST.: 8°=constant).

In order to study vestibular-proprioceptive interaction, we presented our subjects with five different stimulus combinations, keeping peak displacement of the vestibular stimulus constant (8°) while varying that of the proprioceptive stimulus. As can be seen from Fig. 3, one combination represented a repetition of the vestibular stimulus in the frequency series (proprioceptive stimulus: 0°). In the other four combinations, the proprioceptive stimulus had the opposite direction of the vestibular stimulus (indicated by negative sign) and was varied from -4°, -8°, -12° to -16° (note that, in the combination with the -8° proprioceptive stimulus, the trunk was rotated about the stationary feet). These combinations were applied with the five different frequencies of the frequency series (0.025, 0.05, 0.1, 0.2, and 0.4 Hz).

The results are shown in Fig. 3 in terms of a three dimensional plot which gives median gain of the eye response as a function of both, the stimulus combination and frequency. The gain during vestibular stimulation essentially reproduced the above described results, i.e., the gain was on the order of 0.6-0.7 at high frequency and clearly attenuated with decreasing frequency. Addition of the counter-phase proprioceptive stimuli did not change the gain to any considerable degree at high frequencies. At low frequencies, in contrast, attenuation of the response gain was less, the increment depending in a monotonic way on the magnitude of the proprioceptive stimulus, but appeared to saturate with the -12° and -16° stimuli. Similarly, the phase of the response did not show any considerable change at high frequency, whereas the phase lead of the VOR that developed with decreasing frequency (compare Fig. 2A) became less with increasing magnitude of the proprioceptive stimulus (not shown).

Similar as in our previous study (Mergner et al., 1993), leg proprioceptive stimulation exerted a clear influence on subjects' self(trunk)-motion perception in space, which also was recorded (see Methods). In particular, rotation of the foot sled about the stationary trunk evoked an illusion of a trunk rotation in space in the direction trunk-to-foot excursion. The gain (ratio of indicated trunk-in-space rotation versus actual trunk-to-foot rotation) depended on stimulus frequency in a similar way to that of the gain curve of the proprioceptive eye response in Fig. 2A. Also, the vestibular evoked self-motion perception in space was improved in the low frequency range when adding the counter-phase proprioceptive stimuli. The behaviour of the gain (ratio of indicated versus actual trunk-in-space rotation) as a function of stimulus combination and frequency resembled closely the results shown in Fig. 3 for the eye response. These analogies between perceptual and eye responses led us to look for a correlation between the proprioceptive effects in the two data sets, separately for pure proprioceptive stimulation in the frequency series and for the stimulus combinations at 0.025 Hz.

A very weak correlation between perceptual and eye response was found for pure proprioceptive stimulation (r= 0.53). No significant correlation was observed for the stimulus combinations at 0.025 Hz (r= 0.37). Thus, despite the similarity in the median values, there was no considerable correlation with respect to the individual responses.

Discussion

Rotation of the feet relative to the stationary trunk evoked a small, but consistent eye response in intact human adults. The response consisted of both a slow component, which was oriented in the direction of the foot displacement, and saccadic components in the opposite direction. There is a qualitative similarity between this response pattern and that of the VOR. One might ask, therefore, whether the response could possibly have resulted from a small head-in-space rotation associated with the foot rotation. We can largely exclude this possibility, since our subjects' backs and flanks were pressed on back and side rests of the chair, with the head held in fixed alignment with the trunk by means of a chin rest. Also, if the head had been moved slightly in the direction of the foot displacement, the VOR would have produced eye movements opposite in direction to that actually observed. Furthermore, a vestibular origin should have produced a higher response gain at high as compared to low frequencies, which is just the opposite to what we observed.

Therefore, we consider the response to stem from proprioceptive input reaching vestibular centers in the brain, as demonstrated earlier at neuronal levels in animal studies (see Introduction). Also, we take our findings as a confirmation of the "arthrokinetic nystagmus" reported by Brandt et al. (1977; constant velocity stimuli, essentially applied to proximal arm joints). We prefer, however, the term limb proprioceptive eye response instead of arthrokinetic nystagmus, since the input that reaches vestibular neurons in the CNS is likely to stem from muscle spindle receptors rather than from joint receptors. The responses obtained in our experiments resemble those reported by Brandt et al. (1977) with respect to the response pattern, the direction and a clear response saturation at higher stimulus velocities. The velocity saturation most likely accounts for the apparent low-pass behaviour of the leg proprioceptive response observed in our study. This notion is supported by the finding that the phase remained essentially constant across frequency.

In the Introduction we raised the question, to what extent might the leg proprioceptive input "help" the VOR in gaze stabilization. Our experiments show that combining a vestibular stimulus with counter-phase leg proprioceptive stimuli does not raise the gain of the eye response to unity ("normal" gain of VOR: 0.5-0.7). Yet, the vestibular-proprioceptive combination improved the frequency behaviour of the gain, in the sense that the gain attenuation of the VOR at low stimulus frequency became clearly less. Therefore, a possible role of the leg proprioceptive input to the oculomotor system could be to aid gaze stabilization during slow head movements, as they may occur, for instance, during slow spontaneous body sways on the stationary feet.

Interestingly, a similar concept has been suggested by Meiry (1971) for the COR. Meiry observed a COR of considerable gain at low stimulus frequency, where the VOR showed a clear gain attenuation. Upon combined stimulation during head rotation on the stationary trunk, he found the gain of the eye response to remain essentially constant across frequency, and he attributed this to a VOR-COR interaction. As pointed out in the Introduction, however, his findings were not confirmed in later studies, at least in intact human adults. We would like to revive the concept, but to apply it to the leg rather than to the neck proprioceptive eye response.

Neck and leg proprioceptive effects on human self-motion perception are analogous to each other to a large degree (Mergner et al., 1993). Why should neck and leg proprioceptive eye responses then be different? A possible explanation is that a stereotype COR, established to help the VOR during head rotation on the stationary trunk, might be detrimental in other, equally often occurring behavioural conditions. During trunk excursions while walking, for instance, gaze stabilization is obtained in part by a compensatory head movement in addition to the eye movement (VOR). In this situation, where head-on-trunk and head-in-space movements have opposite directions, the COR would counteract the VOR. Compensatory trunk movements in response to motion of the foot support, on the other hand, are certainly rare events as compared to orienting trunk movements, so that one could conceive that the latter condition prevails the former with respect to the adaptation mechanisms that shape the gaze stabilization.

Eye response and self-motion perception during leg proprioceptive stimulation resemble each other with respect to the frequency behaviour of the median gain values. This finding raised the question, to what extent the two responses are interrelated. One could imagine, for instance, that the eye response arises as a consequence of the self-motion perception. On basis of the individual data, however, we found no considerable correlation between the two responses. Therefore, we assume that the perceptual and eye movement responses, although receiving afferents from the same receptor organs, are processed in different channels and thus carry different noise signals. This is similar to the notion of Peterka and Benolken (1992) who found no correlation between the commonly observed imbalances of VOR and of vestibular evoked self-motion perception in the dark.

Acknowledgment

Supported by DFG, SFB 325

References

Bles, W. (1981). Stepping around: Circular vection and Coriolis effects. In Attention and performance, Vol. IX, J. Long and A. Baddeley Eds., Lawrence Erlbaum, Hillsdale, pp. 47-61.

Bles, W., de Jong, J.M.B.V. & Rasmussens, J.J. (1984). Postural and oculomotor signs in labyrinthine-defective subjects. Acta Otolaryngol. (Stockh.) Suppl. 406, 101-104.

Brandt, T., Büchele, W. & Arnold, F. (1977). Arthrokinetic nystagmus and ego-motion sensation. Exp. Brain Res. 30, 331-338.

Bronstein, A.M. & Hood, J.D. (1986). The cervico-ocular reflex in normal subjects and patients with absent vestibular function. Brain Res. 373, 399-408.

Dichgans, J., Bizzi, E., Morasso, P. & Tagliasco, V. (1973). Mechanisms underlying recovery of eye-head coordination following bilateral labyrinthectomy in monkeys. Exp. Brain Res. 18, 548-562.

Huygen, P.L.M., Verhagen, W.I.M. & Nicolasen, M.G.M. (1991). Cervico-ocular reflex enhancement in labyrinthine-defective and normal subjects. Exp. Brain Res. 87, 457-464.

Jürgens, R. & Mergner, T. (1989). Interaction between cervio-ocular and vestibulo-ocular reflexes in normal adults. Exp. Brain Res. 77, 381-390.

Kasai, T. & Zee, D.S. (1978). Eye-head coordination in labyrinthine-defective human beings. Brain Res. 144, 123-141.

Lackner, J.R. & DiZio, P. (1984). Some efferent and somatosensory influences on body orientation and oculomotor control. In Sensory experience, adaptation, and perception, L. Spillmann & B.R. Wooten Eds., Lawrence Erlbaum, Hillsdale, pp. 281-301.

Meiry, J.L. (1971). Vestibular and proprioceptive stimulation of eye movements. In The control of eye movements, P. Bach-y-Rita, C.C. Collins & E. Hyde Eds., Academic Press, New York, pp. 483-496.

Mergner, T., Deecke, L., Becker, W. & Kornhuber, H.H. (1983). Vestibulo-proprioceptive interaction: Neurophysiology and psychophysics. In Multimodal convergences in sensory systems, Fortschr Zoologie. Vol 28, E. Horn Ed., Gustav Fischer, Stuttgart, pp. 241-252.

Mergner, T., Siebold, C., Schweigart, G. & Becker, W. (1991). Human perception of horizontal head and trunk rotation in space during vestibular and neck stimulation. Exp. Brain Res. 85, 389-404.

Mergner, T., Hlavacka, F. & Schweigart, G. (1993). Interaction of vestibular and proprioceptive inputs. J. Vest. Res. 3, 41-57.

Peterka, R.J. & Benolken, M.S. (1992). Relation between perception of vertical axis rotation and vestibulo-ocular reflex symmetry. J. Vest. Res. 2, 59-69.

EFFECTS OF PREDICTION ON SMOOTH PURSUIT VELOCITY GAIN IN CEREBELLAR PATIENTS AND CONTROLS

GU Lekwuwa, GR Barnes, MA Grealy.

MRC Human Movement & Balance Unit, Institute of Neurology, Queen Square, London WC 1N 3BG, UK

Abstract

We have examined 12 cerebellar patients and 10 age-matched controls in a predictive and non-predictive pursuit task using a constant velocity ramp stimulus. The subject sat in total darkness while the target was presented with a pulse duration of 640ms and inter-pulse interval of 1.7s. The target passed through the central position half way through its presentation and executed velocities between 10 and 40 degrees per sec. For the non-predictive mode the direction and timing of the stimulus were randomised. In both the cerebellar patients and controls the eye velocity in the predictive task started to accelerate in the appropriate direction before the onset of target motion. For each group, the eye velocity for equivalent time periods in the predictive task remained significantly above that in the non-predictive task for the first 300ms, although the peak velocities were not significantly different. The eye velocity in the cerebellar patients remained lower than that of controls for equivalent time periods, and peaked at a significantly lower level (p < 0.001). The latency of the visual feedback response was significantly increased in cerebellar patients (p < 0.01). Peak eye velocities were achieved significantly earlier in the controls when compared with cerebellar patients (p < 0.05). The results show that prediction is intact in cerebellar patients although both anticipatory eye velocity and peak velocity are reduced. Prediction conferred a velocity advantage to both controls and cerebellar patients. The increased latency of the visual feedback response in cerebellar patients and the delayed timing of the peak velocity may account for the increased phase lags seen in these patients during sinusoidal pursuit.

Keywords

Eye movement, ocular pursuit, prediction, cerebellar patients.

Introduction

Smooth pursuit eye movements are effected by two main mechanisms (Barnes & Asselman, 1991; Barnes, Donnelly & Eason, 1987; Dallos & Jones, 1963; Bahill & McDonald, 1983). There is a basic mechanism that uses information about retinal error velocity to drive the pursuit system but has a fairly low gain. It is also known to contain a substantial time delay of approximately 100ms (Carl & Gellman, 1987) which would naturally be expected to give rise to a large phase error. The second mechanism functions mainly in the production of predictive activity. The addition of a predictive component to the pursuit response causes the eye movements to become progressively phase advanced, and enhances the gain (Barnes & Asselman, 1991; Barnes & Grealy, 1992).

Eye Movement Research/J.M. Findlay et al. (Editors)

Predictive mechanisms have been postulated to consist of a predictive velocity estimator which samples and holds the gaze velocity information, and a periodicity estimator which controls the anticipatory release of stored waveforms to enhance the gain of smooth pursuit (Barnes, Donnelly & Eason, 1987; Bahill & McDonald, 1983). The neural substrate for prediction in smooth pursuit has not been clearly localised. Evidence from recordings in the flocculus of the cerebellum suggests that this might be an important site for the generation of the predictive component of pursuit (Miles & Fuller, 1975; Lisberger & Fuchs, 1978; Noda & Warabi, 1986; Noda & Warabi, 1987). However previous studies (Waterston, Barnes, & Grealy, 1992) on patients with various forms of cerebellar disease suggest that prediction is preserved in these patients.

Different methods have been used to assess and study the predictive component of pursuit. In one method, repeated transient stimulations are used to demonstrate the temporal characteristics of smooth pursuit eye movements (Barnes & Asselman, 1991; Barnes & Grealy, 1992; Boman & Hotson, 1988). Subjects are instructed to follow the motion of a constant velocity target during brief periods of stimulation that are separated by periods of darkness. The effect of prediction is revealed firstly by the fact that smooth pursuit eye movements become progressively phase advanced with repeated stimulation, and secondly by the observation that anticipatory eye movements occur before the onset of target motion. Prediction has also been assessed by examination of the oculomotor response when, after following a number of cycles of similar stimuli, there is an unexpected change in frequency, direction, or amplitude of target motion. Prediction is demonstrated if the eye movement following the unexpected change has velocity and timing characteristics similar to the preceding eye movements (Keating, 1991; Barnes & Asselman, 1991; Lisberger, Evinger, Johanson, & Fuchs, 1981). The same type of results can also be obtained if a predictable target is unexpectedly blanked out instead of changing the direction, velocity, or amplitude of the target trajectory (Barnes & Grealy, 1992).

In the present study we determined the presence or absence of prediction in cerebellar patients and controls by assessing their ability to build up anticipatory eye movements after repeated runs of transiently illuminated predictable stimuli.

Methods

Subjects

We studied 12 patients with cerebellar disorders of various etiologies, and 10 age matched controls. The cerebellar patients had an age range of 19-74 years with a mean age of 42 +/- 17.91 years. The controls had an age range of 18-74 years with a mean of 40.5 +/- 19.87 years. The clinical characteristics of the cerebellar patients are shown in table 1. Patients with multiple sclerosis and multiple system atrophy were included only if their dominant presenting features were cerebellar. Important cerebellar features in the inclusion criteria were truncal ataxia, limb ataxia, intention tremor, dysarthia, dysdiadokokinesia, dysmetria, hypotonia, and rebound phenomenon. The controls consisted of naive patients admitted into the wards on selective basis for investigation of disorders known not to affect any brain structures. All the subjects were conscious, alert and co-operative. Informed consent was obtained from each subject and experiments were approved by the ethical committee.

Case no.	Age	Sex	Diagnosis	Medication
1	23	m	multiple sclerosis	yes
2	21	f	posterior fossa tumor, (epidermoid cyst)	no
3	58	m	tuberculoma of cerebellar vermis	no
4	44	f	multiple sclerosis	yes
5	50	m	primary cerebellar degeneration	no
6	52	f	multiple sclerosis	yes
7	45	f	cerebellar degeneration	no
8	40	m	multiple system atrophy	no
9	58	f	multiple system atrophy	no
10	19	m	Friedreichs Ataxia	no
11	20	f	multiple sclerosis	yes
12	74	f	multiple system atrophy	no

Table 1. Summary of clinical features of cerebellar patients.

Apparatus

The subjects were seated in the centre of a darkened room in front of a semicircular screen of radius 1.5m. Eye movements were recorded using an infrared limbus reflection technique (Iris 6500 system, Skalar Medical) with a resolution of 10min. or arc, and a linear range of at least +/-20 degrees. The eye movement recorders were mounted on a helmet assembly which was attached firmly to the subject's head. The head position was fixed by padded clamps that fitted snugly on both sides of the head. Subjects were instructed to track the targets actively using only eye movements.

The stimulus consisted of a circle of diameter 50min. of arc with superimposed cross hairs. It was made to move across the screen in the horizontal plane by a mirror galvanometer. The motion of the target was controlled by a computer generated waveform. The waveform of each record was composed of four consecutive sequences of 8 or 9 pairs of alternating velocity sweeps, with respective sequences having differing ramp velocities of 10, 20, 30, or 40 deg/s selected on a randomised basis. At the transition between sequences, the pulse was unexpectedly blanked for one cycle.

In the predictive task the basic target waveform was a regular triangular wave (fig.1) with a period of 3.4s, the target alternating directions at interpulse intervals of 1.7s. The target was exposed for a pulse duration of 640ms in each velocity sweep, and it passed through the central position halfway through each presentation period. In the non-predictive task, the target direction, and the inter-pulse intervals were randomised, although averaged inter-pulse interval remained at 1.7s. Eye movements were calibrated before each record.

All records were analysed off-line by computer (Hewlett-Packard 345) using an interactive computer graphics procedure described by Barnes (1982) to remove the saccadic components and thus obtain the smooth eye velocity.

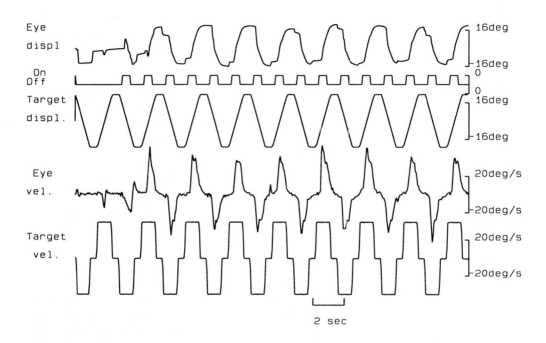

Figure 1. Traces of eye displacement, target displacement, eye velocity, and target velocity in a predictive task. The second trace from the top indicates the transient pulses of target illumination. Fast phase components have been removed from the eye velocity trace. Peak target velocity +/-40deg/s. Pulse duration 640ms; interpulse interval 1.7s. Results from a single control subject.

Results

Predictive versus non-predictive responses.

In the predictive task, controls and cerebellar patients showed anticipatory eye movements in the appropriate direction 350-500ms before the onset of target motion (Figure 2 -PRED condition). In the non-predictive task appropriate eye movements were not seen in both groups until 80-100ms after target onset (Figure 2 -NPR condition). Eye velocities at various time intervals from target onset (100, 200, 300ms) were compared in the predictive and non-predictive tasks (figs. 3A,3B and 3C). In both groups the eye velocity in the predictive task remained significantly above that achieved in the non-predictive task at equivalent time periods (by ANOVA $p < 0.001$ at 100ms, 200ms; $p < 0.05$ for the controls and < 0.001 for the cerebellar patients at 300ms).

Peak velocities.

At every level of stimulus velocity the controls had peak eye velocities significantly greater than the cerebellar patients ($p < 0.001$), (Figure 3D). Comparison of the peak eye velocities attained in the predictive and non-predictive tasks in each group showed that for the controls, the eye velocity in the non-predictive task eventually matched that in the predictive task. Indeed the peak eye velocity of the controls in the non-predictive task was regularly greater than that attained in the predictive task, although this effect was not statistically significant. In contrast, in the cerebellar patients the peak eye velocity attained in the predictive task remained higher than in the non-predictive task for all stimulus velocities (Figure 3D) although this too did not reach statistical significance.

Latency of visual feedback response.

The latency for the visual feedback response was measured from the time of target onset to the time when there was a sharp acceleration response in the eye velocity slope. In the predictive task the visual feedback response was not always well demarcated from the anticipatory response. However, individual records averaged at each velocity level showed that for control subjects this sudden acceleration response to visual feedback occurred 35 - 80ms after target onset (fig. 4A), and for the cerebellar patients this response occurred 70 - 100ms after target onset. In the non-predictive task, the onset of the visual feedback response was better demarcated. For the control subjects it occurred with mean values that ranged from 100 -110ms for the different target velocities, whereas in the cerebellar patients the range was 140 - 180ms from target onset (fig. 4A). This difference in latency between cerebellar patients and controls was significant in the non-predictive task at all the target velocity levels ($p < 0.01$).

Timing of the peak eye velocity.

The timing of the peak velocities in the various groups is shown in Figure 4B. In the controls the peak eye velocity in the predictive task was attained after an overall mean of 389ms irrespective of the target velocity. In the same task the cerebellar patients attained their peak later with an overall mean of 429ms. This difference in the timing of the peak velocity between the two groups was statistically significant ($p<0.05$). For the non-predictive task the timing of the peak velocities was more variable, with an overall mean of 422ms in the controls and 497ms in the cerebellar patients.

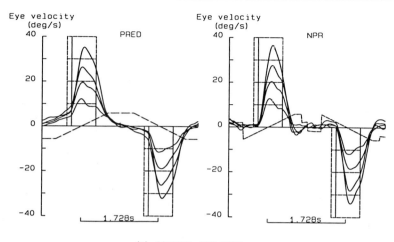

(A) CONTROL SUBJECTS

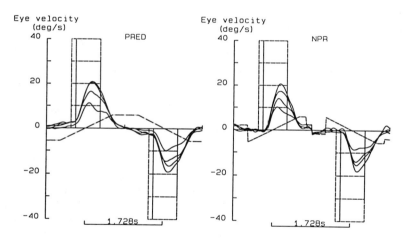

(B) CEREBELLAR PATIENTS

Figure. 2. Eye velocity trajectories during the predictive (PRED) and non-predictive (NPR) tasks. The eye velocity contours correspond to target velocities of 10, 20, 30, and 40 deg/s. respectively in ascending order. The vertical lines represent target onset, the 100ms mark (solid line), and the end of the target pulse or 640ms mark. Note the progressive build up of anticipatory eye movements before the target onset in the predictive paradigm. Averaged responses from 16 ramp velocity sweeps (8 cycles), in (A) 10 control subjects (upper traces) and (B) 12 cerebellar patients (lower traces).

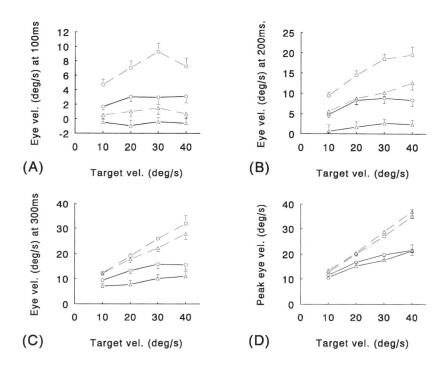

Figure 3. Mean eye velocity as a function of target velocity for all subjects at various time intervals from the onset of target motion - (A) 100ms; (B) 200ms & (C) 300ms. Peak eye velocity is shown in (D). Means from 12 cerebellar patients (solid lines) and 10 controls (broken lines). Circles indicate predictive response; triangles indicate non-predictive response.

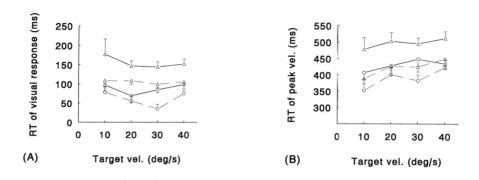

Fig. 4. (A) The reaction time (RT) between the onset of target presentation and the time at which the visual feedback induced an abrupt change in eye acceleration. Mean of 12 cerebellar patients (solid lines) and 10 controls (broken Lines). Note that in the cerebellar patients the response Latency was significantly increased in the non-predictive task. (B). The mean reaction time (RT) between the onset of target presentation and the attainment of peak velocity in 10 controls (broken lines) and 12 cerebellar patients (solid lines). Circles indicate predictive response; triangles indicate non-predictive response.

Discussion

The presence of directionally appropriate anticipatory eye movements in the predictive paradigm in the cerebellar patients indicates that prediction is preserved in cerebellar disease. Previous studies (Waterston, Barnes, & Grealy, 1992) using other predictive paradigms have also indicated that predictive eye movements are preserved in cerebellar disorders. To the extent that the normal visual feedback is preserved, these results suggest that the diminished gain of pursuit in cerebellar patients cannot be attributed to a loss of the predictive component of pursuit. Rather, it appears that visual feedback gain is degraded, and because the predictive component is dependent on the visual feedback, it too is degraded. Although previous neurophysiological recordings (Lisberger & Fuchs, 1978; Noda, 1986; Noda & Warabi, 1986; Noda & Warabi, 1987) in the cerebellum have indicated activity in the cerebellar flocculus before target onset, this activity may not initiate prediction but may represent the effects of earlier activities in other parts of the nervous system. In one of such recordings from the flocculus (Noda, 1986), activities were recorded about 125ms before target onset. Other workers (MacAvoy, Gottlieb, & Bruce, 1991; Gottlieb, MacAvoy, & Bruce, 1989) have recorded predictive neuronal activities in the frontal eye fields (FEF) 500ms before the onset of target motion, and before the eyes reversed to the preferred direction of the FEF neurons that are firing. Ablation of the FEF in monkeys (MacAvoy, Gottlieb, & Bruce, 1991; Keating, 1991; MacAvoy & Bruce, 1989) has also been shown to abolish predictive eye movements.

In the non-predictive task the cerebellar patients had an increased latency of eye movement initiation. Previous works have shown that lateral cerebellar lesions produce a prolonged reaction time and movement time during other voluntary motor actions (Stein & Glickstein, 1992; Becker, Kinnesh, & Freund, 1990).

For each of the 4 sequences of target velocities used the peak eye velocity attained was always greater in the controls than in the cerebellar patients. Impaired smooth pursuit eye movement velocity has been well documented in cerebellar patients (Waterston, Barnes, & Grealy, 1992; Zee, Yamazaki, Butler, & Gucer, 1981; Zee, Yee, Cogan, Robinson, & Engel, 1976), and this has been attributed to lesions of the flocculus, and lobules VI and VII of the vermis (Zee, 1982; Büttner, Waespe, & Henn, 1982; Keller, 1988). In the control subjects the peak eye velocities attained in the predictive task did not differ from those in the non-predictive task. Prediction conferred a velocity advantage only in the earlier parts of pursuit. Previous experiments in normal subjects (Barnes & Asselman, 1991) have shown that when the pulse duration exceeds 240ms, peak eye velocity is frequently attained before the target disappears. In the non-predictive task, the continued increase in eye velocity after the time at which the velocity had peaked in the predictive task annulled the earlier velocity advantage conferred by prediction. In the cerebellar patients this advantage conferred by prediction persisted even at the peak velocities (fig. 3). For the cerebellar patients whose latency of pursuit onset and pursuit velocity are basically impaired, the advantage conferred by prediction becomes very important to ameliorate their deficits, although even with prediction their responses are delayed.

The timing of peak eye velocities varied little with different target velocities in the predictive task. This effect has been noted previously (Barnes & Asselman, 1991) when the pulse duration was kept constant but target velocities varied. However the timing of the peak velocities was significantly later in the cerebellar patients when compared with controls in the predictive tasks. The inability of cerebellar patients to correctly time movement durations and terminations has been well documented in saccades (Vilis & Hore, 1981; Vilis, Snow, & Hore, 1983), and limb movements (Conrad & Brooks, 1974). Abnormal timing of movement durations and terminations incapacitates the ability to perform rapidly alternating movements (adiadokokinesia).

The results in this study show that cerebellar patients are still able to programme and initiate predictive eye movements. But the pursuit eye movements are abnormal in velocity and timing of movement durations. This abnormality in timing of movement durations could be due to an in-built error of calibration in the motor plan of the pursuit movement, or an impaired. on-line execution of the timing in a normal motor plan. The cerebellum has been implicated both in the timing and programming of movements (Stein & Glickstein, 1992) and in the on-line control of eye movements (Keller, 1988). Although the ability to make predictive eye movements is preserved in cerebellar patients, their inability to perform rapidly alternating movements limits their ability to utilise prediction at high frequencies of target motion. This could be one explanation for the increased phase lag at high frequencies of sinusoidal pursuit which has been described in these patients (Waterston, Barnes, & Grealy, 1992).

References

Bahill, A.T. & McDonald, J.D. (1983). Model emulates human smooth pursuit system producing zero-latency target tracking. Biological Cybernetics, 48, 213-222.

Barnes, G.R. (1982). A procedure for the analysis of nystagmus and other eye movements. Aviation Space and Environmental Medicine, 53, 676-682.

Barnes, G.R. & Asselman, P.T. (1991). The mechanism of prediction in human smooth pursuit eye movements. Journal of Physiology (London), 439, 439-461.

Barnes, G.R., Donnelly, S.F. & Eason, R.D. (1987). Predictive velocity estimation in the pursuit reflex response to pseudo-random and step displacement stimuli in man. Journal of Physiology (London). 389, 111-136.

Barnes, G.R. & Grealy, M.A. (1992). The role of prediction in head-free pursuit and VOR suppression. Annals of the New York Academy of Sciences, 656, 687-694.

Becker, W.J., Kinnesh, E., & Freund, H.J. (1990). Co-ordination of multi-joint movement in normal humans and patients with cerebellar dysfunction. Canadian Journal of Neurological Science, 17, 264-274..

Boman, D.K. & Hotson, J.R. (1988). Stimulus conditions that enhance anticipatory slow eye movements. Vision Research, 28, 1157-1165.

Büttner, U., Waespe, W., & Henn, V. (1982). The role of the cerebellum in the control of slow conjugate eye movements. In G. Lennerstrand, D.S. Zee, & E.L. Keller (Eds.), Functional basis of ocular motility disorders (pp. 431-439). Pergamon press..

Carl, J.R. & Gellman, R.S. (1987). Human smooth pursuit: stimulus-dependent responses. Journal of Neurophysiology, 57, 1446-1463.

Conrad, B. & Brooks, V.B. (1974). Effects of dentate cooling on rapid alternating arm movements. J Neurophysiol, 37, 792-804.

Dallos, P.J. & Jones, R.W. (1963). Learning behaviour of the eye fixation control system. IEEE Transactions, Ac-8, 218-227.

Gottlieb, J.P., MacAvoy, M.G., & Bruce, C.J. (1989). Unit activity related to smooth pursuit eye movements in rhesus monkey frontal eye fields. Soc Neurosci Abstr, 15, 1203.(Abstract)

Keating, E.G. (1991). Frontal eye field lesions impair predictive and visually-guided eye movements. Exp Brain Res, 86, 311-323.

Keller, E. (1988). Cerebellar Involvement in Smooth Pursuit Eye Movement Generation: Flocculus and Vermis. In C. Kennard & F.C. Rose (Eds.), Physiological Aspects of Clinical Neuro-Ophthalmology (pp. 341-355). London: Chapman & Hall.

Lisberger, S.G., Evinger, C., Johanson, G.W., & Fuchs, A.F. (1981). Relationship between eye acceleration and retinal image velocity during foveal smooth pursuit in man and monkey. Journal of Neurophysiology, 46, 229-249.

Lisberger, S.G. & Fuchs, A.F. (1978). Role of primate flocculus during rapid behavioural modification of vestibulocular reflex. I. Purkinje cell activity during visually guided horizontal smooth-pursuit eye movements and passive head rotation. Journal of Neurophysiology, 41, 733-763.

MacAvoy, M.G., Gottlieb, J.P., & Bruce, C.J. (1991). Smooth pursuit eye movement representation in the primate frontal eye field. Cerebral cortex, 1, 95-102.

MacAvoy, M.G. & Bruce, C.J. (1989). Oculomotor deficits associated with lesions of the frontal eye field area in macaque monkeys. Soc Neurosci Abstr, 15, 1203.(Abstract)

Miles, F.A. & Fuller, J.H. (1975). Visual tracking and the primate flocculus. Science, 189, 1000-1002.

Noda, H. (1986). Mossy fibres sending retinal-slip, eye and head velocity signals to the flocculus of the monkey. Journal of Physiology (London), 379, 39-60.

Noda, H. & Warabi, T. (1986). Discharges of Purkinje cells in monkey's flocculus during smooth eye movements and visual stimulus movements. Experimental Neurology, 93, 390-403.

Noda, H. & Warabi, T. (1987). Responses of Purkinje cells and mossy fibres in the flocculus of the monkey during sinusoidal movements of a visual pattern. Journal of Physiology (London), 387, 611-628.

Stein, J.F. & Glickstein, M. (1992). Role of the cerebellum in visual guidance of movement. Physiological Reviews, 72(4)., 967-1017.

Vilis, T. & Hore, J. (1981). Characteristics of saccadic dysmetria in monkeys during reversible lesions of medial cerebellar nuclei. J Neurophysiol, 46, 828-838.

Vilis, T., Snow, R., & Hore, J. (1983). Cerebellar saccadic dysmetria is not equal in the two eyes. Exp Brain Res, 51, 343-350.

Waterston, J.A., Barnes, G.R., & Grealy, M.A. (1992). A quantitative study of eye and head movements during smooth pursuit in patients with cerebellar disease. Brain, 115, 1343-1358.

Zee, D.S. (1982). Ocular motor abnormalities related to lesions in the vestibulocerebellum in primate. In G. Lennerstrand, D.S. Zee, & E.L. Keller (Eds.), Functional basis of ocular motility disorders (pp. 423-430). Pergamon press.

Zee, D.S., Yamazaki, A., Butler, P.H., & Gucer, G. (1981). Effects of ablation of flocculus and paraflocculus on eye movements in primate. Journal of Neurophysiology, 46, 878-899.

SACCADE AND FIXATION CONTROL

THE INITIAL DIRECTION AND LANDING POSITION OF SACCADES

Casper J. Erkelens and Ingrid M.L.C. Vogels

Utrecht Biophysics Research Institute, Department of Medical and Physiological Physics, Buys Ballot Laboratory, University of Utrecht, P.O. Box 80000, 3508 TA Utrecht, The Netherlands[1]

Abstract

We studied the trajectories of self-paced saccades in two experimental conditions. Saccades were made between two visual targets in one condition and between the same two, not visible, positions in the other condition. Target pairs were presented which required oblique saccades of 20 or 40 deg. At least 200 saccades were made between each pair of targets. Horizontal and vertical eye movements were measured of the right eye with a scleral coil technique. We computed the angle between starting and end point of each primary saccade (effective direction). We also computed the angle between starting point and eye position when the saccade had covered a distance of 2.5 deg (initial direction). We found that variability in initial directions was two to seven times larger than variability in the effective directions. This effect was found in both experimental directions for saccades made in all tested directions. We conclude that curvedness of saccades is the result of a purposeful control strategy. The saccadic trajectories show that, initially, the eye is accelerated roughly in the direction of the target and subsequently is guided to the target. This behavior cannot be described by present models of saccade generation. We suggest that the coupling between saccadic pulse and step signals is not as tight as generally is accepted in the literature.

Keywords

Cardinal and oblique saccades, Saccadic trajectories, Models of saccade generation

Introduction

A conspicuous feature of eye movements in general and of saccades in particular is that, different from limb movements, their kinematics can hardly be affected by effort. Saccadic eye movements appear to be rather stereotyped behaviors. This observation has made saccades a favourite object for studying human motor behavior, The neural control of saccades is studied as a model of how conversion of sensory

---

[1] This research was partly supported by the Foundation for Biophysics of the Netherlands Organization for Scientific Research (NWO).

Eye Movement Research/J.M. Findlay et al. (Editors)

input into motor action is organized in the human brain. The "humans are machines" and "humans are animals" metaphores (Arbib, 1989) have encouraged researchers to develop engineering models of saccades of which the components can be found in structures of the brains of humans and animals (for reviews see Büttner-Ennever, 1988; Wurtz & Goldberg, 1989; Sparks & L.E. Mays, 1990).

Contempory models (Robinson, 1975; Zee et al., 1976; van Gisbergen et al., 1981; van Gisbergen et al., 1985; Tweed & Vilis, 1985; Scudder, 1988; Grossman & Robinson, 1988; Becker & Jürgens, 1990) of the neural control of saccades are based on an idea proposed by Robinson (1975). This idea, now widely accepted in the saccadic literature, is that saccades are produced by a pulse generator mechanism. The intensity of the pulse determines the saccadic velocity, the time integral of the pulse, called the step, determines the amplitude of the saccade. The basic idea was developed for describing the neural control of horizontal saccades made by a single eye. Experimental data of vertical and oblique saccades (van Gisbergen et al., 1985; Yee et al., 1985; King et al, 1986; Collewijn et al., 1988a; Smit & van Gisbergen, 1990; Smit et al., 1990) made available due to improved recording methods for vertical eye movements, caused extension of one-dimensional of models of saccades to two dimensions (van Gisbergen et al., 1985; Tweed & Vilis, 1985; Scudder, 1988; Becker & Jürgens, 1990). Although two-dimensional saccades created specific problems for the modeling of their neural control, the basic idea of Robinson was never questioned and incorporated in all models.

A further extension of experimental data was obtained from accurate measurements of conjugate (Collewijn et al., 1988a; Collewijn et al., 1988b) and disjunctive (Erkelens et al., 1989) saccades in the two eyes together. The data showed that saccades are not as stereotyped as once has been thought. Binocular saccades showed no fixed relationship between amplitude, velocity and duration of saccades ("the main-sequence parameters").

We have observed from experimental recordings of ourselves and others (see for example Collewijn et al., 1988a; Collewijn et al., 1988b; Becker & Jürgens, 1990; Smit & van Gisbergen, 1990) that trajectories of saccades show a fair amount of variability. Present models predict a fixed trajectory between start and landing position of saccades. This implies that the existing models do not allow variability. The objective of the research described here is to investigate the characteristics of the spatial variability and to test in how far spatial variability can be described by noise in components of existing models for the generation of saccades.

Methods

Subjects

Three subjects participated in the experiments. They had visual acuities of 20/20 or better, with (1 subjects) or without (2 subjects) correction. None of them showed any ocular or oculomotor pathologies. One subject was experienced in oculomotor research. The other subjects were participating in such experiments for the first time.

Informed consent was obtained from all subjects before they embarked on the study.

Apparatus

The position of the right eye was measured with an induction coil mounted in a scleral annulus in an a.c. magnetic field as first described by Robinson, 1963 and modified and refined by Collewijn et al., 1975. The dynamic range of the recording system was d.c. to better than 100 Hz (3 dB down), noise level less than $\pm$ 3 min arc and deviation from linearity less than 1% over a range of $\pm$ 25 deg. The head position of the subjects was restricted by a chin rest and a head support. A large (200 x 250 cm) translucent screen (Marata) was positioned in front of the subject at a distance of 130 cm. Stimuli, generated on a microprocessor (Atari), were back-projected on the screen by using a projection system (Barco Data 800). The microprocessor was also used for data acquisition and for controlling the experiment. Positions of the right eye were digitized on-line at a frequency of 512 Hz with a resolution of 3 min arc. Onsets of saccades were detected on-line by using a simple velocity-threshold criterion. The detection of saccadic onsets was used as a trigger to manipulate the visual stimulus during saccadic eye movements. The period between the detection of a saccade and the completion of the stimulus replacement was about 16 ms. This period was mainly determined by the refreshing rate of the Atari screen (70 Hz). Off-line the data were transferred to an Apollo 10000 computer system that was used for analysis.

Procedure

The experiments were carried out in a darkened room. The duration of each experimental session was limited to about half an hour. The sensitivity of the eye movement recorder was adjusted at the start of each experimental session. A calibration target containing a matrix of 3 x 3 equally spaced fixation marks was presented. The subject fixated in turn each mark while the polarity, gain and offset of the signal for eye position were inspected and roughly adjusted. After these adjustments, voluntary gaze shifts made between the calibration marks were recorded and used for calibration of the eye position signals.

The subjects were asked to make voluntary saccades between two target positions in their own rhythm for periods of 5 min. Both targets (discs, 0.8 deg dia) were located on a circle with a diameter of either 20 or 40 deg centered about the primary position (Fig. 1). Three target configurations were used. In one experiment the targets remained visible at a fixed position during the whole session. Saccades made between such targets are called *V-saccades* (Smit et al., 1987). In a second experiment both targets were displaced during the saccades to positions shown in Fig. 1. The effect of this procedure was that, after a few saccades, the saccades were not directed to the visible position of the target but to its remembered position. Such saccades are called *R-saccades* (Smit et al., 1987). The target positions were chosen such that both V-saccades and R-saccades were made between the same positions. In one experimental session, both types of saccades were made with the targets at three different positions, so that the duration of a session was limited to about half an hour.

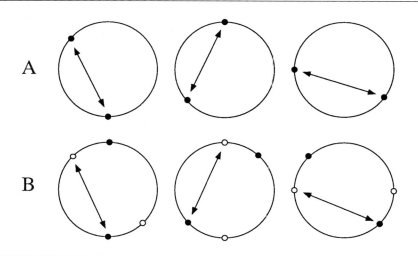

Figure 1: *Target positions for V-saccades (A) and R-saccades (B). Targets switched between the open and filled positions during R-saccades. The arrows indicate the positions from which the saccades were made.*

Data analysis

In the off-line analysis, saccade onset as well as saccade offset were detected by a velocity threshold of 15 deg/s in combination with a required minimum saccade duration of 15 ms. From the recordings we computed amplitude, velocity as a function of time, maximum velocity, initial direction, effective direction, and direction as a function of time for each saccade. The initial direction of a saccade was defined as the angle between its starting point and eye position when the saccade had covered a distance of 2.5 deg. The effective direction was defined as the angle between the positions of the eye at saccade onset and offset. Directional deviation of a saccade was defined as the difference between initial and effective direction. The directions were calculated for primary saccades, secondary saccades were excluded from the analysis. Due to noise of the recording system (noise level less than $\pm$ 3 min arc) the initial direction at a distance of 2.5 deg could be computed with an accuracy of 1.1 deg. Inaccuracy in the computation of the effective direction was negligible. We computed mean and standard deviations of amplitude, maximum velocity, initial direction, effective direction, and directions as a function of time for each set of about 100 repeated saccades. The relationship between amplitude and maximum velocity data was analyzed by computing linear correlation coefficients.

Results

Trajectories of saccades

All the results presented here are of 40 deg saccades made by one subject who has been measured most extensively. The results are representative for the other subjects and also for saccades with amplitudes of 20 deg. During the periods of 5 minutes, in which the subjects made saccades to and fro between two targets, generally about 200 saccades were made meaning about 100 saccades in each direction.

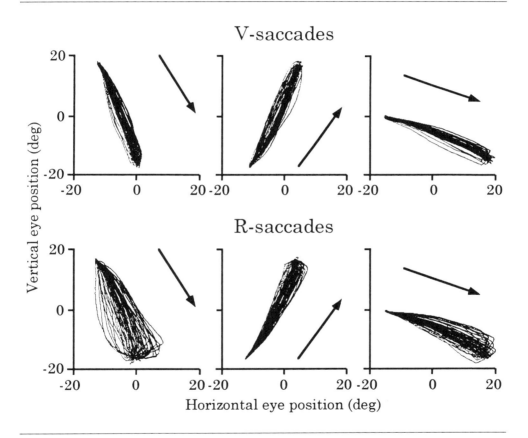

Figure 2: *Trajectories of V-saccades and R-saccades. Arrows indicate the direction in which the saccades are made.*

Fig. 2 shows trajectories of saccades made in various directions in the two experimental conditions. A saccades can have either a clockwise or counterclockwise curvedness. The collection of saccades made to a specific target generally contains both types a saccades. Bundles of trajectories of saccades made to a single target form cigarlike patterns which are generally wider for R-saccades than for V-saccades. Especially for R-saccades, the variability of the trajectories depends heavily on the direction in which the saccades are made. However, the variability does not show a systematic pattern. Directions in which saccades show a large variability in one subject may show little variability in other subjects. A clear relationship between the variability of V-saccades and R-saccades in a specific direction is also not present.

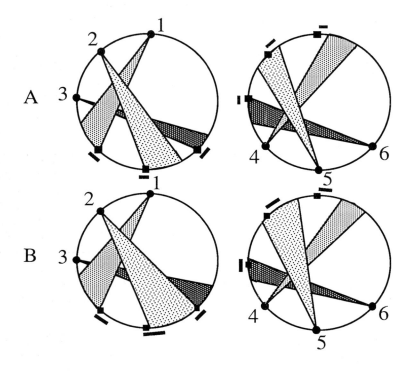

Figure 3: *Variability in initial and effective directions of V-saccades (A) and R-saccades (B). Dots indicate the starting positions of saccades, squares the target positions. Cones, indicating variability in initial directions, have widths of ± 1 SD and are centered about the mean initial directions. Bars, indicating variability in effective directions, have lengths of ± 1 SD centered about the mean effective directions. Numbers indicate the different positions from which saccades were made.*

Variability in initial and effective directions of saccades

The cigarlike shaped patterns, shown in Fig. 2, suggest that variability is larger in the beginning of saccades than near the end. Fig. 3 shows the variability in initial and effective directions of saccades for all the tested directions. Indeed, this figure clearly shows that variabilities in initial directions are considerably larger than those in effective directions. This difference in variability appeared to be a very general feature: it was observed for V-saccades and R-saccades, in all tested directions, and in all the three subjects. Not only the variability was smaller near the end of saccades, the landing position of saccades was also more precise than predicted by the mean initial directions. Landing positions were closer located to the target position than positions suggested by straight extrapolations of the initial directions. Variabilities in initial directions (SDi) varied considerably among directions between about 2 and 6 deg in V-saccades and between 4 and 10 deg in R-saccades (Table 1).

	Direction	SDi (deg)	SDe (deg)	Mdd (deg)	Rh	Rv
V-saccades	1	6.2	1.1	9.5	-0.02	-0.11
	2	5.9	1.4	-6.5	0.50	0.15
	3	2.9	1.6	-4.1	0.01	-0.53
	4	6.5	1.3	8.5	0.32	0.16
	5	5.6	1.4	-3.6	-0.25	-0.04
	6	3.5	1.4	-0.9	0.31	0.33
R-saccades	1	8.1	3.5	11.0	-0.30	-0.40
	2	9.9	5.1	-11.6	0.13	-0.24
	3	4.8	3.4	-4.0	0.13	-0.52
	4	4.3	2.2	11.3	0.32	0.14
	5	9.2	2.1	2.2	-0.44	-0.10
	6	4.8	2.7	-2.3	-0.28	0.44

Table 1: *SD's of initial (SDi) and effective (SDe) directions, mean directional deviations (Mdd), and coefficients of correlation between maximum velocity and amplitude of the horizontal (Rh) and vertical (Rv) components of saccades. The saccadic directions are indicated by numbers which correspond to those shown in figure 3.*

SDe varied between about 1 and 2 deg in V-saccades and between 2 and 5 deg in R-saccades. The ratio between SDi and SDe varied between 2 and 7 in the various directions and conditions. Sizes of paired SDi's and SDe's were not related to each other. In other words, a particular SDi did not predict the magnitude of SDe. Table 1 further shows that mean directional deviations (Mdd) of saccades, i.e. mean curvatures, were observed in both clockwise or counterclockwise directions. Mean curvature had the same sign for V-saccades and R-saccades in a specific direction in most cases, however, the signs were opposite in a few cases (see for example Mdd in direction 5 in Table 1).

In order to be able to relate the present findings to existing models of saccade generation is was important to have details about the relationship between amplitude and maximum velocity of saccades made at one target distance. Table 1 shows that maximum velocity and amplitude were hardly correlated for the horizontal and vertical components of saccades.

Discussion

Curvedness of saccades

The present study shows that human saccades generally have a curved trajectory. We found this for V-saccades as well as for R-saccades. This finding corroborates results of Smit & van Gisbergen (1990a) who reported curvedness of oblique as well as cardinal saccades. Curvedness is most likely a feature of the control mechanism because differences in curvature were found between V-saccades and R-saccades made in the same direction. Another characteristic of human saccades, which was generally present in our subjects, is that variability in initial direction is far larger than variability in effective direction. We showed that effective directions of saccades are more precise and more accurate than can be expected from their initial directions. This implies that curvedness of saccades is most likely the result of a purposefull control mechanism.

   If we accept this conclusion, then there are two obvious explanations for the curvedness of saccades. The first explanation is that the effect of the pulse, which accelerates the eye, is evaluated and corrected during the saccade. The second explanation is that the initial acceleration is based on one source of information and the end position on another, more accurate source. Present models of saccadic generation predict a fixed relationship between initial direction and landing position. We questioned whether the addition of noise to components of existing models for saccade generation can explain the present results. In a quantitative analysis we investigated the effect of noise in two different models. Due to the feedback character of these model noisy components will affect the kinematics but not the amplitudes of saccades. The lack of correlation between maximum velocity and amplitude of the saccadic components (Table 1) suggests that noise, in principle, may be a possible cause for the observed variability effects.

The effect of noise

Presently, there are two acceptable, two-dimensional models for the generation of saccades. Both models are based on the same internal feedback principle but are different in the arrangement of their components. In the cross-coupled pulse generator (CPG) model (Grossman & Robinson, 1988; Becker & Jürgens, 1990), the vectorial error signal firstly is decomposed into two orthogonal error signals which are input for the horizontal and vertical burst generators. In the common source (CS) model (van Gisbergen et al., 1985), the vectorial error signal firstly is fed to a vectorial pulse generator of which the output is decomposed into the horizontal and vertical pulse signals. The CS-model predicts "component stretching" (Evinger & Fuchs, 1978) but cannot produce curved saccadic trajectories. The CPG-model predicts "component stretching" as well as curvedness of saccades, although both features require very different cross-coupling factors between the pulse generators (Smit et al., 1990).

   Simulations of both models showed that independent variation of horizontal and vertical components of model parameters, in most cases, produced realistic distributions of initial directions of saccades. In general, variations of about 20% in the parameters were necessary to produce distributions with SD's of 5 deg. However, time courses of saccades were also affected by variation of the parameter values. Time courses only remained realistic when the presence of high noise levels was limited to the burst generators and the cross-coupling factor. Therefore, noise in these parameters seems to be the likely candidate for creating large variability in initial directions.

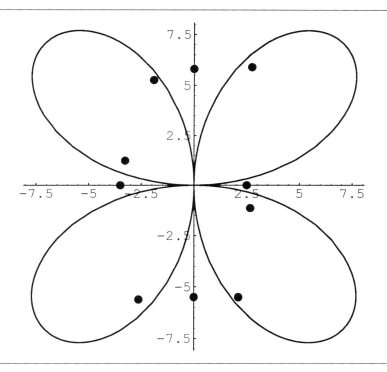

Figure 4: *Polar plot of measured SDi values (dots). The four lobes are model predictions assuming similar variability in the two orthogonal components.*

However, fluctuations in the amplitude of the burst generators and in the strength of the cross-coupling factor have specific consequences for the variability of saccades as a function of the saccadic direction. Fig. 4 shows predictions of both models for the variability in the initial directions when we assume similar amounts of noise in the two orthogonal components. Variabilities in the initial directions of saccades should be symmetrically distributed about the cardinal axes. Variability should zero for cardinal saccades and maximum for oblique saccades in directions of 45, 135, 225 and 315 deg due to the fact that the orthogonal components contribute to saccades made in any direction. Furthermore, variability should be equally large in all four quadrants.

Fig. 4 shows that this relationship between variability and saccadic direction is not present in measured saccades. Variability data computed from V-saccades of the present experiments together with variability data of pure horizontal and vertical saccades, all obtained from one subject, have a more or less elliptical distribution. The long axis of the distributions is directed vertically in this example, however, other subjects showed different directions. The clearly different distributions in experiment and model indicate that noise in orthogonal components cannot describe the spatial variability of saccades.

Quantitative analysis of the models shows that introduction of a major noise source in the generation of saccades must occur before the error signal is decomposed into two orthogonal components. Addition of noise at the position where, in the CS-model, the vectorial error signal is transformed into a vectorial pulse signal, allows saccades with realistic variability in initial and effective directions. Such saccades still have realistic time courses. The amplitude of noise has to be dependent on direction in order to allow for different variabilities in different directions. An additional consequence of noise at this position in the CS-model, which weakens the relationship between pulse and step signals, is that the model is now able to generate curved saccades.

Concluding remarks

Saccades show variability in the temporal (maximum speed, duration) as well as in the spatial (initial direction, curvedness) domain. Initially, models for the generation of saccades have been developed to describe the temporal characteristics of saccades. Later, these models were adapted to describe spatial characteristics as well. Still, Smit & van Gisbergen (1990) concluded that essentially all existing models are un-satisfactory in providing an explanation for the curvature patterns of saccades. The present study shows that these models are also unsatisfactory in describing the initial and effective directions of saccades. Weakening the coupling between pulse and step signals, which effectively destroys the feedback character of the models, appears to be essential for description of the spatial variability. Recently, Enright (Enright, 1992) suggested that pulse and step components of saccadic motoneuron activity may be generated by largely independent processes. From examining disjunctive saccades he proposed that the step component for each eye depends only on that eye's visual input, whereas the pulse components generated for each eye depend on weighted averaging of visual stimuli that impinge on both eyes. The results described here support the view that pulse and step signals are not as tightly coupled to each other as generally is suggested in the literature.

References

Arbib, M.A. (1989). The Metaphorical Brain 2, John Wiley & Sons, New York.

Becker, W. & Fuchs, A.F. (1969). Further properties of the human saccadic system: eye movements and correction saccades with and without visual fixation points. Vision Res. 9, 1247-1258.

Becker, W. & Jürgens, R. (1990). Human oblique saccades: quantitative analysis of the relation between horizontal and vertical components. Vision Res. 30, 893-920.

Büttner-Ennever, J.A. (ed.) (1988). Reviews of Oculomotor Research 2, Elsevier, Amsterdam.

Collewijn, H., van der Mark, F. & Jansen, T.C. (1975). Precise recording of human eye movements. Vision Res. 15, 447-450.

Collewijn, H., Erkelens C.J. & Steinman, R.M. (1988a). Bino cular co-ordination of human horizontal saccadic eye movements. J. Physiol. 404, 157-182.

Collewijn, H., Erkelens C.J. & Steinman, R.M. (1988b). Bino cular co-ordination of human vertical saccadic eye movements. J. Physiol. 404, 183-197.

Enright, J.T. (1992). The remarkable saccades of asymmetrical vergence. Vision Res. 32, 2261-2276.

Erkelens C.J., Collewijn, H. & Steinman, R.M. (1989). Ocular vergence under natural conditions. I. Continuous changes of target distance along the median plane. Proc. Roy. Soc. of London B 236, 441-465.

Evinger, C. & Fuchs, A.F. (1978). Saccadic, smooth pursuit and optokinetic eye movements of the trained cat. J. Physiol. 285, 209-229.

Grossman, G.E. & Robinson, D.A. (1988). Ambivalence in modelling oblique saccades. Biol. Cybern. 58, 13-18.

King, W.M., Lisberger, S.G. & Fuchs, A.F. (1986). Oblique saccadic eye movements of primates. J. Neurophysiol. 56, 769-783.

Robinson, D.A. (1963). A method of measuring eye movements using a scleral search coil in a magnetic field. IEEE Trans. Biomed. Electr. 10, 137-145.

Robinson, D.A. (1975). Oculomotor control signals. In: Lennerstrand, G. & Bach-y-Rita, P. (Eds.) Mechanisms of ocular motility and their clinical implications, Pergamon Press, Oxford, pp. 337-374.

Scudder, C.A. (1988). A new local feedback model of the saccadic burstgenerator. J. Neurophysiol. 59, 1455-1475.

Smit, A., van Gisbergen, J.A.M. & Cools, A.R. (1987). A parametric analysis of human saccades in different experimental paradigms. Vision Res. 57, 1745-1762.

Smit, A. & van Gisbergen, J.A.M. (1990). An analysis of curvature in fast and slow human saccades. Exp. Brain Res. 81, 335-345.

Smit, A., van Gisbergen, J.A.M. & van Opstal, J. (1990). Component stretching in fast and slow oblique saccades in the human. Exp. Brain Res. 81, 325-334.

Sparks, D.L. & Mays, L.E. (1990). Signal transformations required for the generation of saccadic eye movements. Annual Reviews of Neuroscience 13, 309-336.

Tweed, D. & Vilis, T. (1985). A two dimensional model for saccade generation. Biol. Cybern. 52, 219-227.

Van Gisbergen, J.A.M., Robinson, D.A. & Gielen, S. (1981). A quantitative analysis of generation of saccadic eye movements by burst neurons. J. Neurophysiol. 45, 417-442.

Van Gisbergen, J.A.M., van Opstal, A.J. & Schoenmakers, J.J.M. (1985). Experimental test of two models for the generation of o blique saccades. Exp. Brain Res. 57, 321-336.

Wurtz, R.H. & Goldberg, M.E. (1989). (eds.) Reviews of Oculomotor Research 3, Elsevier, Amsterdam.

Yee, R.D., Schiller, V.L., Lim, V., Baloh, F.G. & Honrubia, V. (1985). Velocities of vertical saccades with different eye movement recording methods. Invest. Ophthalmol. Vis. Sci. 26, 938-944.

Zee, D.S., Optican, L.M., Cook, J.D., Robinson, D.A. & King Engel, W. (1976). Slow saccades in spinocerebellar degeneration. Arch. Neurol. 33, 243-251.

MECHANISMS FOR FIXATION IN MAN: EVIDENCE FROM SACCADIC REACTION TIMES

Monica Biscaldi, Heike Weber, Burkhart Fischer, and Volker Stuhr

Department of Neurophysiology, Hansastr 9a
University of Freiburg, D-79104 Freiburg, Germany

Abstract

In this paper we present experimental evidence for the existence of a neural system in man related to fixation and to attention which prevents the saccade system from executing reflexive saccades to suddenly appearing targets. In the first series of experiments, we looked, in a gap task, at the reaction and correction time (measured from target onset and from the end of the first saccade respectively) of secondary corrective saccades in relationship to the amplitude of the primary saccade. We find that corrective saccades following anticipatory primary saccades (in particular direction errors) can be elicited after a very short correction time (about 0 ms) and that their reaction time is in the range of express saccades. We explain these data by the fact that, in case of an anticipatory primary saccade, the retinal error is very large so that the target has not entered the dead zone and, consequently, has not activated the fixation system. In the second part, the data of naive subjects who make spontaneously large numbers of express saccades in the overlap task are illustrated. These subjects have great difficulties in suppressing express saccades to the stimulus in the anti-saccade gap and overlap tasks. Furthermore, in a memory-guided task, they glance reflexively to the target in most of the trials, instead of maintaining fixation until fixation point offset as required. These subjects seem to have a reduced control upon their saccade system in the presence of a target onset: Their fixation system is too weak. Finally, a group of experiments is described in which trained subjects are requested to allocate attention to the "fixation point" shifted to a peripheral location. This instruction reduces the occurrence of express saccades for targets appearing at the same position as the peripheral "fixation point" or at its opposite symmetrical location. The effect decreases progressively for increasing distances between saccade target and attention target. All these results are discussed in terms of cortical and subcortical structures that exert an inhibitory control over the saccade system.

Keywords

express saccade, fixation, attention, anti-saccade task

Introduction

Express saccades - i.e. visually guided saccades with very short reaction times - have been considered reflex-like optomotor reactions to the sudden onset of a visual target in monkey (Fischer & Boch, 1983) and in man (Fischer & Ramsperger, 1984). In recent years the characteristics of these saccades and the experimental conditions under which they are present or absent, have been quite intensively investigated: Express saccades are usually suppressed by attentive fixation of a permanently presented foveal fixation point (Mayfrank et al, 1986). They are also largely absent when the saccade target appears close to the fovea within the

Eye Movement Research/J.M. Findlay et al. (Editors)

socalled dead zone for express saccades (Weber et al.1992). Even non-target stimuli (distractors), which occur simultaneously together with the saccade target, decrease the chance of making an express saccade (Weber & Fischer, in press). Furthermore, instructing the subject to pay attention to a peripherally located attention target also decreases the number of express saccades (Mayfrank et al.1986); (Braun & Breitmeyer, 1988); (Breitmeyer & Braun, 1990).

These experimental results and the neurophysiological finding of collicular neurons specifically responsible for fixation (Munoz & Wurtz, 1993a,b) have led us and others to the hypothesis that - as part of the oculomotor system in primates - there exists a neural subsystem used for actively preventing the eyes from making a saccade. Here we present experimental evidence for the existence of such a subsystem which, when activated, suppresses the occurrence of express saccades (in the result section we will show the data from different single subjects being representative for the others that we have tested). We will consider the effect of foveal (part A and B) versus extrafoveal stimulation (part C) on the activity of this subsystem. Two main hypothesis will be discussed: whether the subsystem has to do with the engagement /disengagement of visual attention, or whether it represents an oculomotor function that simply acts on the saccade system for preventing the eyes from making saccadic eye movement during fixation.

Results

A. Evidence from secondary saccades

Primary saccades to peripheral visual targets often fail to hit the target at once, and a secondary corrective saccade is made to achieve the target position more precisely. The existence of a dead zone for express saccades led us to the idea that this mechanism may affect not only the latencies of primary saccades but also those of secondary saccades (Fischer et al.1993): the reaction time of a secondary saccade will depend on the position that the eye has achieved after the primary saccade, i.e. whether or not the primary saccade has brought the target into the dead zone. One would therefore predict that small corrective saccades can be generated only after latencies well above 130 msec, whereas larger corrective saccades may follow the primary saccades after express time.

Following this idea, we analyzed the time of the occurrence of secondary (corrective) saccades following primary saccades to a single target. The gap task (fixation point offset precedes target onset by 200 ms) was used; target location was randomized between 4° to the right and left. Eye movements were monitored by an infrared light reflection method. The onset time and the size of the first two saccades following target onset were determined. The data of two trained adult subjects and 8 naive children, all making large numbers of express saccades, were investigated.

Fig. 1 shows two different ways of looking at the same data obtained from one of the two adult subjects who had received extensive training in the gap task. The results obtained from the other subjects were essentially the same and they are considered in more details by Fischer et al., 1993.

Reaction and Correction Times of
Primary and Secondary Saccades

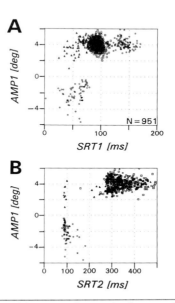

Fig. 1. Data from one trained human subject. A: Scatter plot of amplitudes versus reaction times of primary saccades. Positive values: saccades in the direction of the target: Triangles symbolize anticipatory saccades, squares the express saccades, and circles the fast regular saccades. Circles with negative values indicate direction errors. B: Scatter plot of amplitudes of primary versus reaction times of secondary saccades (measured from target onset). The symbols correspond to the same saccade groups as in A.

In part A of Fig. 1 we plotted the size of the primary saccade (AMP1) versus its reaction time (SRT1). The triangles indicate anticipatory reactions in the direction of the target. The open squares represent the population of express saccades and the open circles (further to the right) represent the fast regular saccades with still longer reaction times. Negative values of AMP1 (open circles) indicate direction errors, where the subject anticipated the occurrence of the target in the wrong direction. In part B of Fig. 1 the same data are replotted. This time the amplitude of the first saccade is plotted versus the reaction time of the secondary saccade measured from the onset of the target. It becomes clear immediately that most secondary saccades following anticipatory saccades, both in the right and in the wrong direction, occur in a vertical band at 100 msec. They represent secondary express saccades made in response to the onset of the target. For these saccades the corresponding intersaccadic interval was extremely short, sometimes close to zero. The latencies of corrective saccades following visually guided ones, on the other hand, were usually in the order of 150 to 250msec. The visually guided saccades to the 4° target usually hit the target quite well, such that the following corrective saccades are small. When we presented the saccade target at large eccentricities (beyond 20°, data not shown), the primary saccade often failed to reach the target by far, and the following corrective saccade was rather large (>2°). In this case it was found that the intersaccadic interval between the primary and the corresponding corrective saccade could have values of about 100ms, which corresponds to the latency of express saccades.

We suggest that the state of the optomotor system, i.e. its "readiness" to generate the next saccade, strongly depends on the conditions of foveal or parafoveal vision right after the preceding saccade: a new period of fixation is initiated only if a visual target has entered the dead zone.

B. Evidence from primary saccades of "Express Saccade Makers"

When trained and naive subjects make saccades to suddenly appearing targets, while the fixation point remains on (overlap task, OVL), the saccadic reaction times are usually long (SRT>150ms) and the mode of express saccades is very reduced or absent in the SRT-distribution. The upper part of Fig. 2 shows an example of a naive subject's reaction times: during overlap trials a peak around 150 msec is obtained. Using the gap task the subject produced a clearly bimodal distribution of express and fast regular saccades to the left target and only a few express saccades to the right target.

The performance of this normal naive subject in the anti saccade task (where a saccade in the direction opposite to the side of presentation of a visual stimulus is required by instruction) for overlap and gap conditions is shown in the lower part of Fig. 2. During overlap trials very few direction errors occur (N = 11 + 3 = 14 prosaccades to the target rather than to the opposite side). During gap trials the number of prosaccades is increased (N = 19 + 15 = 34), but it remains also in this case below 20%.

The hypothesis of a separate fixation system mentioned above implies, however, that there may exist subjects with a selective impairment of active fixation: they may be able to "keep their eyes still", but any target appearing somewhere in their field of view would reflexively trigger a saccade, because the active inhibition of the saccade system would be weak or even missing.

During the course of our experiments over the last years we have recorded the eye movements of subjects who made almost exclusively express saccades even in the overlap task without any previous training ("express saccade makers"). To test their ability of maintaining fixation and of suppressing saccades, the anti saccade task and the remembered saccade task (a saccade to a previously cued position must be made only after fixation point offset) were used. These subjects had considerable difficulties in following the instructions: in the anti saccade gap task they made large numbers (well above 40 %) of direction errors, i.e. saccades to the stimulus (prosaccades). Fig. 3 shows the data collected from such a subject for the normal and anti saccade overlap and gap tasks: the SRT-distribution for trials in the normal tasks are hardly different (in both overlap and gap conditions only express saccades occur). In the anti task he made large numbers of prosaccades (59 in the overlap, 107 in the gap task, respectively 30 % and 70 %). Again, almost all of these prosaccades are of the express type.

If the instruction was given to make saccades always to the right side while the target was presented randomly to the right or left, the subject made anti-saccades of the express type and still a few prosaccades to the left (direction errors). In the remembered saccade task, instead of maintaining fixation, the subject often could not avoid looking at the cue, and many of these saccades were again of the express type.

CONTROL SUBJECT (MR, 32y, female)

Normal Tasks

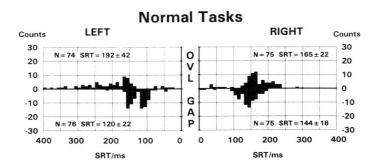

Anti Tasks

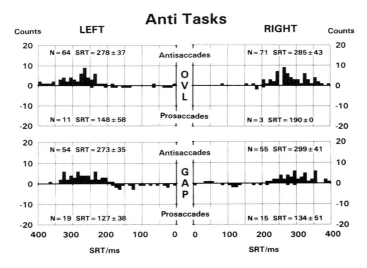

Fig. 2. Data from a naive normal subject (control) obtained in the normal gap (upwards histogram) and overlap (downwards histogram) tasks (top diagram). The two diagrams on the bottom show the saccadic SRT distributions in the overlap and gap anti saccade tasks. The data for left and right directed saccades are mirror drawn in all three diagrams. The inset in each histogram gives the number of saccades and the median SRT.

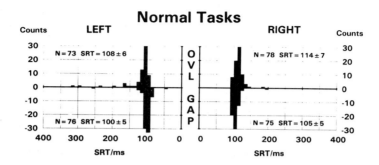

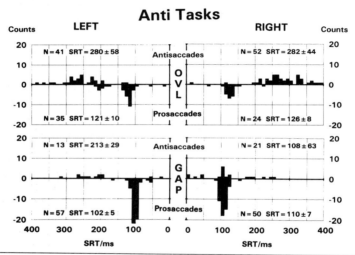

Fig. 3. Same format as for Fig.2. Data from a non-dyslexic express saccade maker.

By now we have seen 7 subjects of this type out of 102 naive and not dyslexic normal subjects, corresponding to about 6% of the population. However, among dyslexic subjects this percentage seems to be increased to above 20%: regardless of the subjects' age, we have seen 15 out of 64 dyslexics (12 males and 3 females), who were classified as express makers on the basis of their performance in the overlap task.

Fig. 4 shows the data from one of these dyslexic subjects. His performance is hardly different from that shown by the non-dyslexic express maker's data shown in Fig. 3. Note that in the gap anti saccade task the dyslexic express maker made 71 out of 77 saccades to the right target. All of these saccades had reaction times below 150 msec with a clear peak of express saccades at 110 msec.

C. Evidence from extra foveal fixation stimuli

So far we have discussed experiments in which a foveal fixation point was used throughout. It is however known, since the experiments of Mayfrank et al. (1986) and of Braun and Breitmeyer (1988, 1990), that attention targets presented in the periphery of the visual field also influence the saccadic reaction time, and in particular the chances of making express saccades. While it had been previously concluded (Posner, 1980) that preattended locations are reached after shorter (manual) reaction times in comparison with non attended ones, the investigators mentioned above found, on the other hand, that saccadic reaction times were delayed for attended target locations.

We have now investigated the spatial selectivity of this phenomenon. Five subjects, all trained to produce high numbers of express saccades in the gap task, were tested in the following experiments: we presented a peripheral attention point at 7 different horizontal positions between 6° to the left and 6° to the right in separate experimental blocks, in order to see the effect of the location of the attention point. Target location was again random at 4° to the left or right. Gaze direction was towards the centre of the screen during the presentation of the peripheral attention point. Fig. 5 shows the data from one of these subjects. The middle two panels give the results obtained when a central fixation point was used, which was extinguished 200ms (gap duration) before the peripheral saccade target appeared. The reaction time distribution of this subject is clearly symmetric: her saccades directed to the right target (right histogram), as well as those to the left target (left histogram) were nearly all of the express type. When the experiment was repeated with the attention point located at 4° to the right, i.e. at the position where the right target was presented (upper two panels of Fig. 5), the number of express saccades towards the right target was drastically reduced in favour of fast and slow regular saccades. Surprisingly, there was in addition an effect with the left directed saccades (note that in this case the target occurred contralateral to the attention point): the number of express saccades was again clearly decreased. This bilateral effect was also observed when the attention point was placed at 4° to the left (lower two panels of Fig. 5). Again express saccades are nearly absent for both target locations. These strong bilateral effects were observed in 3 of the 5 subjects. In the other 2 subjects there was once again a decrease of express saccades with ipsilateral presentation of the attention target, while the contralateral effects were not as clear. Moreover it was found in all subjects that the reduction of express saccades depended from the relative position between attention point and target location: the effect decreased - and the percentage of express saccades increased - with increasing distance between the location of the saccade targets and the attention point.

These results show that it is not foveal fixation nor peripheral attention per se which modulate the occurrence of express saccades; there are clear spatial aspects in the attentional control of the saccade system. The results of these experiments tell us that preattended locations seem to have the least chance of being reached by express saccades, even in the gap task, where the attention target is switched off 100 or 200 msec before the saccade target occurs. It will therefore be also interesting to systematically investigate the effect of the gap duration in order to pursue the time course of the inhibitory effect of the peripheral attention target after its offset.

DYSLEXIC EXPRESS MAKER (BF, 13y, male)

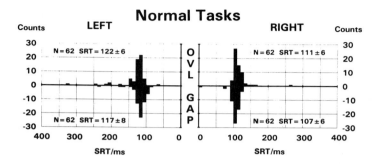

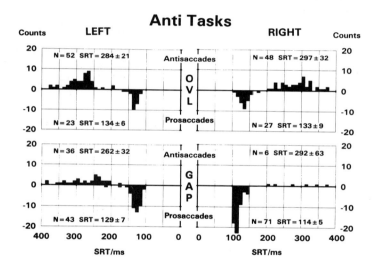

Fig. 4. Same format as for Fig.2 and 3. Data from a dyslexic express saccade maker.

We conclude that the "express saccade makers" have a deficit in their fixational/attentional system, which usually inhibits saccades. This does not necessarily imply that the express makers form a separate group in the population. It is very well possible that the amount of express saccades and the amount of prosaccades varies in a more transitional manner from one extreme to the other between different individual subjects.

Tg random l/r, gap 200
Subject MB

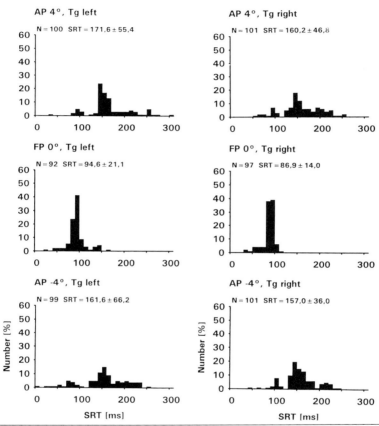

Fig. 5. Data from one human subject with high practice in oculomotor tasks. Upper two panels: Frequency distributions of saccadic reaction times for right (right histogram) versus left (left histogram) target presentation. Location of the attention (fixation) point: 4° to the right; Gap duration: 200ms. Gaze direction was straight ahead (without fixation target). Middle and lower two panels: Same target presentation. Location of the attention point was in the centre of the screen (0°) for the middle two panels, and at 4° to the left for the lower two panels.

Discussion

In this paper we have briefly summarized the results from secondary saccades, from special subjects, who make high numbers of express saccades in the overlap task and high numbers of direction errors in the anti saccade gap task, and from experiments with a peripheral attention point. All these experiments provide evidence for the existence of neural structures in humans exhibiting a suppressive effect on the saccade generating structures. Anatomical and neurophysiological results from several research groups have in fact revealed some cortical and subcortical brain structures involved in attentive fixation and the attentional control of the saccade system. For example it is known that the frontal eye field exerts an inhibitory effect upon the generation of fast saccades. Evidence for this notion comes from neurological patients with lesions of the frontal eye field area, who cannot suppress reflexive express saccades to suddenly occurring visual stimuli in a gap anti saccade task (Guitton et al.1985). But, in contrast to the express saccade makers of our present study, such patients have SRTs in the overlap task which are in the normal latency range (Braun et al.1992). Indeed, microstimulation experiments in the frontal eye field of awake behaving monkeys have shown that the threshold currents needed to elicit a saccade are elevated when the animal fixates a visual stimulus (Goldberg et al.1986). Similar results were obtained by microstimulation of the parietal cortex in monkeys (Shibutani et al.1984). Lesion studies in monkeys have revealed the crucial role played by the superior colliculus in the generation of express saccades (Schiller et al., 1987). Recently, it could be shown by Munoz & Wurtz (1993a,b) that the foveal representation in the rostral part of the superior colliculus is important for maintaining proper fixation on a visual stimulus. Chemical inactivation of these neurons causes monkeys not to be able to fixate a fixation point in a remembered saccade task and to make reflexive saccades to the target: a behaviour similar to that observed in the express saccade makers. The fact that express makers are more frequently found among dyslexic subjects raises the question whether a deficit in one or more of these neural structures is present at least in a subgroup of dyslexics. We know at present that abnormalities in the cytoarchitectonic structure of the cortex of dyslexic subjects can indeed be found at different levels of the magnocellular system, which projects up to the parietal cortex, as reviewed by Breitmeyer (1989).

Our results suggest also that the attentional control upon the saccadic system must be spatially organized. This finding is probably related to the retinotopical organization of most of the brain structures involved in vision and in saccade control. Moreover, the bilateral effects upon the occurrence of express saccades caused by the presence of a unilateral peripheral attention target could have their anatomical basis, for example, in the bilateral projections to the brainstem descending from the frontal eye fields (Schnyder et al.1985). The details concerning the spatial aspects of the attentional control on the saccade system, however, must be further investigated.

Acknowledgment:

This work was supported by the Deutsche Forschungsgemeinschaft, SFB 325, Teilprojekt C5 and C7.

References

Braun, D., Weber, H., Mergner, T., & Schulte-Mönting, J. (1992). Saccadic reaction times in patients with frontal and parietal lesions. Brain 115, 1359-1386.

Braun, D. & Breitmeyer, B. G. (1988). Relationship between directed visual attention and saccadic reaction times. Exp. Brain Res. 73, 546-552.

Breitmeyer, B. G. and Braun, D. (1990). Effects of fixation and attention on saccadic reaction time. North Holland: Elsevier Science Publishers B. V., p.71.

Breitmeyer, B. G. (1989). A visually based deficit in specific reading disability. Irish Journal of Psychology 10, 534-541.

Fischer, B., Weber, H., & Biscaldi, M. (1993). The time of secondary saccades to primary targets. Exp. Brain Res. (in press)

Fischer, B. & Boch, R. (1983). Saccadic eye movements after extremely short reaction times in the monkey. Brain Res. 260, 21-26.

Fischer, B. & Ramsperger, E. (1984). Human express saccades: extremely short reaction times of goal directed eye movements. Exp. Brain Res. 57, 191-195.

Goldberg, M. E., Bushnell, M. C., & Bruce, C. J. (1986). The effect of attentive fixation on eye movements evoked by electrical stimulation of the frontal eye fields. Exp. Brain Res. 61, 579-584.

Guitton, D., Buchtel, H. A., & Douglas, R. M. (1985). Frontal lobe lesions in man cause difficulties in suppressing reflexive glances and in generating goal-directed saccades. Exp. Brain Res. 58, 455-472.

Mayfrank, L., Mobashery, M., Kimmig, H., & Fischer, B. (1986). The role of fixation and visual attention in the occurrence of express saccades in man. Eur. Arch. Psychiatry Neurol. Sci. 235, 269-275.

Munoz, D. P. & Wurtz, R. H. (1993a). Fixation cells in monkey superior colliculus I. Characteristics of cell discharge. J. Neurophys. 70(2), 559-575.

Munoz, D. P. & Wurtz, R. H. (1993b). Fixation cells in monkey superior colliculus II. Reversible activation and deactivation. J. Neurophys. 70(2), 576-589.

Posner, M. I.(1980). Orienting of attention. Q. J. Exp. Psychol. 32, 3-25.

Schiller, P. H., Sandell, J. H., Maunsell, J. H. (1987). The effect of frontal eye field and superior colliculus lesions on saccadic latencies in the rhesus monkey. J. Neurophys. 57, 1033-1049.

Schnyder, H., Reisine, H., Hepp, K., & Henn, V. (1985). Frontal eye field projection to the paramedian pontine reticular formation traced with wheat germ agglutinin in the monkey. Brain Res. 329, 151-160.

Shibutani, H., Sakata, H., & Hyvarinen, J. (1984). Saccade and blinking evoked by microstimulation of the posterior parietal association cortex of the monkey. Exp. Brain Res. 55, 1-8.

Weber, H., Aiple, F., Fischer, B., & Latanov, A. (1992). Dead zone for express saccades. Exp. Brain Res. 89, 214-222.

Weber, H. & Fischer, B. Differential effects of non-target stimuli on the occurrence of express saccades in man. Vision Res., in press.

SACCADE LATENCY TOWARDS AUDITORY TARGETS

Daniela Zambarbieri [a], *Ilaria Andrei* [a], *Giorgio Beltrami* [a], *Laura Fontana* [a], *MaurizioVersino* [b]

[a] Dipartimento di Informatica e Sistemistica, Università di Pavia, Italy

[2] Clinica Neurologica "C. Mondino", Università di Pavia, Italy

Abstract

The latency of saccades, defined as the interval between target presentation and the beginning of eye movement, reflects the time required for the execution of a number of central processes. Several studies have been carried out to investigate the influence of psychophysiological factors on saccade latency, but little attention has been devoted to the reciprocal interaction between the initial eye position in the orbit and target position in space. In order to further investigate these aspects, we have examined saccadic responses evoked by the presentation of visual and auditory targets in the random stimulation protocol. The position of the reference target was changed from trial to trial in order to obtain different initial positions of the eyes in the orbit. In the case of auditory target presentation, saccade latency was found to decrease with the eccentricity of the target with respect to the initial position of the eyes in the orbit.

Keywords

Saccadic eye movements, auditory targets, latency, sound localization, superior colliculus, auditory map.

Introduction

When a subject is asked to orient his eyes toward a target suddenly appearing in space, a goal directed saccade is produced after a time interval from target onset which represents the latency of the response. Saccade latency is the result of a number of sequential and/or parallel processes executed within the central nervous system, such as: the transmission of sensory information from the peripheral receptors, the release of attention from the current fixation point, the localization of target position in the relevant coordinate system, the decision to make a saccade, the generation of an appropriate reference signal for the saccadic execution mechanism.

Saccade latency can be strongly influenced by several psychophysiological factors and by the experimental conditions. The latter can be suitably manipulated in order to investigate the different processes underlying saccade latency. For instance, the gap-overlap paradigm, in which the disappearance of the fixation target and the appearance of a lateral target do not occur at the same instant, has been used to demonstrate the influence of visual attention release on the latency of saccades (Mayfrank et al., 1986).

Eye Movement Research/J.M. Findlay et al. (Editors)

The process of target localization in space can be better investigated when target appearance is completely random, both in time and space, and no warning cues are provided to the subject. In this experimental condition, saccade latency is likely to be related to the sensory signal transformation occurring at the level of superior colliculus (SC) (see Sparks & Hartwich-Young, 1989, for a review).

To further investigate the effect of target localization on saccade latency we have compared saccadic responses evoked by the presentation of visual and auditory targets. Some preliminary results of this study will be presented in this paper. Attention will be focused on the influence of eye position on the latency of saccadic eye movements evoked by the presentation of auditory targets. Targets were always randomly presented to the subjects, but the relative position of the target with respect to the eyes and to head was appropriately manipulated from trial to trial.

Methods

Seven subjects with normal auditory, visual and oculomotor functions were examined. Subjects were seated in total darkness at the center of a circular frame, 220 cm in diameter, supporting visual and auditory targets placed every 5 deg. Visual targets were red light emitting diodes (LED); auditory targets were 5 cm diameter loudspeakers fed with a square wave signal at 15 Hz. During all the experiments, subject's head was restrained by means of a bite board. Eye movements were recorded by conventional electrooculography (EOG).

Following the calibration of eye movements, a visual target was presented to the subject in a fixed position (referred to as "fixation target"). After a random interval, varying betweeen 2 and 4 sec, a lateral auditory target was presented for 2 sec. Then, the LED placed in the same lateral position was switched on to obtain a corrective saccade whenever an error occurred in the fixation of the auditory target. The position of the lateral target was randomly selected among 14 available positions (up to ±35 deg). Each subject was tested in three sessions with different positions of the fixation target (0, +20, -20 deg). The three experimental conditions will be referred to as "C0", "C+20" and "C-20", respectively.

The position of the target evaluated with respect to the subject's midsagittal plane is defined as "target position" (TP). In our experimental conditions, in which the head is restrained in the central position, TP represents, of course, also the position of the target with respect to the head. Moreover, "target offset" (TO) will be used to indicate the position of the target with respect to the position of the eyes in the orbit. Therefore, TO corresponds to the amplitude of target displacement from the fixation position to the lateral position. In other words, the same position of the target can be expressed in a craniotopic (TP) or in a retinotopic (TO) reference system. TP and TO are coincident only in the experimental session with fixation target placed at 0 deg.

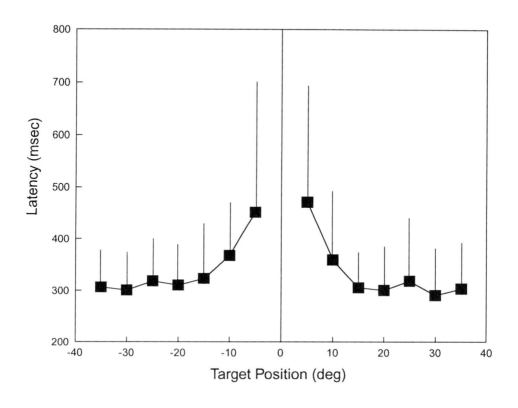

Figure 1. Mean latency versus target position in space of saccades toward auditory targets obtained in condition C0. Vertical bars indicate one standard deviation.

All the experiments were carried on under the control of a personal computer, equipped with a NATIONAL DIO-24 device, that executed the control of the stimulation, and a NATIONAL AT-MIO16 device for the acquisition of the EOG signal. Eye movements were sampled at the frequency of 250 Hz and stored. Saccade parameters were evaluated by means of an interactive program that computed, for each movement, latency (L), duration (D), amplitude (A) and peak velocity (P_V).

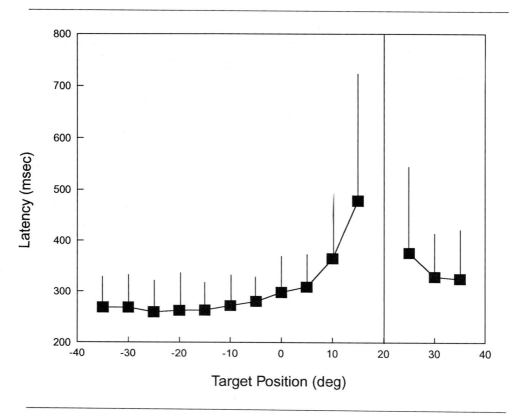

Figure 2. Mean latency versus target position in space of saccades toward auditory targets obtained in condition C+20. Vertical bars indicate one standard deviation.

Results

The dynamic characteristics (A/D and A/P$_v$ relationships) of the responses evoked by auditory target presentation recorded from our population of subjects are perfectly comparable with those reported in the literature (Zahn et al., 1978; Zambarbieri et al., 1982). Saccades evoked by the presentation of auditory targets are characterized by a longer duration and a reduced peak velocity with respect to saccades of the same amplitude evoked by the presentation of visual targets.

Response accuracy was evaluated from the amplitude of the corrective saccade made by the subject at the end of the response, when the auditory target was replaced by the visual one. In all the experimental conditions, a precision of about 95% was observed for all TO.

Concerning the main point of this paper, that is central processing underlying saccade latency, we have obtained the following results. In condition CO, the latency of the primary saccades toward auditory targets decreases as a function of target position in space as shown in Fig. 1. The latency for targets placed at ±5 deg is of about 480 msec, and decreases to about 300 msec for the most eccentric targets. From this diagram it appears that some process prolonging saccade latency occurs for target placed near the midline, but as already mentioned in this experimental condition no difference exists between TP and TO. No evidences can therefore be found on which one between the two parameters of the stimulus actually influences the latency.

Fig. 2 shows the mean latency obtained from condition C+20. The greater values of latency can now be observed for TP of +15 and +25 deg. A mirror-symmetrical diagram has been obtained in condition C-20 where the greater values are observed for TP of -25 and -15 deg. In both experimental conditions it appears that a shift has been introduced that causes the greater values of latency to be centered around the initial position of the eyes in the orbit.

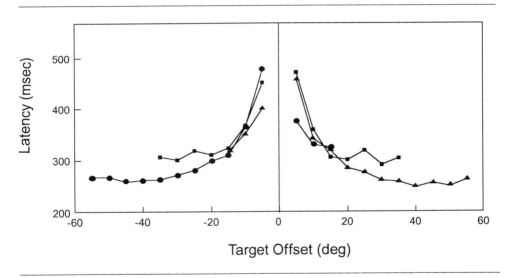

Figure 3. Mean latency versus target offset in the three experimental conditions considered. Squares: condition C0; circles: condition C+20; triangles: condition C-20. For the sake of clearness, standard deviations are not reported in this figure.

It can therefore been concluded that the results obtained in conditions C+20 and C-20 suggest that the latency of auditory saccades are related to TO rather than to TP. This result can be easily appreciated even by eye inspection when the latency is plotted as a function of TO for the three experimental conditions. As shown in Fig. 3, the curves of latency are symmetrical with respect to zero.

Discussion

The significant variation of the latency in saccadic responses toward auditory targets could be first of all ascribed to the process of localization of sound in space. The localization of a source of sound is based primarily on the time and phase differences between the signals perceived by the two ears. When a sound is located near the midline with respect to the head, both time and phase differences are very small and a great level of uncertainty could be supposed to affect the sound localization process.

Based on this assumption, a stochastic model of central processing in saccade generation has been proposed by Schmid et al. (1982). In this model, the uncertainty of sound localization due to the time and phase differences has been represented as a white gaussian noise affecting the incoming sensory signal. The localization of a sound source could therefore be imagined as a process of estimation of a signal affected by noise. The smaller the signal to noise ratio, the longer the time required to reach a threshold level in the estimation process. If the sound localization process is the only reason for a longer latency, it has to be expected that saccade latency does not depend on the position of the eyes in the orbit, since of course eye position has no effect on the performance of the peripheral auditory system.

The results described in this paper, and some other similar results reported in the literature (Zahn et al., 1979, Jay & Sparks, 1990), seem nevertheless to indicate that the main factor affecting saccade latency cannot be that related to sound localization, but that another factor has to exist which is related to the position of the eyes in the orbit. The central processing underlying target localization is likely to consist of several steps: STEP I, corresponding to the transmission of the incoming sensory signals; STEP II, corresponding to the excitation on the relevant SC map, and STEP III, corresponding to the reconstruction of the command signal for the execution mechanism. In a previous study, the simple reaction time to the appearance of visual and auditory targets in space was found to be independent of target position (Zambarbieri et al., 1982). Moreover, the reaction time to auditory targets was found to be about 40 msec shorter than that to visual targets. STEP I seems therefore not to be responsible for the latency profiles shown in Fig. 3. Concerning STEP III, which deals with the efferent signals sent to the premotor structures, a comparison with saccades evoked by visual targets can give some suggestions. In fact, visual stimuli evoke saccades with a latency which is only slightly and positively related to the amplitude of TO. For visual targets confined to about the central 1 deg (that is for TO significantly smaller than those considered in our study) saccade latency increases (Kalesnykas and Hallet, this volume). Since different stimuli reaching the same efferent system display different latency behaviors, the possibility can be excluded that the latency profiles of Fig. 3 are due to the processes performed at the level of STEP III. After having excluded a role of STEP I and III, let us finally consider STEP II. It is reasonable to assume that the latency versus TO relationship originates at the level of the sensory-motor transformation.

At least two hypotheses can be suggested concerning this process. The first hypothesis takes into account the results of a study performed by Jays and Sparks

(1987) on the activity of monkey SC neurons. Monkeys were placed in front of an auditory target located at 20 deg right with respect to the animal's head. The initial position of the eyes was changed from trial to trial and the monkey was trained to make saccades toward the auditory target in darkness. The same neuron, in the SC, that was strongly active when the eyes were initially deviated 24 deg left, has a reduced activity when the eyes were centered in the orbit and finally, it did not respond at all when the eyes were deviated 24 deg right. The authors concluded that "the auditory receptive fields shifted with changes in eye position, allowing the auditory and visual maps to remain in register". It seems therefore that both maps are organized in a "retinotopic frame". If the auditory map shifts with the initial eye position, the receptive field which is excited by a given target in space depends on TO. Depending on the initial position of the eyes different target positions can therefore converge on the same receptive field in the auditory map. Thus the behavorial observation that the latency of auditory responses depends primarily on TO could find its explanation by assuming that the noise affecting the process of localization is related to the resolution of the corresponding receptive field.

The second hypothesis makes reference to the excitation threshold needed to trigger a saccadic eye movement. By assuming that the auditory signal representing target location in space is somewhere compared to the actual eye position, a different level of excitation could be needed, according to the amplitude of the required eye movement, in the case of auditory targets. Lower level of this threshold seems to be required when visual targets are used.

As a matter of fact, from a computational point of view, the hypothesis of an amplitude dependent threshold level gives the same results as the hypothesis of a variable signal to noise ratio, in terms of the time required to produce a signal able to trigger a saccade. The problem remains therefore that of validating the proposed theoretical hypotheses from a physiological point of view. To reach this goal, behavioral data obtained in humans need to be correlated to specific single units recording in the animal SC in order to further investigate the time sequences of the central processing underlying saccade generation.

References

Jay M.F. & Sparks D.L. (1987) Sensorimotor integration in the primate superior colliculus. II. Coordinates of auditory signals. J. Neurophysiol. 57, 35-55.

Jay M.F. & Sparks D.L. (1990) Localization of auditory and visual targets for the initiation of saccadic eye movements. In: Comparative Perception. I. Basic Mechanisms, M. Berkley, W. Stebbins (Eds.), pp.351-374.

Mayfrank L., Mobashery M., Kimmig H. & Fischer B. (1986) The role of fixation and visual attention in the occurence of express saccades in man. Psychiatry and Neurological Sciences 235, 269-275.

Schmid R., Magenes G. & Zambarbieri D. (1982) A Stochastic model of central processing in the generation of fixation saccades. In: Physiological and Pathological Aspects of Eye Movements, A. Roucoux M. Crommelinck (Eds.), Dr W. Junk Publ., The Hague, pp. 301-311.

Sparks D.L. & Hartwich-Young R. (1989) The deep layers of the superior colliculus. In: The Neurobiology of Saccadic Eye Movements, Wurtz R.H., Goldberg M.E. (Eds.), Elsevier Science Publ. B.V., Amsterdam, pp. 213-255.

Zahn J.R., Abel L.A. & Dell'Osso L.F. (1978) Audio-ocular response characteristics. Sensory Process 2, 32-37.

Zahn J.R., Abel L.A., Dell'Osso L.F. & Daroff R.B. (1979) The audio-ocular response: intersensory delay. Sensory Process. 3, 60-65.

Zambarbieri D., Schmid R., Magenes G. & Prablanc C. (1982) Saccadic responses evoked by the presentation of visual and auditory targets. Exp. Brain Res. 47, 417-427.

RETINAL ECCENTRICITY AND THE LATENCY
OF EYE SACCADES

P.E. Hallett[1-4] and R.P. Kalesnykas[1]

Departments of Physiology[1], Zoology[2], Ophthalmology[3] and Institute of Biomedical Engineering[4], University of Toronto, Toronto, Ontario, Canada M5S 1A8.

Abstract

The latency of the first saccade towards a small green or red target on the horizontal hemi-retinal meridian varies as a function of target eccentricity. The function is bowl-shaped with a central latency peak, a minimum plateau from 0.75 to 12°, and a gradual increase in latency towards the periphery. The function is highly reproducible and the central peak is a robust finding. The height of the peak for step sizes of 7.5 to 15 arc min is around 75 and 35 msec, respectively. Manipulations of target intensity and colour show that sensory contributions to the central peak are generally small (5-15 msec) for adequately suprathreshold targets. Beyond 35° in the temporal retina (nasal visual field) latencies become variable, even for bright green targets, and there are direction errors. Frequent direction errors occur over a much wider range of eccentricities for red targets that are 1 log above the foveal threshold for perception - this is attributed to insufficient luminance relative to the peripheral threshold. Although this study examines possible low-level sensory issues, central or motor explanations of the central latency peak are certainly not excluded. For example, the alerting effect of a target-locked auditory tone is absent for small intrafoveal steps.

Keywords

central latency peak, stimulus intensity, direction errors

Introduction

When the retinal image of a fixated target is stepped into the periphery the oculomotor stimulus can be specified either by the amplitude of the step or equivalently in normal subjects given the accuracy and stability of fixation: Steinman, 1965; Snodderly, 1987) by the retinal eccentricity or position stimulated. The literature contains only fragmentary data in any single study on which to base any conclusion as to the relation between the latency of the first or primary saccade to a target and the step size or retinal eccentricity. Findlay's (1983) summary plot allows as one possibility that this relation may be a bowl-shaped function, in the sense that latencies might be increased for eccentricities less than 0.75 or greater than 12°, but the data are scattered by the different experimental protocols and the wide variation in stimulus characteristics. An alternative is that latencies decline almost monotonically from the centre. Given the practical and clinical importance of saccadic eye movements we present new data as part of a standardized study that demonstrate the highly reproducible nature of the latency-eccentricity function for a wide range of eccentricities and several different stimulus conditions. Of interest are a robust central peak of latency for very small target displacements within the foveal region (<0.75°;

Eye Movement Research/J.M. Findlay et al. (Editors)

Wyman & Steinman, 1973; Kowler & Anton, 1987), anomalous responses at the larger eccentricities, and the problems of specifying and matching differently coloured targets.

Methods

Eye Movement Detection

(i) A pupil tracker (Hallett & Lightstone, 1976) was used for the eccentricity range of 3 to 66°. Latencies were measured to the nearest msec. Instrument noise was 5-8 arc min peak-to-peak. Saccades were detected when eye velocity exceeded 20°/sec, so latency was probably overestimated by no more than 1-3 msec. An acrylic dental impression and subject training eliminated head movement.

(ii) A precision two-dimensional corneal-reflex tracker was used for the eccentricity range 0 to 6° with a noise of 3 arc min peak-to-peak (Frecker, Eizenman & Hallett, 1984). Saccade trigger velocity was 7-10°/sec, so any overestimate in latency was presumably <1 msec. A dental impression and a forehead rest provided restraint.

Stimulus Arrangements

Target stimuli consisted of unequally spaced light emitting diodes (LEDs) or an oscilloscope. The lenses of the diodes were ground off and polished flat to provide small targets. Two arcs of differing red and one of yellow-green diodes were used (660, 670 or 565 nm peak, 20-30 nm half-widths), with viewing distances of 57, 114, or 228 cm. Blue-green, 6 arc min targets were provided by a filtered P15 rapid decay phosphor oscilloscope (501 nm, 20 nm half-width) at 57 cm distance.

Saccadic eye movements for the Big Step experiment were recorded using the wide angle Big Step Stimulus Array of Figure 1a and the pupil tracker. The arc (drawn to scale) of radius 57 cm allowed 19 different displacements, left or right, in

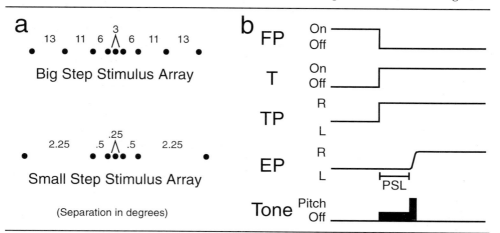

Figure 1. a): *Arrangements of stimuli for Big and Small Step experiments.* b): *Trial schematic.* FP - *fixation point;* T - *target stimulus;* TP - *target position;* EP - *eye position;* R - *right;* L - *left;* PSL - *primary saccade latency.*

the range ±3 to ±66° (displacements of 3, 6, 9, 11, 12, 13, 17, 18, 20, 23, 24, 29, 30, 33, 36, 40, 42, 53, 66°), i.e., 38 retinal eccentricities. Measured target size was 2.0 arc min at 660 nm peak (20 nm half-width) and 8.4 arc min at 565 nm (30 nm hw).

The small angular array of Figure 1a (drawn to scale) and the Purkinje image tracker was used for the Small Step experiments. The arc provided 22 different retinal eccentricities for a 4.2 arc min spot, from ±0.25 to ±6° when the viewing distance was 114 cm (11 left or right displacements of 0.25, 0.5, 0.75, 1.0, 1.5, 2.25, 2.75, 3.0, 3.25, 3.75, 6.0°), and a display of half that angular size at 228 cm. The red LEDs were at 670 nm peak. Corrective spectacle lenses were used if necessary.

Target intensities for a particular colour are specified relative to the dark-adapted foveal threshold FT for perception, e.g., FT+2 log or FT+2 for short. The FT for natural pupils (typically 6.60 mm in diameter) was determined each session after 15 min dark-adaptation from artificial room lighting. Intensities were changed either by altering the supply voltage or with neutral filters. LEDs were maintained in good photometric match.

As in this laboratory's previous studies (e.g., Hallett, 1978) the target step began a soft tone that was truncated by the primary saccade (see Figure 1b), and each saccade was marked by a tone pip; this helped maintain the alertness and confidence of the dark-adapted subject.

General Procedure

The dark-adapted subject viewed the stimuli with the left eye, the right being covered. One of the LEDs was lit at random to provide a point for steady fixation. Extinguishing it and randomly lighting another LED as the new fixation point generated a wide range of target displacements. Note that there is no single fixation point, any diode can be the fixation point; consequently an arc of diodes subtending 66° at the eye can stimulate retinal eccentricities of ±66°.

Figure 1b illustrates the temporal sequence of lighting the fixation point and target, target position and eye position, and the latencies of interest. The subject is in full control of the start and stop of the experiment. Each random amplitude target displacement occurred after a random foreperiod of 1.5 to 2.4 sec. A period of 1 sec from the end of the first or primary saccade was allowed for recording any secondary saccade, and for final accurate foveation of the target prior to the start of the next trial. The sequence then repeated automatically for 800 trials (total time about 1 hour); the subject could rest whenever necessary. The direction though not the distance of the next stimulus is predictable when fixating the end of the array; however, these few saccades made no significant contribution to the data. Any direction errors were analyzed separately. There were up to three training sessions of 800 trials each. The fourth and later sessions confirmed that latencies had stabilized at reduced values by the second or third session. Each experimental session began with 30-50 practice trials.

Subjects

Seven volunteer subjects (aged 18 to 51 years; three male and four female) participated in these experiments. Subjects RPK and HDK were highly experienced, while the others were naive as to the purpose of the experiments.

Results

Figure 2 shows for one subject a latency-eccentricity function measured along the dark-adapted horizontal retinal meridian at five different target intensities. Red targets are shown in the top panel and yellow-green in the bottom. The parameter to the curves in Figure 2 and elsewhere is the intensity of the target, which was varied from 0.5 to 4 log above the perceptual dark-adapted foveal threshold (FT). In general, the latency-eccentricity function shows a local peak within the foveal region for eccentricities of 0.25 to 0.75°, which will be examined in more detail later. Depending on the subject, saccadic latency for bright yellow-green targets tends to remain constant up to 12° or to show a minimum between about 1 to 12°; it then rises and becomes more variable at larger eccentricities, especially in the temporal hemi-retina (nasal field) beyond 35°. Latency is longer at the lower intensities. The large spike in

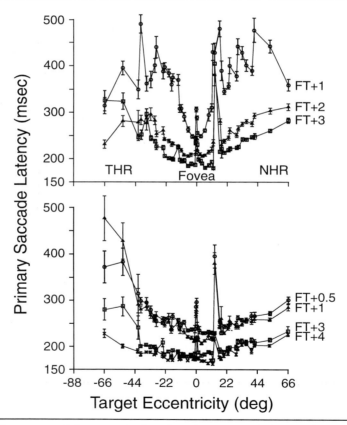

Figure 2.   Primary saccade latency as a function of target eccentricity and stimulus intensity expressed as log units above foveal threshold (FT). Top: Red stimuli. Bottom: Yellow-green stimuli. FT+2 omitted for clarity. ±1 SEM plotted, for this and all other figures.

the nasal hemi-retina represents correctly directed long latency responses to scattered light when the target fell on the optic disk (also Hallett, 1978). Less extensive measurements at ±3 to ±66° for three other subjects confirm the overall trends.

Red stimuli with an intensity of FT+1 log, though moderately bright by foveal viewing, elicit saccades with several peculiarities. At this intensity mean latencies increase steeply beyond ±6° eccentricity and beyond about 15° eccentricity are greatly prolonged (as long as for FT+2 or FT+3 targets at the optic disk) and highly erratic. At FT+1, direction error saccades, normally never seen in simple foveating tasks, are relatively common across much of the retina (Figure 3), especially for the shorter latency subjects HDK and MCP. On the other hand, direction errors for bright red targets (FT+3) are no more frequent than those for bright green ones, being restricted to targets at the optic disk or beyond 35° in the temporal hemi-retina.

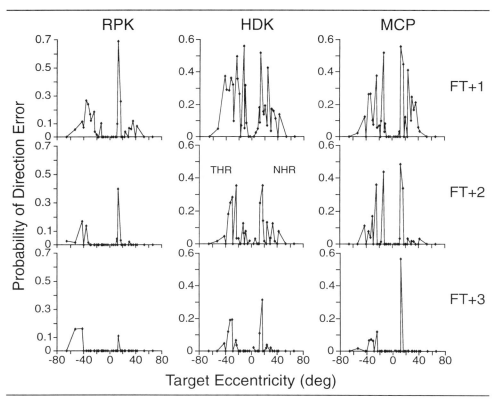

Figure 3. Probability of saccade direction error as a function of target eccentricity and stimulus intensity for red stimuli. Subjects RPK, HDK, and MCP.

Suprathreshold yellow-green targets rarely evoked direction error saccades at any intensity, and when they did the target was either at the optic disk or beyond 40° in the temporal hemi-retina.

The data in Figures 4 and 5 shows the typical narrow form of the latency peak for suprathreshold target intensities. Its height, measured as the difference between the latencies of saccades to targets at 15 and 45 arc min retinal eccentricity (Figure 4), is 35 msec (n=30, sd=7 msec) when pooled across subjects and suprathreshold intensities; measured between 7.5 and 45 arc min it is 75.3 msec (n=6, sd=9.9 msec) (Figure 5).

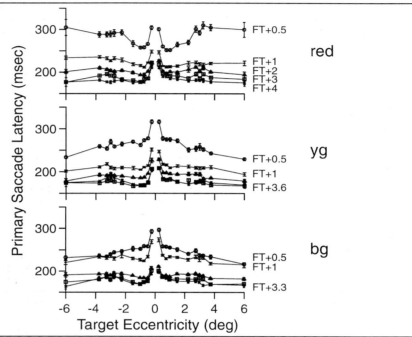

Figure 4.   Primary saccade latency as a function of target eccentricity, stimulus intensity, and colour.

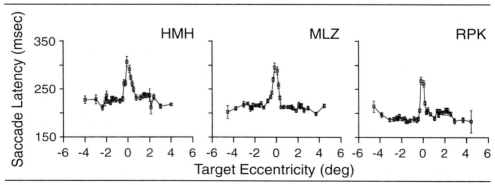

Figure 5.   Mean saccade latency for target displacements from 0.125 to 4.5°. Subjects HMH, MLZ, and RPK. Yellow-green stimuli at FT+3 log.

Figure 6 shows that the height and spatial extent of the central latency peak is greatly increased for blue-green (bg) and yellow-green (yg) targets of an intensity equal to the perceptual threshold FT; latency is increased by 300-400 msec relative to that near 4° eccentricity. These curves are for correctly directed saccades only; the accompanying direction error plots reveal what almost amounts to a central nocturnal scotoma — a region where the probability of a direction error rises from 0 to nearly 0.5. Failing to make a saccade is another possible response but was rare in this particular subject. The correctly directed responses in the central ±1.5° were found to be of fixed mean amplitude and severely hypermetric. This was not the case for adequately suprathreshold targets.

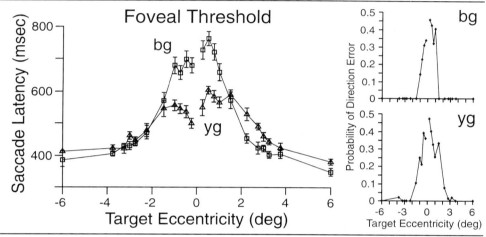

Figure 6. Primary saccade latency and direction errors at foveal threshold.

Figure 7 shows mean latencies when the target array is rotated 90° so that target displacements and saccades are vertical rather than horizontal; the central peak is still present. Figure 7, right, shows one example of a routine plot: primary saccadic

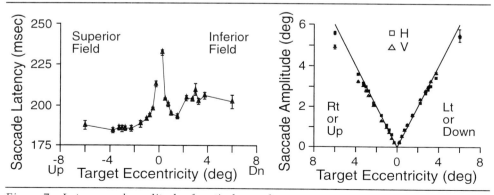

Figure 7. Latency and amplitude of vertical saccades.

amplitudes are typically fairly close to target displacement (unity lines) for targets which are adequately above the foveal threshold, and there is no apparent difference in tracking between vertical and horizontal arrays.

At high lighting intensities (FT+3) the prolonged latencies within the peak persist and are scarcely affected by the presence or absence of the *alerting* tone (Figure 8). Latencies for larger step sizes are less variable (by up to 17 msec in SD) and are reduced in mean by 5 to 46 msec in the presence of the tone (also Hallett & Adams, 1980).

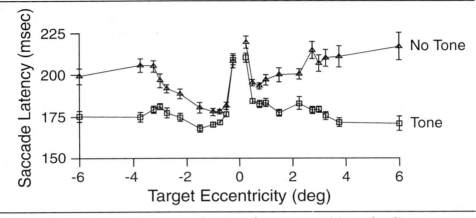

Figure 8. Primary saccade latency as a function of target eccentricity and auditory tone.

Discussion

These data are generally consistent with the hemi-retinal latency-eccentricity function being bowl-shaped, highly replicable in a given subject, and qualitatively similar across normal subjects. The limitations to this function in the temporal retina (nasal field) are not well understood. Le Grand (1957) contrasted a theoretical optical limit of about 99° nasally with a perceptual perimetric limit of at most 60°. We find that saccadic performance tends to be variable beyond about 35° in that direction, with direction errors, whatever the intensity and colour of the target. We stress that our result is not due to obstruction by the nose.

The peculiar responses in the red are plausibly due to limited target luminance. An FT+1 log red target is certainly not dim when viewed foveally, but our subjects reported that it was difficult to see in the periphery. Because of a variety of factors relating to the changes in the photoreceptor population with eccentricity, two targets which are matched for luminance by foveal vision need no longer match at increasing retinal eccentricities. For the present dark adapted state, the threshold for red light tends to increase slowly with eccentricity, while that for yellow-green can be expected to drop appreciably, because of the intervention of the rods, before rising in the farther periphery (e.g., data of Pirenne, 1948). Consequently and FT+1 red target comes closer to the local perceptual threshold as it is marched into the periphery, and some saccadic anomalies are to be expected. As latencies for red FT+1 targets are commensurate with latencies for red FT+2 or FT+3 targets falling on the optic nerve head (Figure 2, top), and as direction errors are often nearly as frequent (Figure 3, top panels), it would seem that the peripheral temporal retina is almost blind to small red targets unless these are quite bright by foveal criteria. Perceptual perimetric fields are also restricted for small dim red stimuli (Traquair, 1957). The many known anomalies for peripheral *colour* vision in the red are not particularly relevant here, because perceptual and saccadic studies have both determined that small high contrast targets are primarily luminance stimuli, not chromatic ones (e.g., King-Smith & Carden, 1976; Doma & Hallett, 1988).

The local threshold is less relevant when the latency response is nearly saturated, i.e., when comparing very bright coloured targets that produce nearly minimal latencies. However, an effect of hue remains conceivable if the various cone mechanisms should have differing latencies (for related work see Nissen & Pokorny, 1977; van Asten et al., 1988). Our earlier studies (Doma & Hallett, 1988, 1989) stressed the general importance of target luminance, rather than hue, but did detect in one subject a small effect of colour as such, for bright yellow and blue lights directly matched by flicker photometry at the retinal eccentricity of the target. The present targets are only matched at the fovea, so there is no need to postulate hue effects in the periphery.

On the basis of limited perceptual observations Wheeless (1965) suggested that the central dead zone for saccades might progressively increase from zero to 4° in extent as the target is dimmed from foveal threshold to an intensity 10 times lower. Our direct oculomotor observations are rather different: the central latency peak remains relatively constant as a target is dimmed from high intensities towards the foveal threshold, increasing quite suddenly in spatial extent and height at just suprathreshold intensities. If we choose to equate a latency of over 500 msec with no effective response, then Figure 5 could be interpreted as indicating a 4° saccadic "dead zone" that comes into existence at an intensity between 1 and 3 times the foveal threshold.

The frequency of seeing at the perceptual threshold setting FT is likely around 0.5-0.8 in the central fovea, while the frequencies of correctly directed (CD) and direction error (DE) saccades both approach 0.5 for the smallest steps. As the CD responses are then severely hypermetric and of constant mean amplitude, the most likely interpretation is that the frequency of *valid* tracking responses at intensity FT for very small displacements is much less than 0.5, if not nearly zero. This is the expectation if the threshold for purely cone-driven saccades is distinctly higher than the perceptual threshold, as previously suggested (Doma & Hallett, 1988, 1989).

Any link between the macular pigment and the central latency peak can be excluded by noting that light from the red and yellow-green targets is not absorbed by this screening pigment (Snodderly et al., 1984a,b), so prolonged central latencies cannot be attributed to locally reduced intensities. It is true that the blue-green target is about 50% attenuated by macular pigment but experiment showed that a 50% dimming shifted the latency-eccentricity function by only 5-15 msec, less than the 25 msec amplitude measured for the peak in that case.

Replotting the present data as primary latency *versus* target intensity, with retinal eccentricity as the parameter, gives functions generally resembling the curves of

Doma and Hallett, but in less detail (not shown). Although Wheeless (1965) suggested that the latency function saturates at FT+2 log, we more typically estimate a higher intensity nearer FT+3. The saturation intensities appear very similar for 0.25 and 3.0° eccentricity, with no naso-temporal difference. Average estimates are FT+2.6 log (n = 10) at 0.25° and FT+2.9 (n = 12) at 3°.

The central latency peak is a very robust phenomenon for all conditions. In addition we have routinely inspected saccadic accuracy (primary saccade amplitude *versus* target eccentricity) and profile (main sequence plots). There was no indication of any peculiarity in the saccadic response to a target at small eccentricity (for a comparable range of data see also Eizenman et al., 1986; Hallett, 1986). Weber et al. (1992) claim a small (10%) excess velocity for comparable small saccades relative to larger ones.

In our method the latencies for a given retinal position are pooled over different orbital positions; this does not appear to be a cause for concern. Our "head shifted" condition controls for this factor, and appropriate plots of the data for our standard condition did not show any orbital effect either (not shown). Also Accardo et al. (1986, 1987) did not find any relevant differences in the latencies of centripetal, centrifugal or mid-line crossing saccades of less than 30° amplitude.

This study concentrates on possible sensory factors underlying the central latency peak, but other factors are not excluded. It is very curious that the alerting tone has little or no latency reducing effect for *small* target displacements within the central foveal region. It may be that the typical tendency is for this tiny region to be always alert, so that it cannot be further alerted by a tone. Reuter-Lorenz et al. (1992) refer to an effect of tones on "express" saccades as an intersensory facilitation. There is also some interest in the role played by "disengagement", or temporary release of attention by the foveal central region, on latencies in general (e.g., Posner, 1980; Kalesnykas & Hallett, 1989; Fischer & Weber, 1993; Mackeben & Nakayama, 1993).

Although the present experiments show very clearly that the central latency peak does not critically depend on low-level sensory factors related to light capture, they do not (with the exception of the tone experiment) address the issue of whether there are more important higher-level sensory, central (e.g., decision or attentional factors), or motor determinants (e.g., Wyman & Steinman, 1973). Further work is in progress.

Acknowledgements
This work was aided by a grant from the Medical Research Council of Canada to PEH.

References

Accardo, A.P., Inchingolo, P. & Pensiero, S. (1986) Influence of eccentricity on saccadic latency. In Developments in Oculomotor Research, IUPS Satellite Symposium, XXX International Congress, Oregon, USA. pg. 24.

Accardo, A.P., Inchingolo, P. & Pensiero, S. (1987) Gaze-position dependence of saccadic latency and accuracy. In Eye Movements: From Physiology to Cognition. Edited by: O'Regan, J.K. & Levy-Schoen, A. Elsevier North Holland, Amsterdam. pg. 150-151.

van Asten, W.N.J.C., Gielen, C.C.A.M. & de Winkel, M.E.M. (1988) The effect of isoluminant and isochromatic stimuli on latency and amplitude of saccades. Vision Res. 28, 827-840.

Doma, H. & Hallett, P.E. (1988) Rod cone dependence of saccadic eye-movement latency in a foveating task. Vision Res. 28, 899-913.

Doma, H. & Hallett, P.E. (1989) Variable contributions of rods and cones to saccadic eye-movement latency in a non-foveating task. Vision Res. 29, 563-577.

Eizenman, M., Frecker, R.C. & Hallett, P.E. (1986) Continuity and asymmetry in amplitude-velocity-duration relations for normal eye saccades. Research in Biological and Computational Vision, Technical Report RBCV-TR-86-13, University of Toronto (available on request from the Department of Computer Science).

Findlay, J.M. (1983) Visual information processing for saccadic eye movements. In Spatially Oriented Behavior. Edited by: Hein, A. & Jeannerod, M. Springer-Verlag, New York. Chap. 16, pg. 281-303.

Fischer, B. & Weber, H. (1993) Express saccades and visual attention. Behav. Brain Res. 16, 553-610.

Frecker, R.C., Eizenman, M. & Hallett, P.E. (1984) High-precision real-time measurement of eye position using the first Purkinje image. In Theoretical and Applied Aspects of Eye Movement Research. Edited by: Gale, A.G. & Johnson, F. Elsevier Science Pub., North-Holland. pg. 13-20.

Hallett, P.E. (1978) Primary and secondary saccades to goals defined by instructions. Vision Res. 18, 1279-1296.

Hallett, P.E. (1986) Eye movements. In Handbook of Perception and Human Performance. Edited by: Boff, K.R., Kaufman, L. & Thomas, J.P. Wiley and Sons, New York. Vol. I, Chap. 10, pg. 78-101.

Hallett, P.E. & Adams, B.D. (1980) The predictability of saccadic latency in a novel voluntary oculomotor task. Vision Res. 20, 329-339.

Hallett, P.E. & Lightstone, A.D. (1976) Saccadic eye movements towards stimuli triggered by prior saccades. Vision Res. 16, 99-106.

Kalesnykas, R.P. & Hallett, P.E. (1989) Human saccadic eye movement: latency and volition? Invest. Ophthalmol. Vis. Sci. 30, Suppl., 184.

King-Smith, P.E. & Carden, D. (1976) Luminance and opponent-colour contributions to visual detection and adaptation and to temporal and spatial integration. J. Opt. Soc. Am. 66, 709-717.

Kowler, E. & Anton, S. (1987) Reading twisted text: Implications for the role of saccades. Vision Res. 27, 45-60.

Le Grand, Y. (1957) Light, colour and vision. London; Chapman & Hall.

Mackeben, M. & Nakayama, K. (1993) Express attentional shifts. Vision Res. 33, 85-90.

Nissen, M.J. & Pokorny, J. (1977) Wavelength effects on simple reaction time. Percept. Psychophys. 22, 457-462.

Pirenne, M.H. (1948/1967) Vision and the eye. (2nd Edition) Science Paperbacks; Chapman & Hall. Figures 3.4 and 3.6.

Posner, M.I. (1980) Orienting of attention. Quart. J. Exp. Psych. 32, 3-25.

Reuter-Lorenz, P.A., Nozawa, G. & Hughes, H.C. (1992) Intersensory facilitation and express saccades. Invest. Ophthalmol. Vis. Sci. 33, Suppl., 1357.

Snodderly, D.M. (1987) Effects of light and dark environments on macaque and human fixational eye movements. Vision Res. 27, 401-415.

Snodderly, D.M., Brown, P.K., Delori, F.C. & Auran, J.D. (1984a) The macular pigment. I. Absorbance spectra, localization, and discrimination from other yellow pigments in primate retinas. Invest. Ophthalmol. Vis. Sci. 25, 660-673.

Snodderly, D.M., Auran, J.D. & Delori, F.C. (1984b) The macular pigment. II. Spatial distribution in primate retinas. Invest. Ophthalmol. Vis. Sci. 25, 674-685.

Steinman, R.M. (1965) Effect of target size, luminance, and color on monocular fixation. J. Opt. Soc. Am. 55, 1158-1165.

Traquair, H.M. (1957) Traquair's Clinical Perimetry. Edited by: Scott, G.I. Published by: Henry Kimpton, London.

Weber, H., Aiple, F., Fischer, B. & Latanov, A. (1992) Dead zone for express saccades. Exp. Brain Res. 89, 214-222.

Wheeless, L. (1965) The effect of intensity on the eye movement control system. PhD Thesis, Univ. of Rochester.

Wyman, D. & Steinman, R.M. (1973) Latency characteristics of small saccades. Vision Res. 13, 2173-2175.

IS SACCADIC ADAPTATION CONTEXT-SPECIFIC ?

Heiner Deubel

Max-Planck-Institut für psychologische Forschung
Leopoldstrasse 24, D-80802 München, Germany

Abstract

An analysis is given of the capability of saccadic gain control to adapt specifically to various context variables such as the orbital starting position of the saccade, the form and color of the saccade target, the presence or absence of a background structure, and the distinction between intentional and stimulus-elicited saccade generation. The data demonstrate that there is no fast position-specific learning, and that visual stimulus features do not trigger the use of specific parameter sets. Intentionally generated and visually triggered saccades, however, can be adapted specifically, suggesting the existence of separate pathways and adaptive control mechanisms for both types of saccades.

Keywords

Saccadic adaptation, context-specificity, endogenous vs. exogenous saccades, orbital eye position specificity

Introduction

The maintenance of saccadic accuracy over the lifetime presumes a continuous monitoring of saccadic performance and, when required, the ability to recalibrate the saccadic response. Evidence for the capability of the saccadic system to compensate adaptively for pathological dysmetria has indeed accumulated from both clinical observations and experimental lesion studies (e.g. Kommerell et al., 1976; Optican & Robinson, 1980). Several findings suggest that parts of the cerebellum, especially the midline cerebellar vermis, are essential for the production of accurate saccades. As a possible mechanism, the cerebellar vermis could act on the saccade generator of the brainstem, e.g. by adjusting the gain of the saccadic internal-feedback loop (Dean et al., 1992).

In the laboratory, saccadic adaptation can be easily induced by systematic displacements of a visual target during saccadic eye movements as shown in Figure 1 (e.g. McLaughlin, 1967; Deubel et al., 1986). In this "double-step" paradigm, the subject has to track steps of a small target; while the eye follows with a saccade, the target is shifted systematically by a small amount, for example into the opposite direction of the saccade. Typically, the saccadic system adapts to this situation quickly, by reducing saccadic magnitude to the

Eye Movement Research/J.M. Findlay et al. (Editors)

required value. This type of conditioning can occur within less than 200 training trials. Interestingly, it has been observed that when, after training, single steps are provided, recalibration to the normal gain value may take longer than conditioning (Deubel et al., 1986; Deubel, 1987). The induced effects are so persistent that they sometimes even show up on the day after the adaptation session (personal observations). Since the gain reductions that can be achieved are considerable, large and frequent saccadic corrections should be expected after adaptation sessions when the precise foveation of small targets is required.

This is not what we found when we occasionally made a subject scan objects in the normal environment outside the laboratory, measuring eye movements with EOG. Rough inspection of the eye movement traces showed that the saccadic responses seemed to have normal accuracy. This surprising observation indicates that the induced adaptation effects are limited to the laboratory context in which the conditioning occurred, implying the existence of context-specific mechanisms and possibly the involvement of higher-level strategies.

I here present five experiments out of a series of studies which investigated whether "context" variables, i.e. stimulus features other than retinal target eccentricity and direction or timing of target presentation, have an effect on adaptation. Potentially, these "context" variables can be manifold. The work presented here studies the effect of visual properties of target and environment, spatial position of the target, and intentional vs. stimulus-guided responses. The findings strongly suggest that visual stimulus features do not trigger specific parameter selection. However, evidence is provided for a selective modification of intentionally generated and stimulus-elicited saccades, demonstrating individual gain adjustment mechanisms for the reactive and the internally triggered saccadic responses which is in line with recent findings by Erkelens & Hulleman (1993). Preliminary reports were presented elsewhere (Deubel, 1993; Deubel, in press).

General Methods

In general, each session consisted of three phases. Initially, saccadic gain (defined as the ratio of primary saccade size and target eccentricity) was determined in the unadapted subject for both types of "context" that were under investigation in the individual experiment. For this purpose, subjects had to follow steps of a visual target, and, in Experiment 3 and 4,

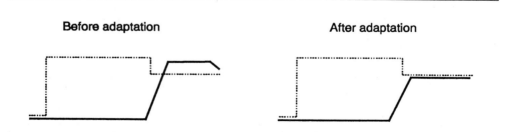

Before adaptation   **After adaptation**

Fig.1 - Adaptation paradigm. Sequence of target shifts and saccade responses before and after "double-step" adaptation. The final target position is taken as the starting point of the next sequence.

had to scan several small objects; step sizes ranged between 4 and 8 deg. The second phase was a long adaptation period (300-500 trials) where the conditioning stimuli occurred only with a certain "context". The conditioning stimuli consisted of systematic, intrasaccadic shifts of the saccade target into the opposite direction of the primary saccade as shown in Figure 1. The sizes of these shifts were 20-33% of the first step, small enough to go largely undetected by the subjects. Finally, saccadic gain was again determined with single step stimuli, and the amount of transfer to saccades occurring under different context conditions was tested.

Eye position was registered with a SRI Generation 5.5 Dual-Purkinje image eyetracker (Crane and Steele, 1985) and sampled at 400 Hz. Vision was binocular and head movements were restricted by a biteboard and a forehead rest. A fast video-graphics system providing a frame frequency of 100 Hz was used for stimulus presentation.

Experiment 1

The first experimental series tested whether different response parameters can be coupled with different visual target features. Since subjects usually select targets based on their color and form, we may hypothesize that two or more specific gain values can be connected to certain targets by means of oculomotor conditioning.

Therefore, in this experiment, the subject had to track targets that were either green crosses or red circles, selected at random. When the subject followed the red circle on the screen, no conditioning intrasaccadic target shifts were provided. When green crosses were given as saccade targets, intrasaccadic target displacements as in Figure 1 occurred. Two of the four participating subjects were informed about the double-step target sequence, the others were naive. Figure 2 shows, on the left, the time course of gain for both target types during adaptation for a typical subject.

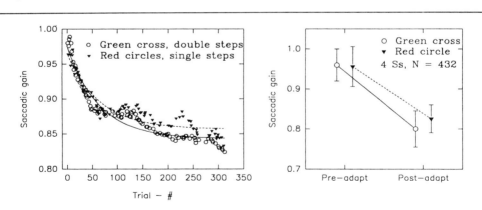

Fig. 2 - Experiment 1. Left: saccadic gain as a function of trial number, for saccades to both target types. Data points represent 10 trial running averages. The conditioning second target displacement occurred only with the green targets. Right: Mean saccadic gain values before and after adaptation, given for both target types.

The data demonstrate that the gain is at any time during learning very similar for both target types. This data pattern is found for the naive subjects as well as the informed subjects. The right diagram presents a comparison of saccadic gain before and after the adaptation. Obviously, saccades to both target types show equally reduced gain values after the adaptation, suggesting that there is no specificity of saccadic gain control with respect to simple visual aspects of the target such as color and form.

Experiment 2

Experiment 2 investigated whether the presence or absence of a background structure might serve as a relevant cue to switch between different sets of response parameters. It is well-known that for space constancy, the presence or absence of a visual background structure is of fundamental importance. With a similar mechanism, prominent landmarks may serve as a frame of reference that allows the system to rapidly recalibrate the response.

Therefore, in this experiment, various types of background structures - one example is shown in Figure 3 - were used. In the first experimental phase, saccadic gain was determined for both the presence and the absence of the background structure. Then, in one type of adaptation session, the conditioning intrasaccadic target displacements occurred while a background was present, in another type of session, the background was absent during adaptation. In the third phase, saccadic gain was again determined with and without background. The results shown in Figure 3 demonstrate, first, that adaptation also occurs with presence of the stationary background. This is interesting per se since the background did not share the intrasaccadic shift with the target implying that the *target shift* alone provides the essential error signal for the adaptive system. Second, and more importantly, the amount of gain reduction does not interact with the presence or absence of a visual background in the test period. This means that presence or absence of a background structure does not act as a relevant context parameter for the switching between different sets of gain parameters.

Experiment 3

The previous experiments failed to find any indication of context specificity. However, one of the characteristics of all these experiments was that the subject's eye movements were completely guided by the stimulus: All these saccades were "sensory" saccades in the sense that they were triggered by an external event, and that the location of where the eye had to go was determined by the location of the single stimulus in the visual field. Most of our everyday saccades are not determined by external events, however; we intentionally decide where to move the eyes next and when to trigger a saccade. The next experiment aimed to mimic a situation where the saccades were not determined by the onset of a single target in an otherwise empty field, but where the subject voluntarily selected a target from several alternatives. Therefore, Experiment 3 investigated the transfer of adaptation induced with stepping targets on the saccadic behavior in a situation where the *scanning* of static items was required.

For adaptation, subjects had to follow a small cross on a video screen that performed steps of 6 deg. As in the previous experiments, the saccade onset systematically triggered a small

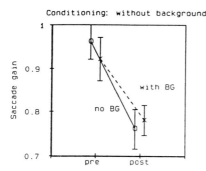

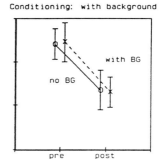

Fig.3 - Experiment 2. Top: Example of one type of background structure used in the experiment. Bottom: Adaptive gain changes for the test conditions with and without presence of a background structure.

secondary target displacement of 1.5 deg, into the opposite direction to the initial step. Experimental blocks were included in which oculomotor scanning behavior was tested by presenting a display of 6 items as shown in Fig. 4. The items consisted either of the complete letter "T" or a version of the letter with three missing pixels. The letters had a horizontal separation of 6 deg. Initially fixating the cross, the subjects had to scan the items (clock-wise or anti-clockwise) in order to report the number of complete "T"s in the display. Since the letters were small, the task required precise foveation of each item. Again, four subjects participated in the experiments.

The right part of Figure 4 presents results from a typical experimental subject. The open circles display the amplitudes of individual saccades to the stepping targets. It can be seen that in the adaptation phase starting after 120 control trials saccadic amplitude has decreased significantly. The small dots show the data from the scanning trials that were intermixed in the session. Obviously, the sizes of the scanning saccades are not affected by the adaptation, remaining approximately constant during the session.

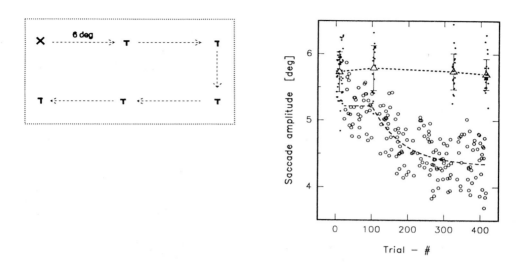

Fig.4 - Experiment 3. Left: Visual display in the "scanning" condition. Right:Saccadic
amplitudes as a function of trial number for the "double-step" (open circles) and the
"scanning" (dots) conditions.

Experiment 4

The previous experiment demonstrated that when stimulus-driven saccades are adapted, no
transfer to the intentionally controlled saccades in scanning occurs. The question arises
whether it is also possible to adapt scanning saccades without affecting the stimulus-driven
responses. In order to investigate this question the scanning paradigm as described before was
again used: starting from the fixation cross, the subject performed a self-paced scanning of
the display of the six items and reported the number of "T"s. In order to induce adaptation,
the whole display was shifted by 25% of the item separation (1.5 deg) during each scanning
saccade. The shifts occurred into the opposite direction to the saccade. The test condition
consisted of several experimental blocks within the session in which the subject had to follow
single steps of a target.

The data from four subjects are displayed in Figure 5. The small dots denote the sizes of
the saccades when scanning the letters, the open triangles display saccade magnitudes to the
stepping targets. Due to the intrasaccadic backward shifts of the scanning display, saccadic
amplitude is reduced quickly, leading to a total mean effect of 18.3% gain reduction. The test
condition with single steps of jumping targets also revealed a significant gain reduction,
however, but the mean amount of gain reduction was only 7.8%. This result demonstrates
that some transfer of adaptation of intentionally controlled saccades to target-elicited saccades
is indeed present, this transfer however being far from complete.

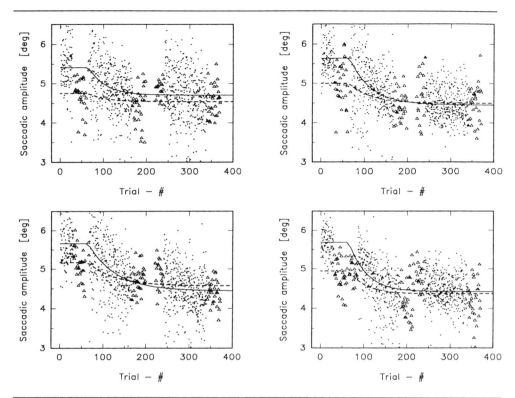

Fig.5 - Experiment 4. Saccade sizes as a function of trial number given for four subjects. Dots denote scanning saccades, open triangles represent saccades to steps of single targets. The curves represent the result of fitting exponentials to the data. Only the scanning saccades received conditioning stimuli.

Experiment 5

The final experiment tested the capability of the saccadic system to adapt specifically for the orbital eye position. It has been reported that saccadic adaptation can be highly specific to the (angular) direction of the eye movement, but (for a given direction) always generalizes with respect to saccade amplitude (Deubel, 1987). Experiment 5 now probed the capability of the saccadic system to specifically adjust the parameters of saccades starting from different orbital eye positions. I here present an example where the conditioning intrasaccadic target step only occurred with *centrifugal* saccades. So, for example, leftward saccades received the conditioning stimulus only when they started from or to the left of the primary position. The upper diagrams of Figure 6 show predicted saccadic gain as a function of starting position, assuming adaptation is perfectly specific to orbital eye position.

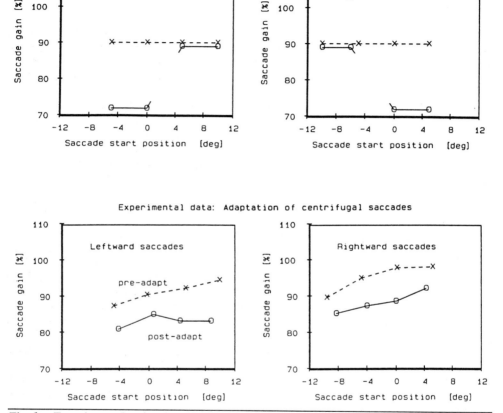

Fig.6 - Experiment 5. The diagrams on top present predicted saccadic gain values as a function of orbital eye position from where the saccade starts. Dashed curves: before adaptation. Solid curves: after selective adaptation of centrifugal saccades. Bottom: Experimental data from one subject before (dashed curves) and after (solid curves) selective conditioning of centrifugal saccades.

The lower diagrams of Figure 6 presents typical experimental data showing saccadic gain before and after more than 400 training epochs as a function of the starting positions of the saccades. It is obvious that the conditioning stimulus has induced a significant gain decrease, but the effect generalizes to a similar amount to all starting positions. In general, these experiments revealed no significant indication of a fast, orbital-position specific gain control.

This also holds for other types of sessions where only centripetal saccades were adapted, and for sessions where only saccades occurring in one hemifield received the conditioning intrasaccadic target step.

Discussion

The data shown here are another demonstration of the amazing adaptability of certain aspects of the saccadic response to the requirements presented by the environment. Some of the adaptations found are extremely fast, even when the system is confronted with the additional demand for disconjugate changes (van der Steen, 1993; Eggert & Kapoula, 1993; Eggert et al., 1994). In considering the general validity of ultra-fast adaptations as truly adaptive processes, it is important to be aware of the possibility that these changes could reflect the switching to a specific strategy rather than the plastic modification of oculomotor parameters. The results from Experiments 1 and 2 however seem to argue against the hypothesis that ultra-fast adaptations represent the cognitively controlled selection of response strategies. They demonstrate that visual properties of target or background cannot be used to switch between different sets of response parameters and suggest that there is no specificity of saccadic gain control with respect to visual aspects of the target. This is further emphasized by the finding that even the informed subjects are not able to voluntarily adjust their saccadic responses to the individual stimulus requirements.

The most important finding from this study is that saccadic adaptation induced when a subject has to track a stepping target does not transfer to saccades which are controlled internally, for example in a situation where the scanning of stationary items is required. Further, when internally controlled saccades are adapted, this conditioning transfers only to a small degree to stimulus-elicited saccades. The latter result is a confirmation of recent findings by Erkelens & Hulleman (1993) who also suggested from their data that internally triggered saccades can be selectively adapted, leaving stimulus-triggered eye movements largely unaffected. In their experiments subjects made saccades between two fixed stimuli. During the primary saccade, the target was systematically replaced by a third stimulus halfway between previous fixation and target. The test condition consisted in saccading to a target that appeared suddenly in the periphery. The adaptive decreases in saccade magnitude were 21% for the condition with the stationary targets, but only 5% in the test condition. Experiment 4 presented above extends these results to a more complex situation where the subjects are even not aware of the intrasaccadic shifts of the scanned items.

Other behavioral studies also hint at a dissociation of endogenously and exogenously controlled saccades. So, Lemij & Collewijn (1990) found that saccades to stationary targets are more accurate than saccades to jumping targets. Collewijn et al. (1988) demonstrated that the amplitudes of endogenous saccades do not show the 10% undershooting that is normally seen in saccades to onset targets. Further, it is known that internally generated saccades such as saccades directed to memorized targets exhibit slightly lower maximum velocities than target-driven saccades (Smit et al., 1987).

The results provide strong arguments that reflexive, stimulus-guided saccades and intentionally elicited saccades are controlled by different neural mechanisms. Separate and largely independent adaptive control systems seem to exist for the protection of both the visuo-motor reflex pathways and the pathways for intentionally induced eye movements.

Dependent on the conditions under which adaptation occurs, either endogenous or exogenous saccades are affected. It is very tempting to speculate that the neural substrate of the reflexive system might be the rather direct retino-collicular pathways, maybe including primary cortex, while the substrate for the intentionally controlled saccades may include the frontal eye fields. The findings provide new insight concerning the neuronal site where the adaptation may take place. They make it unlikely that the adaptive gain modification occurs in lower oculomotor centers such as the brainstem, where the collicular and frontal eye field signals have presumably already converged.

Further, these results raise new questions as to the role of the cerebellum in the *selective* adaptation of both response types. It is well-established that cerebellar structures, especially the cerebellar midline vermis and the fastigial nuclei are essential for the adaptive maintenance of saccadic accuracy (e.g. Ritchie, 1976; Optican & Robinson, 1980). From the experiments presented here the question arises whether these cerebellar mechanisms are responsible for maintaining proper functioning of *both* of the postulated pathways. Recent studies in cerebellar patients indeed suggest that cerebellar lesions may specifically affect only one of both response types (Straube, Deubel & Büttner, submitted).

Due to the inherent nonlinearities of the oculomotor plant, eye muscle innervation for saccadic eye movements strongly depends on the orbital eye position from which the saccade starts and the position to where it is directed. Therefore, saccadic adaptations that repair for lesion of the peripheral oculomotor system must necessarily take the orbital eye position into account (Optican & Robinson, 1980; Inchingolo et al., 1991). For this reason, the finding that position-specific adaptations cannot be induced within the limited time during which conditioning was provided is somewhat unexpected, suggesting that the adaptive mechanism responsible for the fast adaptations does not consider eye position. On the other hand, position-specific adjustments of the saccadic response have been observed in lesion studies that allow for a long-term oculomotor learning. Thus we may hypothesize that while fast adaptive mechanisms provide the means to rapidly adjust the magnitude of saccadic responses individually for each movement direction (Deubel, 1987), a second, much slower adaptive process allows compensation for position-dependent effects (Optican and Robinson, 1980; Inchingolo et al., 1991).

Acknowledgements: This work was supported in part by the "Science" program of the European Communities, contract No. SC1*-CT91-0747. I wish to thank Silvia Hieke for her patience and competence in conducting the experiments.

References

Collewijn, H., Erkelens, C.J. & Steinman, R.M. (1988). Binocular co-ordination of human horizontal eye movements. Journal of Physiology, 404, 157-182.

Crane, H. D. & Steele, C. M. (1985). Generation-V dual-Purkinje-Image eyetracker. Applied Optics 24, 527-537.

Dean P., Mayhew J.E.W. & Langdon P. (1992). A neural net model for adaptive control of saccadic accuracy by primate cerebellum and brainstem. In J.E. Moody, S.J. Hanson, & R.P. Lippmann (Eds.), Advances in Neural Information Processing Systems 4 (pp. 595-602). San Mateo, CA: Morgan Kaufmann Publishers.

Deubel H. (1987). Adaptivity of gain and direction in oblique saccades. In O'Regan J.K. & A. Levy-Schoen (Eds.), Eye Movements: From Physiology to Cognition (pp. 181-190). Elsevier North Holland.

Deubel, H. (1993). Context specificity of saccadic adaptation. Investigative Ophthalmology and Visual Sciences 34, Suppl., 3947.

Deubel, H. (in press) Selective adaptation of intentional and reactive saccades. In A.F. Fuchs, Th. Brandt, U. Buttner & D.S. Zee (Eds), Contemporary ocular motor and vestibular research: A tribute to David A. Robinson. Thieme Verlag, Stuttgart.

Deubel, H., Wolf, W. & Hauske, G. (1986). Adaptive gain control of saccadic eye movements. Human Neurobiology 5, 245-253.

Eggert, T. & Kapoula, Z. (1993) Fast disconjugate adaptations to anisokonia. Neuroscience Abstracts 18, 102.7.

Eggert, T., Kapoula, Z. & Bucci, M.P. (1994). Fast disconjugate adaptations of saccades: Dependence on stimulus characteristics. This Volume.

Erkelens, C.J. & Hulleman, J. (1993). Selective adaptation of internally triggered saccades made to visual targets. Experimental Brain Research, 93, 157-164.

Inchingolo, P., Optican, L.M., Fitzgibbon, E.J. & Goldberg, M.E. (1991). Adaptive mechanisms in the monkey saccadic system. In R. Schmid & D. Zambarbieri (Eds.), Oculomotor control and cognitive processes: Normal and pathological aspects (pp. 147-162). Elsevier Science Publ., Amsterdam.

Kommerell, G., Olivier, D. & Theopold, H. (1976). Adaptive programming of phasic and tonic components in saccadic eye movements, Investigative Ophthalmology, 15, 657-660.

Lemij, H. & Collewijn, H. (1990). Differences in accuracy of human saccades between stationary and jumping targets. Vision Research, 29, 1737-1748.

McLaughlin, S. (1967) Parametric adjustment in saccadic eye movements. Perception and Psychophysics, 2, 359-362.

Optican, L.M. & Robinson, D.A. (1980). Cerebellar-dependent adaptive control of primate saccadic system. Journal of Neurophysiology, 44, 1058-1076.

Ritchie, L. (1976) Effects of cerebellar lesions on saccadic eye movements. Journal of Neurophysiology, 70, 1246-1256.

Smit, A.C., van Gisbergen, J.A.M. & Cools, A.R. (1987). A parametric analysis of human saccades in different experimental paradigms. Vision Research, 57, 1745-1762.

van der Steen, J. (1993). Nonconjugate adaptation of human saccades: Fast changes in binocular motor programming. Neuroscience Abstracts, 18, 102.3.

FAST DISCONJUGATE ADAPTATIONS OF SACCADES: DEPENDENCY ON STIMULUS CHARACTERISTICS.

T. Eggert, Z. Kapoula and M.-P. Bucci

Laboratoire de Physiologie de la Perception et de l'Action,
College de France - CNRS, 15 rue de l'Ecole de Medecine, 75006 Paris, France

Abstract

Recent studies have shown that the conjugacy of saccades can be altered almost immediately. This is achieved by exposing normal subjects to an image which is unequal in the two eyes (aniseikonia). Such disconjugacy is stimulated by the resulting disparity. The present study examines the additional influence of monocular depth cues. The aniseikonia experiment was repeated with three images that contained different types of monocular depth cues: a grid with a clear linear perspective, a random-dot containing no monocular depth cues and a complex image containing a variety of overlapping forms. In all three experiments saccades became unequal in the two eyes immediately (from the first minute). The complex image produced the largest, even overcompensating saccade disconjugacy. The grid produced a saccade disconjugacy that was strongly position dependent. The random-dot produced a disconjugacy that was consistent and appropriate for the disparity. In the experiments with the random-dot and the grid image the disconjugacy persisted under subsequent monocular viewing; the large disconjugacy induced with the complex image did not persist. Hence, both the immediate disconjugacy and the learned, adaptive disconjugacy (evaluated from saccades under subsequent monocular viewing) are modulated by monocular depth cues. This suggests that the vergence oculomotor system is involved in this type of adaptation. Fast changes in saccade disconjugacy could be produced by a natural mechanism of saccade-vergence interaction similar to that occurring when we make saccades between targets that differ both in direction and in depth.

Keywords

Saccades, Disconjugate Adaptation, Disparity, Vergence, Aniseikonia

Introduction

Saccades between targets that do not require any change in vergence angle differ by less than 1 deg in the two eyes (Collewijn, Erkelens & Steinman, 1988). To maintain this remarkable conjugacy the central nervous system has to compensate for the changes and the eventual asymmetries of the oculomotor plants. Furthermore, such asymmetries may require adjustments of the central commands that are specific to the initial and final position of the saccade. Changes in the oculomotor plant during development are relatively slow and can be compensated by slow adaptive mechanisms. It has been recently demonstrated, however, (Van der Steen, 1992; Eggert & Kapoula, 1992) that the conjugacy of saccades can undergo very fast plastic changes. This was demonstrated by exposing normal subjects to large aniseikonia (inequality of the size of the image between the left and right eye). Eggert & Kapoula (1992) used a dichoptic viewing arrangement based on polarizers to provide an image to each eye. The right eye image was uniformly magnified by about 10%. Subjects were asked to saccade between different horizontal or vertical points of the image for 25 min. Because of the size inequality the distance between any two pairs of points was larger in the right eye. To overcome the resulting disparity the eye viewing

Eye Movement Research/J.M. Findlay et al. (Editors)

the larger pattern should make systematically larger saccades. Within 1 to 3 minutes subjects made unequal saccades so that binocular fixation could be obtained immediately afterward. Interestingly, the inequality persisted even under subsequent monocular viewing (in the absence of disparity cues).

An important question is whether the fast induced saccade disconjugacy involves a fast plastic adaptive capability to adjust the saccade pulse-step signals separately to each eye. Alternatively, such disconjugacy could be produced by a natural mechanism of saccade-vergence interaction stimulated by the disparity. Under natural viewing conditions where a large variety of depth cues are available (e.g. disparity, blur, linear perspective, size), quick and flexible saccade-vergence interactions are used to saccade between targets that differ both in direction and in depth (Enright, 1985; Erkelens et al., 1988; Zee et al., 1992). Enright (1985) and Erkelens et al. (1988) showed that even under monocular viewing saccades between such natural targets are disconjugate, indicating that monocular depth cues are involved. The question is to what extent the aniseikonia experiments stimulate the same mechanism. The present study tries to understand better the visual cues that are important for inducing fast changes in the binocular coordination of saccades. It focuses on the role of monocular depth cues. For this purpose we performed three aniseikonia experiments using three images that differed in the type of monocular depth cues they contained: cues based on linear perspective, overlap and object size were very different between images. In contrast the image size inequality was kept the same in all three experiments.

Methods

Adaptation Paradigm

The subject was seated at 1 m from a flat translucent screen. Two projectors were used to provide each eye with its own image. The beams of the two projectors were polarized 90 deg apart. Subjects viewed through filters also polarized 90 deg. The two images were centered on the screen as shown in Fig. 1B. One image subtended 35 deg the other 38.5 deg (10% uniform magnification). In all three experiments the right eye saw the larger image. The subject was asked to saccade back and forth between the center and different points of the image located approximately along the horizontal and vertical meridians. A typical exploring sequence was: from center to middle left, from there to the left border and then back to center; similar for right, up and down. Thus all centrifugal saccades had amplitudes of about 8 deg, all centripetal saccades were about 15 deg. The total duration of this training, adapting period was 25 min. Subjects were asked to change the direction of saccades (horizontal or vertical) every 6 minutes. Eye movements were recorded binocularly not only while the subject viewed the unequal images but also while viewing the smaller image monocularly. This open loop condition (absence of disparity cues) was recorded both before and after training.

The image size inequality produced disparities requiring the following changes in ocular alignment. Fixating the center of an image located at 1 m the vergence angle was 3.4 deg. Due to the aniseikonia all rightward saccades required a divergent disconjugacy and all leftward saccades a convergent disconjugacy. Since the tangent viewing screen was flat and relatively close to the subject (1 m), looking at the periphery required divergence relative to looking at the center. The two requirements added for saccades in the right

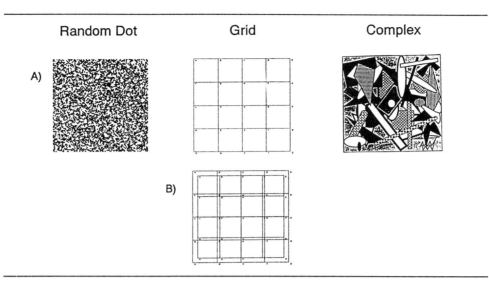

Fig. 1 - A: Images used to perform three experiments. B: In each experiment a dichoptic setup enabled to provide an image to each eye. In all experiments the right eye image was magnified by 10% The two images were centred on the screen as shown for the experiment with the grids.

field but subtracted for saccades in the left field. Thus, on average saccades in the right field (leftward or rightward) required a disconjugacy of 10.7% of the saccade amplitude, while the requirement for saccades in the left field was only 7.4%. This distribution of horizontal disparities is identical to that of a plane tilted around a vertical axis by 58 deg. Nevertheless subjects did not perceive such tilt. Vertical disparities or monocular depth cues could suppress the tilt perception that could have resulted from the horizontal disparities alone.

Fig. 1A shows the three images used in the three experiments. The random-dot image is the simplest stimulus. It contains no large forms, no contours and no monocular depth cues. The grid contains a square structure which is a strong geometric depth cue indicating a frontoparallel plane. The complex image has a variety of monocular depth cues such as overlapping or linear perspective that create a very inhomogeneous distribution of depth; this image was created and first used by F. Miles at the LSR/NIH (personal communication). Thus although all three images contained the same distribution of binocular disparity due to aniseikonia they differed in the type of monocular depth cues.

Subjects

Four subjects with visual acuities of 20/20 or better and perfect binocular vision (TNO test, better than 60 sec of arc) participated each in three experiments testing the three images. Experimental runs were always separated by at least one week. To distinguish between image dependencies and eventual long-lasting training effects the order of experiments was counterbalanced among the subjects. The data presented here were collected in one of these subjects in the following order: random-dot, grid, complex. Only results for horizontal saccades are presented. This subject was unexperienced with eye movement recording and ignored the exact purpose of the experiment. His results are representative of the group.

Eye movement recording - Analysis of data

Stimulus presentation and data collection were directed by REX, a software developed for real time experiments and run on a PC (HP RS/20). Eye movements were recorded with the search coil-magnetic field method (Collewijn et al., 1975; Robinson, 1963). Calibrations were determined for each individual eye when it alone viewed a pair of nonius lines that stepped horizontally. The eye position signals were filtered with a bandwidth of 0-200 Hz and digitized with a 12-bit analog-to-digital converter sampling each channel 500 times per second.

Calibration factors were extracted from saccades to the nonius lines. A polynomial function with five parameters was used to fit the calibration data. Thus, measured eye position data were corrected for the intrinsic sine non-linearity of the coil system and for

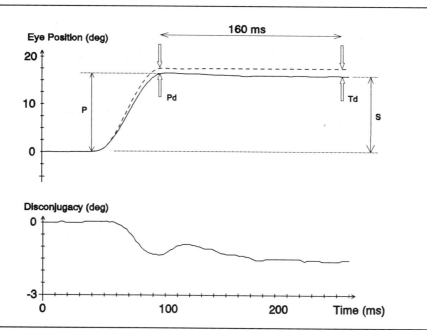

Fig. 2 - Typical binocular recording of a saccade in the presence of unequal images. The dashed line is the right eye, the solid line is the left eye. The distance indicated by P is the pulsatile component of the saccade; the distance indicated by S is the step component. Arrows indicate the end of the saccade and the first 160 ms post-saccadic period. The lower trace shows the disconjugacy of the saccade (left - right eye difference).

the tangent screen effect. Off-line computer algorithms based on standard velocity and acceleration criteria were used to determine the times of saccade onset and offset. Saccade offset was taken as the time when eye velocity dropped below 5 deg/sec. The accuracy of these computer-generated marks was verified by the human editor. The post-saccadic eye drift was determined for a period of 160 ms after saccade offset (Fig. 2). For each individual saccade we measured the disconjugacy (left - right eye difference) of the amplitude of the saccade (Pd) and the total or steady-state disconjugacy after 160 ms (Td). These disconjugacies were expressed as a percentage of the saccade amplitude represented by the value averaged over the two eyes.

Results

Qualitative results

Fig. 3 shows the disconjugacy of saccades during the first six minutes of exposure to unequal random-dot patterns. Saccades are divided into four groups according to their direction. The starting and ending position of each group of saccades are illustrated in the right-up box. Traces with up inflection indicate a convergent disconjugacy, traces with down inflection indicate divergent disconjugacy. In each panel, most of the saccades

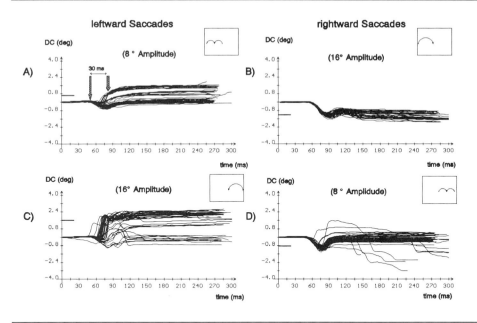

Fig. 3 - Disconjugacy (left-right eye difference) during the first 6 min of exposure to unequally sized random dot patterns. A,B,C,D: Four groups of horizontal saccades starting and ending at the positions illustrated in the right-up square box. For all traces saccade onset is at 50 ms. Arrows in A, indicate the onset and the offset of the saccades. The horizontal line on each ordinate indicates the disconjugacy required for each group of saccades.

exhibit a disconjugacy in the appropriate direction: disconjugacy is convergent for leftward saccades (Figs. 3A and 3C) and divergent for rightward saccades (Figs. 3B and 3D). Its amplitude corresponds quite well to the requirement (indicated by the horizontal line). Convergent disconjugacy is larger but more variable than divergent disconjugacy (Fig. 3A and 3C vs Figs. 3B and 3D). The traces in figure 3A form several distinct populations. A first population shows only a transient divergent-convergent disconjugacy (described for normal saccades by Kapoula et al., 1987; Collewijn et al., 1988; Zee et al., 1992); these saccades start at 8 deg and end at 16 deg to the right. The remaining traces in Fig. 3A are from saccades that start from the center and end at 8 deg to the right; they all show sustained convergent disconjugacy. Thus the induced disconjugacy is position dependent. The variability in the disconjugacy seen in Fig. 3C is due to trial-by-trial fluctuations.

Despite this variability, Fig. 3 shows remarkable disconjugacy in the appropriate direction for the majority of the saccades. This occurs almost immediately, within the first 1 to 3 minutes of exposure to unequal images. Most of the disconjugacy is accomplished during the saccade itself. In Fig. 3A, arrows indicate the onset and offset of the saccade, a duration of approximately 30 ms. The onset of the saccade causes a quick change in ocular alignment which is almost finished at the end of the saccade. Disconjugacy in post-saccadic eye drift is negligible. Thus the quantitative results presented next show the intra-saccadic disconjugacy only.

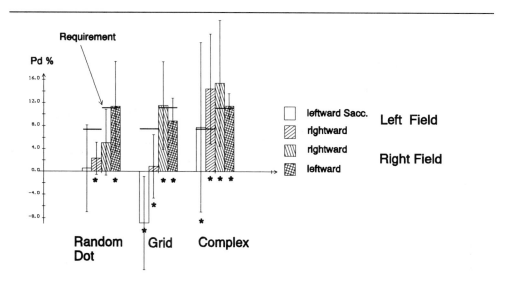

*Fig. 4 - Actual intra-saccadic disconjugacy (Pd in %) while viewing the unequal images. Positive bars indicate a disconjugacy in the appropriate direction. The vertical lines are ± 1 SD. Horizontal lines show the required disconjugacy. All leftward saccades require a convergent disconjugacy, all rightward saccades a divergent disconjugacy. *: significant difference from normal saccades recorded before training (at the P < 0.05 level, t test).*

Quantitative results

Fig. 4 shows the actual mean intra-saccadic disconjugacy for all horizontal saccades recorded during the 25 min of training. For each experiment saccades are divided into 4 groups according to their initial and final position (using the same classification as in Fig. 3). For all experiments and for most groups of saccades the actual disconjugacy is significantly different from normal values; this difference is always in the appropriate direction (indicated by an *). For the experiment with the random-dot image the actual disconjugacy is always in the appropriate direction; its amplitude is considerably smaller for saccades in the left field. For the grid, only saccades to the right field show appropriate divergent or convergent disconjugacy. For leftward saccades in the left field the difference from normal is in the right direction, but the actual disconjugacy is still wrong. Thus the disconjugacy induced with the grid is strongly asymmetric. The complex image produces the largest disconjugacy. For rightward saccades in either field the divergent disconjugacy exceeds the requirement. Such overcompensation is not seen in the other two experiments using the random-dot pattern or the grid. The difference between the required and the actual disconjugacy (for all horizontal saccades) is 0.7 deg for the random-dot image, 0.9 deg for the grid and only -0.2 deg for the complex image. Thus the compensation of the disparity is best for the complex image.

Fig. 5 shows the mean changes in disconjugacy for saccades under monocular viewing (in the absence of disparity cues), recorded before and after training. Interestingly, the large disconjugacy observed in the presence of the complex images does not lead to

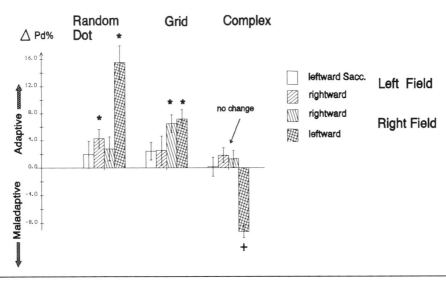

*Fig. 5 - Pre-post training changes in intra-saccadic disconjugacy (Pd in %) for saccades under monocular viewing. Positive bars indicate a change in the appropriate direction. The vertical lines indicate ± 1 SD *: a change from normal which is statistically significant (at P<0.05 level, t test). +: a significant change in the wrong direction.*

consistent changes for saccades under monocular viewing. The pre-post training changes shown in Fig. 5 are not statistically significant or are in the wrong direction (indicated by a +). In contrast, significant changes occur for the experiments with the random-dot and the grid (indicated by an *). Leftward saccades in the right field show the largest change, particularly in the experiment with the random-dot image.

Summary and discussion

The present study demonstrates that exposure to unequal images can induce immediately disconjugacy in the amplitude of the saccades that persists even under subsequent monocular viewing. This suggests the presence of an adaptive mechanism capable to alter the binocular coordination of saccades very quickly. These findings confirm a prior study (Eggert & Kapoula, 1992).

The new findings concern the dependency of the induced disconjugacy on the visual characteristics of the image. Disconjugacy was induced with all three images. Its amplitude and consistency, however, depended on the type of the image. The subject whose data were presented here was first examined with the random-dot image, then with the grid (a week later) and finally with the complex image (6 weeks later). Yet his results in Fig. 4, show larger and more consistent disconjugacy for the first experiment using the random-dot pattern than for the second experiment with the grid. Furthermore, Fig. 5 shows no adaptive persistence of the disconjugacy in the experiment using the complex image which was the last experiment. All subjects showed similar image dependencies. Thus the results presented here are due to the characteristics of the image and not to a bias or transfer from one experiment to another.

How do we explain these findings? Fast disconjugate changes could be achieved by an adaptive mechanism capable to readjust the pulse-step signals of the saccadic innervation separately to each eye. Alternatively, such disconjugacy could be produced by a natural interaction between the saccade and the vergence oculomotor system. Such interaction occurs naturally when we saccade between targets that differ both in depth and in direction.

Fig. 6 shows a block diagram of the saccade-vergence interaction hypothesis. We suggest that the immediate changes in intra-saccadic disconjugacy are due to a fast vergence response to the visual disparity detected in the periphery before each saccade. This immediate vergence command could be influenced by monocular depth cues. To explain the persistence of intra-saccadic disconjugacy in the absence of disparity cues (saccades under monocular viewing), we hypothesize the existence of a learning mechanism using a second vergence circuit. This learning mechanism is supposed to associate with each saccade a preprogrammed, fast vergence command. What is the stimulus driving this learning mechanism? We suggest that post-saccadic fixation disparity which is coupled with the saccade, is the most important error signal used to drive the learning mechanism. Again, monocular depth cues can influence the effectiveness of the disparity error in driving the learning mechanism.

Vergence movements are thought to be slow. However, Zee et al. (1992) reported an acceleration of vergence when it occurs during a saccade (horizontal or vertical). They proposed a model of saccade-vergence interaction based on the idea that the omnipause neurons serve as a link between saccade and vergence bursters. Evidence supporting this

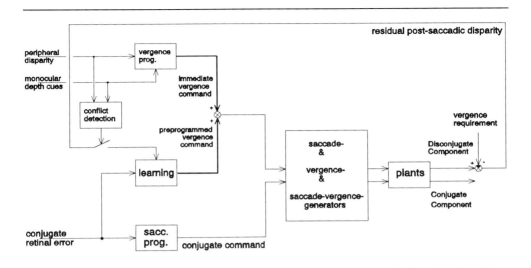

Fig. 6 - Conceptual diagram to explain fast induced changes in saccade disconjugacy. Two different types of vergence commands serve as input to the brainstem saccade, vergence generators and, may be, to a special, saccade-vergence generator, capable to produce fast vergence. The immediate vergence command is driven visually by the disparity detected in the periphery before each saccade. The saccade-initiated, preprogrammed, vergence command is driven by the post-saccadic disparity error detected consistently over time. The preprogrammed vergence command is under adaptive, learning control. The post-saccadic disparity error may become ineffective in case of a sensory conflict between disparity and monocular depth cues.

model comes from the study of Mays and Gamlin (1992). We suggest that both the preprogrammed and the immediate vergence commands use this special saccade-vergence brainstem circuitry to produce the fast vergence change occurring during the saccade. The term saccade-vergence generator in Fig. 6 refers to this idea.

The differences observed between the three types of images can be explained the following way. As already mentioned, the complex image contained a variety of monocular depth cues corresponding to an inhomogeneous distribution of depth. Such cues could enhance the immediate, vergence responses. In the presence of a robust, visually-driven, saccade-vergence capability a learning adaptive process would be unnecessary.

The random-dot image did not contain monocular depth cues. Because of its homogenous structure extraction of disparity did not necessarily involve complex preprocessing such as image segmentation. It could be simply based on the matching of the two retinal images around each fixation point. The evaluation of the disparity would be more accurate and consistent over time. Consequently, the residual disparity error after the saccade would be more consistent and therefore would drive the learning mechanism more efficiently.

The square structure of the grid provides a strong cue for looking at a flat frontal plane. Under natural viewing a tangent flat screen requires divergence for centrifugal saccades and convergence for centripetal saccades. Thus the aniseikonia and the tangent screen disconjugacy requirements were in conflict for saccades in the left field. This conflict was present for all three experiments. We suggest, however, that it was stronger for the grid because of its square structure. This would explain the strong field asymmetry of the disconjugacy found with the grid.

In summary, this study shows that fast disconjugate adaptation of saccades depends on the structure of the image, particularly on monocular depth cues. Monocular depth cues are important in evaluating depth and are used by the vergence system. We suggest that the mechanism producing this adaptation is based on a fast saccade-vergence interaction similar to that occurring when we make saccades to targets that differ both in depth and in direction.

Acknowledgments

The authors thank Prof. H. Colewijn and Dr. J. Van der Steen for suggesting the aniseikonia paradigm. Dr. T. Hain provided the initial PC version of the REX, MARK and PRINT software. This research was supported by the European Community contract SCI*-CT91-0747 (TSTS) and a grant from the french CRAMIF. Maria Pia Bucci was supported by a COMETT LI.SA. grant and then by an ESF/ENP fellowship.

References

Collewjin, H., Van der Mark, F. & Jansen, T.C. (1975). Precise recordings of human eye movements. Vision Res. 15, 447-450.

Collewijn, H., Erkelens, C.J. & Steinman, R.M. (1988). Binocular co-ordination of human horizontal saccadic eye movements. J. Physiol. 404, 157-182.

Eggert, T. & Kapoula, Z. (1992). Fast disconjugate adaptations to aniseikonia. Neurosci. Abstr. 18,102.7

Enright, J.T. (1986). Facilitation of vergence changes by saccades: Influences of misfocussed images and of disparity stimuli in man. J. Physiol. 371, 69-87.

Erkelens, C.J, Steinman, R.M. & Collewijn H. (1989). Ocular vergence under natural conditions. II. Gaze shifts between real targets differing in distance and in direction. Proc. Royal Soc. Lond. B 236, 441-465.

Kapoula, Z., Hain, T.C., Zee, D.S. & Robinson, D.A. (1987). Adaptive changes in post-saccadic drift induced by patching one eye. Vision Research 27, 1299-1307

Robinson, D.A. (1963). A method of measuring eye movements using a scleral search coil in a magnetic field. IEEE Trans. Biomed. Eng. 10, 137-145.

Van der Steen, J. (1992). Nonconjugate adaptation of human saccades: Fast changes in binocular motor programming. Neurosci. Abstr. 18, 102.3

Zee, D.S., Fizgibbon, E.J. & L.M. Optican. 1992. Saccade-vergence interactions in human beings. J. Neurophysiol. 68, 1624-1641.

Mays L.E., Gamlin P.D.R. (1992) Role of omnipause neurons in saccade-vergence interactions. Neurosci. Abstr. 19.2

VISUAL MISLOCALIZATION IN MOVING BACKGROUND AND SACCADIC EYE MOVEMENT CONDITIONS

Hitoshi Honda

Department of Psychology, Faculty of Humanities, Niigata University, Niigata 950-21, Japan

Abstract

There is a possibility that visual mislocalization of targets flashed at the time of a saccadic eye movement is due almost entirely to the shift of the retinal image of the background. To clarify this matter, errors in target localization were analyzed in both saccadic eye movement and moving background conditions. In the latter condition, subject kept fixating and the visual background made a saccadic movement. In both conditions, a horizontal luminous scale was used as the background and the subject reported the position on the scale that the target appeared to occupy. Large localization errors were shown in both conditions. However, the pattern of error for the moving background condition was distinctively different from that for the saccadic eye movement condition, suggesting that the shift of the background image is not sufficient to explain the localization error in the saccadic eye movement condition.

Keywords

saccade, visual stability, image displacement, signal, copy.

Introduction

A visual stimulus flashed at the time near a saccadic eye movement is perceived at a different position from its actual position in space. This phenomenon is observed when visual stimuli are presented on an illuminated structured background (Bischof & Kramer, 1968; Honda, 1993; Mateeff, 1978; O'Regan, 1984) as well as when they are presented in the dark (Honda, 1989, 1990, 1991; Matin, Matin, Pearce, 1969; Matin, Matin & Pola, 1970) In the former condition, the retinal image of the background scene rapidly shifts on the retina with the movement of the eye. Therefore, there is a possibility that the mislocalization of visual stimuli observed in the illuminated background conditions is irrelevant to the eye movement itself, but rather is produced by a complex retinal event generated by quick movements of retinal images contingent upon the eye movement (O'Regan, 1984).

In 1965, Sperling and Speelman reported that the position of a line was mislocalized when it was flashed at a time near the rapid displacement of the background scene (Sperling & Speelman, 1965). Their finding was later confirmed by MacKay (1970). In MacKay's experiment, as was the case in Sperling and Speelman's, a scale pattern was moved horizontally in the visual field, and a small flash stimulus was presented at various points in time before, during, or after the displacement of the scale pattern. The subjects were asked

Eye Movement Research/J.M. Findlay et al. (Editors)

to report the position on the scale that the flash appeared to occupy. According to MacKay, mislocalization was confined chiefly to flashes presented in the 50 msec or so before and during displacement. From this observation, together with the finding by Sperling and Speelman, MacKay argued that the oculomotor system need not be implicated in order to account for mislocalization during saccadic eye movement. O'Regan (1984) also presented a similar interpretation. He argued that mislocalization is caused by the difference between the foveal and the peripheral position sense, and by the smearing of the retinal image of the background.

This kind of explanation is very persuasive because the time course of mislocalization in the moving background condition reported by MacKay seems very similar to that reported for saccadic eye movement conditions by Mateeff (1978). However, it is not conclusive. This is because, in the MacKay and Sperling and Speelman experiments, target stimuli for visual localization were always presented at the midpoint of a moving scale used as a background scene. The subjects were asked to fixate the midpoint of the scale (in the MacKay experiment) or the left side of the scale (in the Sperling and Speelman experiment), and then the scale moved horizontally. If we focus on the target position on the retina, we can easily understand that these experimental situations corresponded to the saccadic eye movement conditions in which a target was always presented at the position of the original fixation point (MacKay's experiment) or at the midpoint between the fixation point and the goal of the saccade (Sperling and Speelman's experiment). However, it should be noted here that, in the saccadic eye movement condition, the size and the direction of mislocalization largely depend on the actual position of the target. Honda (1993), for example, investigated the accuracy in judging the position of targets briefly presented at a time of a saccade at various positions scattered two-dimensionally on a dimly illuminated structured background, and demonstrated that the spatio-temporal pattern of errors in judging the target position largely varied with the actual target position on the illuminated background. For example, when a target was presented at a position beyond the saccade destination, mislocalization occurred exclusively in the direction opposite to the saccade. On the other hand, when a target was presented on the opposite side to the saccade destination across the original fixation point, errors in the saccade direction were dominant.

Therefore, if it is true that mislocalization in the saccadic eye movement condition is caused by retinal image displacements which are not necessarily contingent upon saccadic eye movements, then it will be expected that, irrespective of the target position in space, the pattern of errors in target localization will be almost the same between the moving background and the saccadic eye movement conditions. The present study was conducted to directly examine this prediction, and to explore the mechanism responsible for producing mislocalization in the saccadic eye movement condition.

Methods

Saccadic eye movement condition

A subject was seated with the head fixed by a chin- and forehead rest. Horizontal movements of the right eye were monitored by a photo-electric method. The subject's right eye was illuminated by an i.r. light-emitting diode (Toshiba, TLN103), and the reflected light from the two positions of the lower limbus (iris-sclera boundaries in 4 o'clock and 8 o'clock positions) was collected by a phototransistor (Toshiba, TPS601), and horizontal eye movements were monitored by recording the difference between the two phototransistor outputs.

A horizontal luminous scale with divisions was used as a background scene (Figure 1). Its image [white (12 cd/m^2) on a dark ground (5cd/m^2)] was rear-projected on a screen (50cm x 75cm) placed 57cm from the subject's eye. On each trial, a buzzer warning signal was given, and then a fixation point (red LED, 0.3 deg in dia, 20 cd/m^2) was presented at the zero scale division. The duration of the fixation point varied randomly from trial to trial between 1.0 and 2.0 sec. The subject was asked to binocularly keep watching the fixation point.   At the offset of the fixation point, a small visual cue stimulus for saccades was presented for 20msec, at the position of 8 deg right of the fixation point, i.e., at the eight scale division   The visual cue consisted of two vertically arranged rectangular red LEDs (0.1 deg x 0.3 deg, 18 cd/m^2), the distance between the center of the LEDs being 0.4 deg. The subject was asked to make a horizontal saccade (primary saccade) toward the visual cue. Because the duration of the visual cue was short (20 msec), it disappeared before the beginning of the primary saccade. At various points in time before, during, or after the primary saccade, a vertical rectangular visual stimulus (0.3 deg x 1.5 deg), which was illuminated by an electronic flash tube (Nisshin, HD-100), was presented at the -4, +4, or +12 scale division, and used as a target for visual localization. The subject verbally reported the scale division on which he had seen the target. For example, he reported "minus five", "nine point five", or "I did not see". The fixation point and the visual cue for saccade were set on a black board placed at a different position from the screen, and seen by the subject through a hail-mirror set before the subject's eye. By this method, these stimuli were presented as an optical image on the background scene. To present the target during or after the saccade, the output from the eye movement monitor apparatus was fed into a differential circuit that triggered the electronic flash tube. Targets before the saccade were presented by pre-setting a shorter time interval than a normal saccade latency (200 msec) between the target and the visual cue for eliciting the saccade.

In addition to the saccade condition described above, localization was also examined in a condition in which the target was presented when the eye remained still. In this control condition, either the fixation point or the cue for saccade was presented for 1.8 sec, and the subject was asked to keep watching these stimuli. Just after the offset of these stimuli, a flash target was presented. The subject made a saccade to the target, and reported its apparent position.

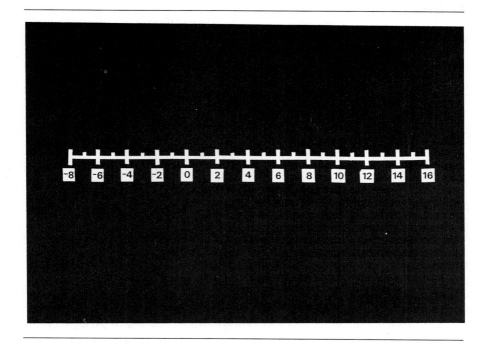

Figure 1. A background scene used in this experiment. The horizontal length of the scale (from the -8 to the +16 scale division) was 24^0 when it was projected on the screen.

Moving background condition

In this condition, the background scene was rapidly displaced by moving a mirror placed between the screen and the projector. The mirror was mounted on a galvanometer, and moved by a computer. As was the case in the saccadic eye movement condition, the subject was asked to keep watching the fixation point. Two hundred msec after the offset of the fixation point, the background scene was rapidly displaced horizontally 8 deg to the left. (The 200msec interval was chosen as an equivalence to normal saccade latencies.) The duration of the displacement was 30 msec, approximately equal to the average saccade duration observed in the saccadic eye movement condition. At various point in time before, during, or after the beginning of the background displacement, a target was flashed at the -4, +4, or +12 scale division. The subject verbally reported the scale division on which the target appeared. Localization accuracy was examined also in a condition in which the background scene did not move. In this control condition, the target was flashed while the subject was watching the fixation point.

Subjects and procedure

Four subjects participated in this experiment. Subject HH was the author and the remaining three subjects were male university students. Each served as a subject for 6 days, 3 days for the saccadic eye movement conditions and 3 days for the moving background condition. On each day, the target's position was restricted to one of the three positions (-4, +4, or +12 on the scale). In the saccadic eye movement condition, eight sessions of the experimental (saccade) condition and one session of the control condition were conducted on each day. Each session consisted of 17 experimental trials or 14 control trials. The timing of the target's presentation was randomized within each session. In the moving background condition, a total of 136 trials, divided into eight sessions, were conducted on each day. In each session, 16 experimental (moving background) trials and one control trial were conducted in random order. The timing of the target's presentation also was randomized within each session.

Results

Saccadic eye movement condition

Primary saccade

The subject's eye movement was analyzed by a high-speed digital storage scope (Iwatsu, DS-6121A). In the saccadic eye movement condition, the subjects sometimes failed in making a saccade in the way required. That is, they made saccades with extremely short (<50msec) or long (>300msec) latencies. In these cases, the target was not presented. The frequencies (percentage) of these expected or delayed saccade responses varied with the subjects ranging from 2.4% (subject HH) to 14.2% (subject HU). These trials were excluded from the following data analysis.

When a target was presented immediately after the presentation of the visual cue for eliciting a primary saccade, the eye sometimes moved directly to the target. In this case, therefore, a saccade to the visual cue did not occur. This type of response was observed in about 10% of the trials. In the remaining trials, the expected primary saccades of about 8 deg were observed. The means of the latency and the duration of the primary saccades were about 200 msec and 32 msec, respectively.

Visual Localization

Figure 2 shows the localization errors, i.e., the discrepancies between the actual target position and the perceptually judged position, as a function of the time interval between the onset of the primary saccade and the occurrence of the target. It is evident that the time course of the error largely varied with the position at which targets were actually presented. When targets were presented at a position opposite to the saccade direction (i.e., on the -4 scale division), all subjects showed a localization error in the saccade direction. The localization error was shown when the target was presented just before or during the primary saccade. When targets were presented at a position located between the original fixation point and the saccade destination (ie., on the +4 scale division), the error was relatively small. At this position, small errors in the saccade direction were shown when the target was presented immediately before the saccade onset, while errors in the direction opposite to the saccade direction were observed when the target was presented at the end of the saccade. Finally, when targets were presented at a position beyond the saccade destination (i.e., on the +12 scale division), all subjects except subject AT showed errors only in the direction opposite to the saccade when the target was presented at the end of the saccade. Subject AT exceptionally showed small errors in the saccade direction as well as errors in the direction opposite to the saccade.

In summary, the results in the saccadic eye movement condition were approximately the same as those I previously reported elsewhere (Honda, 1993), in which it was demonstrated that localization errors were largely dependent upon the actual target's position in the illuminated visual field.

Moving background condition

Eye movements

Eye movement recordings showed that all subjects kept watching the position of the original fixation point during the trials. Usually, the eye moved toward the apparent target position more than 300msec (in subjects AT and HH) or 200msec after the movement of the background (in subjects NN and HU).

Visual localization

Figure 3 shows the results obtained in the moving background condition. When targets were presented on the -4 scale division, errors in both saccade direction and in the opposite direction to the saccade were observed. However, in two subjects, MM and HU, the errors in the saccade direction were small or almost absent. When targets were presented on the +4 scale division, all subjects showed a bipolar pattern of mislocalization consisting of errors in the saccade direction and ones in the direction opposite to saccade. Similar results were shown for the targets presented on the 12 scale division. In this case, however, the size of the error remarkably increased. In three subjects, AT, MN, and HU, the error in the saccade direction appeared about 100msec before the saccade onset.

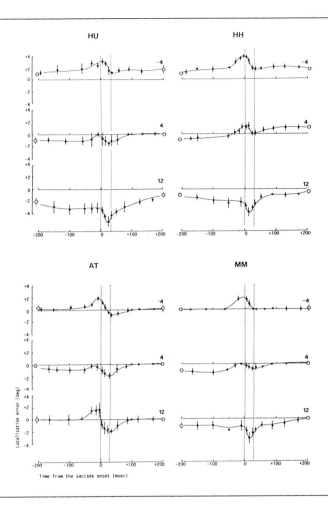

Figure 2. The time course of mislocalization in the saccadic eye movement condition. For each subject. results are shown separately for each actual target position (-4, +4, and +12). The abscissa indicates the time interval (msec) between saccade onset and target presentation. A minus sign in the abscissa shows that targets were presented before the saccade onset. The ordinate indicates the size of mislocalization (deg). A plus sign in the ordinate shows mislocalization in the saccade direction (rightward), and a minus sign mislocalization in the direction opposite to the saccade (leftward). Each dot in the figure represents the average error (and the SD) of about 5-25 trials. The error curves were fitted by eye based on the average errors (dots). Open circles indicate the results on control trials in which the subjects kept watching the fixation point (left circle) or the visual cue for saccade (right circle). Vertical dotted lines indicate the mean duration of the saccades.

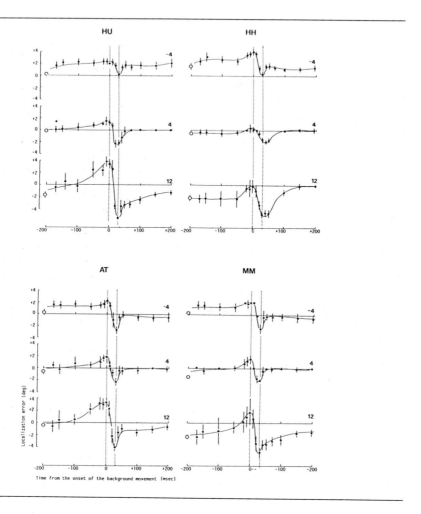

Figure 3. Time course of mislocalization in the moving background condition. Notations are the same as in Figure 3 except the following. The abscissa indicates the time interval between the beginning of background movements and the target presentation. Each dot indicates the average of 8 trials. Open circles in the figure indicate the results in the control condition in which the background did not move. Vertical dotted lines show the movement time of the background.

Discussion

Comparison of saccadic eye movement and moving background conditions

In this study, it was assumed that the retinal event was the same between the two experimental conditions. If the localization error observed in the two conditions was exclusively generated by a retinal event, i.e., rapid displacements of retinal images produced either by saccadic eye movements or by movements of the background scene, then it should be expected that the pattern of errors was the same between the two conditions. However, this was not the case. The results clearly showed that the pattern of errors was not the same between the two conditions. The discrepancy was predominant in particular when the target was presented on the -4 or the +12 scale division. Therefore, it is impossible to explain that errors in the two conditions were generated by the same mechanism: a complex retinal event produced by a rapid displacement of retinal images.

It should be noted here that some authors suggested that a pattern of errors similar to one reported for saccadic eye movement conditions was observed also in moving background conditions (O'Regan, 1984; Sperling & Speelman, 1965). However, their finding is not conclusive. In Sperling and Speelman's study, targets were presented exclusively at the position between the original and the final fixation points (i.e., on the 4 scale division in the present study). As shown in Figures 2 and 3, at this target's position, the shape of error curves was about the same between saccadic eye movement and the moving background conditions. Therefore, Sperling and Speelman's experiment should be replicated using other target positions than they employed in their pioneering In O'Regan's (1984) study, targets were presented only during the displacement of the background scene, and never presented before or after the displacement. Therefore, the exact time course of localization error is not clear, and it is hard to make a comparison between the moving background and the saccade conditions.

Origins of localization error

As described above, the pattern of localization error shown in the saccadic eye moment condition was substantially different from that in the moving background condition. Does this mean that there is a possibility that mislocalization observed in the each of two conditions was caused by a distinctively different mechanism? Before considering this question, it seems necessary to examine the explanations proposed so far in the earlier studies.

As regards mislocalization in moving background conditions, MacKay (1970) explained this as follows. "The location of a flashed image to its background is determined by an interaction between the neural signals generated by each, which interaction takes an appreciable time to complete. If during this time the retinal image of the background shifts to a new position, the integrative process will have two different background signals to cope with, each making its own contribution to the total weight of evidence respect to flash location." For example, "the later the flash comes, before the moment of transition, the greater will be the weight attached to the new scale-position as compared with the old." (MacKay, 1970, p.732) Thus, the apparent position of a flash changes as a function of the timing of flash presentation relative to the background's displacement. Furthermore, another important implication of the MacKay's argument is that the position of the flash is determined by the background scene before and after its shift, not by the background during displacement.

The fact that the image of the background during displacement is irrelevant to mislocalization was established by Sperling (1990). In one of his experiments, the background scene was distinguished during its displacement. Even in this condition without image smearing on the retina, mislocalization very similar to the one in normal moving-background conditions was observed. The idea that mislocalization is produced by the successive appearance of the background in two different positions seems plausible because subjects usually report that they cannot perceive the scale during its shift owing to its smearing. However, this explanation has a serious problem. If this explanation is correct, then it is expected that the pattern of errors will be the same irrespective of the position at which flashes are presented. Strictly speaking, this was not the case in the present study although the basic shape of error curves seems approximately the same.

Next, why were the flashes mislocalized when they were presented at the time of saccadic eye movements? In this condition, too, a rapid displacement of images occurred on the retina. In addition, neural activities of the oculomotor system for generating a saccade were involved. Mislocalization occurs also when targets are presented in the dark (Honda, 1989, 1991; Kennard, Hartman, Kraft, Glaser, 1971; Matin et al., 1969, 1970) as well as when they were presented on an illuminated background. In the case of experiments conducted in the dark, there is almost no retinal event such as image displacements of the background scene. Therefore, the mislocalization in the dark seems to be caused primarily by a sluggish activity of the extraretinal eye position signal (EEPS), resulting in a failure in completely canceling the shift of images of flash targets on the retina by the EEPS.

Then, why was a flash mislocalized when it was presented on an illuminated background on the occasion of saccade generation? Honda (1993) examined the accuracy of localization of flash targets presented at the time of a saccade at various positions scattered two-dimensionally on a dimly illuminated structured background, and argued that localization error is primarily produced by a sluggish activity of the EEPS, and that the error is partially corrected by visual cues from the illuminated background and modified by the subject's selective inattention to image displacements. In addition, it was assumed that the rapid displacement of the retinal image of the background has no substantial role in producing mislocalization. According to this explanation, it is expected that, in the moving background condition without saccadic eye movements, there will be shown no or, if any, small localization errors, because it was speculated that the primary factor for mislocalization is the sluggish activity of the EEPS. However, the present study did not support this prediction; mislocalization was larger in the moving background condition than in the saccadic eye movement condition!

Tentative explanation

Why did the subject mislocalize the flash when it was presented on an illuminated background in the saccadic eye movement condition? It should be noted here that, in both moving background and saccadic eye movement conditions, the image of the background rapidly moved on the retina. That is, the retinal event was the same between the two conditions, and the only difference was the involvement of neural activities of the oculomotor system (EEPS) in the saccadic eye movement condition. This leads us to speculate that the discrepancy in localization errors between the two conditions was produced as a result of these additional neural activities. The present study showed that localization errors were in general larger in the moving background than in the saccadic eye movement conditions. From this finding, we can suppose that, in both conditions, mislocalization was primarily caused by retinal image displacements and that, in the saccadic eye movement condition, the mislocalization was corrected by an involvement of the EEPS. In other words, saccadic eye movements reduced the size of localization error.

Although this explanation sounds reasonable. it has some difficulties. As shown in Figures 2 an 3, when a flash was presented at the -4 scale division, the error in the minus direction observed in the moving background condition disappeared in the saccadic eye movement condition. On the other hand, at the target position of the +12 scale division, the large error in the plus direction observed in the moving background condition was not shown in the saccadic eye movement condition. In short, reduction of error by the EEPS was dependent upon the position where the flash was presented. Although it is evident that mislocalization was reduced in the saccadic eye movement condition, it is not clear why the corrective effect of the EEPS depended on the target position. This is because, if the EEPS has a corrective effect on localization error, the effect should be observed in the same way irrespective of the target position.

To resolve this problem, an additional explanation is necessary. One possible explanation is that a saccadic eye movement generates some kind of cognitive bias such as attentional bias (Bridgeman, 1983) which depends on the spatio-temporal aspects of the saccade, and that this bias produces the corrective effect of the EEPS which works differently from position to position in the visual field. According to this idea, the results of the present study are explained as follows. At the time immediately before the saccade onset and during the first half of the saccade, the corrective effect appears at the positions near or beyond the saccade's destination, but not at the position located in the direction opposite to the saccade. This is because the visual system enhances the efficiency of its cognitive function in a part of the visual field to which the eye gets after the saccade, and in this area localization is not so much influenced by the retinal event produced by rapid image displacements. On the other hand, at the end of the saccade, the corrective effect appears only at the position near the original fixation point. This is because the enhanced cognitive function is brought back to the original fixation point from the saccade destination, or it spreads out all over the visual field. (the latter possibility comes from the fact that, when a flash was presented at the time near the end of the saccade, the reduction of errors was observed also at the position beyond the saccade destination, i.e., on the +12 scale division.)

The explanation described above also has a serious problem which should be answered. My earlier studies (Honda, 1993) showed that mislocalization was larger when a flash was presented in the dark than when it was presented on an illuminated background, and further it was suggested that the large mislocalization in the dark was primarily caused by a sluggish activity of the EEPS. This earlier finding is not consistent with the assumption proposed here that the EEPS corrects the error produced by retinal image displacements. In other words, why does the EEPS, which generates large localization errors in the dark, reduce the errors when a flash target is presented on an illuminated background?

In conclusion, the present study demonstrated that the shape of the error curves in the saccadic eye movement condition was not the same as that in the moving background condition. This finding casts doubt on the idea that mislocalization during saccadic eye movements is primarily caused by retinal image displacements which are not necessarily contingent upon the movement of the eye. It was argued that some kind of additional assumption such as saccade-contingent cognitive bias is necessary to successfully explain the discrepancy between the two conditions.

(This research was supported by a 1992 Grant-in-Aid for Scientific Research from the Ministry of Education, Science and Culture to the author.)

References

Bischof, N. & Kramer, E. (1968). Untersuchung und Uberlegungen zur Richitungswahrnehmung bei Willkurlichen sakkadischen Augenbewegungen. Psychologische Forschung, 32, 185-218.

Bridgeman, B. (1983). Mechanisms of space constancy. In Hein, A. & Jeannerod, M. (eds.), Spatially Oriented Behavior (pp. 263-279). New York: Springer.

Honda, H. (1989). Perceptual localization of visual stimuli flashed during saccades. Perception and Psychophysics, 45, 162-174.

Honda, H. (1990). Eye movement to a visual stimulus flashed before, during, or after a saccade. In Jeannerod, M. (ed.), Attention and Performance (Vol. 13, pp. 567-582). Hillsdale: LEA.

Honda, H. (1991). The time courses of visual mislocalization and of extra-retinal eye position signals at the time of vertical saccades. Vision Research, 31, 1915-1921.

Honda, H. (1993). Saccade-contingent displacement of the apparent position of visual stimuli flashed on a dimly illuminated structured background. Vision Research, 33, 709-716.

Kennard, D.W., Hartman, R.W., Kraft, D., & Glaser. G. H. (1971). Brief conceptual (nonreal) events during eye movement. Biological Psychiatry, 3, 205-215.

Mackay, D. M.. (1970). Mislocation of test flashes during saccadic image displacements. Nature, 227, 731-733.

Mateeff, S. (1978). Saccadic eye movements and localization of visual stimuli. Perception and Psychophysics, 24, 215-224.

Matin, L., Matin, E., & Pearce, D. G. (1969). Visual perception of direction when voluntary saccades occur: I. Relation of visual direction of a fixation target extinguished before a saccade to a flash presented during the saccade. Perception and Psychophysics, 5, 65-80.

Matin, L., Matin, E., & Pola, J. (1970). Visual perception of direction when voluntary saccades occur: II. Relation of visual direction of a fixation target extinguished before a saccade to a subsequent test flash presented before the saccade. Perception and Psychophysics, 8, 9-14.

O'Regan, J. K. (1984). Retinal versus extraretinal influences in flash localization during saccadic eye movements in the presence of a visible background. Perception and Psychophysics, 36, 1-14.

Sperling, G. (1990). Comparison of perception in the moving and stationary eye. In Kowler, E. (ed.), Eye Movements and Their Role in Visual and Cognitive Processes (pp. 307-351). Amsterdam: Elsevier.

Sperling, G. & Speelman, R. (1965). Visual spatial localization during object motion, apparent object motion, and image motion produced by eye movements. Journal of the Optical Society of America, 55, 1576

OCULOMOTOR PHYSIOLOGY

A NEURAL MECHANISM SUBSERVING SACCADE-VERGENCE INTERACTIONS

Lawrence E. Mays and Paul D.R. Gamlin

Department of Physiological Optics, University of Alabama at Birmingham,
Birmingham, Alabama 35294 USA

Abstract

When a saccade occurs during a vergence movement, the amplitudes of the saccades in the two eyes may be markedly unequal (Ono, Nakamizo & Steinbach, 1978; Kenyon, Ciuffreda & Stark, 1980). This results in a large increase in vergence velocity, regardless of the direction of the saccade or vergence movement. The observation of unequal saccades in the two eyes has led to speculation that humans retain circuitry which allows relatively independent control of the movements of the two eyes (Enright, 1984). This hypothesis is notable because, if it were correct, it would require an extensive revision of our basic models of oculomotor control systems. We report experiments which suggest an alternative hypothesis: saccadic facilitation of vergence results from the interruption of inhibition of vergence burst neurons by pontine omnipause neurons, which are involved in initiating saccades.

Keywords

saccade, vergence, neurophysiology, brainstem, omnipause neurons

Introduction

Saccades are high velocity movements in which the two eyes move as if yoked. Looking from a far to a near target requires ocular convergence, in which the two eyes move slowly in opposite directions. In ordinary experience, most gaze shifts between targets at different distances also require changes in direction mediated by saccades. When a saccade occurs during a vergence movement, the amplitudes of the saccades in the two eyes are often different (Ono, Nakamizo & Steinbach, 1978; Kenyon, Ciuffreda & Stark, 1980). These disconjugate saccades result in a large increase in vergence velocity, regardless of the direction of the saccade or vergence movement. The increase in vergence velocity is too large to be accounted for by a linear addition of saccadic and vergence velocities (Ono et al., 1978). The observation of unequal saccades in the two eyes has led to speculation that under certain stimulus conditions, primates, including humans, retain circuitry which allows relatively independent control of the movements of the two eyes (Enright, 1984). We report experiments which suggest an alternative hypothesis: saccadic facilitation of vergence

Eye Movement Research/J.M. Findlay et al. (Editors)

results from the interruption of inhibition of vergence burst neurons by pontine omnipause neurons, which are involved in initiating saccades.

Methods

The experiments reported herein were performed on rhesus monkeys trained to look at visual targets for a fruit juice reward. The movements of both eyes were recorded using implanted eye coils. Visual targets could be moved in the horizontal, vertical and depth dimensions, allowing saccades, vergence, and mixed (vergence with saccades) movements to be elicited. Metal microelectrodes were lowered through chronically implanted recording wells for unit recording and microstimulation. A more detailed description of the methods can be found in Zhang, Mays and Gamlin (1992).

Results

Premotor circuitry for saccades and vergence

Vergence and saccadic eye movements are distinctly different, and although both types of movements share the same motoneurons (Mays & Porter, 1984), their pre-motor neuronal circuitry appears to be separate (Mays, 1984). Figure 1 shows horizontal saccadic (1A), vergence (1B) and vergence plus saccadic (1C) eye movements as a function of time. In Fig. 1A the amplitudes of the saccades in the two eyes are virtually identical and hence there is no net change in vergence angle. The peak velocity of ≈ 300°/sec for the 6° saccade in 1A is an order of magnitude greater than the peak eye velocity seen during vergence movements (1B). In 1C the target stepped 6° to the right and was moved toward the subject in order to elicit a 6° increase in vergence angle. The initial vergence velocity matches that seen in 1B until the point at which a rightward saccade occurs. The rapid rise in vergence velocity associated with the saccade appears to be the result of a larger saccade in the left eye than in the right eye. The discrepancy between the amplitudes of the saccades in the two eyes is too great to be explained by a linear addition of the ongoing symmetrical vergence movement and a conjugate saccade.

The velocity of horizontal saccades is controlled by saccadic burst cells in the pons, which can be either excitatory or inhibitory (Igusa, Sasaki & Shimazu, 1980; Kaneko, Evinger & Fuchs, 1981; Hikosaka & Kawakami, 1977). Excitatory burst neurons provide a pulse of innervation to agonist motoneurons for saccades, while inhibitory burst neurons cause the antagonist motoneurons to cease firing. These burst cells are normally inhibited by pontine omnipause neurons (OPNs), so-called because they cease firing for saccades in all directions (Keller, 1974). It is the cessation of the OPNs that allows saccades to occur (Keller, 1974). For vergence movements in which no saccade occurs (1B), OPNs continue to fire and saccadic burst neurons are silent (unpublished observation).

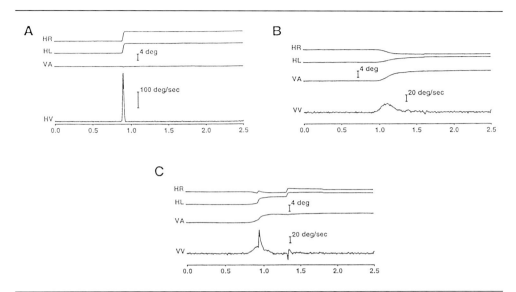

Fig. 1 - Characteristics of typical saccades, vergence eye movements, and their interaction. The horizontal position of the right eye (HR), left eye (HL), and the vergence angle (VA = HR - HL) are shown. Horizontal movements to the right and convergence are shown as upward deflections. Eye movement velocity in the horizontal plane (HV) is shown in 1A, and the vergence velocity (VV) is shown in 1B and 1C. Time scale is in seconds.

Another type of neuron, the midbrain vergence burst cell, has an activity pattern which matches the ongoing vergence velocity (Mays, Porter, Gamlin & Tello, 1986). Both convergence and divergence burst cells have been identified in the midbrain reticular formation. An example of the activity of a convergence burst cell for convergence without a saccade is shown in Fig. 2. Vergence burst cells are believed to provide the input to a neural integrator which constructs a vergence position signal for the motoneurons (Mays et al. 1986). Thus, their role in the vergence system is analogous to that of the saccadic system's burst neurons.

In addition, Mays (1984), and Judge and Cumming (1986) have described midbrain near response cells close to the oculomotor nucleus. Although these cells were characterized as having a vergence position signal, it is now clear that many near response cells carry a vergence velocity signal as well. Figure 3 shows the activity of a near response cell with a position and velocity signal during a symmetrical convergent movement. Zhang et al. (1992) have shown that cells with similar characteristics can be antidromically activated from the medial rectus subdivisions of the oculomotor nucleus. Thus, these cells are presumed to provide a vergence position and velocity signal to medial rectus motoneurons.

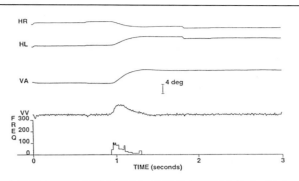

Fig. 2 - *Activity of a midbrain convergence burst neuron for a symmetrical convergent eye movement without a saccadic intrusion. Labels are the same as in Fig. 1. Firing frequency (FREQ) is the reciprocal of interspike interval. These cells appear to encode vergence velocity. They do not fire for conjugate eye movements or for divergent eye movements.*

Activity of midbrain neurons during vergence with saccades

Although the saccadic and vergence systems are thought to be independent (Mays 1984; Gamlin, Gnadt & Mays, 1989), we have observed that a subset of convergence burst neurons displays a burst of activity associated with saccades during convergence. Figure 4A shows the activity of the same convergence burst cell as in Fig. 2 during a mixed movement. It should be noted that this cell fires more vigorously when a saccade occurs during convergence than for convergence without saccades (Fig. 2). Furthermore, convergence burst cells typically do not discharge for saccades in the absence of convergence. It appears that the activity of the convergence burst cell is augmented by the occurrence of a saccade.

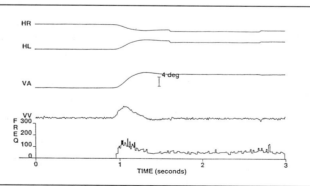

Fig. 3 - *Activity of a midbrain near response cell for a symmetrical convergent eye movement without a saccadic intrusion. Note that this cell appears to encode vergence velocity as well as vergence angle. Near response cells do not change their firing rate for conjugate eye movements.*

A similar effect can be noted on many midbrain near response cells. Figure 4B shows the same cell as in Fig. 3 during a mixed movement. A marked increase in activity just before and during the saccade is evident. Saccade-related activity is not observed on these cells without a vergence movement. The saccade-related increase

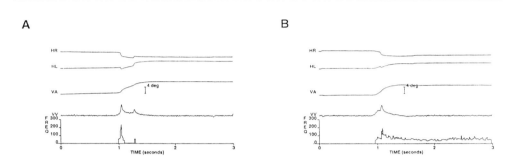

Fig. 4 - *Activity patterns of a midbrain convergence burst cell (4A) and near response cell (4B) for convergence with saccades. The cell shown in 4A is the same as that shown in Fig. 2, and the cell in 4B is the same as shown in Fig. 3. A comparison of these activity patterns with those in Figs. 2 and 3 reveals bursts of activity associated with the saccadic intrusions. These cells show no change in activity for saccades in the absence of a vergence eye movement.*

in activity in these cells is likely to result from the extra saccade-related burst on convergence burst cells. If, as is presumed, the midbrain near response cells provide the vergence signal to the extraocular motoneurons, then this pattern of activity would be responsible for the increase in vergence velocity seen during saccades.

How is the activity of vergence burst cells enhanced by saccades?

The neuronal circuits for generating the horizontal and vertical components of saccades are located in the pons and midbrain, respectively. Omnipause neurons (OPNs), as their name implies, cease firing for vertical and horizontal saccades, and thus constitute an element which is common to both the horizontal and vertical circuitry. The finding that vergence is speeded by the occurrence of saccades in all directions, including vertical (Enright, 1984), suggests the involvement of the OPNs. Furthermore, a role for vergence burst neurons is suggested by Kenyon and Stark's (1983) observations on the importance of vergence velocity for unequal saccades. Recently, Zee, FitzGibbon and Optican (1992) have proposed that the accelerated vergence movements might be the result of the release of OPN inhibition of vergence burst cells during saccades. A schematic diagram of the circuitry for such an interaction is shown in Fig. 5. The right and left medial rectus (RMR and LMR) motoneurons receive saccadic commands from fibers in the medial longitudinal fasciculus. For a leftward saccade, the signal to the RMR is a positive pulse-step (R SAC), while the command to the LMR has the opposite form (L SAC). Omnipause neurons (OPNs) cease firing just before the saccade onset. The pulses required for saccades (both positive and negative) are a consequence of the release of OPN inhibition of saccadic burst neurons (not shown). Convergence burst (CB) neurons have a firing pattern which matches vergence velocity (Mays et al. 1986). It is hypothesized that this signal drives a vergence integrator ($\int$). Vergence velocity and position signals go to convergence burst-tonic (CBT) cells which in turn drive the

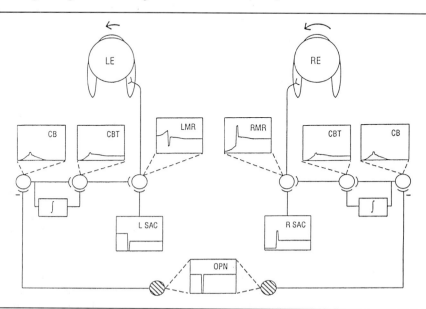

Fig. 5 - Scheme by which omnipause neuron (OPN) inhibition of vergence burst cells could produce unequal saccades in association with vergence eye movements. RE and LE are the right and left eyes with medial and lateral rectus muscles. Circles represent neurons. Labelled boxes connected to the circles represent the neuronal activity pattern during a mixed saccade-vergence eye movement. Circuitry needed for controlling the lateral rectus muscles is omitted for simplicity.

medial rectus motoneurons. The release of OPN inhibition of CB cells during saccades should result in a transient increase (pulse) in CB cells. This pattern (shown in the CB box) will result in the activity pattern shown in box CBT, which is added to the saccadic innervation pattern (L SAC and R SAC) at the medial rectus motoneurons. Because the OPNs also control the saccades, the pulse of activity on the CBT cells will be synchronized with the saccadic pulses. This will result in an augmentation of the pulse-step pattern for the RMR and a diminution of the LMR pulse-step, hence a larger saccade in the right eye than in the left eye. This hypothesis indicates that OPN inhibition of vergence burst cells must be weak enough to allow them to fire during vergence movements in the absence of saccades. Cessation of OPN inhibition would have little or no effect on vergence burst cells in the absence of vergence movements but would allow them to fire more vigorously during mixed vergence-saccadic movements.

A test of the OPN hypothesis

Electrical microstimulation of the OPN region during vergence movements offers a critical test of this hypothesis. OPN stimulation should decrease the velocity of vergence movements by increasing the level of inhibition of vergence burst neurons. Figure 6A shows a control convergent eye movement with an amplitude of ≈ 3°, with no saccadic intrusion during the convergence. This movement had a peak vergence

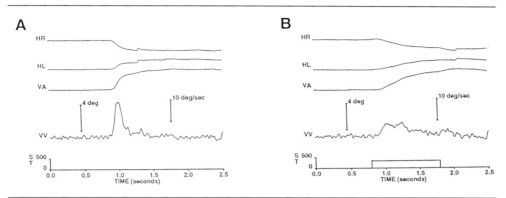

Fig. 6 - Critical test of the OPN hypothesis of saccade-vergence interactions. Eye movement traces are labelled as in Fig. 1.

velocity of ≈ 15°/sec. Figure 6B shows the effect of a one second train of electrical stimulation (300 Hz, 30 µA, 0.2 ms bipolar) of the OPN region beginning just before another convergent eye movement of the same amplitude. OPN micro-stimulation (indicated by the open bar on line ST) results in a 50% reduction in vergence velocity. Because the target was continuously visible, the animal was still able to acquire the target. The OPN region (which was ≈ 1 mm ventral and ≈ 1 mm rostral to the abducens nucleus, on the midline) was identified by recording neurons which paused for all saccades and by observing that electrical micro-stimulation abolished all saccades.

Discussion

Three hypotheses have been offered to explain the interaction between saccades and vergence eye movements. Kenyon and Stark (1983) noted a positive systematic relationship between the size of saccadic inequalities and the vergence velocity at the time the saccade occurred. They suggested that unequal saccade sizes were not the result of unequal saccadic innervation to the two eyes. Their hypothesis, supported by modelling studies, was that eye velocity prior to the saccade influenced the saccade size through non-linear interactions with the peripheral (i.e., muscle and globe) oculomotor system. According to this view, the saccadic and vergence systems generate premotor signals independently, and these signals are combined linearly at the motoneurons. Non-linear properties of the peripheral plant are believed to be responsible for different saccadic amplitudes. An alternative hypothesis, discussed by Enright (1984), is that the saccadic system is capable of independent control of the two eyes under the appropriate stimulus conditions. This hypothesis is notable because, if it were correct, it would require an extensive revision of basic models of the saccadic and vergence control systems. Presumably, this hypothesis would require two populations of excitatory and inhibitory burst cells (one for each eye) on each side of the brain, and would also require that the abducens internuclear neurons and motoneurons, which are intermixed within the same nucleus, receive separate inputs. Our results support a third hypothesis. It appears that the saccadic neural circuitry generates eye movement control signals which are essentially identical for the two eyes, but that the addition of an extra intra-saccadic pulse of innervation from the vergence system to the motoneurons modifies the saccade size. The sharing of OPN inhibition by the saccadic and vergence systems is the key to understanding the interaction between these two systems. OPNs must inhibit vergence burst cells, since the cessation of OPN activity during saccades without vergence has little effect on vergence angle. The shared OPN hypothesis is consistent with the observation that the occurrence of a saccade in any direction always facilitates vergence.

Acknowledgments

We thank David Morrisse and Lorna Mayo for their assistance. This work was supported by National Eye Institute Research Grants to L. Mays (EY 03463) and P.D.R. Gamlin (EY 07558) and by NEI Core Grant EY 03039.

References

Enright, J. Changes in vergence mediated by saccades. *J. Physiol.* **350**, 9-31 (1984).

Gamlin, P.D.R., Gnadt, J.W. & Mays, L.E. Abducens internuclear neurons carry an inappropriate signal for ocular convergence. *J. Neurophysiol.* **62**, 70-81 (1989).

Hikosaka, O. & Kawakami, T. Inhibitory reticular neuroelated to the quick phase of vestibular nystagmus -- their location and projection. *Exp. Brain Res.* **27**, 377-386 (1977).

Igusa, Y., Sasaki, S. & Shimazu, H. Excitatory premotor burst neurons in the cat pontine reticular formation related to the quick phase of vestibular nystagmus. *Brain Res.* **182**, 451-456 (1980).

Judge, S.J. & Cumming, B.G. Neurons in the monkey midbrain with activity related to vergence eye movement and accommodation. *J. Neurophysiol.* **55**, 915-930 (1986).

Kaneko, C.R.C., Evinger, C. & Fuchs, A.F. Role of cat pontine burst neurons in generation of saccadic eye movements. *J. Neurophysiol.* **46**, 387-408 (1981).

Keller, E.L. Participation of medial pontine reticular formation in eye movement generation in monkey. *J. Neurophysiol.* **37**, 316-332 (1974).

Kenyon, R.V., Ciuffreda, K.J. & Stark, L. Unequal saccades during vergence. *Am. J. Optom. and Physiol. Optics* **57**, 586-594 (1980).

Kenyon, R.V. & Stark, L. Unequal saccades generated by velocity interactions in the peripheral oculomotor system. *Math. Biosci.* **63**, 187-197 (1983).

Mays, L.E. Neural control of vergence eye movements: Convergence and divergence neurons in midbrain. *J. Neurophysiol.* **51**, 1091-1108 (1984).

Mays, L.E. & Porter, J.D. Neural control of vergence eye movements: Activity of abducens and oculomotor neurons. *J. Neurophysiol.* **52**, 743-761 (1984).

Mays, L.E., Porter, J.D., Gamlin, P.D.R. & Tello, C.A. Neural control of vergence eye movements: Neurons encoding vergence velocity. *J. Neurophysiol.* **56**, 1007-1021 (1986).

Ono, H., Nakamizo, S. & Steinbach, M.J. Nonadditivity of vergence and saccadic eye movement. *Vision Res.* **18**, 735-739 (1978).

Zee, D.S., FitzGibbon, E.J. & Optican, L.M. Saccade-vergence interactions in humans. *J. Neurophysiol.* **68**, 1624-1641 (1992).

Zhang, Y., Mays, L.E. & Gamlin, P.D.R. Characteristics of near response cells projecting to the oculomotor nucleus. *J. Neurophysiol.* **67**, 944-960 (1992).

EYE POSITION EFFECTS ON PURSUIT RELATED RESPONSES IN AREA LIP OF MACAQUE MONKEY

F. Bremmer and K-P. Hoffmann

Dept. Zoology & Neurobiology, Ruhr University Bochum, 44780
Bochum, FRG

Abstract

Pursuit related activity is a well described neuronal phenomenon for monkey visual cortical areas MT and MST in the superior temporal sulcus (STS). For the posterior parietal cortex (PPC) pursuit related activity has been shown for neurons in areas 7A and VIP, which both receive cortico-cortical projections from area MST. Pursuit related activity has yet not been shown for the Lateral Intra Parietal area (LIP), which is reciprocally connected to areas MT and MST. In an accompanying study we could show that pursuit related responses of neurons in areas MT and MST are modulated by eye position. In the actual experiments we tried to reveal whether also in area LIP neurons pursuit related activity if present is modulated by eye-position. Single unit recordings were made from 2 hemispheres of two monkeys (Macaca mulatta). In a first step neurons were tested for pursuit related activity while the monkey pursued a target which moved randomly into one out of four directions on a translucent screen. For 107 of 269 LIP neurons tested a direction specific pursuit related activity could be found. In order to test whether responsiveness of these neurons was modulated by the position of the eye in the orbit the monkey had to pursue a target which started moving with the same speed into the same direction from different locations on the screen. The majority of cells showed a significant effect of the starting position of pursuit on the mean neuronal response. This response varied mostly linearly with both horizontal and vertical eye position. It is suggested that the observed modulatory effect of eye position on visual neuronal responses in area LIP might be a common phenomenon and that it might be used for the generation of a representation of the visual environment in a non-retinocentric frame of reference.

Keywords

Eye position effect, pursuit, monkey, LIP, coordinate transformation

Introduction

A fundamental problem of the control of visually guided movement consists in the transformation of signals used in different frames of reference in sensory and motor systems (sensori-motor integration). The incoming visual signals are organized in a retinotopic manner in the first stages of the visual system. The outgoing signals (motor commands) have to be related to locations in the surrounding three-dimensional space and therefore have to be organized and encoded with respect to the head, body or an external point in space. If e.g. optical flow, which is the resulting image on the retina during ego-motion through an environment, is used for estimating the direction of heading, the incoming visual information might differ depending on the direction of gaze relative to the direction of heading (e.g. Regan & Beverly, 1982; Warren & Hannon, 1988; Lappe & Rauschecker, 1991). In such a case, the incoming visual information depends

Eye Movement Research/J.M. Findlay et al. (Editors)

on the direction of gaze although the required motor output, which is heading in one specific direction, always keeps the same. The problem then is how to relate the incoming visual signal (retinotopic) with the motor output (cranio-, ego- or allocentric). One hypothesis for solving this obvious problem is a coordinate transformation from retinocentric signals to signals encoded in an ego- or allocentric frame of reference (Andersen et al., 1993; Galletti et al., 1989; Pouget et al., 1993). The Lateral Intra Parietal area (LIP) in the posterior parietal cortex of macaque monkey is supposed to play an important role in providing this required coordinate tranformations for visually guided movement. On the one hand this idea was supported by lesion studies which showed that monkeys had deficits in spatial perception and movement under visual guidance after lesions of the posterior parietal cortex (Lynch, 1980; Andersen, 1989). More direct evidence came from a study showing that the light-sensitive, memory and saccade-related responses of neurons in area LIP were affected by eye-position (Andersen et al., 1990). It was suggested that these eye-position dependent cells could be involved in the construction of an internal map of the visual environment in which the object's position in space instead of the retinotopic position of its image is encoded.

Anatomically it has been shown that area LIP is reciprocally connected with areas MT and MST in the superior temporal sulcus (STS) of macaque monkey. For these areas pursuit related activity is a well known phenomenon. However, pursuit related activity has yet to be shown for area LIP.

In the present experiments we could show that there are neurons in area LIP which show a clear pursuit related activity. Furthermore for most of these neurons neuronal discharge during pursuit is affected by the starting location of pursuit. Discharge mostly varies linearly with horizontal and vertical eye position. Gain fields (Andersen et al., 1985) were computed in order to quantify the observed effect. The strength of the modulatory effect was comparable to that found for saccade related activity of neurons in areas LIP and 7A. Some of the recorded neurons were also tested for modulation of neuronal discharge by eye position during fixation in darkness, a phenomenon already described (Andersen et al., 1990). For two thirds of the neurons tested the same modulatory effect did operate for both fixation and pursuit responses. The observed results are discussed with respect to the remaining question of coordinate transformations.

Material and methods

Single cell recordings were made in two awake, behaving monkeys (Macaca mulatta) performing fixation and pursuit tasks. Recordings were made from 2 hemispheres. All procedures were in accordance with published guidelines on the use of animals in research (European Communities Council Directive 86/609/EEC).

Surgery

A holder   for immobilizing the head, recording   chambers for microelectrode penetrations and scleral search coils for monitoring eye position were implanted under pentobarbital anesthesia and aseptic conditions.

Behavioural paradigm

During training and recording sessions the monkey was sitting in a comfortable primate chair with the head fixed (only during recordings) while performing fixation or pursuit tasks for liquid reward. Rewards were given for keeping the eyes   within an electronically   defined   window centered on the fixation or pursuit target. Targets (1.0^0 diameter) were generated by light emitting diodes back- projected on a translucent tangent screen subtending 90^0 x 90^0 at a viewing distance of 35 cm.

Pursuit paradigm

When a cell was isolated receptive fields (RFs) were mapped using a hand-held projector while the monkey fixated a central target. Quantitative testing of visual properties was performed using a galvanometer-mounted slide projecting system displaying light bars or random-dot patterns of different size (Data Aquisition and Data Analysis (DADA) software).

After initial quantitative characterization of the response characteristics cells were tested for pursuit related activity. Firstly this was done with the step ramp paradigm. In this paradigm after an initial phase of fixation (1000 ms) the central fixation light was extinguished and the pursuit target was switched on in the periphery and moved centripetally with a speed of 15^0/s. Starting points for pursuit were 15^0 away from the center and located on both the horizontal and vertical meridian, respectively. The result of this testing determined the preferred direction of pursuit   of each individual neuron.

In order to test for an eye position effect a 'pursuit- matrix' was defined. Here the monkey, after fixating a central fixation target for 1000ms, had to pursue a target which started moving randomly from one out of five different locations, always   with   the same speed (15^0/s) into the same (preferred) direction. Starting locations for pursuit were 15^0 away from the center ((x,y)_(± 10.6^0, ±10.6^0)) and the center itself. Target movement lasted for 1200ms.

Fixation paradigm

In order to test for the eye position effect during active fixation targets were presented in random order at different locations on the screen in darkness. Locations were the center of the screen plus 8 concentrically allocated points 15^0 away from the center. Presentation of fixation targets lasted for 1000ms after the animals eye position was continuously within the electronically defined window for 300ms.

Histology

At the end of the experimental sessions with the first monkey small electrolytic lesions were made in the recorded hemisphere by passing small direct current through the recording electrode at different depths along the electrode track. The monkey then was given a lethal dose of sodium pentobarbital, and transcardially perfused with formalin. The brain was removed and serial frontal sections were made at 50µm thickness stained alternatingly with thionin for cytoarchitecture, and with the Gallyas method for myeloarchitecture. The second monkey is still being employed in the ongoing experiments. Area LIP here is identified by physiological criteria and the recording depth within the intraparietal sulcus.

Data Analysis

Pursuit paradigm: for analyzing neuronal activity in the pursuit paradigm individual trials normally were aligned to the onset of the initial saccade preceeding the smooth pursuit phase. Trials typically were divided into two characteristic epochs. Epoch 1 was defined as the time window beginning with trial onset (Fixation light ON) until 100 ms before the initial saccade. Activity during this epoch normally was taken into account as background activity. Epoch 2 typically was defined as beginning 100 ms after saccade onset until the end of pursuit target movement. Epoch 2 normally lasted for about 1000 ms. Neuronal discharge rates were computed both as raw activity (epoch 2) and as relative activity defined as raw activity (epoch 2) minus background activity (epoch 1).
Fixation paradigm: in this paradigm the monkey's eye position had to be within the electronically defined window 300 ms before starting of recording of neuronal activity. For analysis of neuronal activity the whole fixation time (1000 ms) was taken into consideration. The effects of horizontal and vertical eye position on neuronal discharge were assumed to be independent and non-interacting. Two-dimensional linear regression analyis (least square estimates) was applied to both kind of responses, raw as well as relative activity. For testing the validation of the planar model for fitting the observed data the F-ratio was computed.

Results

Recordings were made from 2 hemispheres of two monkeys. A total of 320 cells were recorded quantitatively. 269 of these were tested for pursuit related activity, 119 from the first animal and 150 from the second. In the second monkey 83 neurons were tested in the fixation paradigm.

Pursuit Paradigm

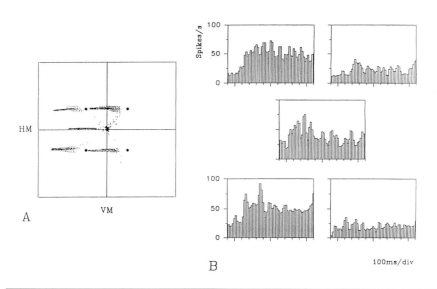

Figure 1: Eye position traces and neuronal discharges during the pursuit paradigm. A: Sampled eye positions when the animal had to pursue a target always moving to the left. This was the preferred direction of the actually recorded neuron. Every 6th sample of recorded eye position is shown. B: The resulting PSTHs are located according to the starting point of pursuit, i.e. the PSTH in the upper left represents the neuronal response when the pursuit target started moving in the upper left etc. Pursuit related response was best when the target started left from the vertical meridian.

Pursuit paradigm: 107 neurons out of 269 tested showed a pursuit related activity. These 107 neurons were predominantly located in the caudal region of area LIP.

A quantitative result for a neuron which preferred pursuit to the left is shown in figure 1. The resulting eye position traces are shown in figure 1a, in figure 1b the resulting pursuit responses are shown. For this individual neuron pursuit response was best when the target started left from the vertical meridian. Pursuit response decreased when the target started from the center of the screen and was lowest when the pursuit started right from the vertical meridian. Furthermore, response was slightly better when the target started above the horizontal meridian in comparison to the trials when the target stared below the horizontal meridian. Figure 2 shows a fitting of a two-dimensional linear regression plane to the resulting data.

Fitting of a Linear Regression Plane

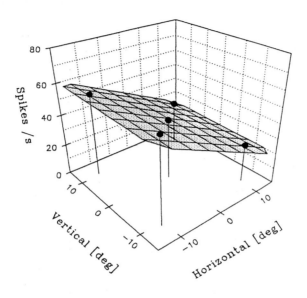

$$z = -1.314x + 0.064y + 40.48$$

$$r = 0.995$$

Figure 2: Two-dimensional linear regression plane fitted to the raw pursuit related activity. The x-y-base represents the central 15 by 15 degrees of the tangent screen where the monkey had to pursue the target. The location of each drop line indicates the starting point of pursuit, the height of each drop line depicts the response for pursuit starting at this location.

The x-y-base of this 3-D-cube represents the central 15 x 15 degrees of the tangent screen where the pursuit targets were presented. Pursuit response is represented in the third (z-) dimension, symbolized by a vertical drop line. The base point of each drop line depicts the starting point of pursuit. The height of the drop line indicates the pursuit response (raw activity). The two-dimensional linear regression function which could be fitted excellently to the relative responses is symbolized by the shaded plane.

107 of the 269 neurons tested showed a pursuit related response. For 88 of them (80%) the discharge was modulated by the starting position of pursuit. For about 58% of them a linear regression plane provided a significant fit (p < 0.1 or better) to the pursuit related discharge.

Slopes and intercepts of the regression planes were fairly evenly distributed in all directions as indicated by figure 3. In this diagram slopes are treated as a 2-D vector. A modulatory effect e.g. which results in a linear regression function with a positive slope in horizontal direction and negative slope in vertical direction is located in the lower right quadrant within this diagram.

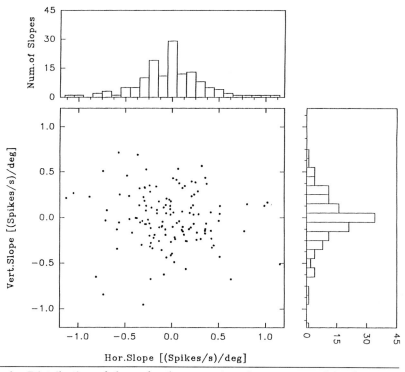

Distribution of Slopes for the Regression planes during the Pursuit Paradigm

Figure 3: Distribution of slopes for the regression planes computed in the pursuit paradigm. The slopes of the regression planes are treated as a two dimensional vector. Each data point represents the 2-dim vector of one plane. Above and right from this scheme the slopes are grouped for only one dimension (bin-width 0.1[(Spikes/s)/deg]).

Fixation paradigm: in order to test whether modulatory effects during pursuit do coincide with possible modulatory effects during active fixation of a dim spot in darkness, 83 neurons also were tested in the fixation paradigm. A typical result is shown in figure 4. For this neuron the activity during fixation without any other visual stimulus was modulated by eye position in the orbit, a phenomenon previously described (Andersen et al., 1990). In this case discharge was largest when the monkey fixated in the upper left of the visual field. The activity gradually decreased for fixation sites located more to the right and below. Consequently, neuronal discharge was lowest for fixation in the lower right. A linear regression plane could be fitted highly significantly to the response strength of this neuron. 38 neurons were tested in the fixation as well as in the pursuit paradigm. For 25 (66%) of them the slopes of the regression planes found during the fixation paradigm did coincide with those found in the pursuit paradigm.

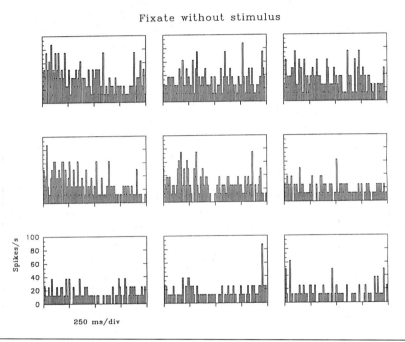

Figure 4: *Neuronal response during fixation of a dim spot in darkness. Discharge is best in the upper left and lowest in the lower right.*

In order to test for any ongoing transformation of visual signals from retinocentric to non-retinocentric frames of reference we computed the population response for the neurons tested in the pursuit as well as the fixation paradigm. Figure 5a shows the sum over all 107 regression planes fitted to the pursuit data. The resulting population plane is fairly flat, indicating nearly no effect of eye position on the population response in the pursuit paradigm. The same is true for neurons tested in the fixation paradigm. Figure 5b shows the cumulated responses of all 83 neurons tested with this paradigm. Again, the resulting population plane is absolutely flat. It thus can be concluded that for most of the cells in area LIP the eye position input produces the same effect on the strength of pursuit and fixation responses. However, the modulatory effect of eye position is balanced out at the population level.

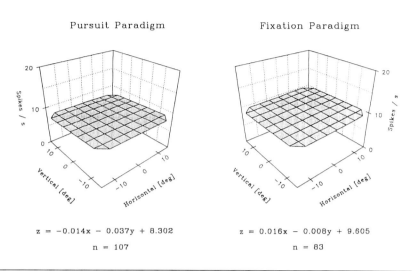

Cumulated Data

Pursuit Paradigm   Fixation Paradigm

z = −0.014x − 0.037y + 8.302   z = 0.016x − 0.008y + 9.605

n = 107   n = 83

Figure 5: Cumulated data of the pursuit and the fixation paradigm. When adding together all the regression planes for neurons tested in the pursuit and fixation paradigms, respectively, the resulting planes turn out to be absolutely flat.

Discussion

Pursuit related activity so far has been shown for monkey visual cortical areas MT and MST (Komatsu & Wurtz, 1988), 7A (Lynch & Mountcastle, 1977) as well as VIP (Colby et al., 1993). Furthermore, anatomical studies (Andersen et al., 1990; Boussaoud et al., 1990) revealed cortical input from areas MT and MST into area LIP as well as strong connections to areas LIP, VIP, and 7a from area MST.

The goal of this study was to investigate whether even in area LIP pursuit related activity can be found. Pursuit related activity could be shown for about 40% of the neurons tested in area LIP. Neurons showing a pursuit related response normally were weak in their light sensitive responses. Some of them gave a brisk response to an onset of a light stimulus, and just a few of them gave a direction selective response when tested with an appropriate stimulus. The discharge of these cells therefore seems to be related mostly to oculomotor behaviour.

A modulatory effect of eye position on the mean neuronal discharge during pursuit could be shown. From the 107 neurons tested in the pursuit paradigm about 80% were modulated in their response by the starting position of the pursuit target. This percentage nearly coincides with the values found for light-sensitive, saccade- and memory related activity in area LIP (Andersen et al., 1990). Furthermore, as for these other kinds of response, about 58% of the modulatory effects turned out to be planar, i.e. neuronal discharges as a function of horizontal and vertical eye position could be fitted significantly by a two- dimensional linear regression plane. It therefore seems, as if the same neuronal processes and inputs, working on saccade and memory related responses as well as light driven activity in area LIP, do also work on its pursuit related activity. The amounts of changes of discharge with change in eye position are approximately equal for all four kinds of responses (saccade, memory, light-sensitive as well as pursuit). This is true for the slopes, found for the fitted regression planes as well as for the intercepts, given also by these planes. In our measurements we could find no bias for maxima of responses or directions of best activity. Maxima and directions seemed to be distributed fairly even.

The same neurons tested in the pursuit paradigm often also showed a modulation of activity during fixation of a dim spot in darkness. 38 neurons were tested in the pursuit paradigm as well as in the fixation paradigm. For about two thirds of them the directions of best responses did coincide. This confirms the hypothesis that the same neural mechnisms influence the activity of the neurons in the different experimental paradigms. We therefore think that the eye position effect is a kind of gating mechnism effecting the neuron. The neuron's discharge is gated without regard of the task the monkey actually is involved in. The source of this gating might be corollary discharge or proprioceptive inputs from eye muscles. However, this question can't be answered by our experiments. Nevertheless, the question remains whether there exists a kind of coordinate transformation performed by these neurons in area LIP.

At the single cell level the effect of eye position during pursuit as well as during fixation for most of the neurons tested was the same, i.e. strength of fixation and pursuit responses covaried with eye position for two thirds of the neurons. However, in none of the experiments done so far on response characteristics of neurons located in monkey cortical areas V3A, V6, 7A, LIP or even PMv (Andersen et al., 1990; Galletti et al., 1989, 1990; Boussaoud et al., 1993) were neurons found which encoded with their discharge a fixed point in space regardless of eye position. As suggested by others, the population of cells located within an area might be capable of transforming the visual signals into a head- or body-centered frame of reference. According to the computations done on population coding for the motor and premotor cortex we computed the population response for the cells tested in the pursuit and the fixation paradigm. What turned out was a little surprising. The eye position effect, found at the level of the single cell, was balanced out when all neuronal responses were taken together. The cumulative regression planes, resulting when taking together all neuronal responses, were absolutely flat. This result led to the hypothesis that sub-populations might do this task. While searching for a more reasonable concept for grouping together neuronal discharges, the idea then was that there might exist a dichotomy of cells: one subpopulation responds better when the task (pursuit or fixation) had to be done on the left side of the vertical meridian, whereas the other responds better when the task has to be done on the right side of the meridian. This hypothesis led to a very interesting result. The summing up of regression planes of these both subpopulations resulted in two planes which had the same slope, however, with an inverted sign. Furthermore, these planes had the same intercept, i.e. they met at the vertical meridian. The analogous result was obtained when both the subpopulation responding better above or below the horizontal meridian were grouped together (see figure 6).

Pursuit Paradigm – Cumulated Data

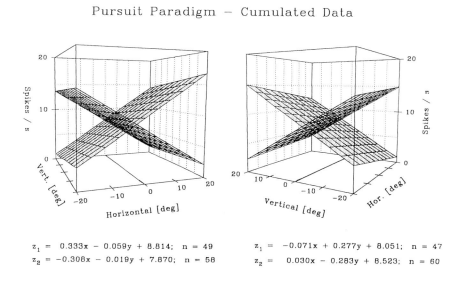

$z_1 = 0.333x - 0.059y + 8.814;\quad n = 49$

$z_2 = -0.308x - 0.019y + 7.870;\quad n = 58$

$z_1 = -0.071x + 0.277y + 8.051;\quad n = 47$

$z_2 = 0.030x - 0.283y + 8.523;\quad n = 60$

Figure 6:   Cumulated regression planes of two subpopulations. Neurons responding better on either side of the meridian (horizontal or vertical) are grouped together into one out of two subpopulations. The resulting regression planes of the two subpopulations do have the same slope, however, with an inverted sign. Furthermore, these planes do have approximately the same intercept, i.e. they meet at the meridian.

With this hypothesis it might be possible to locate points in space with respect to the meridians. Whether this is the meridian with respect to the body or to the head can't be answered by our experiments because in our experimental setup the head of the animal is immobilized and thus the planes always are oriented in the same way with respect to head and body. To dissociate between these two frames of reference, experiments should be conducted in which head and body are moved independently. A comparable experiment already has been done by Andersen and co-workers (1993) which shows that there might be two subpopulations within area LIP, the first one which codes in coordinates with respect to the body, the second one coding in a head centered frame of reference. In summary this could mean that there are two populations of cells in area LIP: the first one which codes with respect to the meridians of the head, the second one which codes with respect to the meridians of the body. However, coding with respect to the meridian is just a hypothesis and has to be verified or falsified in further experiments.

Acknowledgements

We are grateful to Dr.U.J.Ilg for providing the DADA software. Furthermore we thank Dr.U.J.Ilg, A.Thiele, and Dr.W.Werner for surgery. In addition we would like to thank Dr.C.Distler and M.Helming for histology and H.Korbmacher for technical assistance. This work was supported by the ESPRIT Basic Research Action (INSIGHT I) and the Friedrich Ebert Stiftung.

References

Andersen, R.A. (1989). Visual and eye movement functions of the posterior parietal cortex. Annu.Rev.Neurosci. 12, 377- 403.

Andersen, R.A., Asanuma, C., Essick, G. and Siegel, R.M. (1990). Corticocortical connections of anatomically and physiologically defined subdivisions within the inferior parietal lobule. J.Comp.Neurol. 296, 65-113.

Andersen, R.A., Bracewell, R.M., Barash, S., Gnadt, J.W. and Fogassi, L. (1990). Eye position effect on visual, memory, and saccade-related activity in areas LIP and 7A of macaque. J.Neurosci. 10(4), 1176-1196.

Andersen, R.A., Snyder, L.H., Li, C.-S. and Stricanne, B. (1993). Coordinate transformation in the representation of spatial information. Curr.Opin.Neurobiol. 3, 171-176.

Andersen, R.A., Essick, G.K. and Siegel, R.M. (1985). Encoding of spatial location by posterior parietal neurons. Science 230, 456-458.

Boussaoud, D., Barth, T.M. and Wise, S.P. (1993). Effects of gaze on apparent visual responses of monkey frontral cortex neurons. Exp.Brain.Res. 93, 423-434.

Boussaoud, D., Ungerleider, L.G. and Desimone, R. (1990). Pathways for motion analysis: cortical connections of the medial superior temporal and fundus of the superior temporal visual areas in the macaque. J.Comp.Neurol. 296, 462-495.

Colby, C.L., Duhamel, J.-R. and Goldberg, M.E. (1993). The ventral intraparietal Area (VIP) of the macaque: anatomical location and visual properties. J.Neurophysiol. (In Press).

Galletti, C. and Battaglini, P.P. (1989). Gaze-dependent visual neurons in area V3A of monkey prestriate cortex. J.Neurosci. 9, 1112-1125.

Komatsu, H. and Wurtz, R.H. (1988). Relation of cortical areas MT and MST to pursuit eye movements. I: localization and visual properties of neurons. J.Neurophysiol. 60(2), 580-603.

Lappe, M. and Rauschecker, J.P. (1991). A neural network for flow-field processing in the visual motion pathway of higher mammals. Soc.Neurosci.Abstr. 17, 441.

Lynch, J.C. (1980). The functional organization of posterior parietal association cortex. Beh.Brain.Sci. 3, 485- 534.

Lynch, J.C. (1992). Saccade initiation and latency deficits after combined lesions of the frontal and posterior eye fields in monkeys. J.Neurophysiol. 68, 1913-1916.

Lynch, J.C., Mountcastle, V.B., Talbot, W.H. and Yin, T.C.T. (1977). Parietal lobe mechanisms for directed visual attention. J.Neurophysiol. 40(2), 362-389.

Pouget, A., Fisher, S.A. and Sejnowski, T.J. (1993). Egocentric spatial representation in early vision. J.Cogn.Neurosci. 5(2), 150-161.

Regan, D. and Beverley, K.I. (1982). How do we avoid confounding the direction we are looking with the direction we are moving? Science 215, 194-196.

Warren, W.H.. and Hannon, D.J. (1988). Direction of self- motion is perceived from optical flow. Nature 336, 162-163. .

CLINICAL AND MEDICAL ASPECTS
OF EYE MOVEMENTS

PROBLEMS IN MODELLING CONGENITAL NYSTAGMUS: TOWARDS A NEW MODEL

Christopher M. Harris

Department of Ophthalmology, Hospital for Sick Children, London, WC1N 3JH

Abstract

We propose that congenital nystagmus (CN) is due to excessive gain in an internal velocity efference copy loop in the smooth pursuit system around a leaky neural integrator. The model generates accelerating slow phases and pendular waveforms. The CN-null is equivalent to the neural integrator null. The model is compatible with reports on oscillopsia in CN. Further elaboration is needed to replicate all CN waveforms, and null-shifting requires further investigation. We reject the notion of a congenital misrouting as the ultimate cause of CN in favour of a maladaptation to early visual deprivation. Possible causal linkages between poor vision in the first few months of life and the maldevelopment of smooth pursuit gain and the neural integrator are discussed.

Keywords

Nystagmus, smooth pursuit, neural integrator, efference copy, oscillopsia, visual development

Introduction

The mechanisms underlying congenital nystagmus (CN) remain a mystery. Anomalies of the smooth pursuit (SP), fixational, saccadic, and optokinetic systems, and null shifting have all been suggested as causes of CN. Specific models of CN waveforms have employed excessive positive feedback around the horizontal neural integrator (M) (Zee et al, 1980; Optican and Zee, 1984; Tusa et al, 1992). Tusa et al. and Optican and Zee have argued that there is inappropriate positive velocity feedback around the M, which Optican and Zee have suggested results from a congenital neural misrouting.

A good model of CN should not only mimic the observed waveforms parsimoniously, but should also be consistent with, and address the other features of CN. These features have been described frequently and are summarised here.

Onset and Clinical Associations

In spite of the label "congenital", CN does not usually appear until after birth, often in the first 3 months, but very rarely after 6 months (Gresty et al., 1991). Occasionally CN is reported to appear at birth or in the first few days of life (usually based on parents' report).

CN is associated with a wide variety of conditions, such as: achromatopsia, albinism, aniridia, cataracts, colobomata, cone dystrophy, congenital stationary night blindness, corneal opacities, Leber's amaurosis, optic atrophy, optic nerve hypoplasia, retinal dysplasia, and retinopathy of prematurity. Since these associated conditions are genetically unrelated, it is doubtful that CN arises from a single genetic cause. Furthermore, since CN has been reported secondary to neonatally acquired gonorrhoeic infection of the cornea (Ohm, 1958; Kommerel

Eye Movement Research/J.M. Findlay et al. (Editors)

and Mehdorn, 1982), CN cannot be due solely to a congenital malformation. The only apparent common feature among all these conditions is compromised vision in early infancy. In some patients (9%, Weiss and Biersdorf, 1989), however, thorough investigations fail to reveal any abnormalities, and the nystagmus is labelled as congenital idiopathic nystagmus.

Waveforms

Eye movement recordings reveal distinctive waveforms, which are otherwise clinically difficult or impossible to discern (Dell'Osso and Daroff, 1975). The most distinctive is a horizontal accelerating (increasing velocity) slow phase (ASP) nystagmus, which is unique to CN. However, CN is not always so distinctive in the young infant or developmentally delayed child, and the nystagmus is frequently pendular (Reinecke et al, 1988), or roving. There may also be a vertical component (circumrotatory nystagmus), or torsional component. It is now widely believed that there is no relationship between waveform and etiology.

Null Region

CN-patients frequently exhibit a region in their oculomotor range in which the nystagmus is minimal. In this null region the nystagmus may be absent, or have a low amplitude pendular or ASP waveform. When the eyes are eccentric from the null region, the waveform intensity increases and has centrifugal quick phases. There is only one null-region, which may not necessarily be in primary position, although this is frequently the case.

The null region is not always stable. It may slowly alternate spontaneously as in periodic alternating nystagmus (Abadi, this volume). The null can also be shifted by a smooth pursuit or optokinetic stimulus. For example, if an optokinetic stimulus is moved to the left, the null shifts to the right, so that when the patient looks straight ahead during the stimulation, one observes the CN ordinarily seen in left gaze. This has centrifugal quick phases to the left, in the opposite direction to normal OKN quick phases. Thus, this phenomenon has been called "reverse" or "inverted" OKN by some (Halmagyi et al, 1979; Abadi and Dickinson, 1985). However, it is important to recognise that the nystagmus is the patient's own CN, not a true optokinetic response. In the young infant with CN, we frequently observe no OKN response and no obvious null shifting, just deranged nystagmus-like eye movements. Although this can be a useful marker for CN in the sighted young who do not exhibit typical ASP waveforms, it leads us to question whether a null region is common in the very young.

Latent Component

There is often a latent component to CN, where occlusion of one eye induces an intense nystagmus with quick phases directed temporalward for the viewing eye. This may be secondary to strabismus. We call this CN with a latent component. The waveform is different from Latent Nystagmus (whether manifest or not), in which slow phases are decelerating or linear.

Oscillopsia

In spite of sometimes florid nystagmus, it is unusual for patients to complain of instabilities of the perceived visual world (oscillopsia). This should be contrasted with acquired nystagmus, which is often associated with acute or chronic oscillopsia depending on the site of the lesion. However, paradoxically, with retinal image stabilisation the CN persists but patients now perceive oscillations (eg. Leigh et al, 1988). This is very strong evidence for the preservation of extra-retinal signals (efference copy or proprioception). It also shows that CN is not driven by retinal slip, as can also be deduced by the observation that CN persists in the dark in many (but not all) patients.

Fixation Effort and Other Phenomena

The intensity of CN is highly variable both within and between individuals. CN usually intensifies with increased concentration or "fixational effort", but often decreases with convergence. CN may disappear with the eyelids closed. Kelly et al (1989) have reported two unusual cases of CN becoming manifest only during horizontal pursuit. Tusa et al (1992) have reported three patients who could switch their CN on and off at will. We have seen patients who exhibited CN only in the dark or in vertical pursuit; we are also aware of another case of CN becoming manifest only in the dark (Gresty, personal communication).

Other Eye Movement Abnormalities

Most conjugate eye movements are abnormal in CN. As described above OKN is usually absent. Vestibular nystagmus is also usually abnormal (Gresty et al, 1985). The status of the smooth pursuit system is controversial. CN patients can track a moving object with nystagmus superimposed. Although the eye movement is not smooth, it has been argued that the underlying smooth pursuit is nevertheless normal (Dell'Osso, 1986), or even super-normal (Dell'Osso et al, 1972).

The Neural Integrator (NI)

Specific models of CN waveforms have focused on aberrant feedback loops around the horizontal NI (Zee et al, 1980, Optican and Zee, 1984, Tusa et al, 1992). We shall also implicate the NI.

The saccadic, smooth pursuit, optokinetic and vestibular systems all deliver eye velocity commands to the final common path. However, different levels of steady-state innervation to the muscles correspond to different eye positions rather than different eye velocities. To generate the desired eye velocity it is necessary to apply a weighted combination of the velocity command and the mathematical integral of the velocity command to the muscles. With the correct weights and an ideal NI (ie. infinite time-constant), the dominant ocular plant time-constant (t-c) is cancelled and dynamics are determined by residual plant characteristics.

It is generally accepted that there are two anatomical components to the horizontal NI: a brainstem NI with a short t-c of about 1.5 s involving the perihypoglossal region (Cannon and Robinson, 1987), and a cerebellar component involving the flocculus/paraflocculus (Zee et al, 1981), which augments the brainstem NI t-c to its normal human value of about 25 s. Three simple ways of augmenting the brainstem NI t-c are by a cerebellar positive positional feedback, negative velocity feedback, or both (Fig.1)

NI Models

Zee et al. (1980) hypothesised that the accelerating slow phase of CN might be due to excessive *positive positional* feedback around the NI (Fig.1a). If the gain of this feedback were greater than some critical value (the reciprocal of the brainstem time constant, (ie. $k = 1/1.5 = 0.67$) the NI becomes unstable with exponential runaway. However, this model is

C.M. Harris

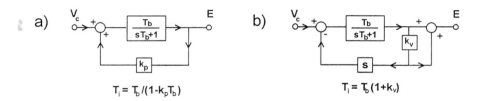

Fig. 1 - *Ways of augmenting the time-constant of the brainstem neural integrator, T_b a) with positive position feedback, k_p b) with negative velocity feedback, k_v. Both feedbacks can be combined.*

clearly inadequate because it cannot generate pendular nystagmus and the direction of the slow phases are always centrifugal, not centripetal as in most CN patients.

An alternative way of augmenting the brainstem NI t-c is by *negative velocity* feedback (Fig.1b). However, a large gain is needed to raise 1.5s to 25s, which can lead to instability. A combination of positional feedback and velocity feedback is a more plausible way to provide a stable NI, since both feedback gains can be kept moderately low. Using combined feedback, Optican and Zee (1984) postulated that the velocity feedback had the *wrong* polarity, namely it became positive. By introducing additional non-linearities in both feedback loops, a variety of CN waveforms could be simulated. Although an improvement over the Zee et al's model, this model also has problems. Positive velocity feedback has no normal function in the context of the NI. To account for the wrong sign, Optican and Zee postulated a congenital neural misrouting [cf. albinism (Collewijn et al, 1985)]. This hardly explains why CN is found in such a wide variety of other congenital and neonatally acquired clinical conditions, whose only apparent common factor is poor vision in early infancy. A second problem is that the model has difficulty in generating large amplitude pendular nystagmus, which is particularly common in infancy. Third, their model essentially generates two null regions, which probably never occurs in CN. To overcome this problem, they required asymmetric non-linearities. Undoubtedly, with tailor-made non-linearities, virtually any waveform can be generated. However, the parsimony of such an approach is brought into question.

Tusa et al (1992) incorporated both positive and negative velocity feedback around the NI to represent an abnormal and a normal fixational mechanisms occurring simultaneously in CN patients. With a velocity input bias, they could model different intensities of ASP nystagmus by varying the gains of these two loops. However, they provided no explanation for the abnormal loop. Because they dispensed with the non-linearities, the model would have a limited repertoire of waveforms. Nevertheless, by reducing the position feedback gain, they were able to induce a positional dependence of the CN. Taken to its logical conclusion, this could account for a single null, as will be described in this report.

The NI time-constant in CN

The NI t-c is usually measured from the centripetal drift in the dark during attempted eccentric fixation. However this is not usually possible in CN patients because the nystagmus occurs in the dark. However occasionally, vestibular nystagmus in the dark can be elicited from CN children, and an alternative method of estimating the NI t-c is to measure the difference in eye velocity immediately before and after a quick phase during induced vestibular nystagmus in the dark, and divide into the amplitude of the quick phase.

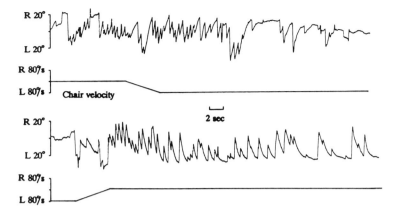

Fig. 2 - Vestibular nystagmus, induced by rotation at 80 deg/s in the dark, recorded with dc-EOG from a 10-year-old boy with congenital idiopathic nystagmus. Note nystagmus in the dark before chair rotation indicates a null in right gaze. Slow phases of induced nystagmus are curvilinear with disparate velocities immediately before and after quick phases indicating a very short neural integrator time-constant of about 1 s. Asymmetry in curvilinearity consistent with integrator null in right gaze.

We have reported an extremely short NI t-c in a patient who exhibited CN in eccentric gaze (Harris et al., 1992). We have also found a short NI t-c in another idiopathic CN patient, who exhibited very curvilinear vestibular nystagmus slow phases in the dark (Fig 2). More recently we have seen similar signs in two infants with presumed CN. These patients demonstrate that the NI t-c is extremely short in CN.

Positive Feedback in Smooth Pursuit

Sites of functional positive feedback are quite restricted in our current understanding of the conjugate oculomotor systems. However, it has been proposed that the near-unity closed-loop gain needed for SP might be achieved with efference copy (Robinson et al, 1986), which we now consider.

Fig. 3 shows a general linear efference-copy model of smooth pursuit. An internal signal representing retinal slip S is given by a linear visual transfer function of retinal image velocity, $v(s)$, which subsumes visual delay, optical magnification, and temporal response function (not modeled here). An efference copy of eye velocity, $c(s)E'(s)$, is added in

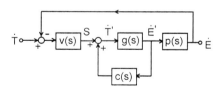

Fig. 3 - General linear model of smooth pursuit system with efference copy, see text.

order to reconstruct an estimate of target velocity, $\dot{T}'(s)$. This signal is then amplified by a forward gain transfer function, $g(s)$, and is passed to the residual plant, $p(s)$. (Except for emphasis, we do not show the dependence on s for clarity.) The overall transfer equation of the system is $\dot{E}=pgv\dot{T}/(1-gc+pgv)$. If the efferent copy is set to $c(s)=p(s)v(s)$, the system becomes balanced and effectively open loop: $\dot{E}(s) = pgv\dot{T}(s)$, and the gain can be set to unity by manipulating the forward gain, $g(s)$.

Accelerating Slow Phase Instability

Efference copy is really positive feedback, and the unbalanced system has the potential for instability if either the visually *independent* internal loop gain, $g(s)c(s)$, or the visually *dependent* external loop gain $p(s)g(s)v(s)$ is large. Consider a more specific model of SP based on a model by Robinson et al (1986) in which the final common pathway is explicit, but without any acceleration emphasis (Fig. 4). If we set the system to be unbalanced, so that the internal loop gain is high and the visual transfer gain is low, then instability occurs with accelerative runaway (Fig. 5a) (a point originally made by Robinson et al, 1986). In our model the direction of the slow phases are determined by post-saccadic drift, which depends on the precise manner in which the saccade switch is modelled and the degree of pulse-step mismatch. In Fig. 5a, a large saccade (which has a high post-saccadic drift rate) tends to set the direction for the ensuing nystagmus.

If v is made relatively larger, then ASP nystagmus in the dark [Fig. 5b(i)] becomes pendular in the light due to too much negative feedback [Fig. 5b(ii)]. In our simple model, this pendular waveform increases in amplitude until a quick phase is triggered. We suspect that in reality a limit cycle would be reached giving rise to a constant pendular waveform. We note that the pendular waveform can easily be made to decrease in amplitude so that the nystagmus disappears in the light (but still remaining ASP in the dark). By mismatching the efference copy and visual delay times, quite bizarre waveforms can be generated. Fig. 5b(iii) illustrates an alternating ASP-type waveform, while Fig. 5b(iv) shows a spontaneous reversal

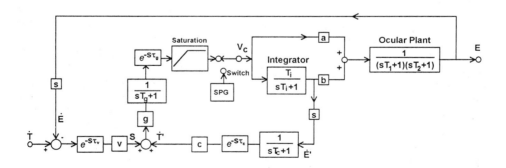

Fig. 4 - Simplified smooth pursuit model based on Robinson et al., (1986), without acceleration component. Normal values: $g=v=1$, $c=0.95$, $\tau_v=\tau_c=50ms$, $\tau_g=80ms$, $T_2=T_c=12ms$, $T_1=250ms$, $a=0.25$, $b=1$, $T_g=80ms$, $T_i=25s$, velocity saturation=90deg/s.

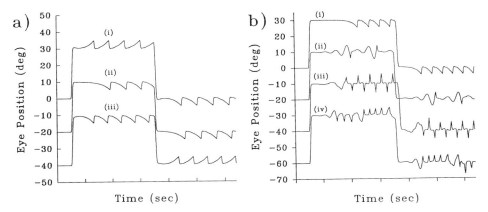

Fig. 5 - Computer simulations of eye position after 30 deg saccades between primary position and 30 deg right gaze. Plots are staggered for clarity. Simulations were carried out using Tutsim software at 1 msec steps. a) accelerating slow phase nystagmus with different pulse-step mismatches: i) over-pulsed, a=0.26; ii) matched, a=0.25; iii) under-pulsed, a=0.24. Other parameters: v=0.5, c=1, g=4, $T_i \rightarrow \infty$. b) Various mismatches: i) accelerating slow phases in the dark: v=0; ii) in the light: v=1; iii) extreme visual delay: τ_v=100ms, τ_c=50ms iv) extreme efferent copy delay τ_v=50ms, τ_c=100ms. Other parameters: c=0.5, g=6, $T_i \rightarrow \infty$,

of beat direction. If we include mismatches of delays, gains, saccadic pulse-step ratios, and the possibilities of non-linear elements and an acceleration SP component, it can be appreciated that there exists a vast parameter space to be explored. There are, however, two salient problems: a) the issue of the saccade switch and b) the lack of positional dependence.

The Saccadic Switch

Optican and Zee incorporated a switch to turn off their positive velocity feedback loop during a saccade in order to avoid distortion of the beginning and end of a saccade. We have also included a switch in which the forward path is switched out by the saccadic pulse (rather than eye velocity), and switched back when eye velocity falls below 5 deg/s. In our model the timing of the switch has only minor effects. However, because the loop delays are longer than the duration of a saccade, signals in the loop that are present at the beginning of a saccade will still be present after the saccade. Thus a saccade would not reset eye velocity to zero as seen in ASP CN. To overcome this we have simply reset signal values to zero during a saccade. However, we raise the question of how the saccadic switch might operate inside feedback loops with delays longer than saccadic durations. (This problem did not arise in Optican and Zee's, and Tusa et al's models because they did not include any delays.)

Efference Copy with a Leaky NI

Since the purpose of efference copy is presumably to reconstruct target motion/position without "artifacts" due to eye-motion, it is important to make the copy as late as possible in the motor system. The role of proprioception is far from clear, and its presence questionable in the young infant (Bruenech and Ruskell, 1993). Instead, we propose that the next best point to make an efference copy is immediately *after* the NI. As described earlier, we have found that the NI can be very leaky in CN. This has two position-dependent effects: a) it introduces a single null, and b) it can distort exponential runaway into a pendular waveform.

A leaky NI does not transfer eye velocity correctly at all gaze positions. Attempted eccentric fixation results in a centripetal *decelerating* drift towards the integrator null. With the NI inside a high-gain internal loop (as we propose), this drift will be amplified and can become accelerative, quasi-linear or stay decelerative depending on the loop gain, NI t-c, and eccentricity. Eventually a centrifugal quick phase will be triggered back toward the eccentric target. When the eyes are at the integrator null there is no centripetal drift and the CN will be minimal (Fig. 6). We propose, therefore, that the CN-null is really the NI-null. We note that, clinically, an acquired leaky NI is associated with cerebellar disease, and is invariably accompanied by poor or absent smooth pursuit. We have only ever found one case of a leaky NI and apparently normal SP, and this patient exhibited CN in eccentric gaze (Harris et al., 1992).

When the eyes are driven further from the null by an ASP in CN and when the NI is leaky, the ocular plant tends to drive the eyes back to the null, and eventually a centrifugal velocity command could not be maintained. Thus when the NI is inside an unstable positive feedback loop, centrifugal exponential runaway will be eventually turned around into the centripetal direction. In other words, provided the NI t-c is less than infinity, ASP nystagmus is ultimately a pendular nystagmus. If the NI t-c is large, the amplitude of the pendular nystagmus will also be large, and never observed because of intervening quick phases (or eventually the limit of gaze). However, if the t-c is very small, pendular nystagmus could be observed, depending on the positional and velocity thresholds for quick phase triggering. It is a general observation in our laboratory that young children tend to make fewer quick phases than adults during induced OKN and vestibular nystagmus. We suspect this reflects higher thresholds for saccadic triggering and could explain why pendular nystagmus is common in the young.

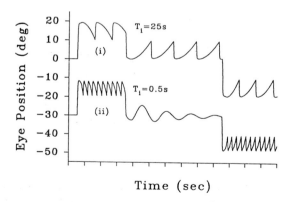

Fig. 6 - Formation of a null region by making the neural integrator leaky. i) waveform virtually independent of position with a normal integrator time-constant: $T_i=25s$; ii) damped pendular waveform around null region with intensified nystagmus eccentric from null: $T_i=0.5s$. Other parameters: $v= 0.4$, $c= 0.7$, $g=4$.

Null Shifting

A striking aspect of CN is that the CN-null is not stable in some patients. We have argued that the CN-null is identical to the NI-null, and therefore focus on the issue of NI-null alignment. It is doubtful that the brainstem has intrinsic veridical knowledge of zero velocity or zero position, but instead, it adapts to external requirements. This leads inevitably to the notion of a velocity null, V0, and a positional null, P0 (Fig. 7).

Velocity Null (V0)

The integrator must integrate relative to some base level, V0, which represents zero-velocity. The stable situation would be for V0 to be equal to desired zero-velocity demand (ie. fixation). However, due to random drift or disease, V0 may not always be aligned with desired zero-velocity, and the NI will integrate the difference over time yielding a vestibular-like nystagmus (Fig. 7b). The NI cannot distinguish between this incorrect outcome and an equivalent desired eye velocity. If we assume that there is no other source of veridical zero-velocity available to the NI, V0 can only be aligned to the long-term average velocity demand. The time-constant of this habituation would have to be long to avoid disrupting normal function of the NI. Since we have placed the NI inside an efference copy feedback loop, this mechanism becomes very similar (if not equivalent) to Leigh et al's (1981) model of periodic alternating nystagmus (PAN). Thus, we predict that patients with CN should also exhibit PAN. It is easy to miss PAN clinically (especially in children) because of the lengthy

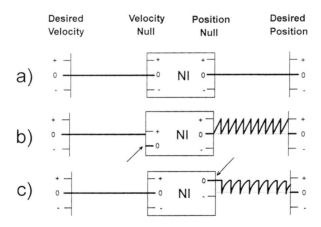

Fig. 7 - Schematic diagram of the effect of velocity and position null misalignment when attempting to fixate (zero desired eye velocity) in primary position (zero desired eye position) a) both nulls aligned. b) velocity null offset (arrow) causes the integrator to integrate the offset and drive the eyes away from desired eye position necessitating resetting quick phases (vestibular nystagmus). c) positional null misalignment (arrow) with a leaky integrator causes the eye to drift away from desired eye position with decelerating slow phases requiring resetting quick phases (gaze-paretic nystagmus).

observation time, but intriguingly Abadi (this volume) reports a high incidence of PAN in albinism. We note, however, that the converse should not be true. Patients who acquire PAN also have poor smooth pursuit (Leigh et al., 1981), and so could not demonstrate CN according to our model.

Position Null (P0)

When the NI is leaky, there is a centripetal drift back to some null position, P0. Without any NI, P0 would be at the equilibrium position of the extraocular muscles (the plant null). However, with a NI, P0 could be at any position. Assume that V0 is correctly aligned with desired zero-velocity, but that P0 is not aligned with the desired fixation position (Fig. 7c). Because the NI is leaky, the eyes will drift away from desired eye position towards P0 with an exponentially decreasing velocity. Saccades will be required to realign the eyes with desired eye position, resulting in gaze-paretic nystagmus (GPN). To overcome this misalignment, we propose that P0 also adapts to the prevailing eye position. Evidence for this is seen in patients with vestibulocerebellar disorders who exhibit GPN and rebound nystagmus (RN) (eg. Zee et al., 1976; Harris et al., 1993), and in normals with end-point nystagmus and RN (Shallo-Hoffman et al, 1990).

To shift P0, it is necessary to shift the NI positional output and to simultaneously subtract the velocity of the same shift from the NI input, so that there is no net eye drift. Hypothetically, there are a variety of ways that this null shifting may be accomplished. We speculate that P0 is driven by comparing efference copy with desired eye velocity (Fig. 8a).

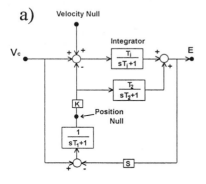
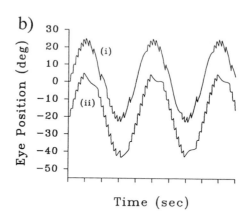

Fig. 8 - a) Hypothetical model of positional null alignment driven by the difference between desired eye velocity and efference copy. Note positional null shifting signal applied to both output and input of integrator to yield ideally a null shift without any net eye movement. b) Simulation of tracking a smooth pursuit target with i) integrator null fixed in primary position; note symmetry of nystagmus. ii) null shifting according to (a); note null moves in opposite direction to target, K=3, T_1=0.5s, T_2=3s, v=0.4, c=0.7, g=2, a=0.23, T_i=0.5s.

For example, in attempted eccentric fixation there is a zero reconstructed target velocity (in a balanced system) but with a leaky integrator there will be a non-zero centripetal efference copy velocity which will give rise to a positional shift towards the eccentric fixation, thereby gradually nulling the centripetal drift. In an unbalanced system with too much internal loop gain, this alignment could become deranged. Fig. 8b illustrates a model of tracking with i) the null fixed at primary position and ii) with the null allowed to shift according to the scheme in Fig. 8a. As can be seen the null moves in the opposite direction to target velocity. Unfortunately, during OKN this model is not satisfactory because the model moves P0 wildly and not necessarily in the opposite direction to the stimulus. According to this model, null shifting becomes very complex and possibly chaotic in an unbalanced system. Clearly, this needs further investigation.

Latent Phenomena

Although the mechanism for null shifting is unclear, it seems plausible that the latent phenomena (LN and CN with a latent component) are also manifestations of null shifting brought about by monocular suppression of vision. In latent LN, there is no nystagmus in primary position with both eyes open. With monocular occlusion, a decelerating slow phase nystagmus appears with quick phases temporalward for the viewing eye. This waveform is similar (if not identical) to that seen in GPN. However, upon monocular nasalward viewing, the nystagmus is greatly dampened, and unlike GPN, LN shows either a small or no reversal in direction. We doubt that the NI is exceptionally leaky in LN, because in our laboratory eccentric gaze holding in the dark is not always impaired. It appears as if P0 has shifted far nasalward, so that even with a moderate NI t-c, GPN (end-point) nystagmus occurs because primary position is now very eccentric from the nasal P0. In CN with a latent component, the excessive internal loop gain causes the decelerating LN slow phases to become accelerating according to our model.

The cause for the null shift is unknown, but there may be a huge tonic imbalance in the velocity command due to naso-temporal tracking asymmetry becoming manifest on suppression of vision in one eye (whether voluntary or otherwise).

Oscillopsia

Efference copy is an integral part of this model. By adding an efference copy of eye velocity to retinal slip, the target motion in space is reconstructed (we are assuming a still head):

$$\dot{T}'(s) = (c - vp)\dot{E}'(s) + v\dot{T}(s) \tag{1}$$

When the system is balanced [ie. $c(s) = v(s)p(s)$], reconstructed target velocity depends only on true target velocity and the visual transfer function, and is independent of efferent copy eye velocity $\dot{E}'(s)$. If the system goes unstable such that $g(s)$ becomes an oscillator $O(s)$ (dependent or independent of its input), reconstructed target velocity is unaltered:

$$\dot{T}'(s) = v\dot{T}(s), \quad \dot{E} = pO(s) \tag{2}$$

Assuming perceptual processes can also reconstruct target motion from involuntary eye movements (Bedell et al, 1989), there would be no perception of motion, and it would be expected that CN patients do not complain of oscillopsia. If the system were not perfectly balanced (ie. $c \neq vp$), then some illusory oscillation would be expected, with an amplitude of $(c/p-v)$ times the actual eye oscillations. However, provided the magnitude is below detection threshold, which may itself be elevated (Dietrich and Brandt, 1987), oscillopsia would not be reported. In retinal stabilization, target velocity is made equal to eye velocity, $T(s)=\dot{E}(s)$, and reconstructed target velocity and eye velocity become:

$$\dot{T}'(s) = cO(s), \qquad \dot{E} = pO(s) \tag{3}$$

so that under these artificial conditions reconstructed eye velocity will also oscillate, and the patient would perceive an illusory oscillatory motion of the target, as has been reported in patients with CN (Leigh et al., 1988). However, the magnitude of the illusory oscillations have been consistently reported to be less than the actual eye movement [50%, Kommerel (1986); 75%, Leigh et al (1988); 76%, Bedell and Currie (1993)], implying incomplete cancellation. However, from (3), we see that the ratio of reconstructed target velocity to eye velocity is given by:

$$\dot{T}'(s) \div \dot{E}(s) = c(s) \div p(s) \tag{4}$$

Thus perfect correspondence would only occur if $c(s) = p(s)$. Even in a balanced system ($c/p=v$), this would be unlikely because the visual system does not transfer retinal slip perfectly [$v(s)<1$], and we would expect the transfer to decrease with increasing retinal slip velocities. Thus, unless we know the CN patient's $v(s)$ and perceptual threshold for motion at the nystagmus frequency, it is difficult to interpret the degree of oscillopsia under retinal stabilization.

Discussion

According to this model, CN results from too much positive velocity feedback in the smooth pursuit system. How does this come about in the young infant? As discussed earlier, the circumstantial evidence implicates early visual deprivation. Could poor central vision (on which the SP system usually depends) lead to a maldeveloped SP with excessive gains?

It is clear from Fig. 3 that the system can never know $v(s)$ from its internal variables, even with extraocular muscle proprioception. Hence the system cannot prescribe the correct parameter values for a balanced system, and must somehow adapt/learn them via some neural network. (We suspect that balancing efference copy is a priority since this is required for spatial constancy, as well as for other eye movements.)

Perhaps the young plastic visuomotor system always interprets poor central retinal acuity (whether due to poor acuity or a pre-existing nystagmus) as a failure of poor smooth tracking, and attempts to compensate by raising the gain. However, raising the gain eventually causes the system to go unstable, which in turn degrades central visual acuity. This in turn leads to further gain increase, and so on, until full-blown nystagmus is entrenched and the visuomotor system is at the limit of its adaptive range. If efference copy is sufficient to prevent oscillopsia, the system would have no internal information about the instability - just poor acuity. The difficulty, though, is why do not all infants develop CN?

Another possibility is that the smooth tracking system is driven not by target motion per se, but by the difference between target and background (full-field) motion. This has the advantage of yielding a veridical estimate of zero velocity in space because the whole visual world never moves (except in an earthquake). Assume that there are two channels which average velocity over retinal space: a central process with a limited central area of velocity summation, and a surround mechanism which averages velocity over the entire retina (the fast optokinetic system?). Because of the preferential weighting of the central visual field the central mechanism normally outweighs the surround mechanism. If the central visual field has abnormally poor motion detection, the difference between the central and surround mechanism will be less than normal, and require a greater than normal forward gain, which could lead to CN. Such a system is reminiscent of the ocular following response described by Miles and Kawano (1987).

Variable gain

In order to keep models relatively simple and quasi-linear, it is usual to consider only fixed gain systems (as we have done). However this is probably overly simplistic. The ultimate proving ground for the smooth pursuit system is presumably how well visual details are perceived. There are two requirements: a) to keep retinal image motion low and b) to maintain foveal registration. The latter is particularly important since more precise gain control is needed for finer visual details. We rule out direct negative visual feedback in the SP system [although others do not (Goldreich et al, 1992)], since this cannot give rise to CN in the dark. However, the forward gain, $g(s)$, could be under negative visual feedback. Thus when the image of a visual detail of interest falls behind or ahead of the fovea, the gain is turned down or up. Such a system is inherently non-linear and could lead to instabilities by itself, as well as inducing too high a gain for the SP system, as we have described here. Clearly, the gain of this gain control would depend on the nature of the visual detail, and would have to be under voluntary/attentional control. This has some intuitive appeal, since introspection tells us that we engage greater tracking or fixation "effort" when we wish to see a finer visual detail. Thus, depending on the visual task, for some CN patients, the whole gain range may fall above the critical gain for instability. In others the range may straddle the critical gain, and in normals the range is always below critical gain. Thus, how the patient responds to tracking a moving target, fixating a stationary target, viewing in the dark, or shutting the eyelids, will dictate the degree of nystagmus. A full description of this type of model is beyond the scope of this report, but it can be seen that it would become necessary to consider a range of gains in the SP system, which may be set in infancy.

Why is the integrator leaky in CN?

According to our model the CN-null is equivalent to the NI positional null, P0. Since P0 only becomes manifest when the integrator is leaky, we predict that a leaky NI should be common, if not universal among CN patients with a null region. We suggest two plausible explanations for a leaky NI.

One possibility is that in the normal infant the NI is originally leaky and the t-c develops along with the need for greater image stability to support emerging foveal vision in the first few months of life. If foveal vision never develops properly because of some congenital or early-onset insult, the NI t-c does not develop either. Second, the NI t-c in CN appears (so far) to be extraordinarily short, suggesting an active shortening of the t-c rather than a failure of development. Robinson et al (1984) have shown that Alexander's law results from a leaky integrator, and proposed that the NI t-c is actively shortened in the face of chronic vestibular imbalance. A similar argument could be recruited for CN patients (Harris et al, 1992). Because of incessant oscillations, the NI is made leaky thereby creating a null region and improving acuity. The shorter the t-c, the more dampened is the CN, but at the price of a narrower null region and more intense nystagmus outside the null region.

Indeed these two scenarios are not mutually exclusive, and when combined, they could lead to an insidious cycle of maldevelopment: the NI fails to develop, which leads to increased SP gain, which leads to nystagmus, which reduces the NI t-c leading to further SP gain increase, and so on.

Acknowledgements

I thank the charities *Help a Child to See,* and the *Iris Fund* for their support. I also thank A. Kriss, F. Shawkat, D. Taylor, and P. West for their help.

References

Abadi, R.V. & Dickinson, C.M. (1985) The influence of preexisting oscillations on the binocular optokinetic response. *Ann Neurol, 17,* 578-586.

Abadi, R.V. (1994) this volume

Bedell, H.E., Klopfenstein JF & Yuan N (1989) Extraretinal infromation about eye position during involuntary eye movement : optokinetic afternystagmus. *Percept Psychophys, 46,* 579-586.

Bedell, H.E. & Currie, D.C. (1993) Extraretinal signals for congenital nystagmus. *Invest Ophthalmol Vis Sci, 34,* 2325-2332.

Bruenech, J.R. & Ruskell, G.L. (1993) Extraocular muscle spindles in human infants. *Proc Vis Sci, 15.*

Cannon, S.C. & Robinson, D.A. (1987) Loss of the neural integrator of the oculomotor system from brain stem lesions in monkey. *J Neurophysiol 57,* 1383-1409.

Collewijn, H., Apkarian, P., & Spekreijse, H. (1985) The oculomotor behaviour of human albinos. *Brain, 108,* 1-28.

Dell'Osso, L.F., Gauthier, G., Liberman, G. & Stark L (1972) Eye movement recordings as a diagnostic tool in a case of congenital nystagmus. *Amer J Optom Arch Acad Optom, 49,* 3-13.

Dell'Osso, L.F. & Daroff, M.D. (1975) Congenital nystagmus waveforms and foveation strategy. *Doc Ophthalmol, 39,* 155-182.

Dell'Osso, L.F. (1986) Evaluation of smooth pursuit in the presence of congenital nystagmus. *Neuro-ophthalmology, 6,* 383-406.

Dietrich, M. & Brandt, Th. (1987) Impaired motion perception in congenital nystagmus and acquired ocular motor palsy. *Clin Vision Sci, 1,* 337-345.

Goldreich, D., Krauzlis, R.J. & Lisberger, S.G. (1992) Effect of changing feedback delay on spontaneous oscillations in smooth pursuit eye movements of monkeys *J Neurophysiol, 67,* 625-638.

Gresty, M.A., Barratt, H.J., Page, N.G.R. & Ell JJ (1985) Assessment of vestibular-ocular reflexes in congenital nystagmus. *Ann Neurol, 17*, 129-136.

Gresty, M.A., Bronstein, A.M. & Page, N.G. (1991) Congenital type nystagmus emerging later in life. *Neurology, 41*, 653-656.

Halmagyi, G.M., Gresty, M.A. & Leech, J. (1979) Reversed optokinetic nystagmus (OKN): mechanism and clinical significance, *Ann Neurol,7*, 429-435.

Harris, C.M., Jacobs, M., Shawkat, F. & Taylor, D. (1992) Human ocular motor neural integrator failure. *Neuro-ophthalmology, 13*, 25-34.

Harris, C.M., Walker, J., Wilson, J., & Russell-Eggitt, I. (1993) Eye movements in a familial vestibulocerebellar disorder. *Neuropediatrics, 24*, 117-122.

Kelly, B.J., Rosenberg, M.L., Zee, D.S. & Optican, L.M. (1989) Unilateral pursuit-induced nystagmus *Neurology, 29*, 424-416.

Kommerell, G. & Mehdorn, E. (1982) Is an optokinetic defect the cause of congenital and latent nystagmus? In: Lennerstrand G, Zee DS, Keller EL, eds. *Functional basis of ocular motility disorders*. New York, Pergamon Press, pp 159-167.

Kommerel, G. (1986) Congenital nystagmus: control of slow tracking movements by target offset from the fovea. *Graefes Arch Clin Exp Ophthalmol, 224*, 295-298.

Leigh, R.J., Dell'Osso, L.F., Yaniglos, S.S. & Thurston, S.E. (1988) Oscillopsia, retinal image stabilization and congenital nystagmus *Invest Ophthalmol Vis Sci, 29*, 279-282.

Leigh, R.J., Robinson, D.A. & Zee, D.S. (1981) A hypothetical explanation for periodic alternating nystagmus: instability of the optokinetic-vestibular system. *Ann NY Acad Sci, 374*, 619-635.

Miles, F.A. & Kawano, K. (1987) Visual stabilization of the eyes. *Trends Neurosci, 10*, 153-158.

Ohm, J. (1958) Nystagmus und Schielenm bei Sehschwachen und Blinden. Enke, Stuttgart.

Optican, L.M. & Zee, D.S. (1984) A hypothetical explanation of congenital nystagmus. *Biol Cybern, 50*, 119-134.

Reinecke, R.D., Guo, S. & Goldstein, H.P. (1988) Waveform evolution in infantile nystagmus: an EOG study. *Binoc Vision 3*, 191-202.

Robinson, D.A., Gordon, J.L. & Gordon, S.E. (1986) A model of the smooth pursuit eye movement system *Biol Cybern, 55*, 43-57.

Robinson, D.A., Zee, D.S., Hain, T.C., Holmes, A., & Rosenberg, L.F. (1984) Alexander's Law: Its behavior and origin in the human vestibulo-ocular reflex *Ann Neurol, 16*, 714-722.

Shallo-Hoffmanm, J., Schwarze, H., Simonsz, H. & Mühlendyck, H. (1990) A reexamination of end-point nystagmus in normals. *Invest Ophthalmol Vis Sci 31*, 388-392.

Tusa, R.J., Zee, D.S., Hain, T.C. & Simonsz, H. (1992) Voluntary control of congenital nystagmus *Clin Vision Sci, 7*, 195-210.

Weiss, A.H. & Biersdorf, W.R. (1989) Visual sensory disorders in congenital nystagmus. *Ophthalmology, 96*, 517-523.

Zee, D.S., Yee, R.D., Cogan, D.G., Robinson, D.A. & Engel, W.K. (1976) Ocular motor abnormalities in hereditary cerebellar ataxia. *Brain 99*, 207-234.

Zee, D.S., Leigh, R.J. & Mathieu-Millaire, F. (1980) Cerebellar control of ocular stability. *Ann Neurol 7*, 37-40.

Zee, D.S., Yamazaki, A., Butler, P. & Gucer, G. (1981) Effects of ablation of the flocculus and paraflocculus on eye movements in the primate. *J Neurophysiol, 46*, 878-899.

EYE MOVEMENT BEHAVIOUR IN HUMAN ALBINOS

R.V. Abadi[1] and E. Pascal[2]

1. Department of Optometry and Vision Sciences, UMIST, PO Box 88, Manchester, M60 1QD, UK,

2. Department of Vision Sciences, Glasgow Caledonian University, Glasgow, G4 OBA, Scotland.

Abstract

Twenty five oculocutaneous albinos and seven ocular albinos participated in the study. Eye movements were monitored using infra-red oculography and six features of the nystagmus were examined: amplitude, frequency, waveform, metrics of the slow phase, beat direction and the temporal nature of the cycle. All the subjects had a nystagmus; twenty having either a jerk with extended foveation and/or a pseudocycloid oscillation in primary gaze, Over one third of the subjects exhibited a periodic alternating nystagmus which proved to be due to the temporal shifting of the null zone,

Keywords

Albino, periodic alternating nystagmus, spatial and temporal null zones.

Introduction

Congenital nystagmus (CN) is a consistent feature of all forms of albinism. The binocular involuntary oscillations are conjugate and occur predominantly in the horizontal plane (Apkarian et al 1983, St. John et al 1984, Collewijn et al 1985, Abadi and Dickinson 1986, Abadi et al 1989). As with other forms of CN there is much intersubject and individual variation in the amplitude, frequency and waveform of the oscillations (Abadi and Dickinson 1986). The purpose of this paper is to describe the results of some of the studies we have carried out into the nature of nystagmus in albinism. In particular the unexpectedly high incidence of periodic alternating nystagmus, previously thought to be quite rare, is described and discussed.

Eye Movement Research/J.M. Findlay et al. (Editors)

The albino state

Albinism is a congenital, inherited disorder of melanin pigment and can be found throughout the animal kingdom. It is due to a metabolic error whereby the enzyme tyrosinase is either absent or inactive (Witkop 1979). This results in hypopigmentation of the hair, skin and eyes (oculocutaneous albinism - OCA) or it may be limited to the eyes (ocular albinism - OA). Considerable heterogeneity exists within albinism: at least ten different forms of OCA and four different types of OA have been described (Taylor 1978, Witkop 1979, O'Donnell and Green 1981, Van Dorp 1985, Kinnear et al 1985, Abadi and Pascal 1989). The identification of each variety of albinism is made using histological, genetic, biochemical and clinical criteria.

The two most prevalent forms of OCA are tyrosinase negative OCA (TNOCA) and tyrosinase positive OCA (TPOCA). In the former no pigment is produced, indicating a total lack of the enzyme tyrosinase, whereas in the latter melanin gradually accumulates in the hair, skin and eyes as the person becomes older. In OA the skin and hair appear similar to those of unaffected family members whilst the eyes show hypopigmentation. OCA and OA are characterised by similar clinical ocular signs: translucent irides, pale fundi with clearly visible vasculature, poor foveal differentiation, a high incidence of large refractive errors, strabismus and congenital nystagmus. The main clinical characteristics of the albino groups described in this paper are given in Table 1.

The motor characteristics of nystagmus in albinos

Despite the term congenital, many albino infants do not seem to exhibit a nystagmus until several weeks after birth and even then the oscillations are very different from those exhibited in adulthood (Abadi and Dickinson 1986). In infancy the eye movements are typically pendular with a large amplitude, especially if the ambient light levels are high. With increasing age changes in the intensity and waveform of the nystagmus occur, such that in adulthood the oscillations are indistinguishable from those of congenital idiopathic nystagmus. However, unlike the albino the idiopath shows no apparent abnormality of the eyes and visual pathway.

Subjects and methods

Recently we investigated the CN exhibited by 32 albinos aged between 8 and 57 years. Horizontal eye movements were recorded using infra-red oculography (Abadi and Dickinson 1986). Each subject was instructed to fixate a stationary white target (0.5°) projected onto a large uniform hemicylindrical screen measuring 172° horizontal and 50° vertical. The target had a luminance of 4.1 cd/m^2 whilst the low internal room illumination provided background screen illuminance of about 0.4 cd/m^2. Five directions of gaze were examined (-20°,-10°,0°,+10°, and +20°) for time periods of up to 8 minutes for each gaze position. In addition each subject was instructed to change

Characteristics of various forms of albinism

Type of albinism		TNOCA	TPOCA	XOA	AROA
Clinical appearance	Skin	White — no tan	Cream → Pink → can tan	May be paler than sibs	May be paler than sibs
	Hair	White	White → yellow	Normal range	Normal range
	Irides	Pale grey → blue	Blue → hazel	Blue → brown	Blue → brown
Pigment change		None	Increase with age	Darkening of eyes with age	None
Fundus	Macula	Hypoplasia	Hypoplasia	Hypoplasia	Hypoplasia
	Periphery	No retinal pigment	Some retinal pigment	As TNOCA	As TPOCA
Vision	VA	6/60 or less	3/60 → 6/18	3/60 → 6/15	3/60 → 6/15
	Photophobia	+ + + +	+ + to + + +	+ + to + + +	+ + to + + +
Inheritance		autosomal recessive	autosomal recessive	X-linked	autosomal recessive
Incidence		1 : 37000	1 : 31000	1 : 50000	1 : 50000
Biochemical tests	Tyrosine	No pigmentation	Pigmentation	Pigmentation	Pigmentation
	Tyrosine and cysteine	No pigmentation	Pigmentation	Pigmentation	Pigmentation
Ocular features		Nystagmus, aberrant visual pathway, squint	As TNOCA	As tTNOCA	As TNOCA

TNOCA - tyrosinase negative oculocutaneous albinism
TPOCA - tyrosinase positive oculocutaneous albinism
XOA   - x-linked ocular albinism
AROA - autosomal recessive ocular albinism

Table 1 - The characteristics of the four most common forms of albinism.

gaze position in response to single step changes in target position. Head movements were minimised using a chin rest and a forehead restraint. Five features of the nystagmus were examined: amplitude, frequency, waveform, the metrics of the slow phase and beat direction (Dell'Osso and Daroff 1975, Abadi and Dickinson 1986, Abadi and Worfolk 1989, Abadi et al 1990).

Binocular and monocular visual acuities were recorded using a high contrast (90%) Bailey-Lovie chart located in the primary position. Visual acuity was recorded as the log of the minimum angle of resolution (log MAR) from 1.0 (Snellen equivalent 6/60) to 0.0 (Snellen equivalent 6/6).

Results

Table 2 gives a summary of some of the pertinent features of the 32 subjects. A marked attenuation of CN intensity with convergence was found in 6 of the albinos, while 7 of the 32 displayed a definite spatial null zone. A further 9 subjects exhibited some variation in the nystagmus at different gaze angles. Three individuals had both convergence and spatial nulls. Abnormal head postures were adopted by 23 of the albinos, especially if a visually demanding task was being performed. As previously reported (Abadi and Whittle 1991), not all of the compensatory head postures were correlated with the null zone, where the nystagmus intensity was at a minimum.

Identical twins with TNOCA

Amongst our albino sample was a pair of 16 year old monozygote twin girls with TNOCA. In primary gaze both displayed conjugate horizontal CN which was similar in waveform (Jerk with extended foveation, Jef) and frequency (2.0Hz:1.9Hz). The mean amplitudes (6.8°:3.7°) and foveation periods, (28% : 68%) defined as the % time CN slow phase velocity $\leq 10°$/sec were, however, markedly different. In addition, the variation in CN parameters with gaze position was dissimilar: for one twin the Jef waveform persisted in most positions of eccentric gaze whereas the other twin exhibited considerable waveform variability, including bidirectional eye movements between 5° and 25° left gaze and dual jerk (DJ) oscillations beyond 25° right gaze. In fact on the basis of their eye movements these identical twins were no more alike than any pair of unrelated albinos. Since monozygote twins have identical genotypes the ocular motor differences must have arisen either from intra-uterine or other environmental factors. A similar conclusion was reached after examining monozygote twins with idiopathic CN (Abadi et al 1983).

The waveform

Most of the waveforms originally described by Dell'Osso and Daroff (1975) were exhibited by the subjects during primary gaze (Figure 1). However, each type did not occur with equal frequency. Instead, the Jef and pseudocycloid (PC) oscillations, either

	TNOCA	TPOCA	XOA	AROA
No. Subjects	16	9	5	2
Age Range	12-57 yrs	13-52 yrs	8-40 yrs	15-38 yrs
VA Range	0.52-1.00	0.34-0.92	0.70-1.12	0.80-0.86
Strabismus	16	9	5	2
CN Waveform				
Jef	4*	3*	2**	1*
PC	5**	3**	1	1*
Jdv	1	-	-	-
P	1	-	-	-
Pfs	2	-	-	-
DJ	1	1	-	-
AP	1	-	-	-
PC+P	-	-	2	-
Jef+DJ	-	1*	-	-
PC+DJ	1*	-	-	-
Torsion	-	1	-	-
PAN	4	4	2	2
Spatial Null Zone	4	2	1	0
Convergence Null	3	2	1	0
Abnormal Head Posture	10	6	5	2
Head Nodding	4	3	1	1

Table 2 - Summary of the eye movement behaviour of the original 32 albino subjects. The waveforms were either Jef (jerk with extended foveation), PC (pseudocycloid), Jdv (jerk with decreasing velocity), P (pendular). Pfs (pendular with foveating saccades), DJ (dual jerk) or AP (asymmetric pendular). The asterisks (*) identify the 12 subjects who were found to have a PAN. Visual acuity is given in log MAR.

alone or in combination with others, were exhibited in the primary position by 75% of the albino population. The PC waveform was the most common, accounting for 41% of the subject pool. There appeared to be no correlation between the CN waveform and the different types of albinism.

Nystagmus and visual acuity

The effect of CN on visual performance is dictated by the retinal slip velocities that comprise each slow phase and, in particular, the duration of the period in which the retinal slip velocity is reasonably slow (Dell'Osso 1973, Abadi and Sandigkcioglu 1974, Dell'Osso and Daroff 1975, Abadi and Dickinson 1986, Abadi et al 1989, Abadi and Worfolk 1989, Abadi et al 1990,). In 1989 Abadi and Worfolk demonstrated that for a group of subjects with idiopathic nystagmus there was a significant correlation between the log of the minimum angle of resolution (log MAR) and the percentage of slow phase time spent at, or below, 10 deg/sec. Figure 2 illustrates this relationship for 22 of the 32 albinos in our group (10 TNOCA, 7 TPOCA, 3XOA AND 2 AO). From the graph it can be seen that the visual acuity improves as dwell time (the proportion of time spent at low velocities) increased. However, unlike the linear relationship previously described by Abadi and Worfolk (1989) for the idiopaths, the albino data were best fitted by an exponential curve. Thus acuity improved as dwell time increased only up to a resolution limit of 0.52 (about 6/18). Any further increase in the duration of low retinal slip velocities did not result in better visual acuity.

Therefore, whilst visual resolution in albinos is affected by the ongoing nystagmus, there must be other factors which are responsible for the absolute limit in performance. Other studies support the view that the lack of a normally differentiated fovea is the limiting factor in albinism (Wilson et al 1988(a), Wilson et al 1988(b), Yo et al 1989,Abadi and Pascal 1991, Abadi and Pascal 1993).

Periodic alternating nystagmus

Periodic alternating nystagmus (PAN) is a conjugate, horizontal, jerk oscillation in which regular reversals in the direction of the fast component are separated by brief "quiet" intervals. As it is considered to be an ocular motor rarity (Davis and Smith 1981, Daroff and Dell'Osso 1974, Baloh et al 1976, Oosterveld and Rademakers 1979, Cross et al 1982, Guyer and Lessell 1986), we were surprised to discover that 12 of the 32 albinos (37.5%) exhibited PAN. These 12 subjects came from all four categories of albinism (4 TNOCA, 4 TPOCA, 2 XO AND 2 AROA). Although each of the subjects had a strabismus none exhibited a latent or manifest latent nystagmus. Generally the intensity of the nystagmus was low at the start of each jerk phase; it gradually built up to a mid-cycle maximum then decreased until the quiet phase was reached, as shown in Figure 3.

During the PAN jerk phases, either Jef or PC waveforms were encountered, whilst during the quiet phases triangular, bidirectional and pendular oscillations were evident. The duration of each PAN cycle displayed inter- and intra-subject variation.

TITLE	WAVEFORM	DESCRIPTION
Pure Pendular P	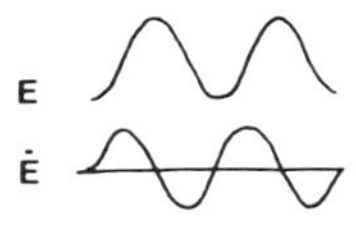 E Ė	Symmetrical sinusoidal oscillation (ie. bidirectional slow phase), biased such that the fovea rests on the target at one of the peaks.
Pendular with foveating saccades Pfs	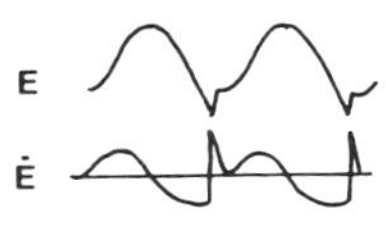 E Ė	Sinusoidal oscillation with small braking saccades that halt the slow phase and realign the retinal image with the fovea.
Pure Jerk J	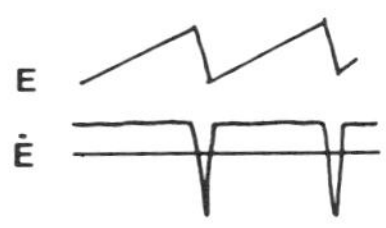 E Ė	Simple linear slow phase followed by a saccadic fast phase that recaptures the target on the fovea.
Jerk with extended foveation Jef	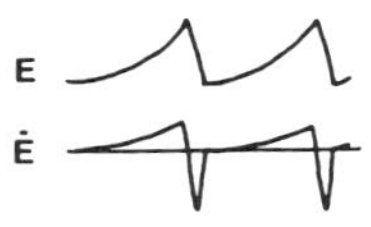 E Ė	Accelerating slow phase corrected by a refoveating saccadic fast phase. The time the retinal image remains motionless is variable.
Pseudocycloid PC	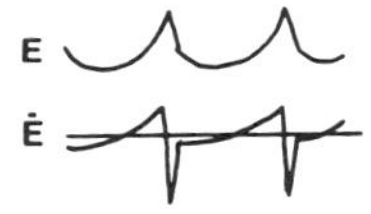 E Ė	Accelerating slow phase terminated by a braking saccade. Following slow phase achieves foveation.
Jerk with decreasing velocity Jdv	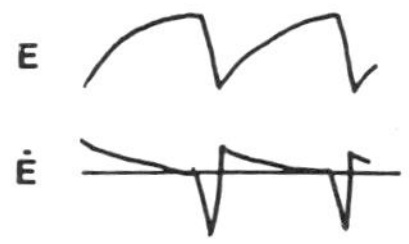 E Ė	Decelerating slow phase off target corrected by a refixation saccade. With slow deceleration, slow phase may seem linear.
Dual Jerk DJ	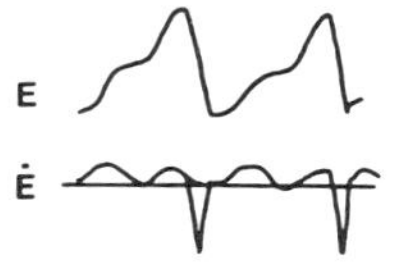 E Ė	Superimposition of rapid, small amplitude sinusoidal oscillation on a slower, larger amplitude jerk waveform.

Fig. 1. - *Schematic illustration and description of the seven CN waveforms seen amongst our albino subjects. E = eye position; Ė = eye velocity.*

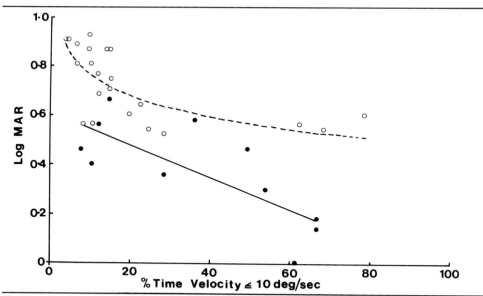

Fig. 2. - The relationship between the log of the minimum angle of resolution (log MAR) and the percentage time that the nystagmus slow phase eye velocity was less than or equal to 10 deg/sec. Open symbols (o) represent the data from 22 albino subjects; with a best fit exponential curve. Solid symbols (o) represent data from 11 subjects with idiopathic nystagmus. The straight line shows the least squares best fit to the data points (r = -0.76, p<0.01).

Across the subjects mean cycle length was 271s with maximum and minimum times of 430s and 141s respectively. Although most of the PAN cycles were fairly symmetrical, some were markedly asymmetric. For example, one ocular albino displayed a right-beating phase of 282s and a left beating phase of 126s during a single cycle. None of the subjects with PAN demonstrated a convergence null or an abnormal head posture.

The effect of gaze changes on PAN
In order to test the hypothesis that PAN results from a temporal shift in the null zone, eye movements were monitored at different gaze positions. Attempts were made to "track" the null zone so that if a subject entered into a quiet phase following a period of right-beating nystagmus, then by looking into the right gaze (ie. to where the null should then move) the quiet phase should be prolonged. This is illustrated in Figure 4. Similarly, it should also be possible to completely abolish or significantly alter the PAN cycle by instructing the subject to maintain fixation on a horizontally eccentric stimulus. In this way one jerk phase would be prolonged at the expense of the other. Figure 5 illustrates for 8 subjects (7 albinos and 1 idiopath) the PAN cycle was indeed modified by eccentric gaze.

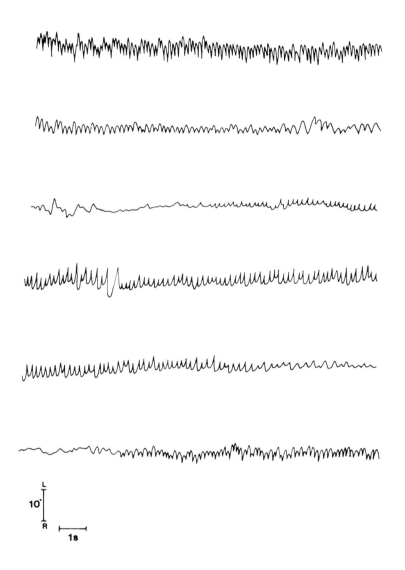

Fig. 3. - Eye movement recordings from a TNOCA with congenital PAN. This trace (which is not continuous in time) illustrates the temporal change in nystagmus waveform, amplitude and frequency as the horizontal, involuntary oscillation goes through its cycle. The left and right beating phases are separated by quiet phases.

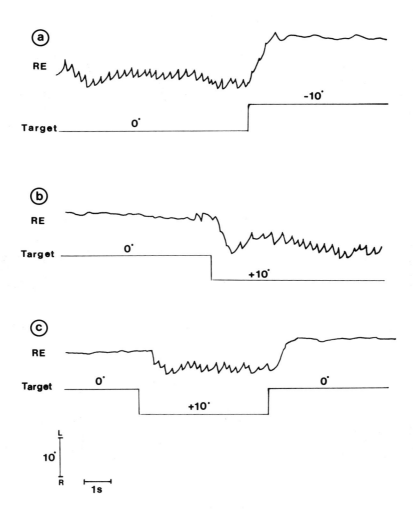

Fig. 4. - *Eye movement recordings from the right eye of a TPOCA subject which support the hypothesis that PAN results from a temporal shift in the null zone. In (a) a change in gaze position from the centre (0°) to the left (-10°) resulted in the reappearance of the quiet phase. In (b) a right beating nystagmus was evoked when gaze position was altered from the centre (0°) to the right (+10°) at the start of a left beating phase. In (c) right gaze (+10°) elicited a right beating oscillation when a quiet phase was present in central gaze (0°).*

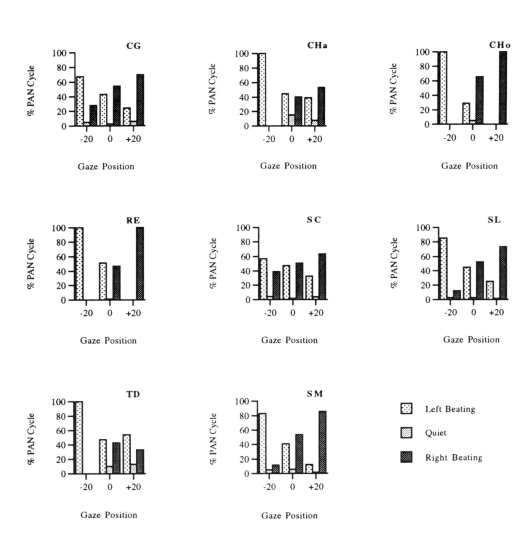

Fig. 5. - The effect of gaze position on the percentage of right beating and left beating nystagmus per PAN cycle. Histograms for seven albinos and one idiopath (subject SM) are shown.

Discussion

To date there have been few published reports of subjects with congenital PAN (Daroff and Dell'Osso 1974, Baloh et al 1976, Oosterveld and Rademakers 1979, Davis and Smith 1981, Cross et al 1982, Collewijn et al 1985, Abadi and Dickinson 1986, Guyer and Lessell 1986, Abadi et al 1989), and in only five papers has PAN been described in association with albinism (Davis and Smith 1981, Collewijn et al 1985, Abadi and Dickinson 1986, Guyer and Lessell 1986, Abadi et al 1989). Although the incidence of albinos with PAN has in the past appeared to be very small, it is clear from this study that PAN is far more common in human albinos than was previously thought. A possible explanation for such a high incidence (37.5%) is that in our laboratory we routinely monitor fixational eye movements over long continuous periods. Other published studies on albinos may have missed the periodicity because the recording sessions were too short to reveal any spontaneous changes in the nystagmus beat direction.

Although the variation in PAN timing with gaze position has been mentioned in several other smaller studies this has invariably been in a qualitative manner. Our present research conclusively supports the hypothesis originally put forward by Daroff and Dell'Osso (1974) that PAN results from a temporal shifting of the null zone. It is also interesting to note that accurate foveation during the minimum velocity period of the slow phase of each waveform can still be achieved even with this form of nystagmus periodicity (Abadi et al 1989). In a parallel study we have also examined idiopaths with congenital PAN and the spatial and temporal nature of the oscillations appear to be indistinguishable from those of the albino group. Thus any attempt to elucidate the mechanisms behind PAN cannot rely on features that are unique to the albino.

Whilst the aetiology of PAN has received some attention, no firm conclusions have been reached. Leigh and his colleagues described a hypothetical model of PAN which relied on an instability in the neural mechanisms that generate vestibular and optokinetic linear slow phases, combined with an inability to process retinal velocity error signals (Leigh et al 1981). More recently Harris (in this volume) has proposed that CN is due to an excess gain in the internal efference copy positive feedback loop. The efference copy being a velocity feedback signal from the output of the common neural integrator of the ocular motor system (ie. the position signal output of the neural integrator is differentiated and fed back). Future work will aim to explore these models with our PAN subjects.

References

Abadi R.V. and Dickinson C.M. (1986) Waveform characteristics in congenital nystagmus. Doc Ophthalmol; 64: 153-167.

Abadi R.V., Dickinson C.M., Lomas, M.S. and Ackerley R. (1983) Congenital nystagmus in identical twins. Br J Ophthalmol; 67: 693-695.

Abadi R.V., Dickinson C.M., Pascal E. and Papas E. (1990) Retinal image quality in albinos. Ophthalmic Paediat Genet; 11: 171-177.

Abadi R.V. and Pascal E. (1989) The recognition and management of albinism. Ophthal Physiol Opt; 9: 3-15.

Abadi R.V. and Pascal E. (1991) Visual resolution limits in human albinism. Vision Res;   31: 1445-1447.

Abadi R.V. and Pascal E. (1993) Incremental light detection thresholds across the central visual fields of human albinos. Invest Ophthalmol Vis Sci; 34: 1683-1690.

Abadi R.V., Pascal E., Whittle J. and Worfolk R. (1989) Retinal fixation behaviour in human albinos. Optom Vision Sci; 66: 276-280.

Abadi R.V. and Sandikcioglu M. (1974) Electro-oculographic responses in a case of bilateral idiopathic nystagmus. Br J Physiol Opt; 29: 73-85.

Abadi R.V. and Whittle J.   (1991) The nature of head postures in congenital nystagmus.   Arch Ophthalmol; 109: 216-220.

Abadi R.V. and Worfolk R. (1989) Retinal slip velocities in congenital nystagmus. Vision Res; 29: 195-205.

Apkarian P., Spekreijse H. and Collewijn H. (1983) Oculomotor behaviour in human albinos. Doc Ophthalmol Proc Series; 37: 361-372.

Baloh R.W. and Honrubia V., Konrad H.R. (1976) Periodic alternating nystagmus. Brain; 99: 11-26

Collewijn H., Apkarian, P. and Spekreijse H. (1985) The oculomotor behaviour of human albinos. Brain; 108: 1-28.

Cross S.A., Smith J.L. and Norton E.W.D. (1982) Periodic alternating nystagmus clearing after vitrectomy. J Clin Neuro-Ophthalmol; 2: 5-11

Daroff R.D. and Dell'Osso L.F. (1974) Periodic alternating nystagmus and the shifting null. Can J Otolaryngol; 3: 367-371

Davis D.G. and Smith J.L. (1971) Periodic alternating nystagmus: A report of eight cases. Am J Ophthalmol; 72: 757-762

Dell'Osso L.F. (1973) Fixation characteristics in hereditary congenital nystagmus. Am

J Optom. Arch Am Optom; 50: 85-90.

Dell'Osso L.F. and Daroff R.B. (1975) Congenital nystagmus waveforms and foveation strategy. Doc Ophthalmol; 39: 155-182.

Guyer D.R., and Lessell S. (1986) Periodic alternating nystagmus associated with albinism. J Clin Neuro-Ophthalmol; 6: 82-85

Harris C.M. (1994) Problems in modelling congenital nystagmus. Eye movement research: Processes, Mechanisms and Applications. Findlay, J.M, Kentridge, R.W. and Walker, R. (eds). Elsevier Press.

Kinnear P.E., Jay B. and Witkop C.J. (1985) Albinism. Surv Ophthalmol; 30: 75-101.

Leigh R.J., Robinson D.A., and Zee D.S. (1981) A hypothetical explanation for periodic alternating nystagmus: Instability in the optokinetic-vestibular system. Ann NY Acad Sci; 374: 619-635

O'Donnell F.E. and Green W.R. (1981) The eye in albinism. In Duane T.D. (ed): Clinical Ophthalmology 4. New York: Harper and Row: 1-23.

Oosterveld W.J. and Rademakers W.J.A.C. (1979) Nystagmus alternans. Acta Otolaryngol; 87: 404-409

St. John R., Fisk J.D., Timney B. and Goodale M.A. (1984) Eye movements of human albinos. Am J Optom Physiol Opt; 61: 377-385.

Taylor W.O.G. (1978) Visual disabilities of oculocutaneous albinism and their alleviation. Trans Ophthalmol Soc U.K.; 98: 423-445.

Van Dorp D.B. (1985) Shades of grey in human albinism. Holland Ophthalmic Publishing Centre, Amersfoot: 1-173.

Wilson H.R., Mets M.B., Nagy, S.E. and Ferrera V.P. (1988a) Spatial frequency and orientation tuning of spatial visual mechanisms in human albinos. Vision Res; 28: 991-999.

Wilson H.R., Mets M.B., Nagy, S.E. and Kressel A.B.(1988b) Albino spatial vision as an instance of arrested visual development. Vision Res; 28: 979-990.

Witkop C.J. (1979) Depigmentations of the general and oral tissues and their genetic foundations. Ala J Med Sci; 16: 331-343.

Yo C., Wilson H.R., Mets M.B., and Ritacco D.G. (1989)Human albinos can discriminate spatial frequency and phase as accurately as normal subjects. Vision Res; 29: 1561-1574.

SMOOTH PURSUIT RESPONSES TO STEP RAMP STIMULI IN PATIENTS WITH DISCRETE FRONTAL LOBE LESIONS

N.A.R. Munro[1], B. Gaymard[2], S. Rivaud[2], J.F Stein[1] and Ch. Pierrot-Deseilligny[2].

[1]University Laboratory of Physiology, Parks Road, Oxford.
[2]Laboratoire INSERM 289, Hôpital de la Salpêtrière, 47 Bvd de l'Hôpital, 75653 Paris cédex 13. France.

Abstract

We recorded smooth pursuit in 4 patients with small, discrete unilateral frontal lobe lesions and in 13 normal subjects. In all cases the lesions affected the frontal eye fields (FEF). The use of step ramp stimuli enabled the measurement of different parameters. Initial acceleration was measured to assess the open loop response, maximum and mean gain was an index of the closed loop response and inaccuracy was used to assess the ability of the saccadic system to refoveate. Patients showed a decreased gain towards the side of the lesion, with significant asymmetry. In contrast, acceleration was reduced in both directions without significant asymmetry. In the monkey, it has been shown that the middle temporal area (MT) codes motion in the contralateral side of space in both directions equally, whereas the posterior parietal cortex (PPC) contains cells which encode velocity asymetrically. Our findings would, therefore, be consistent with the hypothesis that there are two pathways from the homolog of MT in the occipito-tempero-parietal pit to the FEF; a direct pathway and an indirect pathway, via the PPC. The direct pathway might encode target velocity in a retinotopic reference frame without directional asymmetry and be responsible for initial acceleration. The indirect pathway might encode target velocity in a non-retinal frame and be responsible for the predictive mechanisms for maintaining pursuit.

Keywords

Pursuit, Eye Movements, Step-ramp, Human, Prediction, Frontal cortex

Introduction

Smooth pursuit allows the foveate animal to keep the images of small objects moving across a complex background almost stationary on the fovea. The image moving across the retina reduces acuity more than poor foveation (Eckmiller, 1987; Barnes & Smith, 1981). The pursuit system, therefore, achieves optimum visual acuity by minimising retinal slip velocity. Thus we might expect that the pathways for pursuit to reflect the anatomy of the pathways of motion which is represented by the magnocellular system (Zeki, 1990; Merigan et al. 1991; Lisberger & Pavelko, 1989). This begins with the 'Y'-like ganglion cells of the retina and passes through layers 1 and 2 of the lateral geniculate nucleus (LGN) and onto layers $4c\alpha$ and 4β of V1 and thence to the cytochrome rich stripes of V2 and areas V3. The magnocellular pathway

Eye Movement Research/J.M. Findlay et al. (Editors)

then passes through the traditional cortical areas associated with the control of pursuit: the middle temporal area (MT), the medial superior temporal sulcus (MST) and the posterior parietal cortex (PPC) (Ungerleider & Desimone, 1992; Eckmiller, 1987; Seltzer & Pandya, 1989; Van Essen et al., 1981). Lesions in these areas in monkeys result in significant impairment in smooth pursuit(Newsome et al. 1988; Komatsu & Wurtz, 1988; Dursteler et al. 1988; Newsome et al. 1985). In man, impaired smooth pursuit is frequently attributed to damage to the tempero-parietal regions(Leigh, 1989; Thurston et al. 1988; Leigh & Tusla. 1985; Morrow & Sharpe, 1990). A greatly simplified diagram, loosely based on published diagrams of the visual hierarchy in the macaque, is shown in figure 1. The areas on the left of the diagram contain largely magnocellular elements and it may be seen that the frontal eye fields (FEF) and prefrontal cortex (PFC) represent a pinnacle of the hierarchy at which parvo and magnocellular systems meet.

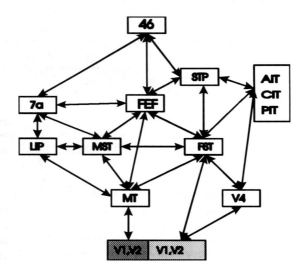

Fig. 1 - A simplified diagram of the visual hierarchy. (AIT, CIT, and PIT denote Anterior, Central and Posterior Inferotemporal, STP, Superior Temporal Polysensry and FST Floor of Superior Temporal, other abbreviations as in text.)

It is now well recognised that the FEF are an important component of the smooth pursuit pathway(Eckmiller, 1987; Sharpe et al. 1989). Abnormalities in smooth pursuit following lesions of the FEF in monkeys have been well documented(Lynch & Allison, 1985; Lynch 1987; Keating et al., 1985; Keating 1991, MacAvoy et al. 1991). There have been a few reports of patients with frontal lobe lesions who demonstrated smooth pursuit asymmetry when tracking targets moving continuously in sinusoidal motion(Pyykkö et al. 1984; Morrow & Sharpe, 1990). We report abnormalities of smooth pursuit in four patients with discrete lesions affecting the frontal lobes. By using step ramp stimuli we were able to study different parameters reflecting both the initiation and maintenance of smooth pursuit. We propose that the different pathways from MT to the frontal lobe structures evident in the visual hierarchy subserve different functions in the generation of smooth pursuit eye movements.

Methods

Subjects

We measured smooth pursuit in 13 normal control subjects (mean age, 42.5; sd, 7.1) and 4 patients (mean age, 48.0; sd, 14.2) whose clinical details are set out in figure 2. All subjects had normal uncorrected vision (with the exception of mild presbyopia) and no visual field defects. There were 3 patients with right sided lesions and 1 with a right sided lesion.

Patient	Age	Delay	Radiological site
L1	35	28 days	FEF and PFC
R1	44	21 days	FEF
R2	41	8 days	FEF
R3	72	5 days	FEF and PFC

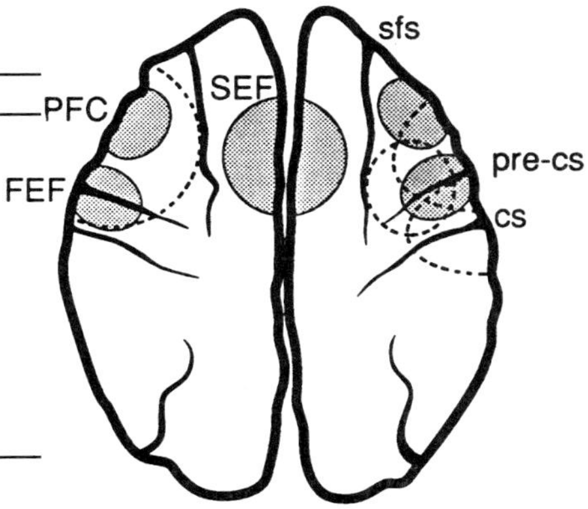

Fig. 2 - The clinical details of the 4 patients indicating their age, the delay between their stroke and examination and the site of the lesion.

Eye movement recording

Horizontal movements were recorded from the right eye by infrared reflectometry (IRIS system, Scalar Ltd) with a resolution of 0.01°. The pursuit stimulus was an array of LEDs sited 80cm from the subject. Each LED was separated from its neighbour by 0.16°. The target was stationary at 0° for a period of 1.8, 2.0, 2.2 or 2.4 seconds, then stepped to one side. The target then moved at a constant speed in the direction opposite to the step to a final position at ±14°. 14 Speeds between 6°/sec and 28°/sec were employed. The size of the step was adjusted so that the target was moving past the midline 150 msec after the onset of target motion. Thus for trials with latencies in the region of 150 msec the target was foveated at the onset of pursuit. This provided the optimum conditions for tracking to be initiated with a smooth pursuit acceleration rather than a saccade. Seven point calibration was performed every 60 seconds. Eye position was digitized on line using a 386 personal computer with a 12 bit analogue to digital converter (Metrabyte inc).

Analysis

An interactive computer program was used to remove saccades, which were identified using velocity and acceleration criteria and checked visually. Gaps left by saccade removal were filled by interpolation of the velocity trace. Further analysis was performed after digitally filtering with a 5 pole Butterworth filter at 35 Hz. Typical traces are shown in figure 3. The top trace shows a normal subject pursuing a target moving to the right at 16°s/sec. The "staircase" line represents the position of the target and the continuous line represents the eye position. It can be seen that pursuit begins with a smooth initial acceleration at about 160 msecs with several "catch-up" saccades. The middle trace shows the velocity of the eye with time. It represents the differential of the top trace, but the saccades have been removed and a parabola has been fitted through the first velocity peak. Because of the high noise inevitable in the velocity trace using the infrared technique numerical averaging and curve fitting techniques were used to obtain consistent measurements of parameters. The lower trace is taken from a patient with a right frontal eye field lesion also pursuing a target moving to the right. The decreased pursuit velocity is evident. Latency was obtained initially by inspection and was then corrected by fitting a straight line through the first 100 msecs and measuring its intersection with the axis. The initial acceleration was obtained from the gradient of the straight line fitted through the first 100 msec of the velocity trace. Maximum gain, (g_{max}) was obtained from the first peak velocity, divided by the target velocity. The mean gain, (g_{mean}) was obtained by averaging eye velocity between the times t_{peak}, the time of the first eye velocity peak, and t_{final}, the time when the target stopped. We chose mean error, (E_{mean}) and root mean square error, (E_{rms}) as the measures of the ability of the saccadic system to commpensate for poor foveation. They are calculated between the times t_{peak} and t_{final} and are defined formally as follows, s and e, denoting target and eye position respectively,

$$E_{rms} = \sqrt{\frac{1}{t_{final}-t_{peak}} \int_{t_{peak}}^{t_{final}} (s-e)^2 dt} \tag{1}$$

$$E_{mean} = \frac{1}{t_{final}-t_{peak}} \int_{t_{peak}}^{t_{final}} (|s|-|e|) dt \tag{2}$$

Statistical Considerations

Each subject was presented with a balanced set of 96 target speeds and directions. 34.7% of these trials were eliminated during offline analysis as blinks, multiple saccades or failure to attend to the target rendered the data unanalysable. This resulted in imbalance as some target speeds were not represented. By having a large number of target speeds with a small interval (2°/sec) it was possible to overcome

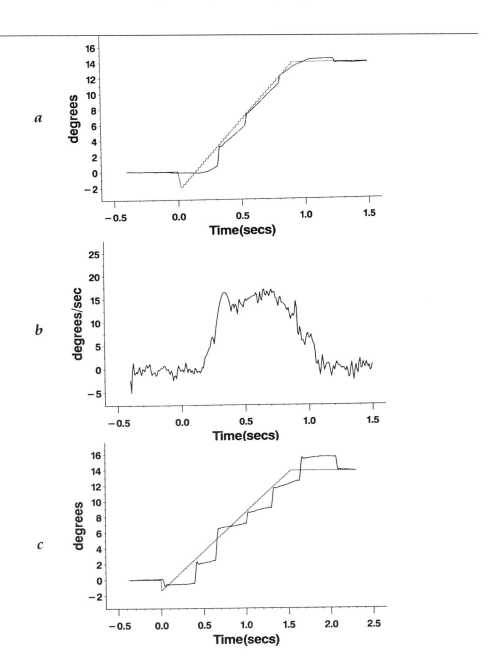

Fig 3 a) Trace of a normal subject pursuing a target moving to the right at 16°/sec; the target trace is a staircase. b) the same trace differentiated and c) the trace of a patient (R2), with a frontal eye field lesion on the right.

imbalance by treating target speed as a continuous rather than discrete variable and using analysis of variance for an unbalanced design utilising the SAS statistical software package (Cary, NC). Bias which could have arisen from rejected trials being more frequent amongst higher target velocities was avoided using this method. "Least square means" which represent the midpoint of regression lines, rather than simple means were therefore used as summary statistics. These statistics are generated as a by product of the analysis of variance.

Results

Figure 4 shows, in graphical form, the least square means for maximum and mean gain, latency, initial acceleration and mean and root mean square error for individual patients and the aggregate of controls. Error bars indicate standard error. The small standard error bars on the control subjects reflect the large number of trials, approximately 1000, over which the averages have been taken. Table 1 shows the least square means of the various parameters in aggregate for patients versus controls. Patients perform less well than controls in all six parameters to a high level of significance.

	Controls	Patients	p value
mean gain	0.987 (0.007)	0.472 (0.014)	< 0.0001
max gain	0.684 (0.005)	0.405 (0.010)	< 0.0001
latency	174.8 (1.7)	207.6 (3.4)	< 0.0001
acceleration	69.0 (1.4)	28.6 (2.9)	< 0.0001
mean error	0.69 (0.03)	1.71 (0.05)	< 0.0001
rms error	1.09 (0.02)	2.14 (0.05)	< 0.0001

Table 1 - Pursuit parameters of patients and controls. The "least square means" averaged over both directions are shown with standard errors in parentheses.

Table 2 shows the values for patients when pursuing a target moving towards (ipsilataleral) and away (contralateral) to the side of the lesion. The significance levels are calculated from the interactive terms of the analysis of variance. For mean and maximum gain and latency performance is much worse towards the side of the lesion witha statistical significance of p < 0.0001. In contrast there is no statisticaly significant asymmetry in initial acceleration and mean and rms error.

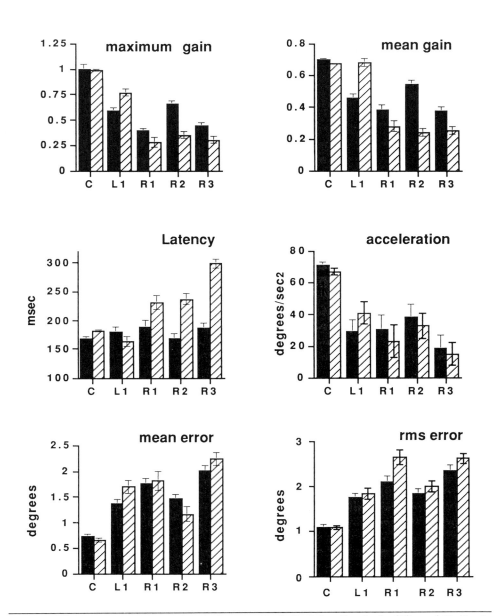

Fig. 4 - The various parameters of pursuit plotted for Controls (C) and Patients (L1, R1,R2 and R3). Dark bars indicate motion in a rightward direction and shaded bars represent motion in a leftward direction.

	Ipsilateral	Contralateral	p value
mean gain	0.308 (0.013)	0.502 (0.013)	< 0.0001
max gain	0.382 (0.019)	0.570 (0.019)	< 0.0001
latency	237.8 (4.7)	177.2 (4.7)	< 0.0001
acceleration	25.0 (4.1)	32.1 (4.1)	0.3448
mean error	1.67 (0.08)	1.74 (0.08)	0.0748
rms error	2.24 (0.07)	2.04 (0.07)	0.0247

Table 2 - *The various parameters in patients pursuing targets moving ipsi and contralateral to the lesion*

Discussion

Step ramp stimuli in patients with frontal lesions were used for the first time here, allowing us to analyze both the maintenance and initiation of smooth pursuit. Abnormalities of smooth pursuit are usually attributed to damage to the Superior Temporal Sulcus (STS) and PPC in monkeys (Komatsu & Wurtz, 1988; Newsome et al. 1988; Dursteler et al. 1987), and the tempero-parietal association cortex in humans(Leigh, 1989; Thurston et al. 1988). Abnormalities of pursuit attributed to damage to FEF in monkeys have been reported (Lynch & Allison, 1985, Lynch, 1987; Keating et al. 1985; Keating, 1991). Reports of slow eye movements following cortical stimulation(Bruce & Goldberg 1985; Huerta et al. 1987; MacAvoy et al. 1988 & 1991; Gottlieb et al., 1993) and the occasional finding of neurons in the FEF responding solely to pursuit eye movements had indicated a possible role for pursuit eye movements in the FEF. The importance of the FEF for smooth pursuit control has been confirmed in a study of 3 rhesus monkeys who underwent surgical ablation of the FEF and were required to track ramps moving at 20°/sec, 30°/sec and 40°/sec(Lynch, 1987). Recordings were made with electrooculography so measurements other than pursuit gain could not be measured. There was significant loss in gain, dropping in one monkey from 1.1 before surgery to 0.15 one week after surgery recovering to 0.75 after 47 days. MacAvoy et al (1988 & 1991) and Gottlieb et al. (1993) lesioned and performed microstimulation and single unit recordings in macaque monkeys. The cortex in the depths of the anterior bank and fundus of the arcuate sulcus was critical for smooth pursuit. They reported only ipsilateral deficits affecting both gain and initial acceleration. In particular, they noted that predictive tracking of sinusoidal target motion was ipsilaterally impaired. Recordings were made from stimulation sites and responded to eye movement in the same direction as stimulated movement and tended to respond to predictive movements, anticipating, for example, changes in direction of a target moving in sinusoidal motion. In humans Pyykkö et al (1984) demonstrated impaired maximum pursuit gain in a group of patients with dysphasia and presumed lesions of the frontal eye fields. Morrow and Sharpe(1990) demonstrated asymmetric smooth pursuit to a target moving in sinusoidal motion in two patients with FEF lesions.

MT connects principally with MST but there are also direct connections to the FEF and the prefrontal cortex (PFC, area 46). MST connects directly with FEF and PFC and indirectly via areas LIP and 7a (Boussaoud et al. 1990; Livingstone & Hubel, 1983; Maioli et al. 1983; Maunsell & Van Essen, 1983; Ungerleider & Desimone, 1992; Van Essen et al. 1981). Lesions in monkeys confined to MT have shown to produce a deficit independant of the direction of the target but dependant on the position of the target in the visual field (Newsome et al., 1985). This is in contrast with lesions of MST or area 7a which are independant of position but do depend upon target direction (Newsome et al., 1988; Komatsu & Wurtz, 1988, I & III). Since neurons in MT are active before neurons in MST it has been postulated that MT is principly responsible for the initiation of pursuit. We address the question as to whether the different functions of MT, MST, and PPC are reflected in the different pathways from MT to the frontal lobe. Specifically we propose that asymmetry in smooth pursuit arises from an asymmetry in the representation of motion or velocity space in the lesioned area.

It is customary to refer to abnormalities of pursuit which depend on the position of the target on the retina as "retinotopic". More generally, the terms "retinotopic" and "spatiotopic" refer to the reference frame in which a body is represented and may apply to all measurements made in that frame be they position, velocity, acceleration or angular rotation. In this discussion and in figure 5, we use $\mathbf{r}$ and $\mathbf{v}$ refer to the position and velocity of the target and $\Re$ and Σ are subscripts which refer to retinotopic and spatiotopic reference frames. An apostrophe denotes asymmetry of representation. Single cell studies in MT confirm that velocity is represented in both directions but target position is only represented contralateral to the visual field(Newsome et al., 1988; Komatsu & Wurtz, 1988, I & III). More formally, in MT, retinotopic velocity space is not lateralised. Using the above notation cells in MT represent $\{\mathbf{r}_\Re{'},\mathbf{v}_\Re\}$. The lack of smooth pursuit asymmetry in monkeys with isolated lesions in MT is consistent with the absence of the lateralisation of velocity space. The observation that lesions in MST and PPC result in pursuit asymmetry leads to the expectation of lateralisation of velocity space: an imbalance in the sensitivity of cells representing direction toward the lesion. In the parietal cortex single cells studies have confirmed such an asymmetry (Steinmetz et al., 1987; Sakata et al, 1988). In addition 79% of visual tracking neurons also showed equal activity during fixation suggesting that they representated non-retinotopic velocity space, $\mathbf{v}_\Sigma$. In PPC the velocity sensitive cells have receptive fields that extend in to the opposite visual field - a decrease in the lateralisation of position. Thus we may represent some of the cellular activity which represents velocity space in PPC, as $\{\mathbf{v}_\Sigma{'},\mathbf{v}_\Re{'}\}$. In MST, asymmetry is less clear. There are cells with small receptive fields representing target velocity and cells with larger receptive fields active to motion in the opposite direction which probably encode the moving background and therefore eye velocity. As in PPC there are also cells that are eye velocity independant (Komatsu & Wurtz, 1988, I & III). Thus the pathway to the FEF from MT via MST and PPC would appear to involve the lateralisation of velocity space in a retinotopic frame and the transform of velocity space into a non retinotopic frame. Prediction requires the representation of the velocity to be non-retinotopic and this pathway might be useful for the maintenance of pursuit.

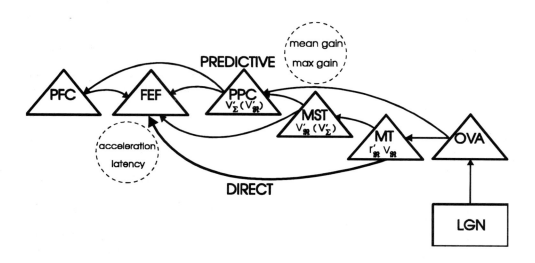

Fig. 5 - Proposed "direct" and "predictive" intracortical pathways from MT to FEF for the initiation and maintenance of pursuit.

We suggest that maximum and mean gain are measures of pursuit maintenance and are mediated via the PPC. Velocity signals reaching FEF are lateralised and result in asymmetry gain.. We propose that acceleration is mediated directly in a retinotopic frame from MT. Signals reach FEF from this pathway are not strongly lateralised accounting for the relative lack of asymmetry in initial acceleration. It may be that smooth pursuit is initiated by the detection of a moving target in the peripheral field and is thus strongly dependant of retinotopic *position*, accounting for the asymmetry of latency. Mean and rms error are defined in the retinal frame. The relative lack of asymmetry of error would suggest that the positional signals in a retinal frame in FEF are not strongly lateralised. It is possible that the increased latency observed in patients might bias these results by causing the target initally to be non-foveate. One might expect this, however, to bias the results in the direction of increased asymmetry of initial acceleration.

Acknowledgement

This work was supported by the McDonnell-Pew foundation and Wellcome Trust.

References

Barnes, G.R. & Smith, R. (1981). The effects on visual discrimination of image movement accross the stationary retina. Aviat Space Environ Med. 52, 466-472.

Boussaoud, D., Ungerleider, L.G. & Desimone, R. (1990). Pathways for motion analysis: Cortical connections of the medial superior temporal visual areas in the macaque. J.Comp.Neurol. 296, 462-495.

Bruce, C.J. & Goldberg, M.E. (1985). Primate frontal eye fields I. Single neurons discharging before saccades. J.Neurophysiol. 53, 606-635.

Dursteler, M.R. & Wurtz, R.H. (1988). Pursuit and optokinetic defects following chemical lesions of cortical areas MT and MST. J.Neurophysiol. 60, 940-964.

Dursteler, M.R., Wurtz, R.H. & Newsome, W.T. (1987). Directional pursuit deficits following lesions of the foveal representation within the superior temporal sulcus of the macaque monkey. J.Neurophysiol. 57, 1262-1287.

Eckmiller, R. (1987). Neural control of pursuit eye movements.Physiol.Rev 67, 797-857.

Gottlieb, J.P., Bruce, C.J. & MacAvoy, M.G. (1993). Smooth eye movements elicited by microstimulation in the primate frontal eye field. J.Neurophysiol. 69(3), 786-799.

Huerta, M.F., Krubitzer, L.A. & Kaas, J.H. (1987). Frontal eye field as defined by intracortical microstimulation in squirrel monkeys, owl monkeys and macaque monkeys. J.Comp.Neurol 265, 332-361.

Keating, E.G. (1991). Frontal eye field lesions impair predictive and visually guided eye movements. Exp.Brain Res. 86(2), 311-323.

Keating, E.G., Gooley, S.G. & Kenny, D.V. (1985). Impaired tracking and loss of predictive eye movements after removal of the frontal eye fields. Soc Neurosci Abstr 11, 472.

Komatsu, H. & Wurtz, R.H. (1988). Relation of cortical areas MT and MST to pursuit eye movements. I. Localization and visual properties of neurons. J.Neurophysiol. 60, 580-603.

Komatsu, H. & Wurtz, R.H. (1988). Relation of cortical areas MT and MST to pursuit eye movements. III. Interaction with full-field visual stimulation. J.Neurophysiol. 60, 621-644.

Leigh, R.J. (1989). The cortical control of ocular pursuit movements. Rev Neurol (Paris) 145, 605-612.

Leigh, R.J. & Tusa, R.J. (1985). Disturbance of smooth pursuit caused by infarction of occipitoparietal cortex. Ann.Neurol. 17, 185-187.

Lisberger, S.G. & Pavelko, T.A. (1989). Topographic and directional organization of visual motion inputs for the initiation of horizontal and vertical smooth-pursuit eye movements in monkeys. J.Neurophysiol. 61, 173-185.

Livingstone, M.S. & Hubel, D.H. (1983). Specificity of cortico-cortical connections in monkey visual system. Nature 304, 531-534.

Lynch, J.C. (1987). Frontal eye field lesions in monkeys disrupt visual pursuit. Exp Brain Res 68(2), 437-441.

Lynch, J.C. & Allison, J.C. (1985). A quantitative study of visual pursuit deficits following lesions of the frontal eye fields in rhesus monkeys. Soc.Neurosci.Abstr. 11, 473.

MacAvoy, M.G., Bruce, C.J. & Gottlieb, G.L. (1988). Smooth eye movements elicited by microstimulation in the frontaleye fields region of alert monkeys. Soc.Neurosci.Abstr. 14, 956.

MacAvoy, M.G., Gottlieb, G.L. & Bruce, C.J. (1991). Smooth-pursuit eye movement representation in the primate frontal eye field. Cerebral Cortex 1, 95-102.

Maioli, M.G., Squatrito, S., Galletti, C., Battaglini, P.P. & Sanseverino, E.R. (1983). Cortico-cortical connections from the visual region of the superior temporal sulcus to frontal eye field in the macaque. Brain Res. 265, 294-299.

Maunsell, J.H. & Van Essen, D.C. (1983). The connections of the middle temporal visual area (MT) and their relationship to a cortical hierarchy in the macaque monkey. J.Neurosci. 3, 2563-2586.

Merigan, W.H., Byrne, C.E. & Maunsell, J.H. (1991). Does primate motion perception depend on the magnocellular pathway? J.Neurosci. 11, 3422-3429.

Morrow, M.J. & Sharpe, J.A. (1990). Cerebral hemispheric localization of smooth pursuit asymmetry. Neurology 40, 284-292.

Newsome, W.T., Wurtz, R.H., Dursteler, M.R. & Mikami, A. (1985). Deficits in visual motion processing following ibotenic acid lesions of the middle temporal visual area of the macaque monkey. J.Neurosci. 5, 825-840.

Newsome, W.T., Wurtz, R.H. & Komatsu, H. (1988). Relation of cortical areas MT and MST to pursuit eye movements. II. Differentiation of retinal from extraretinal inputs. J.Neurophysiol. 60, 604-620.

Pyykkö, I., Dahlen, A.I., Schalën, L. & Hindfelt, B. (1984). Eye movements in patients with speech dyspraxia. Acta Otolaryngol.(Stockh) 98, 485-489.

Sakata, H., Shibutani, H.& Kawano, K. (1983). Functional properties of visual tracking neurons in posterior parietal cortex of the monkey. J.Neurophysiol. 49, 1364-1380.

Seltzer, B. & Pandya, D.N. (1989). Intrinsic connections and architectonics of the superior temporal sulcus in the rhesus monkey. J.Comp.Neurol 290, 451-471.

Sharpe, J.A., Morrow, M.J. & Johnston, J.L. (1989). Smooth pursuit: anatomy, physiology and disorders. Bull.Soc.Belge.Ophtalmol. 237, 113-144.

Steinmetz, M.A., Motter, B.C., Duffy, C.J. & Mountcastle, V.B. (1987). Functional properties of parietal visual neurons: radial organization of directionalities within the visual field. J.Neurosci. 7(1), 177-191.

Thurston, S.E., Leigh, R.J., Crawford, T., Thompson, A. & Kennard, C. (1988). Two distinct deficits of visual tracking caused by unilateral lesions of cerebral cortex in humans. Ann Neurol 23, 266-273.

Ungerleider, L.G. & Desimone, R. (1992). Cortical connections of visual area MT in the macaque. J.Comp.Neurol.

Van Essen, D.C., Maunsell, J.H. & Bixby, J.L. (1981). The middle temporal visual area in the macaque: myeloarchitecture, connections, functional properties and topographic organization. J.Comp.Neurol 199, 293-326.

Zeki, S. 1990 The motion pathways of the visual cortex. In: Vision: coding and efficiency. C. Blakemore, Ed. Cambridge University Press, Cambridge, pp. 319-360.

SMOOTH PURSUIT EYE MOVEMENT ABNORMALITIES IN PATIENTS WITH SCHIZOPHRENIA AND FOCAL CORTICAL LESIONS.

Crawford T.J., Lawden M.C., Haegar B., Henderson L. and Kennard C

Academic Unit of Neuroscience, Charing Cross & Westminster Medical School, London, W6 8RF.

Abstract

Smooth pursuit eye movements have been widely used in clinical research in attempts to clarify the neural mechanism underlying various brain diseases. However, many of these studies are subject to two major weaknesses: a failure to control for neuropharmacological factors and an inadequately defined visual context against which the smooth pursuit tracking is measured. This paper addresses both of these issues in patients with schizophrenia or focal cortical lesions. In the first study we compared smooth pursuit eye movements in the dark in medicated and non-medicated patients fulfilling the DSM-IIIR criteria for schizophrenia, and a group of age-matched control subjects. Relative smooth pursuit eye velocity (i.e. gain) was reduced in both schizophrenic groups; however the effect was significantly greater in the neuroleptically medicated group. In the second study smooth pursuit, with and without a structured background, was compared in patients with discrete cortical lesions and normal subjects. The analysis revealed a cohort of patients manifesting a large inhibitory effect of a structured background on pursuit eye movements. Examination of CT scans showed that two regions are of particular importance in this effect: an area of parietal cortex lying within the architectonic boundaries of Brodmann's area 40 (Brodmann, 1909); and an area of white matter close to the lateral ventricles containing cortico-cortical connections. These data point strongly to the critical importance of neuropharmacological factors and the visual background conditions in studies of smooth pursuit eye movements.

Keywords

smooth pursuit, neuroleptics, schizophrenia, cortical lesions, inferior parietal.

Introduction

The primary function of smooth pursuit eye movements in the service of vision is to maintain a relatively stable image of a moving target on the retina. In order to achieve efficient and accurate pursuit, this remarkable oculomotor system must solve a number of neuro-computational problems encompassing selective attention, transduction of target velocity (see Thurston et al, 1988) and the formation of a program of appropriate motor output to the extraocular muscles. In normal subjects efficient smooth pursuit can also break down when, for example, the target velocity exceeds

Eye Movement Research/J.M. Findlay et al. (Editors)

the performance range for smooth pursuit (Lisberger & Westbrook, 1985). In circumstances where this range is exceeded visuo-motor mechanisms may switch to saccadic eye movements in order to monitor the target, but there will usually be a high cost in target visibility. Saccadic and smooth pursuit systems therefore interact to achieve the final objective of maintaining a stable target image at the fovea and together they provide a prime example of an integrated visuo-motor network involving many levels of brain function. In view of the complexity of these neurocognitive and sensorimotor interactions it is not surprising that smooth pursuit abnormalities have been observed in a wide range of neurological conditions including Alzheimer's disease (Fletcher & Sharpe, 1986), Parkinson's disease (Shibasaki et al, 1979; Sharpe et al, 1987; Gibson, Pimlott & Kennard, 1987) and a variety of brain lesions (Leigh & Zee, 1991).

Diefendorf and Dodge (1908) were the first to recognize that patients with schizophrenia were unable to generate normal smooth pursuit eye movements. This finding has since been extensively confirmed in many laboratories (see Clementz & Sweeney, 1990). Interest in the possible genetic basis of the abnormality was heightened by the demonstration that a smooth pursuit abnormality is present in approximately 50% of first degree relatives of schizophrenics (Holzman et al, 1974). However, there has been relatively little advance in understanding the abnormal neural mechanisms giving rise to this eye movement abnormality. A major problem in the interpretation of previous research has been the confounding factor of neuroleptic medication since most studies have failed to include a cohort of unmedicated patients. Furthermore, many studies have used idiosyncratic measures of smooth pursuit performance combined with the noisy and unstable electro-oculography (EOG) methodology. In the first study we attempted to determine the role of neuroleptic medication by investigating smooth pursuit eye velocity using infra-red oculography in both neuroleptically treated and untreated schizophrenics patients and age-matched normal controls.

Methods

Apparatus: The smooth pursuit stimulus was a bright red laser spot, back-projected onto a translucent screen which was placed 1.5 metres in front of the subject. The target oscillated horizontally with a triangular waveform of amplitude 22.5°. visual angle. Four velocities were used: 10, 20, 30, and 36°/sec. In each block 6 full cycles were recorded at each velocity employed. Subjects were comfortably seated 1.5 metres from the screen. Head movements were constrained by the use of an adjustable head rest. The experiment was conducted in the dark. Eye movements were recorded using an infra-red limbus-reflection device (Skalar IRIS) with a linearity range of ±15°. A hardware anti-aliasing filter (cut off frequency 200Hz) was used to filter eye position and the sampling rate was 500Hz. Blinks were monitored using EOG with electrodes placed above and below one eye. The stimulus display and data sampling were controlled by a PDP 11/73 computer. Smooth pursuit analysis was performed using a smooth pursuit data analysis package (Eyemap, Amtech GmbH, Heidelberg). Saccadic movements were identified and excluded from the analysis. In each half-cycle the portion of smooth pursuit eye movement having the highest velocity was identified and expressed as peak velocity gain (peak eye velocity / target velocity). Smooth pursuit gain was analysed from at least 12 half cycles at each of the four target velocities.

Study 1 Neuroleptics

Subjects

Three subject groups were constituted as follows: 17 schizophrenic patients on neuroleptic treatment (TrSZ), (mean age =40.7, sd =3.0 years); 10 schizophrenic patients off neuroleptics for at least six months (UTrSZ), (mean age =39.3, sd =4.1 years); 17 normal controls (mean age =41.4, sd =2.7 years). All schizophrenic subjects fulfilled the appropriate DSMIII-R criteria. Patients were assessed clinically using the Schedule for Affective Disorders and Schizophrenia, an assessment of negative (SANS) and positive (SAPS) symptoms and the Mini-Mental State Examination. The control subjects reported no history of psychiatric illness. Informed consent was obtained from all the subjects, the study having been approved by the Tower Hamlets Ethical Committee.

Results

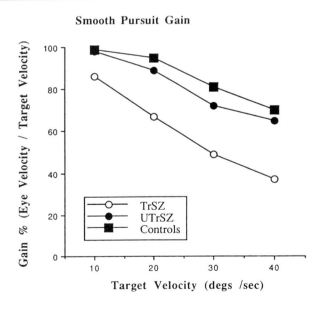

Figure 1. Figure showing smooth pursuit gain as a function of target velocity.

The results were analysed in a mixed two-factor ANOVA using the Statistical Package for the Social Sciences (SPSS), with subject group as the between groups factor and target velocity as the repeated measures factor. There was significant decrease $(F(3,126)=100.77, p<0.001)$ in smooth pursuit gain as the target velocity increased from 10°/sec to 36°/sec for all subject groups. However, whereas UTrSZ and normal control subjects had a similar monotonic function relating smooth pursuit gains and target velocities, compared to other groups the TrSZ patients had significantly reduced smooth pursuit gains $(F(2,42)=7.57, p<0.005)$. There was also a significant group by velocity interaction $(F(6,126)=3.72, p<0.005)$ showing that the difference between pursuit gain in the TrSZ and other groups was greatest at the highest target velocity (see figure 1). Analyses of the psychiatric data showed that TrSZ and UTrSZ patients did not differ clinically in terms of either SAPS $(F(1,24)=0,ns)$ nor SANS $(F1,24) =3.03, ns)$.

Discussion

Smooth pursuit abnormalities in schizophrenia are unlikely to be entirely contingent on a previous history of neuroleptic medication since the original description of the abnormality preceded the emergence of neuroleptics (Diefendorf & Dodge, 1908). However, the current study has shown that the effect of the disease effect is small in comparison to the additional effect of neuroleptic medication. The neuroleptic medication effect cannot be explained in terms of a ceiling effect on pursuit velocity. Pursuit gain was clearly reduced in neuroleptic patients at the lowest target velocity 10°/sec, where the mean absolute eye velocity was 8.6°/sec whereas the mean absolute eye velocity at the 36°/sec target was 13.6°/sec. The effect of neuroleptics on pursuit gain does not seem therefore, to be one of a simple saturation limit on pursuit velocity, but the effect seem to affect the specific neural pathways relating eye to target velocity.

These findings support the results of Kufferle et al (1990) showing that schizophrenics who had taken neuroleptic medication in the preceding 2 years had significantly more disturbed smooth pursuit than controls, affective patients and untreated SZs.

Study 2. Cortical Lesions.

Many of the early studies on smooth pursuit both in schizophrenia and other diseases were conducted under photopic conditions in normal room lighting conditions with many potentially distracting visual targets. Pursuit eye movements under these conditions will produce a complex pattern of optical flow across the retina, providing a powerful stimulus for attentional capture (i.e. distractibility). When a subject pursues an object moving against a background, most retinal flow will be in a direction opposite to the target's motion. This background retinal flow might be expected to provide a powerful stimulus to the optokinetic reflex, counteracting the effect of target retinal slip, and imposing a background "drag" upon the pursuit eye movement.

Indeed, as the area of retina stimulated by the background will ordinarily be much larger than that stimulated by the target, the background motion must be suppressed in order to make smooth pursuit possible.

In this study we aimed to investigate the effect of high contrast background patterns upon smooth pursuit eye movements, both in normal control subjects and in a group of patients who had suffered discrete cerebral lesions. In particular, we wished to investigate whether the ability of the pursuit eye movement mechanism to filter out a distracting background pattern is localised to a specific cortical area.

Methods

Smooth pursuit eye movement data were collected against both dark and structured backgrounds at four target velocities: 10, 20, 30, and 36 °/sec. The latter were produced by front-projection of a high contrast random square pattern, each square subtending 1.9° x 1.9° visual angle, onto the screen from a slide projector mounted above the subject's head. In order that the target spot should remain clearly visible and of constant contrast, it moved within a dark stripe (2.7°), which divided the background pattern horizontally.

The effect of a background pattern upon smooth pursuit gain was quantified as the *background effect*, which was the mean of the differences between gains when tracking against dark and patterned backgrounds at each target velocity. We quantified the effect of a structured background by subtracting the mean peak velocity gain attained with that background from that attained in the dark at each velocity, and then averaged across the four velocities to arrive at a single figure. For the control subjects the range of this *background effect* was -0.07 to 0.19 (mean 0.02; sd ±0.08). When analysing the results obtained with patients we adopted a criterion *background factor* of 0.18. This figure was derived from the mean value of the *background effect* measured in controls plus twice its standard deviation. Patients for whom the value of *background factor* was greater than 0.18 were judged to exhibit an abnormally large degree of inhibition of smooth pursuit in the presence of a background pattern.

Subjects

Recordings were made in 26 patients and in 9 healthy controls. Of the patients 22 had suffered from ischaemic strokes, 1 from a closed head injury, and 3 from spontaneous intracerebral haemorrhages. All had lesions visible on CT brain scans. None showed clinical evidence of dementia or receptive dysphasia sufficiently severe as to compromise their understanding of the task. The patients' ages ranged from 25-81 years (mean 61.2 years); the controls' ages ranged from 28-77 years (mean 59.4 years).

Results

Of the 26 patients examined, 14 showed a significant effect of structured background upon smooth pursuit with values of *background factor* > 0.18 in at least one direction, while in 12 there was no such effect. Lesions visible on CT brain scans were transferred onto standard templates (Damasio & Damasio, 1989).   As similar numbers of lesions were found in each hemisphere in this group (6 right; 8 left) we transposed all lesions onto a single hemisphere to obtain better delineation of the brain areas that seem to be important in producing a strong background effect.   Two "hot spots" of multiple overlaps appeared: one in the anterior inferior portion of Brodmann's area 40 in the parietal lobe cortex and one in white matter containing cortico-cortical connections, close to the lateral ventricle (see figure 2).

   In principle, lesions in patients without a prominent background effect provide at least as much information concerning possible anatomical location of neural mechanisms involved in the background effect as do those in patients manifesting the effect.   If an area of brain is necessary to prevent degradation of pursuit by a distracting background, then that area should never be included in the lesions of patients who do not manifest such degradation.   Figure 2 shows the cumulative overlap of 10 of the 12 patients without a background effect.   Also plotted are the "hot-spots" found in patients with the effect.   These areas should not overlap, and indeed there is a good measure of agreement.

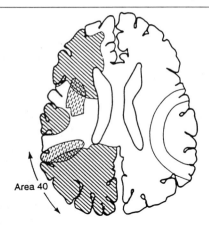

Degree of overlap: Background Effect

No background effect

Figure 2. Section showing the lesion overlaps in 10 of the 12 patients in whom there was no large background effect (transposed onto a single hemisphere). There is good agreement with the areas of maximal overlap found in patients with a large background effect. Two regions are of particular importance: an area of parietal cortex lying within the architectonic boundaries of Brodmann's area 40; and an area of white matter close to the lateral ventricles containing cortico-cortical connections. Two patients (lesions plotted on the right side of the section) had weak background effects despite having lesions which encompassed the parietal "hot spot" of the affected patients.

Discussion

The task of pursuing the target against a structured background requires one to focus attention upon the target and ignore the background. Can it be assumed that the background-sensitive patients simply lack the ability to focus attention on the target and thus become easily distracted by the background? Several workers have demonstrated that pursuit gain to a moving stimulus increases markedly in normal subjects (Cheng & Outerbridge, 1975; Dubois & Collewijn, 1979; Barnes & Hill, 1984) and schizophrenics (Shagass et al, 1976) when subjects actively attend to the image movement. Subjects with parietal lobe lesions are known to have deficits in their ability to direct the focus of visual attention in space (Posner *et al.*, 1984) and a recent PET study has demonstrated parietal activation during tasks involving a covert shift of visual attention (Corbetta *et al.* 1993). However, in both of these studies the parietal areas of importance seemed to lie in the superior parietal lobule (Brodmann areas 5 and 7) rather than in area 40, which is in the inferior parietal lobule. By contrast, Robertson *et al.* (1988) demonstrated deficits in the ability of human subjects to allocate attention to local or global features of a visual target after lesions affecting the inferior parietal lobule.

Summary and Conclusions.

1. Smooth pursuit eye velocity is dramatically reduced in schizophrenic patients on neuroleptic medication compared to age-matched patients off neuroleptics and normal control subjects. This impairment is not due to a simple velocity saturation process in the mechanisms that determine smooth pursuit eye velocity in proportion to the target velocity.

2. The precise neuropharmacological mechanism is unclear since neuroleptics affect a wide range of neural transmitter systems. However, the hypothesis that dopamine pathways may play a central role in the control of smooth pursuit gain is supported by reports of small decreases in smooth pursuit gain in patients with early Parkinson's disease (Gibson et al, 1987), a degenerative disease process affecting dopaminergic pathways in the basal ganglia.

3. We have demonstrated that a group of patients with isolated lesions of the cerebral hemispheres who have particular difficulty maintaining smooth pursuit of a target moving against a high contrast background pattern; a deficit that is either absent or much less prominent when the target moves against a dark featureless background.

These studies have identified two important factors that would mitigate against efficient smooth pursuit performance in patients with schizophrenia: 1. Neuroleptic medication (although as we have stated, this is unlikely to be the sole factor, as the original description of the smooth pursuit abnormality (Diefendorf and Dodge, 1908) preceded the emergence of neuroleptic therapy). 2. Attentional distractibility in the presence of relative background motion.

Interestingly, in a recent study measuring regional cerebral blood using single photon emission tomography (SPET) we found, in schizophrenic patients who are unable to inhibit reflexive saccadic eye movements, in the anti-saccade task reduced cerebral blood flow in a number of brain regions including Brodmann's area 40 (Crawford et al, 1994). This is precisely the same cortical area shown be critical to the effect of a distracting background on smooth pursuit eye movements. One neurocognitive mechanism that would link these findings together in a coherent hypothesis would be in terms of abnormal function within a neural network for selective visual attention that is manifest within the saccadic and smooth pursuit eye movement systems across various brain neuropathology.

Acknowledgements: This work was supported by grants from the Wellcome Trust.

References

Barnes, G.R. & Hill, T. (1984).   The influence of display characteristics on active pursuit and passively induced eye movements. Exp. Brain Res. 56, 438-447.

Brodmann K.   Vergleichende Lokalisationslehre der Grosshirnrinde. Leipzig. Barth, 1909.

Cheng, M. & Outerbridge, J.S. (1985). Optokinetic nystagmus during selective retinal stimulation. Exp. Brain Res. 23: 129-139.

Clementz, B.A. & Sweeney, J.A. (1990). Is eye movement dysfunction a biological marker for schizophrenia ? A methodological review', Psychol. Bull. 108, 1, 77-92.

Corbetta, M. Miezin, F.M., Shulman, G.L. & Petersen, S.E. (1993). A PET study of visuospatial attention. J. Neurosci. 13, 1202-1226.

Crawford, T.J., Puri, B.K., Kennard, C. & Lewis, S. (1994). Regional cerebral blood flow in schizophrenic patients with abnormal inhibition of reflexive saccadic eye movements. Psychiat. Res. 11, 2, 171-172.

Damasio, H. & Damasio, A.R. (1989). Lesion Analysis in Neuropsychology. Oxford University Press, New York.

Diefendorf, A.R. & Dodge, R. (1908). An experimental study of the ocular reactions of the insane from photographic records. Brain, 31, 451-489.

Dubois, M.F.W. & Collewijn, H. (1979). Optokinetic reactions in man elicited by localized retinal motion stimuli. Vis. Res. 19, 1105-1115.

Fletcher, W. & Sharpe, J.A. (1986). Saccadic eye movement dysfunction in Alzheimer's disease. Ann. Neurol. 20, 464-471.

Gibson, J.M., Pimlott, R., & Kennard, C. (1987). Ocular motor and manual tracking in Parkinson's disease and the effect of treatment. J. Neurol. Neurosurg. Psychiat. 50, 853-860.

Holzman, P.S, Protor, L.R., Levy, D.L., Yasillo, N.J., Meltzer, H.Y. & Hurt, S.W. (1974). Eye-tracking in schizophrenics patients and their relatives. Arch. Gen. Psychiat. 1, 143-151.

Kufferle, B., Friedmann, A., Topitz, A., Foldes, P., Anderer, P., Kutzer, M. & Steinerberger, K. (1990). Smooth pursuit eye movements in schizophrenia:

influences of neuroleptic treatment and the question of specificity. Psychopathology, 23, 106-114.

Leigh, R.J. & Zee, D.S. (1991). The Neurology of Eye Movements, 2nd edition. FA Davis Company. Philadelphia.

Lisberger, S.G. & Westbrook, L. E. (1985). Properties of visual input that initiate horizontal smooth pursuit eye movements in monkeys. J. Neurosci., 5, 1662-1673.

Posner, M.I., Walker, J.A., Friedrich, F.J. & Rafal, R.D. (1984). Effects of parietal injury on covert orienting of visual attention. J. Neurosci, 4, 1863-1874.

Robertson, L.C., Lamb, M.R. & Knight, R.T. (1988). Effects of lesions on temporo-parietal junction on perceptual and attentional processing in humans. J. Neurosci, 8, 3757-3769.

Shagass, C., Amadea, M., & Overton, D.A. (1976). Eye-tracking performance and engagement of attention. Arch. Gen. Psychiat. 33, 121-125.

Sharpe, J.A., Fletcher, W.A, Lang, A.E. & Zackon, D.H. (1987). Smooth pursuit during dose-related on-off fluctuations in Parkinson's disease. Neurology, 37, 1389-1392.

Shibasaki, H., Tsuji, S. & Kuroiwa, Y. (1979). Oculomotor abnormalities in Parkinson's disease. Arch. Neurol. 36, 360-364.

Thurston, S., Leigh, J., Crawford, T.J., Thompson, A. & Kennard, C. (1988). Two distinct deficits of visual tracking caused by unilateral lesions of cerebral cortex in man. Ann. Neurol. 23, 266-273.

PEAK SACCADE VELOCITIES, SHORT LATENCY SACCADES AND THEIR RECOVERY AFTER THERAPY IN A PATIENT WITH A PINEAL TUMOR

L.J. Bour, D. van't Ent and J. Brans

Graduate School of Neurosciences Amsterdam, Department of Neurology, Clinical Neurophysiology Unit, Academic Medical Centre H2-214, AZUA, Meibergdreef 9, 1105 AZ Amsterdam, The Netherlands.

Abstract

Peak velocities of visually elicited saccades and latencies of saccades have been studied in a patient with a large tumor in the pineal gland area that exerted compression in the direction of the corpora quadrigemina and underlying structures. Clinically it effected, among other symptoms, a severely impaired vertical upward gaze. Four repeated examinations were performed from initial radiation therapy to 12 months post therapy; normal data were derived from ten control subjects. Following therapeutic intervention, upward and downward gaze normalized almost completely, but peak saccade velocities in vertical directions remained markedly reduced, although there was a slight improvement in upward direction only.

To examine whether the patient could generate short latency 'express' saccades, latencies of horizontal and vertical eye movements were also measured with a 'gap-paradigm' which included a 200 ms gap between peripheral target onset and fixation point offset. Surprisingly, the amount of 'express' saccades was significantly reduced only in the vertical plane. After one year post-treatment a slight decrease in saccade latencies was found in downward direction.

Keywords

Express saccades, Parinaud's syndrome, saccade latency, saccade peak velocity, superior colliculus.

Introduction

Parinaud's syndrome is usually defined as a defect of voluntary conjugate upward gaze. Accompanying symptoms may include loss of the pupillary reaction to light, retraction of the upper lids and paralysis of convergence (Brain & Walton, 1977) Sometimes a paralysis of downward gaze is also present (Ranalli et al., 1988)

The syndrome is almost exclusively caused by mesencephalic lesions (Pierrot-Deseilligny et al., 1982). The mesencephalic reticular formation (MRF) contains premotor structures for the control of vertical eye movements. An important structure is the posterior commissure (PC) which contains fibers that mediate upward gaze. Lesions in the PC or the PC nuclei result in elimination of upward saccades and pursuit (Pasik et al., 1969; Ranalli et al., 1988). Another important structure is the rostral interstitial nucleus of the medial longitudinal fasciculus (riMLF) situated rostral to the tractus retroflexus and the nucleus of Cajal and ventral to the nucleus of Darkschewitsch; this nucleus participates in the direct

Eye Movement Research/J.M. Findlay et al. (Editors)

premotor control of vertical eye movements (Büttner et al., 1977; Büttner et al., 1982).

Pineal gland tumors, or dilation of the supra pineal recess due to aqueduct stenosis can cause Parinaud's syndrome (Lerner et al., 1969; Moffie et al., 1983). In this condition, a space occupying process presses on the quadrigeminal region. Oculomotor abnormalities can be explained by pressure exerted via the colliculus upon mesencephalic structures such as the PC or riMLF and their afferent and efferent fibres, or by second order effects such as brain edema and ischemia due to vascular compression.

However, a dysfunctioning superior colliculus caused by pressure effects can also account for some of the concomitant oculomotor deficits. In monkey, superior colliculus ablation causes an increase in saccade latencies (Mohler & Wurtz, 1977) and reduces fixation accuracy, saccade frequency and saccade velocity (Albano et al., 1982; Schiller et al., 1980). The superior colliculus also seems to play an essential role in the chain of events generating 'express' saccades in monkey. Schiller et al. (1987) found in monkey that 'express' saccades were abolished in the absence of the superior colliculus, whereas saccades with longer latencies were maintained.

Experiments with GAP-paradigms have shown in monkey (Fischer & Boch, 1983) and humans (Fischer & Rampsperger, 1986) that latencies of saccades can be strongly reduced. Often a bimodal latency distribution is found with in humans an initial peak at about 100 ms and a second peak at about 150 ms. Fischer & Boch (1983) called the first peak the 'express' peak because of its short latency characteristics, and saccades belonging to this peak were called 'express' saccades. Wenban-Smith & Findlay (1991) reported in a study on 'express' saccades in humans that short latency visually guided saccades were found, but a clear bimodal distribution of latencies did not exist.

In the majority of reports concerning Parinaud's syndrome, longitudinal examinations were not performed and the abnormalities of eye movements were only assessed by clinical observation. As a result, vertical gaze palsy has not been extensively quantified in terms of degrees of upward and/or downward limits, nor have peak velocities and latencies of saccades been investigated. To accurately measure vertical eye movements, electromagnetic methods such as the magnetic search coil (Collewijn et al., 1975) and the double magnetic induction (DMI) method (Bour et al., 1984) should be used. For example, with a scleral search coil technique Ranalli et al. (1988) described vertical eye movements in a patient with a distinct midbrain lesion. The accurate quantification of vertical eye movements in the Rannali study enabled them to establish correlations between pathophysiologic findings and abnormal oculomotor responses.

In the present study, with the DMI method, we measured peak velocities and latencies of saccades in a case with a pineal gland tumor. We expected that the superior colliculus would be compressed. Since, to our knowledge, it has not been investigated whether the superior colliculus plays a role in the generation of 'express' saccades in humans, we also examined if the patient could still make 'express' saccades.

The results of this case study show that one year after radiation therapy, an initial upward gaze paralysis was normalized clinically. However, peak velocities of saccades in upward and downward direction remained significantly reduced compared to control values. During the entire period of examination the patient was able to produce 'express' saccades in the horizontal plane. However, in the vertical plane virtually no 'express' saccades were found during the early phases of treatment; slight improvement was noted after one year post-treatment, especially in downward direction.

Methods

Case-history

A 24 years old formerly healthy computer technician visited the ophthalmologist because of a slowly progressive diplopia lasting for two weeks. Double images appeared in the vertical plane and arose upon leftward and more prominently upon rightward gaze. On examination the pupils were symmetrical, 4 mm each. In both eyes the pupillary light reflex was absent, but constriction on accommodation was normal (light-near dissociation). A supranuclear upgaze paresis was present, with normal vestibulo-ocular reflexes. On attempting upward gaze and on eliciting vertical optokinetic nystagmus, a retraction-convergence nystagmus occurred, especially in the left eye. When testing eye movements to the left, the abducting eye moved faster than the adducting eye. Visual acuity was 1.0 in both eyes. Fundoscopy showed bilateral papilloedema. With the Maddox-rod test, a non-concomitant divergent strabismus with a clear A pattern was found. Further, general and neurological examination was normal.

Blood examination was normal, including the tumor markers α-fetoprotein, β-HCG and CEA. A CT-scan of the brain showed a calcified pineal gland surrounded by a hyperdense mass which enhanced after intravenous contrast injection. The MRI showed an inhomogeneous process in the region of the pineal gland with enhancement after intravenous gadolinium (Figure 1). The tumor compressed the corpora quadrigemina, cerebral aqueduct and vermis cerebelli and an obstructive hydrocephalus was present. A stereotactic biopsy was performed and during the same session an Omaya resevoir was placed and cerebral spinal fluid (CSF) was extracted for examination. Neuropathologically a germinoma was found. CSF contained tumor cells, but tumormarkers α-fetoprotein, β-HCG and CEA were absent. Treatment consisted of radiation therapy, 10 fractions of 2 Gy, on the tumor region, followed by radiation, 19 fractions of 1.6 Gy, on the craniospinal axis. After two months there was a slight improvement. Diplopia diminished, but it remained troublesome. A CT-scan of the brain showed a remarkable tumor regression.

After six months the light-near dissociation and non-concomitant divergent strabismus were unchanged, but the upgaze paresis had disappeared completely. After one year the patient still reported mild diplopia with vertical image separation upon lateral gaze. A light-near dissociation was still present. Fundoscopy was normal. A CT-scan of the brain did not show any recurrence of the tumor.

One week after radiotherapy was started the first examination of the saccadic eye movements was performed. The subsequent examinations were performed one month, six months and twelve months after radiotherapy.

A control group consisted of 10 healthy individuals aged 21 to 62 years (mean 35 years). The patient and the control subjects were naive to our experiments and gave informed consent.

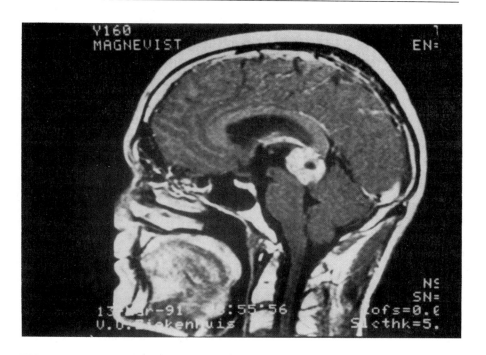

Figure 1 - MRI after gadolinium showing enhancement of a mass in the pineal region com-pressing the lamina quadrigemina and the aqueduct.

Experimental procedure

The eye movements of the subject's right eye were measured with the DMI method (Bour et al., 1984). Within the range of -15° to +15° raw eye signals do not deviate more than 5% from linearity. The average resolution is 5 minarcs. Vision was binocular. The subject's head was stabilized by a head tie and a chin rest in order to reduce drift and head movements. Horizontal and vertical eye position signals were low-pass filtered (-3db at 150 Hz), digitized with a sample frequency of 250 Hz and stored in the computer.

The subject was asked to track as accurately as possible a single red circular laser-spot that was back projected on a tangent white translucent screen by means of a scanning mirror device. The spot diameter was 0.4° and its luminance was 20 cd/m² . A dimly illuminated background was used. The laser spot moved either in horizontal or vertical direction and never exceeded -10° or +10° from the point on the screen that corresponded with gaze straight ahead.

At the beginning of each experimental session the subject was asked to fixate several successive positions of the spot in horizontal and vertical direction. Raw eye position signals obtained from these known positions were used to calibrate the data.

Three experimental paradigms were used. In the first paradigm (RAN) the subject had to track the laser-spot in either horizontal or vertical direction. The laser-spot jumped around gaze straight ahead position so the direction of the jumps was not randomized. The time between successive jumps was randomized between 500 and 1500 ms and the amplitude between 3° and 20°. In the second paradigm (GAP) the laser-spot was first projected at gaze straight ahead position. The subject was asked to fixate the spot. Then after a random period between 600 and 1200 ms the laser beam was disrupted by a shutter and the spot on the screen disappeared. After a temporal blanking period of 200 ms, called the gap, the laser-spot was again projected on the screen. The onset time of this peripheral target was determined by a photocell placed in a beam splitted from the target beam. Direction of the projection measured from gaze straight ahead position was selected randomly to the left or to the right when the paradigm was used in the horizontal plane, and above or below when used in the vertical plane. The peripheral target's eccentricity always was 5°. After 500 ms the laser spot was again projected at gaze straight ahead position to start a new sequence. The third paradigm (NOGAP) was exactly the same as the second paradigm, except that no temporal blanking period was present.

By means of an interactive program, eye movements were displayed on a monitor in a time-window, subtending 1000 ms and centered around target onset. Onset and offset of the saccades to this stimulus were indicated manually by placing hair-lines and subsequently amplitude, gain, duration, peak velocity and latency of the saccade were computed. Velocity of the saccade was computed digitally by means of a second order low-pass (65 Hz) differentiation algorithm (Usui & Amidror, 1982). Saccadic latency was computed as the time between peripheral target onset and saccade onset. The probability that the subject could guess the direction correctly was 50%. Visually guided saccades were separated from anticipatory saccades by using the direction error criterion as proposed by Kalesnykas & Hallett (1987). Histograms were made of the latency distributions for visually guided saccades, using a binwidth of 10 ms. The non-parametric Mann-Whitney U test was used to detect significant differences between latency distributions of the patient and the control group and between latency distributions from the four examinations of the patient.

Saccades obtained from the experiments using the RAN paradigm and which had latencies larger than 100 ms were used for a quantitative evaluation of the amplitude/velocity relationship for visual saccades. Using an algorithm that minimizes the residual sum of squares an exponential relationship $V_m (A_s) = V_a [1 - \exp(-A_s/A_o)]$ was fitted to the data of visual elicited saccades in each direction, i.e., to the right, to the left, upwards and downwards (Bahill et al., 1975). In this equation $V_m (A_s)$ represents estimated peak velocity of a saccade with amplitude A_s; V_a and A_o are constants.

Results

In Figure 2 peak saccade velocity in four different directions is plotted against saccade amplitude. Every symbol in this plot represents a saccade and different symbols are used to indicate the four separate examinations. The upper curves represent exponential amplitude/velocity relationships fitted to mean peak saccade velocities from a control group of ten subjects. The lower curves relate to mean peak velocity minus 2 times the standard deviation. Thus, the grey

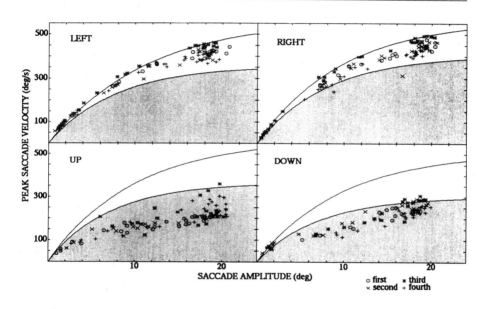

Figure 2 - Peak velocity as a function of amplitude for visually guided saccades in each direction, i.e., to the left, to the right, upwards and downwards. Different symbols are used to indicate the four separate examinations. The upper curves in each quadrant are main-sequence relationships fitted to mean peak velocities of saccades with amplitudes ranging from 0° to 20° computed from the control group. The lower curves are the main sequences fitted to these velocities minus 2 SD. The grey area indicates significant reduction of peak saccade velocity.

shaded area indicates the area of significant reduced peak saccade velocity (p<0.05). Peak velocities of horizontal saccades obtained on the four examination dates fall well within the limits of the control group (with minor exceptions). Peak velocities of vertical saccades are reduced significantly. In upward direction at the third examination, an increase of peak velocity for three saccades with amplitude larger than 15° was observed and at the last examination a considerable number of saccades in this amplitude range showed relatively increased peak velocities. However, the velocities still did not reach the limits of the control group. An interesting observation with respect to the upward directed saccades was that to generate these saccades the patient used an unusual combination of the extraocular muscles. In Figure 3 (a) sampled data points are plotted for an upward and a downward directed 20° saccade of the right eye. The small arrows indicate the direction of the saccades. The time between successive samples is 4 ms. The downward directed saccade reaches the target in about a straight line.

In contrast, the upward directed saccade is accompanied by an initial abduction of the eye. After 50 ms the eye reaches a maximum horizontal eccentricity of about 5°, and consequently during 90 ms the eye makes a glissade back to the plane of the target jumps, and eventually reaches the target. This phenomenon was present also in the left eye but it was not seen with vertical pursuit for either eye.

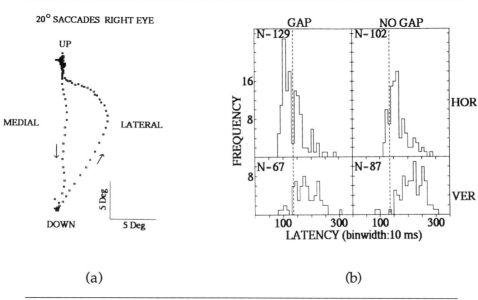

(a)   (b)

Figure 3 (a) - Scanpath of an upward and a downward directed saccade made by the patient, the arrows indicate the direction of the saccades. The sampled data points, obtained with a sample frequency of 250 Hz, are plotted

Figure 3 (b) - Latency distributions of visually guided saccades, made by the patient in the horizontal and the vertical plane obtained with the GAP and the NOGAP paradigm. The dotted line indicates the separation between the two peaks of the distribution in the horizontal plane obtained with, the GAP paradigm (upper left). The line is drawn at the same latency in the other distributions.

Figure 3 (b) shows the latency distributions for visually guided primary saccades of the patient obtained with the GAP and the NOGAP paradigm. The upper and lower two panels refer to horizontal and vertical saccade latencies, respectively. The distributions represent pooled data of the first three examinations.

In the horizontal plane a bimodal distribution was found when the GAP paradigm was used. The first peak has a latency of 120 ms and the dashed vertical line falls between the two peaks of the distribution (135 ms). When the NOGAP paradigm was used for horizontal saccades, the latency distribution shifted to the right as can be seen in the upper right panel. Only a small peak is observed before the dashed line. This result is in agreement with the findings reported by Fischer & Ramsperger (1986) and Deubel & Frank (1991).

In the vertical plane latency distributions were shifted to longer latencies compared to the distributions in the horizontal plane. In the GAP paradigm a small first peak at about 100 ms was observed but most latencies were longer than 130 ms. In the NOGAP paradigm all saccades except 3 had latencies longer than 150 ms (N=87).

To obtain a quantitative measure for the amount of 'express' or short latency saccades in the GAP paradigm, the percentage of the total latency distribution that falls within a window ranging from 90-130 ms, in both horizontal directions (W_1, W_r) and both vertical directions (W_u, W_d) was computed, respectively. The mean and standard deviation of the total latency distributions in each direction also were calculated. Data of the patient obtained during four

		PATIENT							
HORIZONTAL	**ind**			direction					
			left				right		
		N	mean	sd	W_l	N	mean	sd	W_r
			(ms)	(ms)	(%)		(ms)	(ms)	(%)
	1	32	118.8	30.8	62.5	23	135.7	34.3	52.2
	2	14	123.4	43.0	71.4	12	165.3	43.0	25.0
	3	25	147.2	47.3	44.0	23	138.3	25.4	52.2
	4	30	124.4	38.5	63.3	29	123.2	18.9	55.2
		CONTROL SUBJECTS							
	subj			direction					
			left				right		
		N	mean	sd	W_l	N	mean	sd	W_r
			(ms)	(ms)	(%)		(ms)	(ms)	(%)
	A	53	104.8	21.2	71.7	37	115.4	17.7	89.2
	B	45	125.2	36.6	57.8	31	123.1	22.7	61.2
	C	58	118.1	25.3	53.4	40	121.2	32.8	77.5

Table 1 - *Ind and subj relate to the four examinations of the patient and the three control subjects respectively, N equals number of saccades in the latency distribution; W_1 and W_r equals the percentage of saccades in a latency window between 90 and 180 ms in leftward and rightward direction respectively; mean and sd refer to mean (ms) and standard deviation.*

successive examinations and of three control subjects are summarized for the horizontal and vertical plane in Table 1 and Table 2, respectively. In the horizontal plane, rightward saccade latencies in the patient, obtained by pooling the first three examinations, were increased compared to pooled

VERTICAL

	ind	PATIENT direction							
		up				down			
		N	mean (ms)	sd (ms)	W_u (%)	N	mean (ms)	sd (ms)	W_d (%)
	2	13	175.4	34.6	7.7	13	168.6	54.9	0
	3	15	195.7	35.3	6.7	26	175.4	48.9	7.7
	4	28	185.1	36.9	10.7	37	153.1	35.0	16.2

CONTROL SUBJECTS

	subj	direction							
		up				down			
		N	mean (ms)	sd (ms)	W_u (%)	N	mean (ms)	sd (ms)	W_d (%)
	A	33	123.6	16.0	60.6	51	103.6	11.5	88.2
	B	36	136.8	28.1	36.1	56	133.4	39.1	37.5
	C	30	129.1	21.6	50.0	39	135.0	36.4	64.1

Table 2 - Ind and subj relate to the three examinations of the patient (on the first examination the GAP paradigm was not used in the vertical plane) and to the three control subjects respectively, N equals number of saccades in the latency distribution; W_u and W_d equals the percentage of saccades in a latency window between 90 and 130 ms in upward and downward direction respectively; mean and sd refer to mean (ms) and standard deviation.

latencies of the control subjects (p<0.0001). Comparing the patient's data and the control subjects there were no significant differences for leftward saccade latencies. On the fourth examination, the percentage of saccades in the latency window (W_r) increased only slightly but more rightward saccades had shorter latencies compared to the previous examinations as demonstrated by the reduction of the mean and standard deviation of the latency distribution. Consequently no difference was found anymore between the patient and the control subjects comparing rightward and leftward saccade latencies. In the NOGAP paradigm, latencies of leftward and rightward saccades were not increased significantly compared to the control values. In the vertical plane in both paradigms the upward and downward latency distributions of the patient were shifted to longer latencies compared to the distributions of the control subjects (p<0.0001) on all four examinations. The patient's reduced ability to generate short latency 'express' saccades in the vertical plane is demonstrated by the strong reduction of W_u and W_d , compared to W_l and W_r , in the patient and compared to W_l ,W_r , W_u and W_d in the control group.

With respect to intra-individual differences, latencies were reduced significantly on the fourth examination of the patient compared to the pooled latencies of the previous examinations both for the GAP and the NOGAP paradigm, for downward saccades (p<0.05 and p<0.001, respectively). The percentage of 'express' saccades increased especially in downward direction on the fourth examination compared to the previous examinations.

Discussion

In the vertical plane peak velocity of up and down saccades of the patient with Parinaud's syndrome, was slow and slightly improved, after one year, in upward direction only. In contrast, the initial upward gaze paralysis completely resolved. The patient also had increased latencies of upward and downward saccades and was virtually unable to make vertical 'express' saccades with the GAP paradigm. In contrast with peak saccade velocity, latencies of saccades decreased primarily in downward direction, and a slight increase in the percentage of vertical 'express' saccades was found especially for downward 'express' saccades.

In the horizontal plane no significant eye movement abnormalities have been demonstrated. Peak saccade velocity was within normal limits. Latencies did not show an increase as with vertical saccades and it was remarkable that the patient always was able to make 'express' saccades in leftward and rightward directions.

The discrepancy in peak velocity as well as in latency between horizontal and vertical saccades points to the localization the defects in the vertical pulse generator as it is largely separated from the horizontal pulse generator. Lesions located in the mesencephalon can lead to a selective paralysis of upward or downward saccades and as such, the pulse generators for up and down saccades are also, at least partly, separated. Output pathways of the pulse generator for upward saccades and output pathways for upward pursuit decussate through the posterior commissure (Ranalli, 1988). Pressure on this commissure or its nuclei probably caused the initial upward gaze paralysis seen in the patient. The resolving of this paralysis and the slight increase in peak velocity of upward saccades can be explained by the relief of pressure and partial or complete recovery of the posterior commissure function. Damage to afferent or efferent fibers of the up and down pulse generators, or to structures such as the riMLF that contain burst neurons for up and down saccades could cause the remaining deficit in peak velocity of vertical saccades. Peak velocities of horizontal saccades remained within normal limits because the horizontal pulse generators are more caudally situated in the brainstem.

Evidence has been presented that the superior colliculus input to the pulse generators also encodes dynamic properties of saccades (Waitzman et al., 1991). After injecting muscimol in the intermediate layers of the superior colliculus, Hikosaka & Wurtz (1985) measured decreased peak velocities of saccades corresponding with the saccades that were initially elicited by electrical stimulation of the injection site. Reduction of saccade velocity after ablation of the superior colliculus is often not found. However, in most of these experiments the intervals between the time of the ablation and the eye movement recording were considerable and compensation could have occurred for the velocity deficits. Because our first examination of the patient was made within weeks after the tumor was identified and we measured a strong reduction in peak velocity of vertical saccades which only slightly improved, it is unlikely that a dysfunctioning superior colliculus is the main cause of the velocity reduction.

A dysfunctioning of the pulse generators for upward and downward saccades could also, as for the reduced peak velocity, explain the exclusive increase of vertical saccade latencies. However, peak saccade velocity increased in upward direction only, while saccade latency improved most significantly in downward direction. Moreover, experimental studies in monkey have shown that in contrast to reduced saccade velocity, ablation of the superior colliculus frequently leads to an increase in saccade latency (Mohler & Wurtz, 1977; Wurtz & Goldberg, 1972) and results in an inability to make short latency 'express' saccades to the contralateral side of the ablation (Schiller et al., 1980). These latency deficits show little recovery with time. Finally, Robinson (1972) has found that stimulation of the medial area of the colliculus leads to contralateral saccades with an upward component whereas stimulation of the lateral area produced contralateral saccades with a downward component. Thus, it is expected that, with a large centrally placed pineal tumor excerting pressure more on the medial than on the lateral area of the superior colliculus, recovery of latencies will occur primarily for downward saccades. So these considerations give support to the possible role of the superior colliculus in the generation of vertical short latency saccades also in humans.

Horizontal short latency saccades were not involved. Therefore, a role of the superior colliculus in the generation of short latency saccades in at least the horizontal direction remains questionable in humans. However, there are some alternative explanations. Only one colliculus is needed to generate a horizontal saccade. Purely vertical saccades on unilateral stimulation in experimental studies are never found, requiring cooperation of both colliculi. When, in the patient, pressure is exerted on the commissure of the superior colliculus, the colliculi may be partly disconnected and deficits primarily may be expected in vertical saccades. It also could be possible that the superior colliculus has more neurons involved with horizontal than with vertical saccade components. If so, latencies of saccades in the vertical plane could increase more significantly simply because they are more sensible to loss of neurons.

References

Albano J.E., Mishkin M., Westbrook L.E. & Wurtz R.H. (1982) Visuomotor deficits following ablation of monkey superior colliculus. J. Neurophysiol. 48, 338-351.

Bahill A.T., Clark M.R. & Stark L. (1975) The main sequence: A tool for studying eye movements. Math. Biosci. 24, 191-204.

Bender M. (1980) Brain control of conjugate horizontal and vertical eye movements. Brain 103, 23-69.

Büttner U., Büttner-Ennever J.A. & Henn V. (1977) Vertical eye movement related unit activity in the rostral mesencephalic reticular formation of the alert monkey. Brain Res. 130, 239-252.

Büttner-Ennever J.A., Büttner U., Cohen B. & Baumgartner G. (1982) Vertical gaze paralysis and the rostral interstitial nucleus of the medial longitudinal fasciculus. Brain 105, 125-149.

Boch R., Fischer B. & Ramsperger E. (1984) Express saccades of the monkey: reaction time versus intensity, size, duration and eccentricity of their targets. Exp. Brain Res. 55, 223-231.

Boch R. & Fischer B. (1986) Further observations on the occurrence of express saccades in the monkey. Exp. Brain Res. 63, 487-494.

Bour L.J., van Gisbergen J.A.M., Bruyns J. & Ottes F.P. (1984) The double magnetic induction method for measuring eye movements. IEEE Trans on BME. BME-31 No.5.

Brain W.R. & Walton J.N. (1977) Brain's Diseases of the Nervous System. 8th ed. Oxford University Press, Oxford.

Collewijn H., van der Mark F. & Jansen T.C. (1975) Precise recording of human eye movements. Vision Res. 15, 447-450.

Deubel H. & Frank H. (1991) The latency of saccadic eye movements to texture defined stimuli. Oculomotor Contr. and Cogn. Processes, 369-384.

Dehaene I. & Lammens M. (1991) Paralysis of saccades and pursuit, a clinicopathologic study. Neurology 41, 414-415.

Fischer B. & Boch R. (1983) Saccadic eye movements after extremely short reaction times in the monkey. Brain Res. 260, 21-26.

Fischer B. & Ramsperger E. (1986) Human express saccades: effects of randomization and daily practice. Exp. Brain Res. 64, 569-578.

Hatcher M.A. & Klintworth G.K. (1966) The sylvian aqueduct syndrome. Archs Neurol. 15, 215-222.

Henn V., Hepp K. & Vilis T. (1989) Rapid eye movement generation in the primate. Physiology, pathophysiology, and clinical implications. Rev. Neurol. (Paris) 145, 540-545.

Hikosaka O. & Wurtz R.H. (1985) Modification of saccadic eye movements by GABA-related substances. I. Effect of muscimol and bicucciline in monkey superior colliculus. J. Neurophysiol. 53, 266-291.

Kalesnykas R.P. & Hallett P.E. (1987) The differentiation of visually guided and anticipatory saccades in gap and overlap paradigms. Exp. Brain Res. 68, 115-121.

King W.M. & Fuchs F. (1979) Reticular control of vertical saccadic eye movements by mesencephalic burst neurons. J. Neurophysiol. 42, 861-876.

Lerner M.A., Kosary I.Z. & Cohen B.E. (1969) Parinaud's syndrome in aquaduct stenosis: its mechanism and ventriculographic features. Br. J. Radiol. 42, 310-312.

Moffie D., Ongerboer de Visser B.W. & Stefanko S.Z. (1983) Parinaud's syndrome. J. Neurol. Sci. 58, 175-183.

Mohler C.W. & Wurtz R.H. (1977) Role of striate cortex and superior colliculus in visual guidance of saccadic eye movements in monkeys. J. Neurophysiol. 401, 74-94.

Pasik T., Pasik P. & Bender M.B. (1966) The superior colliculi and eye movements. Archs Neurol. 15, 420-436.

Pasik T., Pasik P. & Bender M.B. (1969) The pretectal syndrome in monkeys I.-Disturbances of gaze and body posture. Brain 92, 521-534.

Pierrot-Deseilligny C.H., Chain F., Gray F., Serdaru M., Escourolle R. & Lhermitte F. (1982) Parinaud's syndrome, Electro-Oculographic and anatomic analysis of six vascular cases with deductions about vertical gaze organization in the premotor structures. Brain 105, 667-696.

Ranalli P.J., Sharpe J.A. & Fletcher W.A. (1988) Palsy of upward and downward saccadic, pursuit and vestibular movements with a unilateral midbrain lesion: patho-physiologic correlations. Neurology 38, 114-122.

Robinson D.A. (1972) Eye movements evoked by collicular stimulation in the alert monkey. Vision Res. 12, 1795-1808.

Schiller P.H. (1977) The effect of superior colliculus ablation on saccades elicited by cortical stimulation. Brain Res. 122, 154-156.

Schiller P.H., Sandell J.H. & Mansell J.H.R. (1987) The effect of frontal eye field and superior colliculus lesions on saccadic latencies in the Rhesus monkey. J. Neurophysiol. 574, 1033-1049.

Schiller P.H., True S.O. & Conway J.L. (1980) Deficits in eye movements following frontal eye-field and superior colliculus ablations. J. Neurophysiol. 446, 1175-1189.

Usui S. & Amidror I. (1982) Digital low-pass differentiation for biological signal processing. IEEE Trans on BME. BME-29 No.10, 686-693.

Waitzman D.M., Ma T.P., Optican L.M. & Wurtz R.H. (1991) Superior colliculus neurons mediate the dynamic characteristics of saccades. J. Neurophysiol. 66, 1716-1737.

Wenban-Smith M.G. & Findlay J.M. (1991) Express saccades: is there a separate population in humans? Exp. Brain Res. 87, 218-222.

Wurtz R.H. & Goldberg M.E. (1972) Activity of superior colliculus in behaving monkey. IV. Effects of lesions on eye movements. J. Neurophysiol. 35, 587-596.

EYE MOVEMENTS AND COGNITION

EVIDENCE RELATING TO PREMOTOR THEORIES OF VISUOSPATIAL ATTENTION

Timothy L. Hodgson & Hermann J. Müller

Department of Psychology, Birkbeck College, University of London

Abstract

Two experiments examined whether saccadic eye movements and visuospatial attention can be decoupled and whether orienting of attention to a location is equivalent to programming a saccadic eye movement to that location. Experiment 1, produced evidence against decoupling of attention and eye movements. Experiment 2, which used peripheral cues, produced no evidence for the separate and hierarchical reprogramming of saccade direction and amplitude as predicted by Rizzolatti et al. (1987). Further, there were dissociations between saccadic and simple manual responses such as a differential preference for targets in the upward and downward directions. The results suggest that, whenever a saccadic eye movement is about to be executed, the direction of visuospatial attention is constrained to be compatible with the direction of the eye movement. However, while there is thus a close association between saccadic eye movements and visuospatial attention, it does not seem to be the case that attentional orienting is identical to a program for an eye movement.

Keywords

Saccades, manual responses, visuospatial attention, premotor theory, vertical asymmetries.

Introduction

When a spatial cue, indicating the likely location of a target stimulus, is followed shortly afterwards by a target at the cued location, visual processing of that target is enhanced in comparison with both a neutral (spatially non-informative) cueing condition and uncued locations - even in the absence of head and eye movements (e.g. Eriksen & Colegate, 1971; Downing, 1988; Posner, Snyder & Davidson, 1980; Müller & Humphreys, 1991). This enhancement has been shown to hold across a variety of tasks ranging from detection of luminance increments to visual form discrimination, with measures varying from simple reaction time to sensitivity parameters in terms of signal detection theory. The cue signal may be central, for example an arrow at the fixation point indicating the likely target location, or peripheral, for example a flash at the indicated location.

Posner (1980) referred to the allocation of attention in the absence of head and eye movements as "covert orienting", distinguishing it from "overt orienting" which involves such movements. There has been some debate concerning the nature of the relationship between saccadic eye movements and attentional orienting (see Posner, 1980 and Müller and Hodgson, 1993). Central to this debate is the question as to

Eye Movement Research/J.M. Findlay et al. (Editors)

whether a covert shift in visuospatial attention always involves, or is equivalent to, the preparation of a saccade. "Premotor accounts" of attentional orienting (e.g., Rizzolatti, Riggio, Dascola & Umilta, 1987; Wurtz & Mohler, 1976) maintain that attention and eye movements cannot be dissociated. However, there is evidence suggesting that this assumption may not be valid (Klein, 1980).

Experiment 1 reinvestigated this issue, producing evidence against decoupling of eye movements and visual attention. This is consistent with premotor accounts. Experiment 2 examined a particular recent premotor model, proposed by Rizzolatti et al. (1987). Their model assumes separate and hierarchical programming of saccade direction and amplitude parameters, predicting a characteristic pattern of RT costs for saccadic and manual responses. This pattern, characterised by "meridional" and "distance" effects (see below), was not observed in Experiment 2. The implications of these findings are discussed for premotor accounts of visuospatial orienting.

Experiment 1

One controversial issue is whether eye movements and attention can be dissociated (see Klein, 1980; Müller & Hodgson, 1993). Premotor accounts suggest that attention and eye movements are intrinsically linked, and the act of preparing an eye movement to a location is equivalent to orienting attention to that location. Therefore the two should not be separable, and covert orienting of attention should always precede overt movements of the eyes.

In agreement with this, Nissen, Posner and Snyder (1978) found that, when a saccadic eye movement was summoned by a peripheral cue, the simple manual RT to a luminance increment target at the cued location was facilitated 50 to 100 msec after cue onset - i.e. well before the saccade was executed (latencies of around 240 msec) (see also Remington, 1980). A similar finding was made by Shepherd, Findlay and Hockey (1986), who used central cues to direct both attention and eye movements.

In contrast, Klein (1980) reported data which appear to show that under certain circumstances visuospatial attention and eye movements can be decoupled (see also Remington, 1980). In one of his experiments, the subjects had to perform one of two tasks on a given trial, one likely and one unlikely, in response to the appropriate signal. The likely task was to make a saccade in a prespecified (blocked) direction in response to a peripheral asterisk that was equally likely to appear on either side of fixation. The unlikely task was to give a simple manual response to a peripheral brightening that also appeared with equal probability on either side of fixation. Klein expected that, if subjects prepared an eye movement in advance in the prespecified direction, then simple manual RTs should be faster to targets in that direction. However, Klein found no effect on simple manual RT, while saccade latencies were faster when the asterisk appeared in the movement-compatible direction. This seems to show that voluntary preprogramming of an eye movement in a particular direction does not necessarily facilitate simple manual responses to targets in the direction of the saccade.

However, there is an alternative account of Klein's data. The subjects may not have preprogrammed an eye movement in the prespecified direction. Rather, they may have relied on a stimulus-driven strategy (Henderson 1993). This would allow the peripheral response signal (brightening/asterisk) to reflexively activate both the oculomotor and visuospatial attention systems (Müller & Rabbitt, 1989), improving response signal discrimination at the stimulus location. If an eye movement is not prepared in advance of the peripheral signal, manual RTs would not be affected by whether the manual response signal (brightening) is compatible or incompatible with the prespecified eye movement direction. All that is required in both cases is the replacement of any reflexive saccade by a manual response. However, for eye movement RTs there would be an additional cost with signals incompatible to the prespecified direction, due to the effort of suppression of the saccade directly to the peripheral response signal and the programming of an anti-saccade.

There are two reasons why subjects may have relied on a reflexive strategy rather than active preparation of an eye movement in the prespecified direction on all trials. Firstly, 50% of eye movement signals were incompatible to the specified direction of the eye movement. The effort required to prepare an eye movement in advance would therefore be wasted on these incompatible trials and subjects may have refrained from active preparation of a saccade because of this (see Müller & Rabbitt, 1989). Secondly, eye movement direction was blocked, providing a further disincentive to the preparation of an eye movement (see Posner, Snyder & Davidson, 1980).

This alternative account predicts that by varying the likelihood of the eye movement signal appearing in the compatible or incompatible direction, subjects should have more or less incentive to prepare an eye movement. In addition, by varying the required direction of the eye movement from trial to trial as opposed to blocking the direction, subjects would be more likely to preprogram a saccade. In the present experiment, the direction in which an eye movement is to be prepared is specified with a central arrow cue before the start of each trial. We also varied the "validity" of this cue with respect to the direction of the peripheral eye movement signal.

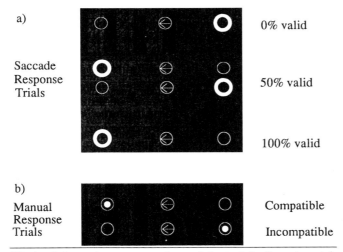

Fig.1 - Schematic representation of the task in Experiment 1, showing: a) Example stimuli appearing on saccade trials in the three validity conditions, b) example compatible and incompatible manual response trials (for simplicity, the central arrow instruction signal is only shown pointing to the left in this figure).

Method

The display, presented on a 16 inch monitor via a Macintosh Quadra computer, consisted of a black background with three dimly lit circles serving as location markers, one positioned centrally and the others 6 degrees to the left and right of the centre. On each trial, an arrow appeared in the central circle pointing either to the left or right location. Subjects were instructed to prepare an eye movement in the direction of the arrow. The arrow was displayed for 1000 msec. After a further interval of 500 msec, one of two possible response signals was presented. On 75% of the trials, one of the peripheral circles was brightened for 1000 msec (eye movement response signal), instructing the subjects to make a saccade in the direction of the previously presented arrow . On 25% of the trials, a small spot appeared in the centre of one of the peripheral circles (manual response signal), instructing the subjects not to make an eye movement, but to produce a manual button press as quickly as possible. Eye movement and manual response signals were presented in randomized order.

There were three experimental conditions, each defined by a particular "validity" of the eye movement signal. In the "0% condition", the eye movement signal always appeared opposite to the direction of the arrow. In the "50% condition", the eye movement signal appeared equally often on the same and the opposite side to the direction of the arrow. In the "100% condition", the eye movement signal always appeared in the direction of the arrow. In all three conditions, the manual response signal was equally likely to appear on either side of fixation. Trials on which the arrow and response signals were in the same and opposite direction will be referred to as compatible and incompatible trials, respectively. The three experimental conditions were presented in separate blocks of trials, the order of which was counterbalanced across subjects and sessions. Horizontal eye movements were recorded using a limbus tracker device (Skalar IRIS). 10 subjects participated in Experiment 1. Figure 1 summarizes the method schematically.

Results

Simple manual RTs. In contrast to Klein (1980), Experiment 1 produced a significant effect of compatibility on simple manual RTs (2-way ANOVA, $F(1,9) = 5.16$, $p < .05$) (see Figure 2a). RTs to manual targets in the arrow-compatible direction were faster than RTs to targets in the arrow-incompatible direction: 511 msec, on average, as compared to 520 msec. Unexpectedly, however, there was no reliable effect of arrow validity. The validity x compatibility interaction was also not significant. The RT advantage for arrow-compatible over incompatible manual response signals was not affected by the probability with which the arrow indicated the location of saccadic response signal. If anything, the effect appeared greatest under the 50% condition (compatible-incompatible: 524-536 msec), which also appeared to produce the slowest RTs (0-50-100%: 511-530-507 msec).

Saccade latencies. Also unlike Klein (1980), saccades in response to signals in the arrow-compatible direction were not reliably faster than saccades in response to signals in the arrow-incompatible direction (compatible-incompatible: 338-343 msec). Further, saccades were significantly faster in the 100% (all compatible) and 0% conditions (all incompatible) (318 and 325 msec, respectively) than in the 50%

condition (357 and 361 msec for compatible and incompatible trials, respectively) (2-way ANOVA, F(1,9) ,= 5.98, p < .05).

Response Errors. Three types of error were examined. Errors on eye movement trials consisting of (1) the execution of a manual response instead of the required saccadic response, and (2) the execution of a saccade to the peripheral eye movement signal rather than the location prespecified by the arrow. (3) Errors on manual response trials consisting of subjects making a saccade to one of the target locations rather than maintaining fixation and giving a manual response.

The subjects only very rarely made a manual response when an eye movement response was required (type 1 errors = 0.3% of total trials), so these errors were not further analysed. The second type of error on eye movement trials, fixating the eye movement signal instead of the prespecified location, was more common (0.8%). This is not surprising given that, when the eye movement signal occurs opposite to the direction of the arrow, a reflexive saccade to that signal must be suppressed and an anti-saccade must be executed.

a)   b)

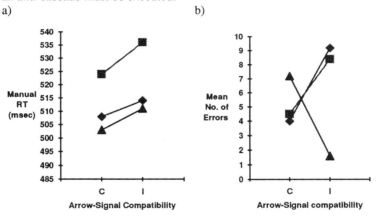

Fig. 2 - a) Graph of inter-subject mean manual RTs against arrow-response signal compatibility (C = compatible, I = incompatible), for the three validity conditions in experiment 1. b) Graph of mean number of saccade errors (type 3) on manual response trials across subjects for the three validity conditions. (diamonds = 0%, squares = 50%, triangles = 100%).

The pattern of saccade errors on manual response trials (type 3 errors) was more interesting. The overall rate of such incorrect eye movement responses was relatively high (12% of manual response trials), and 98% of all such responses consisted of saccades in the direction of the arrow signal. This type of error showed a significant cross-over interaction between arrow validity and arrow-saccadic response signal compatibility (2-way ANOVA, F(2,18) = 27.44, p < .0001) (see Figure 2b). In the 100% validity condition, more saccade errors were made on trials on which the arrow and manual response signals were compatible. In contrast, in the 0% and 50% conditions, this effect was reversed, with more errors occurring on incompatible trials.

Discussion

The task in Experiment 1 can be regarded as requiring the execution of several processing stages: (1) Encode arrow direction -> (2) Prepare an eye movement -> (3) Detect and locate response signal -> (4) Discriminate response signal -> (5) Select Response -> (6) Execute response. The complexity of this task is reflected in relatively long overall response times (saccade latencies: 344 msec; simple manual RTs: 511 msec), and this may have obscured any strong effects on the response times. Nevertheless, an understanding of the manual RT data may be reached by taking the error rate effects (type 3 errors) into account.

In the *100% condition* (arrow and saccadic response signal always compatible), all signals incompatible with the arrowed direction required a manual response. Thus, stimulus discrimination was only required with compatible response signals. On incompatible trials, a manual response could be executed directly (based on the detection of an incompatible response signal alone). This would facilitate the manual RTs on incompatible trials (while also reducing the error rate). This would result in a reduced compatibility effect on the RTs in this condition. This might explain why the compatibility effect was not larger in the 100% condition when compared to the 50% validity condition. The high error rates on compatible trials presumably reflects failures to suppress a saccadic response to peripheral signals in the prepared direction.

In the *0% condition* (arrow and saccadic response signal always incompatible), all compatible response signals required a manual response. Thus, detection of a compatible signal was sufficient to directly make a manual response. However, signals incompatible to the direction of the arrow require discrimination. This would explain the trend towards faster RTs on compatible compared to incompatible trials. It would also explain the increased number of errors in the incompatible direction, due to failures in discrimination of the response signal.

In the *50% condition*, discrimination of the response signal was required with both compatible and incompatible stimuli. So differential task demands cannot account for the faster manual RTs for compatible when compared to incompatible signals. In addition to this RT effect, the error rates were lower for compatible than for incompatible signals. This presumably reflects enhanced stimulus processing at the location indicated by the arrow. Recall further that 98% of errors consisted of saccades in the direction of the arrow, indicating that subjects were indeed utilizing the arrow information to prepare the direction of the subsequent eye movement.

In summary, 1) manual responses were faster to signals presented at the target location of the eye movement; 2) eye movement errors (due to mistaking the manual response signal for a saccadic signal) were less frequent for that location in the 50% valid condition; 3) if an eye movement error was made, it was virtually always a saccade to that location. Taken together, this pattern strongly suggests that attention was allocated to the prespecified target location of the eye movement, and that the direction of attention cannot be decoupled from that of the eye movement.

One finding that seems at odds with this proposal is that the compatibility effect on the manual RTs was not accompanied by a strong and reliable effect on saccadic latencies. It is possible that because subjects actively prepared an eye movement, the effect of signal compatibility on saccade reaction time was not as strong as if the subjects had relied on a reflexive strategy. However, further work would be needed to investigate this possibility.

Together with the previous evidence, Experiment 1 argues that, when a voluntary eye movement is prepared in response to a central arrow signal, or when a reflexive saccade is programmed in response to a compelling peripheral stimulus, the direction of visuospatial attention is constrained to be compatible with the direction of the eye movement. This is consistent with premotor accounts of attention in which the direction of an eye movement and the direction of attention cannot be decoupled.

Experiment 2

One particular premotor model has been put forward by Rizzolatti, Riggio, Dascola and Umilta (1987) to account for the "costs", in terms of simple manual RT, observed when targets appear at uncued locations. Typically, costs are found to increase as a function of the distance between the cued and target location (e.g., Downing & Pinker, 1985), but often an additional cost is observed when cue and target appeared on opposite sides of the vertical or horizontal meridian of the visual field. These findings have been termed distance and meridional effects, respectively. The vertical meridional effect might be explained in terms of the transfer of activation from one hemisphere to the other across the corpus callosum (Hughes & Zimba, 1985), but this account fails for the horizontal meridian. In contrast, the premotor model of Rizzolatti et al. could accommodate both the (vertical and horizontal) meridional and distance effect.

According to this account, making a simple manual response to a target involves the programming a saccade to the location of the target. If the target appears at an unexpected location. the saccade program must be modified, and the degree of modification is reflected in the final manual RT. Rizzolatti et al. (1987) suggested that reprogramming of the direction of the saccade involves greater modification than reprogramming its amplitude (following the "hierarchical" model of motor programming proposed by Rosenbaum, Inhoff and Gordon, 1984). The meridional effect would correspond to reprogramming the direction, and the distance effect would correspond to reprogramming the amplitude of the saccade.

It follows from the model of Rizzolatti et al. that hierarchical programming of saccade parameters ought to be observable directly from the latencies of saccadic eye movements. Reuter-Lorenz and Fendrich (1992) failed to find meridional effects on saccadic latencies or manual RTs for peripherally cued targets; but they found meridional effects for both manual and saccadic responses using central arrow cues. Other workers have reported similar findings (saccade latencies: Abrams & Jonides, 1988; Crawford & Müller, 1992; manual RTs: Egly & Homa 1991), although some studies have reported meridional effects with peripherally cued targets (manual RTs: Hughes & Zimba, 1985; Crawford & Müller, 1992).

Hughes and Zimba (1987) and, more recently, Gawryzewski et al. (1992) have reported that meridional costs for manual responses were greater when crossing oblique meridians. This may be explained in terms of the reprogramming of two orthogonal components of target position rather than a change in the absolute direction of the preprogrammed movement. Note, though, that the methodology behind these conclusions has been criticised (Klein & McCormick, 1989; Henderson 1993), and Henderson (1991) failed to find increased manual RT costs on oblique meridians with peripheral cues.

Experiment 2 attempted to resolve this issue by examining responses to directly cued targets in a two-dimensional framework. Previous failures to find meridional costs for saccadic latencies in response to peripherally cued targets may be due to the fact that only the horizontal meridian has been thoroughly examined. The effect might be more apparent for oblique meridians. Experiment 2 also compared manual and saccadic responses to precued stimuli in two dimensions to assess whether there are qualitative differences indicating possible dissociations of the attentional systems subserving the two responses.

Method

16 target locations were marked by dimly illuminated boxes on a black background. The locations were arranged along 4 axes corresponding to the horizontal, vertical, 45 degree oblique and 135 oblique axes. Target locations were at 3, 6 and 9 degrees eccentricity from the centre of the central box which served as the fixation point. Precue signals consisted of a brief brightening of one of the peripheral boxes. These were followed after 150 msec by the presentation of a bright spot target. The target signal appeared at the cued location on 50% of the trials (valid trials). On 35% of the trials, the target appeared at one of the other locations (excluding the central location) along the same axis (invalid trials). On 10% of the trials, the cue appeared at the central location. This indicated that the target was equally likely to appear at any of the target locations (neutral trials). Finally, on 5% of the trials, a cue signal appeared without being followed by a target (catch trials).

In the first part of Experiment 2, the subject's task was to make a saccade to the target as quickly as possible. In the second part, the subjects had to refrain from making an eye movement to the target; instead, they responded by making a manual button press response. Figure 3 summarizes the method schematically.

7 subjects participated in both parts of the experiment. In addition, 3 subjects took part in the saccade condition only, while 3 different subjects took part in the manual response condition only.

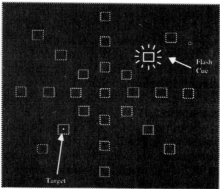

Fig. 3 - Schematic representation of the task in Experiment 2.

Results

The results showed a RT advantage for valid trials over neutral and invalid trials with both manual and saccadic responses (manual: F(2,18) = 5.02, p< 0.02; saccadic: F(2,18) = 16.56, p < 0.0001). There was also a strong relationship between RT and the absolute distance between the cue and the target, with both manual and saccadic RTs increasing as cue-target distance increased (manual: F(6,48) = 3.247, p < 0.01; saccadic: F(6, 54) = 17.12, p < 0.0001).

There was no evidence of meridional effects, for either orthogonal (horizontal and vertical) or oblique (diagonal) meridians. Meridional effects were assessed by comparing trials on which the cue-target distance was equivalent but the target appeared in the same or the opposite direction to the cue, relative to the fixation point. For saccadic RTs, same and opposite direction trials did not differ significantly. However, for manual responses, opposite side trials were significantly faster (rather than slower) than same side trials (F(1,9) = 14.517, p < 0.005).

The data were also analyzed in terms of target direction (with respect to the fixation point). For saccadic responses, there was a large increase in RTs to targets presented in the downward direction on neutral trials. For invalid trials, this up/down asymmetry was reduced; and for valid trials, it was abolished (trial type x target direction interaction: F(14,98) = 1.932, p < 0.05). Manual RTs also showed an up/down asymmetry. However, it was targets directly above fixation which gave rise to increased RTs. Again, this asymmetry was greatly reduced when the target was preceded by a valid cue (trial type x target direction interaction: F(14, 126) = 2.37, p <0.01).

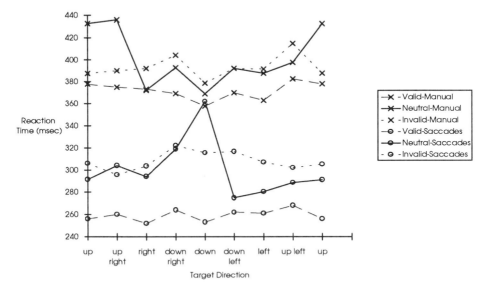

Fig. 4 - Graph of saccadic and simple manual RTs against the absolute direction of the target stimulus in experiment 2. Inter-subject means are shown for valid, neutral and invalid trial types.

There was another dissociation between saccadic and manual responses in Experiment 2, which has been noted elsewhere (Reuter-Lorenz & Fendrich, 1992). For simple manual responses, the RTs were significantly slower when an invalid target appeared more peripheral (far from fixation) to the cued location than when it appeared more central (near fixation). However, there was no evidence of such a difference with saccadic responses (manual/saccadic response x near/far target interaction: $F(1,6) = 20.53, p < 0.005$).

Discussion

In summary, Experiment 2 produced little evidence for the hierarchical programming of saccade direction and amplitude in response to peripheral cues. In particular, there was no indication of meridional effects predicted by the premotor account of Rizzolatti et al., no matter whether cue and target were presented on orthogonal or oblique axes. Since meridional effects are relatively consistently found with central cues, the data of Experiment 2 suggest that responses to peripherally cued targets (reflexive) may involve different spatial coding mechanism than do responses to centrally cued targets (voluntary) (Shepherd & Müller, 1989; Reuter-Lorenz & Fendrich, 1992).

Perhaps the most interesting findings of Experiment 2 concerned the qualitative differences between saccadic and manual response characteristics. Previous reports have described vertical (up/down) asymmetries separately in eye movements (Honda & Findlay, 1992) and manual responses (Rizzolatti et al., 1987; Gawryszewski et al., 1987; see also Previc, 1991). Experiment 2 went further by demonstrating them concurrently in the same visuospatial cueing task.

The differential up-down RT asymmetry and near-far RT costs for manual and saccadic responses have implications for the premotor account of Rizzolatti et al. (1987). If RTs for saccadic eye movements differ qualitatively from RTs for simple manual responses on the same attentional task, is it still plausible to maintain that covert visuospatial orienting is equivalent to (i.e., involves the same mechanisms as) the programming of an overt oculomotor orienting response? The answer would appear to depend on the locus of the dissociations. Do the asymmetries occur at an attentional level? Or are they due to more motoric factors? The differential up/down asymmetry with saccadic and manual responses was found to be dependent on cue validity (cue validity x target direction interaction). This would suggest that these differences must be premotor in nature as valid trials did not show vertical asymmetries in RT. However, further work will be required to clarify this issue. In particular, the same effect also needs to be demonstrated for centrally cued responses.

Summary and Conclusions

Taken together, the results of Experiments 1 and 2 suggest the following. Whenever a saccadic eye movement is about to be executed, the direction of visuospatial attention is constrained to be compatible with the direction of the eye movement. This applies to both reflexive and voluntary saccades. However, while there is thus a close association between saccadic eye movements and visuospatial attention, it does not seem to be the case that attentional orienting is identical to a program for an eye movement. With peripheral cues, there is no evidence that saccade parameters are programmed in a hierarchical fashion (with direction being specified first and amplitude second). Further, there are dissociations between directed saccadic and simple manual responses. Although the nature of these dissociations is not well understood, they would at least caution against an account of simple manual RT effects in terms of oculomotor programming. With central cues, there may be a greater degree of "equivalence" between saccadic and simple manual response processes (e.g., Reuter-Lorenz & Fendrich, 1992). However, possible dissociations need yet to be explored systematically.

More work is required to examine to what extent the particular response (motor) system that is used to index "attentional" processes, contributes to the effects obtained in visuospatial cueing experiments. It is not implausible that visuospatial attention and eye movements are closely associated (i.e. are subject to the same constraints) while being functionally relatively independent of each other.

References

Abrams, R.A. & Jonides, J. (1988). Programming saccadic eye movements. Journal Of Experimental Psychology: Human Perception and Performance, 14, 428-443.

Crawford, T. & Müller, H.J. (1992). Temporal and spatial effects of spatial cueing on saccadic eye movements. Vision Research, 32, 293-304.

Downing, C. (1988). Expectancy and visual-spatial attention: Effects on perceptual quality. Journal of Experimental Psychology: Human Perception and Performance, 14, 188-202.

Downing, C. & Pinker, S. (1985). The spatial structure of visual attention. In M.I. Posner & O.S.M. Marin (Eds.), Attention and Performance XI (pp. 171-187). Hillsdale, NJ: Erlbaum.

Egly, R. & Homa, D. (1991). Reallocation of visual attention. Journal of Experimental Psychology: Human Perception & Performance, 17, 142-159.

Eriksen, C.W. & Colegate, R.L. (1971). Selective attention and serial processing in briefly presented visual displays. Perception & Psychophysics, 10, 321-326.

Gawryszewski, G., Riggio, L., Rizzolatti, G. & Umilta, C. (1987). Movements of attention in three spatial dimensions and the meaning of neutral cues. Neuropsychologia, 25, 19-29.

Gawryszewski, L., Faria, R.B., Thomaz, T.G., Pinheiro, W.M., Rizzolatti, G., Umilta, C. (1992). Reorienting visual spatial attention: Is it based on Cartesian co-ordinates? in Lent, R. (Ed.) The Visual System: from Genesis to Maturity, Birkhäuser, Boston.(pp.267-281).

Henderson, J.M. (1991). Stimulus discrimination following covert attentional Orienting to an exogenous cue. Journal of Experimental Psychology: Human Perception and Performance, 17, 91-106.

Henderson, J.M. (1993). Visual attention and saccadic eye movements. In G. d'Ydewalle & J. Van Rensbergen (Eds.), Perception and Cognition: Advances in Eye Movement Research. Amsterdam: Elsevier (pp. 37-58).

Honda, H. & Findlay, J.M. (1992). Saccades to targets in 3-D space: Dependence of saccade latencies on target location. Perception & Psychophysics, 52, 167-174.

Hughes, H.C. & Zimba, L.D. (1985). Spatial maps of directed visual attention. Journal of Experimental Psychology: Human Perception and Performance, 11, 409-430.

Hughes, H.C. & Zimba, L.D. (1987). Natural boundaries for the spatial spread of directed visual attention. Neuropsychologia, 25, 5-18.

Klein, R. (1980). Does oculomotor readiness mediate cognitive control of visual attention? In R.S. Nickerson (Ed.), Attention and Performance VIII. Hillsdale, NJ: Erlbaum (pp. 259-276).

Klein, R. & McCormick, P. (1989). Covert visual orienting: Hemi-field activation can be mimicked by zoom lens and mid location placement strategies. Acta Psychologia, 770, 225-250.

Müller, H.J. & Hodgson, T.L. (1993). On the relationship between visuospatial attention and saccadic eye movements. Unpublished manuscript.

Müller, H.J. & Rabbitt, P.M.A. (1989). Reflexive and voluntary orienting of visual attention: Time course of activation and resistance to interruption. Journal of Experimental Psychology: Human Perception and Performance, 15, 315-333.

Müller, H.J. & Humphreys, G.W. (1991). Luminance-increment detection: Capacity-limited or not? Journal of Experimental Psychology: Human Perception and Performance, 17, 107-124.

Nissen, M.J., Posner, M.I. & Snyder, C.R.R. (1978). Relationship between attention shifts and saccadic eye movements. Paper presented at the Psychonomics Society, San Antonio, Texas.

Posner, M.I. (1980). Orienting of attention. Quarterly Journal of Experimental Psychology, 32, 3-25.

Posner, M.I., Snyder, C.R.R. & Davidson, B.J. (1980). Attention and the detection of signals. Journal of Experimental Psychology: General, 109, 160-174.

Previc, F.H. (1990). Functional specialization in the lower and upper visual field in humans: Its ecological origins and neurophysiological implications. Behavioral & Brain Sciences, 13, 519-575.

Remington, R.W. (1980). Attention and saccadic eye movements. Journal of Experimental Psychology: Human Perception and Performance, 6, 726-744.

Reuter-Lorenz, P.A. & Fendrich, R. (1992). Oculomotor readiness and covert orienting: Differences between central and reflexive precues. Perception & Psychophysics, 52, 336-344.

Rizzolatti, G., Riggio, L., Dascola, I. & Umilta, C. (1987). Reorienting attention across the horizontal and vertical meridians: Evidence in favour of a premotor theory of attention. Neuropsychologia 25, 31-40.

Rosenbaum, D.A., Inhoff, A.W. & Gordon, A.M. (1984). Choosing between movement sequences: A hierarchical editor model. Journal Of Experimental Psychology: General, 113, 372-393.

Shepherd, M., Findlay, J.M. & Hockey, R.J. (1986). The relationship between eye movements and spatial attention. Quarterly Journal Of Experimental Psychology, 38, 475- 491, 146-154.

Shepherd, M. & Müller, H.J. (1989). Movement versus focusing of visual attention. Perception & Psychophysics, 46,146-154.

Wurtz, R.H. & Mohler, C.W. (1976) Organization of monkey superior colliculus: enhanced visual response of superficial layer cells. Journal of Neurophysiology, 39, 745-765.

Acknowledgements

This work was supported in part by grants from the Central Research Fund of the University of London and from the Science and Engineering Research Council (GR/4/54966). T.L.H. is supported by a Medical Research Council studentship.

VISUAL ATTENTION AND SACCADIC EYE MOVEMENTS: EVIDENCE FOR OBLIGATORY AND SELECTIVE SPATIAL COUPLING

Werner X. Schneider & Heiner Deubel

Ludwig-Maximilians-Universität and
Max-Planck-Institut für psychologische Forschung
München, Germany

Abstract

The spatial interaction of visual attention and saccadic eye movements was investigated in a task that combines stimulus-elicited saccades with letter discrimination. Subjects had to saccade to locations within horizontal letter strings left or right from a central fixation cross. The performance in discriminating between the letters "E" and "mirror-E" presented tachistoscopically within surrounding distractors before the saccade was taken as a measure of attentional orienting. The data showed that discrimination performance is best when discrimination target and saccade target referred to the same object. The results strongly argue for an obligatory coupling of saccade programming and visual attention allocation to one common object. Further, the discrimination performance was better when the saccade is directed to a target located more foveally of the discrimination target than when the saccade target was located more peripherally. Similar asymmetries have been found in lateral masking studies.

Keywords

Visual attention, Saccades, Obligatory coupling, Lateral masking

Introduction

The issue whether and how visual attention and saccadic eye movements are related has been systematically investigated in Cognitive Psychology since the late seventies (e.g. Posner, 1980; Klein, 1980; Shepherd, Findlay, & Hockey, 1986). Visual attention is usually

Eye Movement Research/J.M. Findlay et al. (Editors)

operationalized as differences of reaction times or recognition performance (see, Van der Heijden, 1992, for an overview). The general finding is that the allocation of visual attention to an object or an location in visual space improves both performance measures compared to conditions where visual attention is allocated elsewhere. The standard interpretation of this finding is that attention leads to the space-based priorizing of processing of the attended object (e.g. Schneider, 1993).

That this allocation of visual attention does not depend on the overt initiation of an eye movement was a major findings of spatial precueing paradigms (e.g. Eriksen & Hoffman, 1973; Posner, 1980; Van der Heijden, 1992). If the inverse is true, however, i.e. whether the eye could move by a saccade to a visual target without a concomitant, obligatory shift of attention, is still an open question. Is visual attention strictly spatially and temporally coupled to the saccade target location during the process of "programming" the saccade, or can visual attention be "shifted" to one location while the saccade is programmed to another location? Several investigators have tried to tackle this question since the late seventies (Posner, 1980; Klein, 1980; Remington, 1980; Shepherd, Findlay, & Hockey, 1986; Crawford & Müller, 1992; Reuter-Lorenz & Fendrich, 1992). The study by Shepherd et al. (1986) eliminated some of the methodological problems that were present in the earlier investigations. They combined a simple reaction task - reaction time served as attentional measure -with a saccade task. Their subjects had to saccade to one of two boxes left or right from the fixation point indicated by a central arrow. Shortly before or after the saccade a second stimulus appeared in one of these boxes. The subject had to react with a key press as fast as possible. The basic finding of this study was that the manual reaction time was much faster when saccade target location and the location of the stimulus to which the subject had to react coincided, as compared to the condition where saccade target and imperative stimulus appeared in opposite hemifields. This relationship held when the stimulus for the key press appeared *before* the overt saccade has been initiated.

These findings strongly suggested that the allocation of visual attention and saccade programming are spatially coupled. Since Shepherd et al. (1986) manipulated saccade target and manual reaction target between hemifields, these data are not conclusive about how spatially selective this coupling is, i.e., whether the coupling refers to the hemifield or whether it refers to one common target location *within* a hemifield. The second hypothesis would predict a reaction time increase when the saccade and manual reaction are "directed" to two spatially close but different objects within one hemifield.

A further question is to what extend this coupling is obligatory, i.e. whether subjects are unable to move their eyes to one location and attend to another. Concerning hemifields Shepherd et al. (1986) found evidence that strongly favors the conception of an obligatory coupling. In one of their experimental conditions, subjects had to saccade to a location within one hemifield while being informed that the manual reaction time target would appear with a high probability in the opposite hemifield. In this condition subjects had a high incentive to decouple visual attention allocation and saccade programming. The experimental results show reaction time costs for this condition compared to a condition where the saccade target and the manual key press target were in the same hemifield. Again, the spatial selectivity of this effect is still unclear, i.e. whether it is possible to attend to an object while saccading to a nearby location in the same hemifield.

In order to elucidate these questions we developed a dual-task paradigm where a target-directed saccade was combined with a discrimination task. A cue indicated a saccade target

(ST) consisting of one of the items out of a horizontal string of letters. Before the eye movement, a discrimination target (DT) always appeared at the central position of the item string while the saccade target position was varied. This should present an optimal condition for the decoupling of visual attention and saccade control. In case of a strict coupling discrimination performance should strongly depend on the intended ST location.

Presumably, attentional allocation to one of the items in the string is facilitated when perceptual segmentation into the item elements is easy (Schneider, submitted). Therefore, we also varied the spacing between the items in the string.

Methods

Experimental Set-up

The subject was seated in a dimly illuminated room. The visual stimuli were presented on a fast color monitor (CONRAC 7550) providing a frame frequency of 100 Hz at a resolution of 1024*768 pixels. Active screen size was 40 by 30 cm; the viewing distance was 80 cm. The video signals were generated by a freely programmable graphics board (Kontron KONTRAST 8000), controlled by a PC via the TIGA (Texas Instruments Graphics Adapter) interface. The stimuli appeared on a grey background which was adjusted at a mean luminance of 2.2 cd/m2. The luminance of the stimuli was 25 cd/m2. The relatively high background brightness is essential for avoiding the effects of phosphor persistence, ensuring that the target luminance decays to subthreshold values within less that 20 msec (Wolf & Deubel, 1993). In order to further exclude the possibility that persistence effects allow for target discrimination *after* the saccade, we performed a pilot study in which dark targets were used on a bright background. The results were identical.

Eye movements were recorded with a SRI Generation 5.5 Dual-Purkinje image eyetracker (Crane and Steele, 1985) and sampled at 400 Hz. Head movements were restricted by a biteboard and a forehead rest. The experiment was completely controlled by a 486 Personal Computer. The PC also served for the automatic off-line analysis of the eye movement data in which saccadic latencies and saccade start and end positions were determined.

Procedure

Each session started with a calibration procedure in which the subject had to sequentially fixate 10 positions arranged on a circular array. A block of 216 experimental trials followed the calibration procedure. An example for the sequence of stimuli presented in each trial is given in Figure 1. Each trial started with the presentation of a small fixation cross in the center of the screen. After a variable delay ranging from 500 to 1200 msec, two item strings appeared left and right of the central fixation, each consisting of three "8"s. The width of each item was 0.52 deg of visual angle, its height was 1.05 deg. The central item of the three letters was always presented at an eccentricity of 5 deg. From trial to trial, the inter-item distances were randomly selected from values of 1 deg ("small"), 1.5 deg ("medium"), and 1.9 deg ("large"). After 60 msec, two vertical lines indicating the saccade target (ST) appeared at one of the item positions. The side and the item position where this cue appeared was varied randomly. After a cue lead time of 80 msec, five of the premask items in both strings were replaced by distractors ("**2**" or "**5**"). At the middle position of the string in the cued direction, the mask was replaced by the discrimination target (DT) which was either "**E**" or "**3**". The DT and the distractors disappeared after 60 msec. Thus, the discrimination target was no longer available 140 msec after the onset of the saccade target. As a result of the stimulus timing sequence most of the saccades were initiated long after the disappearance of target and distractors. Saccades with latencies shorter than 160 msec were discarded from the analysis. After the saccade the subject had, without time pressure, to indicate the identity of the discrimination target by a button press (2-alternative forced choice). The central fixation cross reappeared after the subjects discrimination and the next trial was initiated by the computer. The subjects were instructed to saccade to the position indicated by the line cue "as fast and as precisely as possible". They were informed that the discrimination target always appeared at the middle position of the string, in the same direction as that of the elicted saccade.

In a control task ("fixation task"), the subject was required to keep fixation on the central cross. The stimulus sequence was identical to that described before except that line cues appeared at all three item position on one of both sides. Thus, the cues indicated the hemifield where the target would appear.

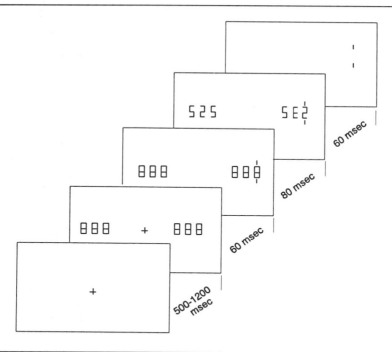

Fig.1 - Sequence of stimulus presentation

Subjects

Five subjects aged 20-27 years participated in the experiments. All had normal vision and were experienced in performing other types of experiments related to oculomotor research. Each subject performed at least 3 sessions.

Results

Figure 2 shows mean discrimination performance (averaged over the five subjects) for the various conditions, in the left diagram (Figure 2a) as a function of saccade target location and in the right diagram (Figure 2b) as a function of spacing. Figure 2b additionally exhibits discrimination performance for the "fixation" condition. It can be clearly seen that discrimination performance strongly depends both on the saccade target position and inter-item spacing. For all spacings, discrimination is best when ST and DT location coincide, i.e. when the saccade is programmed to the discrimination object (Figure 2a). Performance steeply decreases both when the saccade was directed to an object more foveally or more peripherally. For a given ST location, discrimination of DT is facilitated with increased spacing (Figure 2b). Discrimination performance for the fixation condition (Figure 2b) is basically identical to the condition when the eye is cued to the inner item (ST1).

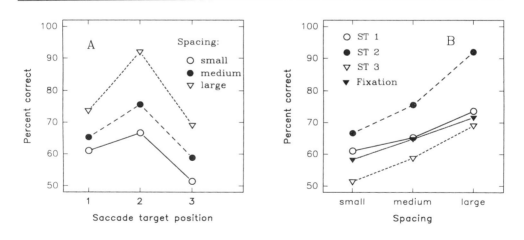

Fig.2 - Discrimination performance (percent correct identification of the target letter), plotted as a function of the saccade target (ST) position (A) and of the inter-item spacing (B). Additionally, Figure 2B shows discrimination performance with fixating eye.

An analysis of variance (repeated measures) shows a significant main effect of ST position ($F_{(2,8)} = 38.3$; $p < 0.01$) and of distance ($F_{(2,8)} = 15.4$; $p < 0.01$), with a non-significant interaction ($F_{(4,16)} = 0.8$; $p > 0.05$). Newman-Keuls tests revealed significant differences of discrimination performance between ST1 and ST2 as well as between ST2 and ST3 ($p <$

0.01), averaged acrossed all inter-item distances. Furthermore, when the first and the third ST positions (ST1 and ST3) are compared, the data showed an asymmetry with performance being worse when the eye is directed more peripherally (p < 0.01).

The saccade latencies for the different experimental conditions are shown in Figure 3a. Saccade latencies were not influenced by the distance manipulation and only very modestly influenced by the ST position. So, an analysis of variance (repeated measures) revealed no significant effects of distance on saccade latencies $(F(2,8) = 0.04; p > 0.05)$ but a significant effect of ST position $(F(2,8) = 7.39; p < 0.05)$ and of the interaction $(F(4,16) = 7.3; p < 0.05)$.

Finally, Figure 3b displays the saccade amplitudes in the different experimental conditions. Mean saccade end positions vary according to the saccade target position demonstrating that the subjects hit the indicated targets with reasonable accuracy. For the near target (ST1), the saccade tends to overshoot the indicated target position, while the peripheral target (ST3) produces systematic undershoots.

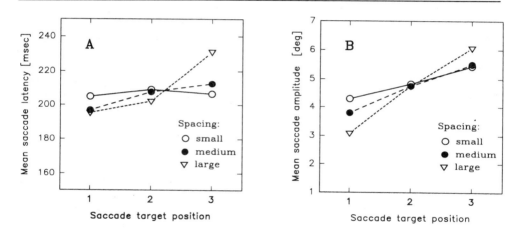

Fig.3 - A: *Mean saccadic latency as a function of saccade target position, given for the various inter-item spacings. B: Mean saccadic amplitudes as a function of the saccade target positions.*

Discussion

The main finding of this study is that the discrimination performance depends strongly on the intended saccade target position, producing best performance when the eye is directed to the object that has to be identified. Discrimination performance declines steeply when the ST and DT refer to items at different locations. This argues against the decoupling hypothesis, i.e. the ability to direct visual attention to one location while simultaneously "programming" a saccade to another location. Instead, the results strongly suggest that both processes are strictly coupled, both temporally and spatially.

The amount of spatial selectivity shown in our data is surprising. Investigations using other paradigms (e.g. reaction time) often demonstrated a rather broad gradient of attentional focussing as an effect of peripheral cueing (e.g. Downing & Pinker, 1985). Results from a related set of experiments (Deubel & Schneider, 1992; Deubel & Schneider, submitted) suggest that an important prerequisite for spatial selectivity present in our experiment is that the visual field can be segregated into individual visual objects. This implies that in our experiment attentional focussing is object-related rather than purely location-specific (see, Schneider, 1993). In other words, when an object in the visual periphery is selected as a saccade target, only this item experiences priorized processing by the attentional system.

The fixation condition where three simultaneous cues indicated the hemifield of the DT yielded a discrimination performance as in the saccade condition "ST1". This finding may indicate an attentional scanning process that always starts with the object closest to the fovea.

The manipulation of the distance between items did not change the pattern of superior discrimination performance when ST and DT locations were identical. Rather, the distance effect was additive to the ST factor. Therefore, even when the spacing between items was large the visual attention could be optimally allocated to the item at DT location the ST position still matters and strongly influences the DT performance.

Interestingly, a spatial asymmetry of discrimination performance between the first (more foveal) and the third (more peripheral) ST position emerged. This resembles a spatial asymmetry effect found in lateral masking experiments (e.g. Chambers & Wolford, 1983). In their experiments, subjects had to report the identity of a briefly presented target letter surrounded in close proximity by a single distractor letter. The results showed that a distractor located more foveally to the target is less harmful than a distractor located more peripherally. The interesting aspect of the spatial asymmetry effect in our experiment is that we did not manipulate the distractor position - like in lateral masking experiments - but instead the ST position while keeping the physical stimulus structure of target and distractors constant.

To indicate the saccade target we used a peripheral cue. There is evidence that this type of cue probably attracts the visual attention mechanism obligatory (see, e.g. Yantis & Jonides, 1984; Müller & Rabbitt, 1989). Therefore, it is not so surprising to find a coupling between this peripheral saccade target and attention-dependent discrimination performance. The question arises whether this coupling also holds for a centrally cued ST. Data from experiments with a somewhat similar experimental paradigm (Deubel & Schneider, submitted) using central cues to indicate the ST position show that central and peripheral ST cues have very similar effects.

Besides collecting further data on the relationship between saccade control and visual attention more precise models for both types of processes are required. They have to take into account not only the specific data patterns on the relationship between eye movements and attention but also data on the neurocomputational architecture and dynamics of the visual attention system and the connected object recognition, saccade and other space-based motor action systems (Schneider, submitted). We are convinced that the saccadic system offers an informative key of investigating the "input" and "output" control of this attentional system.

References

Crane, H. D. & Steele, C. M. (1985) Generation-V dual-Purkinje-Image eyetracker. Applied Optics, 24, 527-537.

Crawford, T. D., & Müller, H. J. (1992). Spatial and temporal effects of spatial attention on human saccadic eye movements. Vision Research, 32, 293-304.

Chambers, L., & Wolford, G. (1983). Lateral masking vertically and horizontally. Bulletin of the Psychonomic Society, 21, 459-461.

Deubel, H. & Schneider, W. X. (1992). Spatial and temporal relationship of saccadic eye movements and attentional orienting. Perception, 21, Suppl., 106.

Deubel, H., & Schneider, W. X. (submitted). Visual attention and saccadic eye movements are coupled spatially and temporally.

Downing, C.J. & Pinker, S. (1985). The spatial structure of visual attention. In M.I. Posner & O.S.M. Martin (Eds.), Attention and Performance, XI (pp. 171-187), Hillsdale, N.J.: Lawrence Erlbaum Associates Inc.

Eriksen, C. W., & Hoffman, J. E. (1973). The extent of processing of noise elements during selective encoding from visual displays. Perception & Psychophysics, 1, 155-160.

Klein, R. (1980). Does oculomotor readiness mediate cognitive control of visual attention? In R. Nickerson (Ed.), Attention and performance, VIII (pp. 259-276). Hillsdale, NJ: Lawrence Erlbaum Associates.

Müller, H. J., & Rabbitt, P. M. (1989). Reflexive and voluntary orienting of visual attention: Time course of activation and resistance to interruption. Journal of Experimental Psychology: Human Perception and Performance, 15, 315-330.

Posner, M. I. (1980). Orienting of attention. Quarterly Journal of Experimental Psychology, 32, 3-25.

Remington, R. W. (1980). Attention and saccadic eye movements. Journal of Experimental Psychology: Human Perception and Performance, 6, 726-744.

Reuter-Lorenz, P. A., & Fendrich, R. . (1992). Oculomotor readiness and covert orienting: Differences between central and peripheral cues. Perception & Psychophysics, 52, 336-344.

Schneider, W. X. (1993). Space-based visual attention models and object selection: Constraints, problems, and possible solutions. Psychological Research, 56, 35-43.

Schneider, W. X. (submitted). VAM: A neuro-cognitive model for visual attention control of segmentation, object recognition and space-based motor actions.

Shepherd, M., Findlay, J. M., & Hockey, R. J. (1986). The relationship between eye movements and spatial attention. Quartely Journal of Experimental Psychology, 38A, 475-491.

Van der Heijden, A. H. C. (1992). Selective attention in vision. London, GB: Routledge.

Wolf, W. & Deubel, H. (1993) Enhanced vision with saccadic eye movements: An artefact due to phosphor persistence of display screens? In Proceedings of the 15th Annual International Conference of the IEEE Engineering in Medicine and Biology Society (pp. 1383-1384), IEEE, New York.

Yantis, S., & Jonides, J. (1984). Abrupt visual onsets and selective attention: Evidence from visual search. Journal of Experimental Psychology: Human Perception and Psychophysics, 10, 601-620.

WHY SOME SEARCH TASKS TAKE LONGER THAN OTHERS: USING EYE MOVEMENTS TO REDEFINE REACTION TIMES

Gregory Zelinsky[a] and David Sheinberg[b]

Cognitive and Linguistic Sciences, Brown University, Providence RI, USA

Abstract

Instead of simply describing search in terms of the time needed to detect a target, this behavior might also be understood as an interaction between two underlying processes: the number of search movements preceding a judgment (the *variable number* model) and the time taken to initiate these movements (the *variable duration* model). We introduce these models to address a specific question: why do some search tasks take longer than others? Search times may take longer in a difficult task due to either an increased number of movements or because of longer durations between search movements. Since a reaction time measure lacks sufficient resolution to answer this question, the oculomotor variables of saccade number and fixation duration were used to estimate how these processes contribute to search behavior. Subject's eye movements were recorded as they searched for a target among either 4 or 16 distracters in an easy "parallel" task or a harder "serial" task. The results showed more eye movements and longer initial saccade latencies in the serial task relative to the parallel. Display size differences were also observed for both oculomotor variables in the parallel task, but only for initial latency in the serial condition. These findings provide partial support for both *variable number* and *duration* interpretations of search. They also suggest that eye movements may be a useful means in which to study the spatio-temporal dynamics of search behavior.

Keywords

saccades, fixation durations, parallel/serial processes, visual attention

Introduction

Everyone is familiar with the frustrating task of trying to find a particular book on a cluttered shelf or locating a specific car in a crowded parking lot. For most people, the only relevant concern during such search tasks is the time that it takes to find the desired object, especially if you are late for work in the morning and the object is your ring of keys. During these frustrating moments, fewer people stop to think about how they go about searching for these objects. They may not notice how many books they glance over before acquiring the desired text, and they are probably unaware of how long it takes to reject a particular car before moving their search to

---

[a]Address reprint requests to: Gregory Zelinsky, Dept. of Computer Studies, University of Rochester, Rochester NY, 14627, USA.
[b]Currently at the Baylor College of Medicine, Division of Neuroscience, Houston TX, 77030, USA.

another vehicle in the parking lot. Still, these variables are essential components of search and are legitimate areas of interest in their own right.

This emphasis on the time needed to find an object can also be found in the experimental study of search behavior. In a standard visual search task, subjects are asked to make speeded judgments about the presence or absence of a designated target in displays containing one or more nontarget distracters. These judgments are typically expressed in the form of button presses, which in turn are recorded as manual reaction times (RTs). So, instead of seeing how many objects are inspected in a scene and the time spent looking at each of these objects, this measure compresses the description of search into a single number. From this single measure of RT, entire theories are constructed to explain how we search for a target in multi-element displays (Treisman and Gormican, 1988; Duncan and Humphreys, 1989; Wolfe and Cave, 1990).

Given the widespread use of this measure to study visual search, it is important then to consider exactly what information is conveyed by RTs. Essentially, a RT measure indicates the time needed to make some judgment about a target. By comparing these search times it then becomes possible to rank order the difficulty of various search tasks. For example, if the time needed to locate a target in search task *A* is longer than the time taken in search task *B*, then task *A* is by definition more difficult. If this assessment of relative difficulty is all you wish to know about a search task, then the simplicity of a RT measure makes it the appropriate dependent variable. However, if you want to understand what is happening *during* search, then a RT measure seems less suited for this purpose. Search is much more than the time taken to press a button in response to a target. It is instead a highly complex behavior that has both a spatial and a temporal component. Because RTs mark only the completion of a search task, it is understandably difficult for this measure to reveal how the process of search progresses over time. In order to address this question, what is needed are dependent measures that vary with these spatio-temporal processes. We propose that oculomotor variables provide such dependent measures.

An oculomotor analysis broadens the study of search along both temporal and spatial dimensions while preserving all of the information available from a RT measure. For example, suppose that a search task yielded a RT of 400 ms. This RT is ambiguous in terms of oculomotor variables. It may have resulted from either two 200 ms fixations, or four fast 100 ms fixations. In the case in which no eye movements occur, the oculomotor redefinition of search would simply degenerate into a single 400 ms initial fixation reflecting the RT response. Since all three of these scenarios are consistent with the 400 ms response, the actual spatio-temporal dynamic underlying this search task is lost when using only a RT measure.

However, with the added resolution made available by oculomotor measures, questions may be asked about how these underlying processes interact during search. This study will focus on one such question which is fundamental to any discussion of search: why do some search tasks take longer than others? Figure 1 illustrates this problem in greater detail. The *variable number* model (the top half of Figure 1) would predict that increases in search task difficulty should result in more search movements (indicating by the number of circles), but little or no change in the duration of these movements.[1] Alternatively, according to the *variable duration* model (the bottom half of Figure 1), as the difficulty of a search task increases more processing time is devoted to each search movement (indicated by the darker circles) but the overall number of search movements should remain about the same.

What factors underlie an increase in search time?

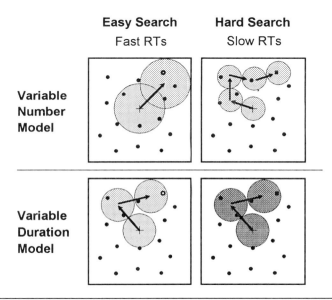

Figure 1: Increased task difficulty may result in more search movements or longer durations.

Since search times would be expected to increase under both the *variable number* and *variable duration* models, it would be difficult to differentiate between these hypotheses using only a RT measure. Fortunately, predictions from these models map rather nicely onto familiar oculomotor variables. If longer RTs are a result of a *variable number* process, then the number of saccades made in the difficult task should increase while individual fixation durations remain unaffected. Conversely, longer fixation durations and constant saccade number would be consistent with a *variable duration* interpretation of search.

---

[1]An additional assumption of this model is that search area (indicated by the diameter of the circles in Figure 1) is large for an easy task and smaller for a more difficult task.

The suggestion that eye movements may be used to study visual search is not a new idea. In fact, there have even been several studies which looked at the number of saccades occurring during search and their fixation durations (Gould and Dill, 1969; Luria and Strauss, 1975; Engel, 1977; Megaw and Richardson, 1979; Scinto *et al.*, 1986; Jacobs, 1986, 1991), the very same oculomotor variables to be considered in this analysis. What is new about this work is the attempt to relate these eye movement results back to the original RT data. The decomposition of search into more simple spatio-temporal processes forces one to question definitions of search based solely on RTs. Is search simply the time needed to find your keys in the morning, or is it a much more sophisticated sequence of directed movements interrupted by pauses of variable duration? By addressing this question, it is our hope that search might be considered as a process rather than merely as an event.

Methods

The stimuli used in this experiment were taken from a well-studied class of search tasks which have been repeatedly shown to yield the range of RTs required for this analysis (Treisman and Gormican, 1988). What distinguishes these search tasks is how they interact with the number of items in a display. Performance in one of these tasks is relatively independent of how many distracters accompany the target, a response pattern consistent with a *parallel* search process. However, detection times in the other task is highly dependent upon distracter number. With each distracter added to these displays, a roughly constant amount of search time is also added to the overall response. This pattern of results is consistent with a *serial* self-terminating search strategy.

The specific target/nontarget combinations used in this experiment are shown in Figure 2. In the case of the parallel task (top panels), the target was a "Q-like" element embedded in a field of "O" distracters. The roles of the target and nontarget were simply reversed in the serial task (bottom panels). The diameters of both elements subtended 2/3° of visual angle, with the "Q-like" element having an additional 2/3° line segment originating at the center of the circle and extending vertically upward. The elements were white ($\approx$ 20 cd/m^2) and presented on an otherwise dark background ($\approx$ 0.1 cd/m^2). Target and nontarget elements were positioned pseudorandomly in the display with the following constraints. Elements appeared within 3° and 6° of initial central fixation, and target elements were further limited to positions at 4° eccentricity.

Of the 128 stimulus configurations per subject per search task, trials were evenly divided into two display sizes (5 and 17 elements) and positive (target present) and negative (target absent) conditions. The two display sizes were randomly inter-leaved throughout the experiment, but the search task trials were blocked. Each of the four subjects saw the same 128 target/nontarget configurations in both the parallel and serial search tasks in order to eliminate position biases arising due to differing element placements. The eye movements of these subjects were monitored (using an AMTech eye tracker sampling at 100 Hz). Subjects were instructed simply to indicate the presence or absence of the designated target by pressing one of two computer mouse buttons. No eye movement instructions were provided other than an instruction to return gaze to a central fixation cross between trials in order to establish a baseline oculomotor starting position.

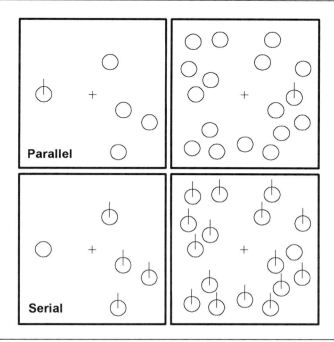

Figure 2: Samples of the parallel and serial stimuli at both the 5 and 17 item display sizes.

Results and Discussion

Reaction Times

Figure 3 shows the subject's marginal mean RTs as they performed either the easy parallel task (solid lines) or the more difficult serial task (dashed lines).[2] The positive data indicates target present trials, and the negative data indicates trials in which a

---

[2]The error bars in this and future figures represents within-subjects variability. This error term is defined as the variance of the difference scores divided by the number of subjects and raised to the ½ power.

target did not appear in the display.[3] A 2×2 repeated-measures ANOVA confirmed a couple of trends visible in the figure. First, RTs in the serial task were longer than those in the parallel task for both the positive (p = 0.042) and negative (p = 0.048) data. Secondly, serial RTs increased with display size while parallel RTs remained unaffected by the number of distracter elements (p = 0.047 for positive trials, p = 0.058 for negative trials).   Both the Search Task × Display Size interaction and the nearly 2:1 ratio of positive to negative serial slopes suggest that these tasks adequately capture the standard search asymmetry commonly reported in the literature.

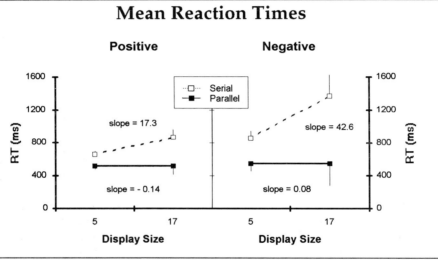

Figure 3: Mean RTs for parallel and serial tasks are plotted as a function of display size.

Saccade Number

What isn't often reported in the search literature is the number of eye movements made while performing these tasks. Figure 4 shows the mean number of saccades initiated before subjects responded with a button press.   As in the case of the previous RT results, a repeated-measures analysis of this data revealed that subjects made more saccades in the serial task relative to the parallel (p = 0.049 for the positive trials, p = 0.030 for the negative trials). But despite this similarity, the Search Task × Display Size interaction which characterized the RT data is very different for the saccade number measure. Instead of the steep serial slopes observed for RTs, mean saccade number remained fairly constant across display size in the serial search condition (p = 0.594 for positive trials; p = 0.301 for negative trials).   The

---

[3]Trials in which errors occurred (less than 2% of the data) were not included in the RT analysis or in any of the oculomotor results.

parallel data was just as discrepant. Whereas RT slopes were flat in the parallel task, subjects made significantly fewer saccades at the larger display size in both the positive (p = 0.010) and negative (p = 0.003) trials.

Mean Number of Saccades

Positive

Negative

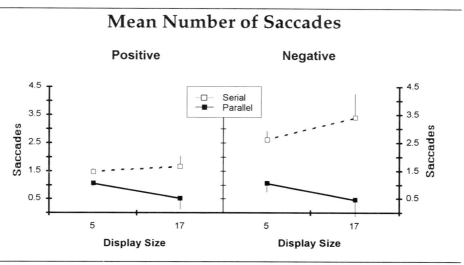

Figure 4: Mean number of saccades occurring during parallel and serial search as a function of display size. Note the decrease in saccade number for 17 item displays in the parallel task.

This decreasing number of eye movements in the parallel task can be largely attributed to the greater percentage of zero-saccade trials at the larger display size. In the case of the positive data, subjects failed to move their eyes in 54% of the 17 item trials but only 7% of the 5 item trials. A similar difference was observed in the negative data (60% of the 17 item trials compared to 20% of the 5 item trials). It is unclear why subjects made fewer eye movements at the larger display size, but one explanation might be that oculomotor search is facilitated by some form of segmentation process. The increased presence of neighboring display elements in the 17 item trials may have enhanced the salience of the target, allowing it to be localized more easily. It should be noted however that this would be a very atypical instance of segmentation, which is an effect usually measured in terms of faster RTs. This facilitation appears to exist only in oculomotor variables.

Fixation Durations

Recall that this analysis is attempting to redefine RTs into oculomotor components. Such a redefinition has certain implications for relationships between these variables. For example, the only way that parallel RT slopes could remain flat despite a reduced occurrence of saccades in the 17 item trials is if fixation durations were to increase with display size. This relationship is precisely what is illustrated in

Figure 5. When subjects saw a 17 item display, they took dramatically longer to make their initial saccades than when the display contained only 5 items ($p = 0.001$ by repeated-measures ANOVA for both the positive and negative trials). This latency increase also interacted significantly with search task ($p = 0.012$ for positive trials, $p = 0.038$ for negative trials). Despite almost identical latencies between tasks at the smaller display size, subjects viewing the 17 item displays took nearly 100 ms longer to make their initial eye movements in the serial task relative to the parallel.

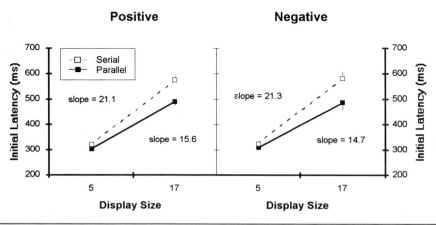

Mean Initial Saccade Latency

Figure 5:   Mean initial saccade latencies are shown for both search tasks and display sizes. Note that latencies increase with display size for both the parallel and serial tasks.

A similar effect of display size on initial saccade latency has been discussed in the context of motor preparation (Inhoff, 1986) and center-of-gravity fixation tendencies (Ottes *et al.*, 1985), but this finding may be new to the eye movement search literature. Contrary to parallel process models which predict no affect of display size on search, this increase in initial saccade latency implies that adding distracters to even a parallel task may influence an oculomotor measure of search behavior.

Figure 6 shows results from a similar analysis performed on all of the remaining fixations occurring during search. When initial saccade latencies were excluded from the data, fixation durations remained relatively unaffected by increasing numbers of distracters. No reliable differences were found between either parallel and serial search conditions ($p = 0.300$ for positive trials, $p = 0.388$ for negative trials) or the two display sizes ($p = 0.157$ for positive trials, $p = 0.330$ for negative trials). This negative finding supports models which postulate additional processing demands only for the initial fixation of a search task.

Mean Non-Initial Fixation Durations

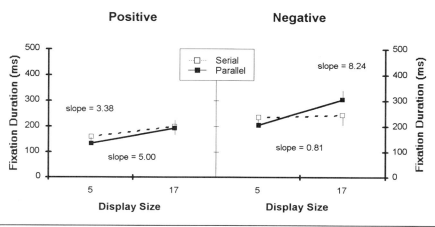

Figure 6: *Fixation durations from all non-initial eye movements. Note the absence of a search task or display size effect in this data.*

To summarize the results then, parallel/serial task differences were observed for the number of saccades occurring during search and the latency of the initial eye movements. To see how well these oculomotor variables can predict the actual search times, saccade number and initial latency were correlated with the RT data from individual trials. Figure 7 shows these raw RT scores plotted against estimated RTs from a multiple linear regression using the two eye movement measures. The diagonal line running through these estimates indicates a theoretically perfect predictor of search times for comparison. It is clear from this figure that the time taken by subjects to make their target judgments can be reliably predicted by the number of eye movements that they use to search the display and the latency of their initial saccades. The correlation coefficient describing this relationship was 0.92, meaning that about 85% of the variance in the individual RT data can be described by just these two oculomotor variables.[4]

Conclusions

RTs might increase with search difficulty according to either a *variable number* or a *variable duration* model. Although either model alone could account for the longer RTs, evidence from oculomotor measures suggest that both processes underlie search

---

[4]When saccade number was partialed out of the regression equation, initial saccade latency was found to account for only 7% of the remaining variance. Although significant ($p < 0.001$ by F test), this contribution illustrates the importance of saccade number relative to the latency measure.

G. Zelinsky & D. Sheinberg

differences observed using O and Q-like stimuli. Evidence for a *variable number* process appeared in the greater number of saccades occurring in the more difficult serial task. Similarly, the *variable duration* model was partially supported by the finding of longer initial saccade latencies in both the more difficult task and at the larger display size. This influence of saccade latency however was largely limited to the first eye movement, with non-initial fixation durations remaining relatively unaffected by manipulations to search difficulty. Figure 8 gives an example of how *variable number* and *duration* processes might interact to produce longer search times. Subjects engaged in an easy search task may make very few eye movements, although the latency of the initial saccade would still be long relative to subsequent fixation durations. However, when the target is less conspicuous, as in the case of the harder search task, subjects may make many more eye movements and take even longer to launch their initial saccades.

Actual × Predicted Reaction Times

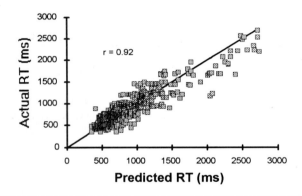

Figure 7: Actual RTs are plotted as a function of search times predicted from saccade number and initial latency measures.

But the oculomotor variables of saccade number and initial latency can do more than simply indicate mechanisms underlying increased search difficulty. Because these variables are so highly correlated with RTs, they can actually parse this traditional measure of search into simpler and more meaningful components. Specifically, parallel and serial search slopes may be redefined by oculomotor functions relating saccade number and initial latency to display size. Figure 9 makes this point more concrete. Serial search slopes may be decomposed into a relatively flat saccade number function and an initial latency function that increases dramatically with display size. In the case of the parallel data, an increasing initial latency function may be offset by a decreasing saccade number function, resulting in the characteristically flat search slope.

Why search times increased with task difficulty

Easy Search   Hard Search

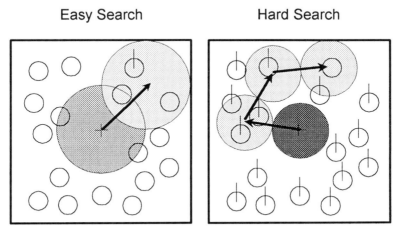

Figure 8: Increasing the difficulty of a search task may result in more search movements and longer initial search latencies.

Components of Search

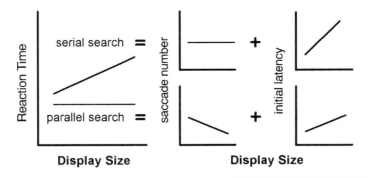

Figure 9: Search times may be meaningfully parsed into two oculomotor variables: saccade number and initial saccade latency.

Given the fact that RT slopes may be redefined into more detailed spatio-temporal processes, it no longer seems sufficient to characterize search as simply either serial or parallel. A parallel search task was found to have a highly serial component (initial latencies), and a parallel component (saccade number) was observed in what appeared to be a serial task. The inability of RTs to resolve these underlying search processes may therefore make this measure not only an incomplete description of search, but also one that is potentially misleading. A better classification scheme would be more sensitive to how *variable number* and *duration* components interact during search. Because eye movements may covary with these component processes, categorization of search behavior based on oculomotor measures becomes an attractive alternative.

References

Duncan, J., & Humphreys, G. (1989). Visual search and stimulus similarity. Psychological Review, 96, 433-458.

Engel, F. (1977). Visual conspicuity, visual search and fixation tendencies of the eye. Vision Research, 17, 95-108.

Gould, J. & Dill, A. (1969). Eye-movement parameters and pattern discrimination. Perception & Psychophysics, 6, 311-320.

Inhoff, A. (1986). Preparing sequences of saccades under choice reaction conditions: Effects of sequence length and context. Acta Psychologica, 61, 211-228.

Jacobs, A. (1986). Eye-movement control in visual search: How direct is visual span control? Perception & Psychophysics, 39, 47-58.

Jacobs, A. (1991). Eye movements in visual search: A test of the limited cognitive effort hypothesis and an analysis of the search operating characteristic. In R. Schmid & D. Zambarbieri (Eds.), Oculomotor control and cognitive processes (pp. 397-410). Amsterdam: North-Holland.

Luria, S., & Strauss, M. (1975). Eye movements during search for coded and uncoded targets. Perception & Psychophysics, 17, 303-308.

Megaw, E. & Richardson, J. (1979). Target uncertainty and visual scanning strategies. Human Factors, 21, 303-315.

Ottes, F., Van Gisbergen, J., & Eggermont, J. (1985). Latency dependence of colour-based target vs nontarget discrimination by the saccadic system. Vision Research, 25, 849-862.

Scinto, L., Pillalamarri, R., & Karsh, R. (1986). Cognitive strategies for visual search. Acta Psychologica, 62, 263-292.

Treisman, A., & Gormican, S. (1988). Feature analysis in early vision: Evidence from search asymmetries. Psychological Review, 95, 15-48.

Wolfe, J., & Cave, K. (1990). Deploying visual attention: The guided search model. In A. Blake & T. Troscianko (Eds.), AI and the eye (pp. 79-103). Chichester, England: Wiley.

EYE MOVEMENTS AND RESPONSE TIMES FOR THE DETECTION OF LINE ORIENTATION DURING VISUAL SEARCH

S. Mannan, K.H. Ruddock and J.R. Wright

Dept. Physics (Biophysics), Imperial College, London SW7 2BZ, UK.

Abstract

We have examined visual search performance made in response to a line target which differs in orientation from the reference or distractor elements. The eye movements made by the subject in the course of the experiments were monitored continuously with an infra-red device. Search performance was assessed by measurement of the manual response time, $T_{1/2'}$, required to register detection of the target, and of the ocular response time, T_0, taken to foveate the target. We also measured the number of saccades made in achieving fixation of the target. Our choice of visual stimulus was guided by the empirical model for pre-attentive detection of line orientation, derived from their experimental data by Foster and Ward (1991 a,b). The principal aim of the experiments was to test the applicability of their model under stimulus conditions which differed markedly from those used to obtain their experimental data. We show that our observations are not consistent with the predictions of the model and we examine the reasons for the divergence between the results of our experimental study and the performance of the empirical model.

Keywords

Visual search, Line orientation, Pre-attentive vision, Parallel processing.

Introduction

Visual search experiments are designed to investigate detection of a target in the presence of multiple non-target elements, known variously as distractors or reference elements. Detectability depends on the nature of the target and on the nature and number of the reference elements, and is usually measured in terms of the response time required for signal detection. In simple search tasks, with the target distinguished by a single parametric difference from a set of identical reference elements, response time is independent of the number of distractors, and is achieved by parallel processing of the stimulus (Neisser, 1963; Egeth *et al.*, 1972; Shiffrin and Gardner, 1972; Treisman, 1988). Under other conditions, including conjunction of different parameters, such as colour and orientation, limitingly small differences between the target and the reference elements, and the presence of more than one class of reference elements, parallel search is frequently not observed (Treisman and Gelade, 1980; Egeth *et al.*, 1984; Treisman and Souther, 1985; Javadnia and Ruddock, 1988; Alkhateeb *et al.*, 1990 a, b). In such cases, response times increase as the number of reference elements increases, giving rise to so-called serial search. The dichotomy between serial and parallel search has been modelled in

Eye Movement Research/J.M. Findlay et al. (Editors)

terms of successive stages of visual processing, involving initial automatic detection of changes in a given image feature, such as colour, and subsequent localisation and conjunction of different features, such as colour and orientation, through which image specification is achieved (Treisman, 1988; Treisman and Gormican, 1988; Cave and Wolfe, 1990). The distinction between parallel and serial search is not, however, always clear cut (Egeth *et al.,* 1984; Kleiss and Lane, 1986) and Duncan and Humpheys (1989) have analysed visual search without invoking this classification of responses. Other authors have attempted to identify the properties of the mechanisms involved in target detection. Binello *et al.* (1993a) have determined, experimentally, the parametric band-widths, discrimination capacities and specificities of visual channels involved in several search tasks.

Foster and Ward (1991a,b) have derived a two filter model which predicts their experimental observations on pre-attentive detection of line orientation. Their proposed detection mechanisms (Fig. 1b) are tuned to near vertical and near horizontal directions, and both have band-widths at half maximum sensitivity of approximately $30°$. Alkhateeb *et al.* (1990a) performed experiments to determine response times for detection of line orientation in the presence of heterogeneous reference elements. Their data are illustrated by Fig. 2, in which manual response times, $T_{1/2}$, for detection of a target line, embedded in reference elements orientated in one of two alternative directions, are plotted against the number of reference elements, N. Values are given for three different pairs of reference orientations, and in each case the target element was orientated so that its angular direction lay approximately midway between those of the two classes of reference elements. Comparison values are given for fields containing only one of the two classes of reference elements. The $T_{1/2}$ values for measurements with a single class of reference elements are always independent of N, that is, parallel processing occurs. The introduction of a second class of reference elements orthogonal to the first has little effect on the response patterns, except that $T_{1/2}$ is increased slightly. With a smaller angular difference between the two classes of reference elements, however, $T_{1/2}$ values increase as N increases and become significantly greater that those measured with either class of reference elements alone. These results demonstrate that line orientation detection mechanisms have sensitivity ranges extending some $30°$ either side of the target orientation. The broad-band characteristics of Foster and Ward's (1991a) filters are, therefore, consistent with the experimental data obtained by Alkhateeb *et al.* (1990a), even though the latter's experimental techniques were very different from Foster and Ward's. Although line orientation is a particularly efficient stimulus for the generation of parallel search, similar responses are observed with other stimulus parameters such as orientation of 2-D elements, magnification and flicker frequency (Alkhateeb *et al.,* 1990a; Binello *et al.* 1993a). Foster and Ward's model may, therefore, have broad applicability in the interpretation of parallel, or pre-attentive, visual search. The principal aim of the study to be described was to test their model against the results of visual search experiments involving the detection of line orientation.

Response times provide a restricted description of visual search performance and in order to investigate more closely performance in the search experiments, we have traced the eye-movements made by our subjects during their observation of the search stimuli. In a previous study, we showed that the ocular search time, that is, the time taken to achieve fixation of the target, is closely correlated with the traditional, manual, search time recorded by pressing a response button (Binello *et al.*, 1993b). Other characteristics of the eye movements, such as the number of saccades made in response to a given target are, however, dependent on the nature of the stimulus and reveal more complex response patterns. We include in our results data for the ocular response times and the number of saccadic eye movements made in response to the visual stimulus.

METHODS

Equipment
Stimuli were generated on a high resolution, Hewlett-Packard display screen (1280 x 1024 pixels), viewed from 1.5m to give a field approximately 13 deg. horizontally by 10 deg. vertically. The display was controlled by a Hewlett-Packard processor (Vectra RS/25c) which also stored and processed the eye movement signals. The latter were measured with an infra-red eye tracking device, with a 20ms temporal resolution limit, the P-scan system (Barbur *et al.*,1987).

Stimulus patterns
The stimulus patterns consisted of a single target embedded in a selected number, N, of reference elements which were either identical or were divided into two sub-groups of N/2 identical elements. The screen locations of all elements were randomised between each presentation, with the restriction that all elements were non-overlapping. The elements were green (C.I.E. 2° chromaticity co-ordinates x =0.293; y=0.607; luminance 79 cd m^{-2}), and were presented against a dark background (luminance 0.1 cd m^{-2}). The combinations of target and reference elements examined in these experiments are illustrated in Fig. 1a. Each experiment comprised 100 stimulus presentations, 50 with a target and 50 without, presented in a random sequence. The stimuli were selected with reference to the response characteristics of the filters described by Foster and Ward (1991a), as is illustrated in Fig. 1b. According to Foster and Ward's (1991a) model, the line orientations corresponding to the maximum filter responses are +4° and -83°, and those corresponding to equal filter responses are +58° and -50° (see Fig. 1). In order to test the validity of the model under our altered experimental conditions (that is, longer stimulus durations), we investigated the properties of line orientation discrimination tasks in which these key orientations were used for background and target line elements. We set up paradigms where the model predicted clear differences between data for experiments using one and two classes of background elements, or predicted no difference, and compared our results to the model's predictions.

Experimental Procedure
During the experimental measurements the subject's head was held steady by clamping it at the temples and supporting it with a chin-rest on a fixed mount. Each presentation commenced with the subject's eye fixated for 1s on a small cross at the centre of the screen, and during this time, the position of the eye was calibrated. The

cross disappeared and 60 ms later the stimulus was presented for a fixed time. The subject was instructed to press a response button as soon as the target was detected and, additionally, was asked to fixate as quickly as possible on the target and to maintain fixation until the stimulus disappeared. The central cross then re-appeared and was fixated prior to the presentation of the next stimulus. The duration of each stimulus presentation depended on the difficulty of the search task and was set long enough to ensure that detection could be achieved without causing discomfort for the subject, who had to suppress blinks during stimulus presentation. The distance between the centre of rotation of the eyeball and the pupil centre was calibrated at the start of each experiment.

Data analysis

The eye movements were analysed off-line by computer, using an algorithm written to identify fixations and to extract other features of the traces. Fixations were defined as a period of 60 ms or greater during which the trace was localised to an area <0.5°. Traces including blinks and saccades between two consecutive fixations which overlapped spatially were both excluded from the analysis. Occasionally, the subject failed to maintain fixation on the target and returned gaze to the centre of the screen before the stimulus disappeared, and these final saccades were also discounted. The ocular response time, T_0, was defined as the mean time from stimulus onset required to fixate the target to within 0.5°. Response times for detection of the target , $T_{1/2}$, was the mean time required to press the response button. The spread of the values given in the figures correspond to $\pm$ 2 standard errors. The dead-time for manual responses was just under 200 ms. Any experiment in which there occurred greater that 10% erroneous responses (false positives and negatives) was discarded.

Subjects

Data are given for two female subjects, both aged 24 years, who are authors. Both had normal 6/5 Snellen visual acuity and some preliminary data (Fig. 2), taken from Alkhateeb et al. (1990a), are given for another female subject who also possessed normal visual acuity.

Results

Data for two classes of reference elements orientated at +58° and -50° to the vertical, both of which provide equal input to the two filters of Foster and Ward's (1991a) model (Fig. 1b), are given in Fig.3 together with data for a single class of reference elements, orientated at -50°. The manual response times, $T_{1/2}$, the ocular response times,T_0, and the probabilities of fixating the target with fewer than two, or fewer than three saccades are all plotted against N, and values are given for two subjects. Each subject responds similarly, with longer response times for the stimuli which contain two classes of reference elements, and correspondingly lower probabilities of target fixation of a given number of saccades. The differences in the response times recorded with two stimulus configurations for subject JW are, in most cases, not significant. Data for a target orientated at -50°, and for two classes of

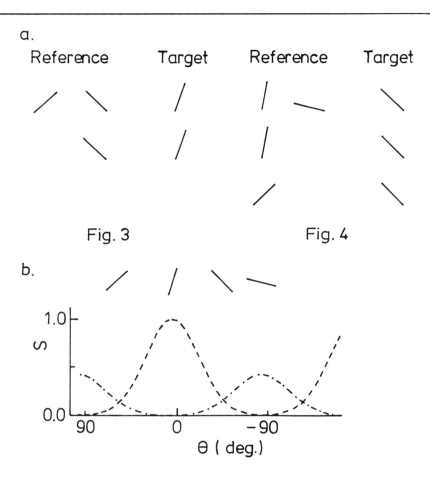

Figure 1 *a) The combinations of reference and target elements used in the experiments, data for which are given in Figures 3 and 4, as noted. When two classes of reference elements were used, there were equal numbers of each. The lines were 1° in length.*

b) The relative sensitivities, S, of the two filters proposed by Foster and Ward (1991a), expressed as a function of q, the angular orientation of the line elements. q is measured in degrees from the vertical, positive clockwise and negative anticlockwise. The lines above the sensitivity functions show the orientations of lines used in our experiments, one each at the angles for peak response of each filter, and two at the angles at which the filter responses intersect, to give equal responses.

reference elements, one orientated along each of the directions for peak response of
Foster and Ward's (1991a) two filters (+4° and -83°), are given in Fig.4, together with
comparison data for only the +4° reference elements. The response times for the
mixed reference elements are greater than those for the single class, and the
probabilities for detection with a given number of saccades are lower, but the
differences between the two types of stimuli are less marked than for the data of Fig.
3. Also plotted in Fig.4 are data measured with the same target, but with a single
class of reference elements, orientated at +58°, so that it provides equal input to the
two filters. These are generally similar to those obtained with reference elements
orientated at +4°.

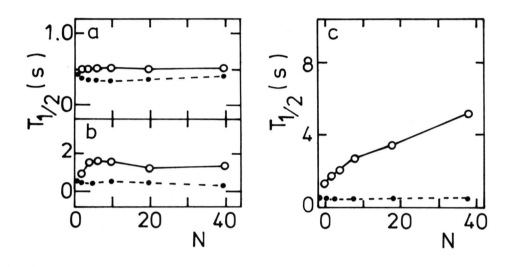

Figure 2 *Response times, $T_{1/2}$, for detection of a line target plotted against the number of
reference elements, N. a). Target orientation 45°; reference orientations 0° and 90° (open
circles), or 0° (full circles) b). Target orientation 30°; reference orientations 0° and 60° (open
circles), or 0° (full circles) c). Target orientation 22°; reference orientations 0° and 40° (open
circles) or 0° (full circles). Subject WA (after Alkhateeb et al., 1990a).*

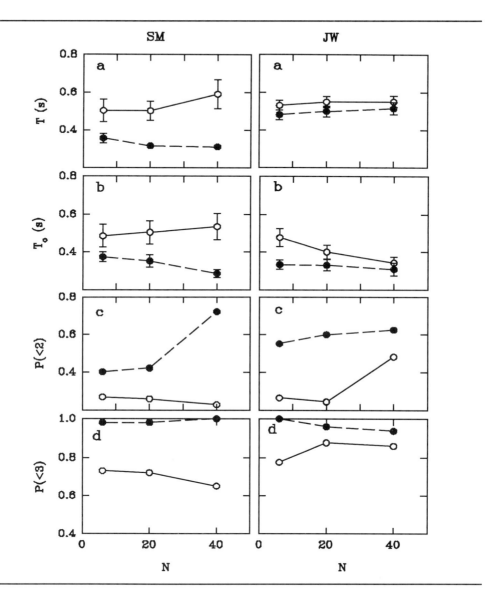

Figure 3 *Response times $T_{1/2}$ and T_0 and the probabilities $P(<2)$ and $P(<3)$ of fixating the target in fewer than two saccades and fewer than three saccades respectively, plotted against N, the number of reference elements. Open circles; target orientation +4°, reference orientation +58° and -50°; full circles; target orientation +4°, reference orientation -50°. Error bars show ± 2 standard errors. Data for SM (left column) and JW (right column)*

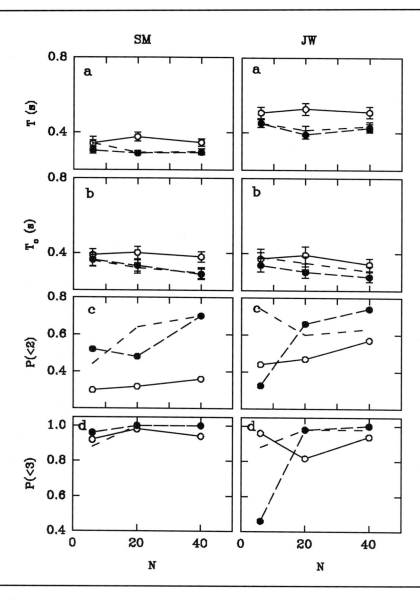

Figure 4 *As Fig. 3, but for different line orientations. Open circles; target orientation - 50°, reference orientations +4° and -83°. Full circles; target orientation -50°, reference orientation +4°; dotted lines, target orientation -50°, reference orientation +58 °.*

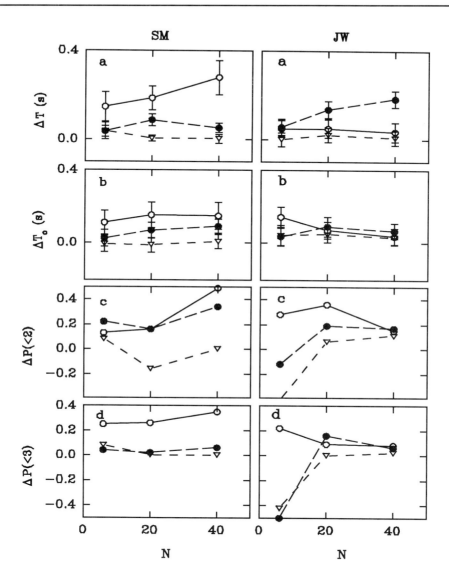

Figure 5 *Difference values taken between the two sets of data points in Fig. 3 (open circles), and between the open and full circles of Fig. 4 (full circles) and between the full circles and the points joined by the dotted lines of Fig. 4 (triangles).*

Discussion

The data of Fig. 2, taken from Alkhateeb *et al.* (1990a), demonstrate that the presence of reference elements orientated orthogonally to each other (vertically and horizontally) has relatively little effect on manual response times compared to values for vertical reference elements alone. When two classes of reference elements differ in their orientations by 60° or 40°, however, response times increase significantly and search becomes serial rather than parallel. Alkhateeb *et al.* argued that the detection mechanisms for line orientation must respond over an angular range of at least ± 30° around the direction of the target orientation, in order to be influenced by the presence of two class of reference elements which differ in orientation by 60°. This compares well with the ± 30° half-widths of Foster and Ward's (1991a) filters, which suggests that the same mechanisms may be active in both detection tasks, despite the many differences in stimulus conditions. In the new experiments reported by us, the orientations of the line elements were chosen to be +4° or -83°, such that they stimulate only one or other of the two filters, or + 58° and -50°, such that they stimulate both filters equally. Thus, the two classes of reference elements used to obtain the data of Figure 3 both provide equal inputs to the two filters and should be entirely equivalent to each other. According to the model, substitution of one class of reference elements for the other should not affect the responses, but experimentally, response times are longer with the two classes of reference elements than with one (Fig. 3a,b). Correspondingly, the probabilities of achieving target fixation with less than three saccades are greater for the single class of reference elements (Fig. 3c,d). The elements of the mixed reference field used to obtain data presented in Fig. 4 (open circles) are orientated so that they provide input to only one or other of the filters, whereas the target stimulates both equally. Removal of one of the two classes of references elements should, therefore, facilitate target detection by the filter which no longer receives an input from the reference elements. The experimental data show the expected effects (Fig. 4) as the response times for the mixed reference elements (Fig.4a,b open circles) are greater than those obtained when one class is removed (Fig. 4, a.b., full circles) and the probabilities of detection within a given number of saccades are fewer (Fig. 4c,d). For SM, however, the differences (Fig. 5, full circles) are smaller those found in the previous experiment and for JW, the differences in the manual response times, $T_{1/2}$ are consistently greater than the corresponding values found in the previous experiment.

The final comparison is made between two sets of data for the same target (orientated at 50°), with the reference elements orientated either at +4° (full circles, Figure 4) or at +58° (dotted lines, Figure 4). The model predicts that detection of the target should be easier in the former case because one of the filters receives inputs from the target, but not from the reference elements. In practice, the differences between the two sets of data (Fig. 5, crosses) are inconsistent, and often zero. In summary, two of the three experiments used to test the model yield results contrary to its predictions, there being consistent differences between results for a pair of stimulus configurations for which there should be no difference (Fig. 3, Fig. 5 open circles) and no consistent differences between those for a pair which should yield them (Fig. 4, full circles and dotted lines; Fig. 5 crosses). Our results show that detection is always more difficult with two classes of reference elements than with one, regardless of relative orientations. This is not predicted by Foster and Ward's model, but is consistent with the subjective impression that a reference field containing two classes of elements is more 'cluttered', thereby rendering target detection more difficult.

Foster and Ward (1991a) used a 20° x 20° field, with the target restricted to an annulus between 3° and 8° from the central fixation, and stimuli were presented for 40ms, followed by 60 ms inter stimulus interval and a 500 ms masking field. The target in our experiments, however, could lie anywhere within the 13° x 10° field, and the stimulus was visible for a fixed period of some 2s to 3s. The former arrangement elicits strictly pre-attentive vision whereas in the latter, the subject is aware of the content of the visual images. We believe that as a consequence, mechanisms other than those involved in the early, pre-attentive stages of vision contribute to the responses we observe, and cause our data to diverge from those predicted on the basis of Foster and Ward's model.

Acknowledgement

We are particularly grateful to Professor D.H. Foster, of the University of Keele, for his advice on the design of the experiments, and for general discussions. This work was supported by a grant awarded to KHR by the Defence Research Agency (No CB/RAE/9/4/2037/384/RARDE), and JW was supported by a Prize Research Studentship awarded by The Wellcome Trust.

References

Alkhateeb W.F., Morris R.J. and Ruddock K.H. (1990a) Effects of stimulus complexity on simple spatial discriminations. Spatial Vision 5, 129-141.

Alkhateeb W.F., Morland A.B., Ruddock K.H. and Savage C.J. (1990b). Spatial, colour and contrast response characteristics of mechanisms which mediate discrimination of pattern orientation and magnification. Spatial Vision 5, 143-157.

Barbur J.L., Forsyth P.M. and Thompson D. (1987). A new system for the simultaneous measurement of pupil size and two-dimensional eye-movements. Clin. Vision Sci. 2, 131-142.

Binello A, Mannan S. and Ruddock K.H. (1993a). Control of visual search by different stimulus parameters: the hierarchical organisation of processing. Visual Search 3 (ed. A. Gale and K. Carr). In press.

Binello A., Mannan S. and Ruddock K.H. (1993b) Eye movements in visual search performed with multi-element fields. Visual Search 3 (ed A. Gale and K. Carr) In press.

Cave K.R. and Wolfe J.M. (1990) Modeling the role of parallel processing and visual search. Cognitive Psychol. 22, 225-271.

Duncan J. and Humphreys G.W. (1989). Visual search and stimulus similarity. Psychol. Rev. 96, 433-458.

Egeth H., Jonides J.and Wall S. (1972). Parallel processing of multielement displays. Cog. Psychol. 3, 674-698.

Egeth H., Virzi R.A. and Garbart H. (1984). Searching for conjunctively defined targets. J. Exp. Psychol. Human Percept. Perform. 10, 32-39.

Foster D.H. and Ward P.A. (1991a) Asymmetries in oriented-line detection indicate two orthogonal filters in early vision. Proc. Roy. Soc. London B, 243, 75-81.

Foster D.H. and Ward P.A. (1991b). Horizontal-vertical filters in early vision predict anomalous line orientation identification frequencies. Proc. Roy. Soc. London B, 243, 83-86.

Javadnia, A. and Ruddock, K.H. (1988). The limits of parallel processing in the visual discrimination of orientation and magnification. Spatial Vision 3, 97-114.

Kleiss, J.A. and Lane, D.M. (1986). Locus and persistence of capacity limitations in visual information processing . J. Exp. Psychol: Human Percept. Perform. 12, 200-210.

Neisser, U. (1963). Decision time without reaction-time. Experiments in visual scanning. Am. J. Psychol. 76, 376-385.

Shiffrin, R.M. and Gardner, G.T. (1972). Visual processing capacity and attentional control. J. Exp. Psychol., 93, 72-83.

Treisman. A. (1988). Features and objects. The fourteenth Bartlett memorial leture. Q.J. Exp. Psychol. 40A, 201-237.

Treisman A. and Gelade, G. (1980). A feature-integration theory of attention. Cog. Psychol. 12, 97-136.

Treisman A. and Gormican, S. (1988). Feature analysis in early vision; Evidence from search asymmetries. Psychol. Rev. 95, 15-48.

Treisman A. and Souther, J. (1985). Search asymmetry; A diagnostic for preattentive processing of separable features. J. Exp. Psychol. Gen. 114, 285-310.

CHRONOMETRY OF FOVEAL INFORMATION EXTRACTION DURING SCENE PERCEPTION

Paul M. J. van Diepen, Peter De Graef and Géry d'Ydewalle

Laboratory of Experimental Psychology, University of Leuven,
Tiensestraat 102, B 3000 Leuven, Belgium

Abstract

The moving mask paradigm was applied to scene perception, to determine whether there exists a fixed, privileged period for foveal information extraction at the beginning of each fixation. Subjects freely explored line drawings of realistic scenes in the context of a search task during which their eye movements were recorded. During each fixation, foveal information was masked whenever the fixation lasted longer than a preset mask onset delay. Scene inspection time was used as a global measure of task difficulty. An asymptotic masking curve was observed, with longer mask onset delays producing a decrease in task difficulty, asymptoting to base level performance at onset delays of 45-75 ms. This suggests that most of the foveal scene information can be encoded within this interval. Contrary to earlier findings in reading research, mean fixation durations, a local measure of encoding difficulty, did not increase when the mask onset delay decreased. An analysis of fixation duration distributions, however, did reveal a time-locked effect of mask onset on fixation duration. These findings are discussed in terms of their implications for estimates of the chronometry of visual information encoding in scene perception.

Keywords

Foveal Masking, Scene Perception, Fixation Duration.

Introduction

Reading research has successfully employed two eye-contingent display-change techniques to study chronometric and spatial characteristics of the processing of written text (for a review, see Rayner & Pollatsek, 1989, chapter 4). In the moving window paradigm, all text within a predefined window moving in synchrony with the eye is left unaltered while all text outside the window is changed in various ways. In the moving mask paradigm, the situation is reversed and only the text within a predefined area around the current fixation position is masked while everything else remains visible. By manipulating mask onset delays as measured from the start of the fixation, Rayner, Inhoff, Morrison, Slowiaczek, and Bertera (1981) observed that most of the foveal information acquired in the course of a single fixation is encoded during the first 50 ms of the fixation: Detrimental masking effects on global and local measures of information processing efficiency decreased to an asymptote as the mask onset delay approached the 50 ms mark. Globally, the effective reading rate for a text and the mean number of forward fixations on the text exhibited this asymptotic function; locally, the same function was found for the mean duration of individual fixations. The shape of these functions is quite similar to that of Type A backward masking

Eye Movement Research/J.M. Findlay et al. (Editors)

as found in classical visual masking experiments (Breitmeyer & Ganz, 1976). Notwithstanding differences in stimuli and procedure this suggests that each individual fixation in the moving mask paradigm can be regarded as a single trial in a traditional backward masking experiment: As the target-mask onset asynchrony increases, target and mask information are less likely to be integrated in the same encoding channel. Hence, the Rayner et al. (1981) results appear to validate fixation durations as a measure of ease of visual information encoding.

In the domain of scene perception, several authors have defended object fixation times as a reliable measure of the ease with which an object is identified (Antes & Penland, 1981; De Graef, 1992; Friedman, 1979; Henderson, 1992a; Rayner & Pollatsek, 1992). In addition, effects of scene context on object fixation times have been interpreted as effects on ease of object encoding (De Graef, Christiaens, & d'Ydewalle, 1990; De Graef, De Troy, & d'Ydewalle, 1992; Loftus & Mackworth, 1978). These claims are not unchallenged and it has been argued that object fixation times are simply too long to reflect fast object encoding processes (Biederman, Mezzanotte, & Rabinowitz, 1982). By the same token, context effects on object fixation times are ascribed to changes in interest value of the object, or to difficulties in integrating the object in a coherent scene interpretation (Boyce & Pollatsek, 1992).

In view of this controversy, research on scene perception could greatly benefit from an application of the moving mask paradigm to the study of object fixation times. Thus far, technical difficulties have prevented such an application, but recently van Diepen, De Graef, and Van Rensbergen (in press) developed a new display-change technique that enables use of the moving mask and window paradigms in scene perception. As a result, the present experiment is a first step towards unravelling the nature and chronometry of processing reflected in object fixation times. As a working hypothesis, we assumed that Henderson's (1992b) general description of eye movement control in reading also applied to scene perception: During the first part of a fixation, foveal information is extracted from the stimulus. Once this information is "understood", attention is shifted from its foveal location to an extrafoveal location and a saccade towards that location will be programmed and executed, unless attention is re-allocated prior to a given point of no return in saccade programming. Selection of a new fixation location is supposedly governed by salient low-level stimulus characteristics such as contour density or discontinuity. If information is not understood prior to a saccade programming deadline, attention will not be shifted and the foveal stimulus will be refixated.

Our main goal was to manipulate onset delays of foveal masks to determine whether a privileged period of foveal information acquisition could also be identified in object fixation times. If so, future research of scene context effects on object encoding would have to demonstrate that this specific time interval is affected by contextual manipulations. We attempted to probe scene perception under conditions that imposed minimal demands in terms of post-perceptual processes. Subjects were asked to freely explore line drawings of scenes in search of non-objects, i.e. closed, meaningless figures with a part-structure and size-range comparable to that of real objects. Presentation of each scene was self-terminated, following which subjects indicated the number of non-objects they had found.

During scene presentation, eye movements were measured continuously which allowed us to display foveal masks at various delays following the start of a fixation. In addition to Mask Onset, we also manipulated Mask Size: One mask was sufficiently large to completely cover the average object when fixated in its center, while the second mask covered only part of the object. Finally, a No Mask condition was included to obtain a control measurement of normal performance. Processing efficiency was measured globally by the total time subjects spent on a given scene, and locally by the average duration of all individual fixations. These measures could in principle also include off-object fixations, but earlier research indicated that objects constitute the perceptually most relevant fixation locations in natural scenes (Antes, Singsaas, & Metzger, 1978; Metzger & Antes, 1983). In line with the reading research, we hoped to find a masking effect on both measures at shorter mask onset delays which would dissolve and approximate No Mask performance as delays became longer. While the manipulation of Mask Size could have an overall effect on fixation and inspection times and thus would indicate how resistant object identification is to partial occlusion, it was primarily intended to determine whether viewers could develop strategic saccadic behavior to cope with the mask. Specifically, subjects could try to circumvent the mask by making several small saccades within the mask, thus producing a series of "snapshots" of the object covered by the mask. An increase of the relative frequency of small saccades in the mask conditions would indicate such a strategy. Alternatively, if saccades are indeed fundamentally governed by low-level visual saliency then saccadic amplitude should increase as a direct function of mask size, provided that the presented masks are not smaller than the normal minimal saccade length in scene exploration.

Method

Stimuli. 31 black-on-white line drawings of different real-world scenes were assembled, most of them adapted from the set described in De Graef et al. (1990). All scenes contained several easily recognizable objects, whose presence, size, and position were consistent with everyday experience. 12 of the scenes, selected on grounds of richness and clarity of detail, were designated as targets for the subsequent analyses, while the remaining 19 served as fillers. From this stimulus pool, 12 trial blocks were constructed of 27 stimuli each. Every block contained the 12 target scenes and a random selection of 15 fillers. Stimulus order in each block was randomized, with the restriction that the first 3 scenes in a block had to be fillers. All filler scenes were populated with a randomly chosen number of non-objects, varying between 0 and 6. Target scenes contained from 0 to 3 non-objects, with a .2 probability that 1, 2, or 3 non-objects were present, and a .4 probability that no non-object was present. Which non-objects would be present was also randomized. By this distribution of non-objects over target and filler scenes, we hoped to disrupt normal scene-processing as little as possible in the target scenes, while maintaining subjects' commitment to extensively search each scene for non-objects even if the scene was presented more than once.

The masks employed in this experiment were elliptic noise patterns of pixels with random brightness ranging from black to light-grey. Luminance of each mask approximated the luminance of scenes. Two mask sizes were used: one that would cover only part of the average object when centered on it (1.5° wide x 1.0° high), and one that would cover the whole object (2.5° x 1.7°). The delay between the start of a new fixation and the onset of a mask was 15, 45, 75, or 120 ms.

Subjects. 8 subjects from the University of Leuven subject pool were paid to participate in the experiment. All subjects had normal or corrected to normal vision. Two of the subjects had previous experience with eye-movement research in unrelated experiments.

Procedure and design. Upon arriving for the experiment, subjects were told that they were to participate in a series of experiments on the speed and accuracy with which certain kinds of information could be extracted from complex visual displays. In the present experiment, they would have to search and count non-objects in line drawings of everyday scenes that were presented on a display in front of them. The notion of non-objects was explained and illustrated with several examples (see De Graef et al., 1990). Subjects were told that the number of non-objects would vary from stimulus to stimulus and that not all stimuli would contain non-objects. They were instructed to explore each stimulus as quickly and as accurately as possible, after which they could terminate the stimulus by a button-press and give their answer. Subjects were told that their eye movements would be recorded during their exploration of the scene in order to determine whether they effectively spotted all the non-objects.

Following these instructions, subjects were seated at 125 cm from the stimulus display, with their heads stabilized by a head-rest and a bite bar. Once the eye-tracker was successfully calibrated for 9 points along the diagonals of the stimulus field, the first block of trials was initiated. Each trial consisted of the following events: First, a fixation spot appeared in the center of the display and subjects were instructed to fixate this point. Once the eye-tracker determined that subjects were steadily gazing at the fixation point, a stimulus was presented. Subjects freely explored the stimulus, and terminated the display when they thought they had found all the non-objects. Displays were terminated by pressing the right of two response buttons which brought a counter on the screen which subjects could increment by pressing the left response button. Once the counter displayed the detected number of non-objects, the trial was terminated by pressing the right button again.

All subjects saw all 12 blocks of 27 scenes in the same order over a two-day period. Each day consisted of 6 blocks: an unmasked presentation of the stimuli in the first and last block, and 4 masking blocks in between, each with a different mask-onset delay. All masks presented on the same day had the same size: Half of the subjects started with the large mask on day 1, and went on with the small mask on day 2, while the order was reversed for the other half. The order of the 4 mask-onset delays was counterbalanced across blocks within each group of 4 subjects that received the same mask size at the same time. Following the first block on each day, subjects were warned that in the next 4 blocks their task would be made more difficult by the appearance of a blur at the position they were looking at. Between each block, subjects were given a brief rest. Typically, the whole experiment took about 1 hr 45 min. on each day.

Apparatus. Eye movements were recorded with a Generation-V dual-Purkinje-image eye-tracker (Crane & Steele, 1985). This system has an accuracy of 1 min of arc and a 1000 Hz sampling rate. It was interfaced to an Intel-386 PC, storing every sample of the eye's position. For each sample, the computer made an on-line decision about the eye state: fixation, saccade, blink, or signal loss. Eye state and position were fed into a second Intel-386 PC, in control of stimulus presentation (for an extensive description of this dual-PC eye-tracking system see Van Rensbergen & De Troy, 1993). The second PC contained a Truevision ATVista Videographics Adapter, connected to an identical graphics board in a third PC. This interface enabled use of the moving overlay technique (van Diepen, De Graef, & Van Rensbergen, in press), which allows for fast mask-movement across a stable scene. Timing of the mask's appearance at the current fixation position was initiated whenever a new fixation was detected, i.e. whenever the eye moved less than 5 min of arc over a 5 ms period. At that time the computer determined on what frame the mask should be enabled in order to display the whole mask in one frame and approximate the requested mask-onset delay as closely as possible. This resulted in mean onset delays of 17.1, 46.4, 76.4, and 121.4 ms for the 4 onset conditions, with a standard deviation of 5.5 ms. Stimuli appeared in 60-Hz interlaced NTSC mode on a Barco 6351 CRT with a 756 x 486 resolution, and subtended approximately 16 x 12 degrees of visual angle.

Results

All subsequent analyses pertain to the data for the 12 target scenes only. For each scene presentation we computed total scene inspection time, average fixation duration and average saccadic amplitude. Scene inspection time ranged from .65 to 45.4 s. in all but one (excluded) case, with a mean of 5.5 s, as measured from computer-determined scene-onset to subject-determined scene-offset. Analyses excluded all fixations and saccades which were followed or preceded by periods of signal loss or eye blinks. The remaining fixations and saccades accounted for an average 79.2% of the total scene inspection time.

Preliminary plots of the dependent variables as a function of trial block (from 1 to 12), showed an exponential decrease for inspection time and a linear decrease for average fixation duration. Therefore, base level performance in the No Mask condition was estimated by averaging all 4 control blocks (1, 6, 7, and 12). The impact of practice on the effects of Mask Size (large vs. small) and Mask Onset (15, 45, 75, 120 ms) was eliminated by counterbalancing, but did inflate the variability of these effects across individual subjects. Since this greatly reduced the power of tests for the within-subjects Mask Size and Mask Onset effects, it was decided to remove the practice effects by means of linear regression. For this purpose, inspection time was first linearized by applying a log transformation. This transformation also normalized the frequency distribution of inspection times (see also Bush, Hess, & Wolford, 1993). Fixation duration and saccadic amplitude distributions did not approach the normal distribution any better when transformed, and were therefore left unaltered. Subsequently, separate regression equations were computed for each day of testing, thus taking into account between-day discontinuities in the practice effect (see Figure 1). Regressions were computed on the data from the masking conditions, and were used to obtain residual inspection times and fixation durations. To preserve between-day differences in performance and comparability with the No Mask condition, all residuals were augmented with the overall uncorrected mean for the

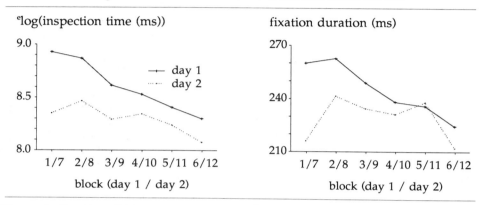

Figure 1: Practice effects were apparent on log-transformed inspection time (left) and mean fixation duration (right).

corresponding day. Note that these corrections only reduce error variance and in no way affect mean performance levels for the various conditions.

Following the correction for practice effects, all data were entered in a repeated-measures analysis of variance (ANOVA). All 8 Mask Size x Mask Onset conditions and the No Mask condition were treated as within-subjects levels. Mask Order (large/small vs. small/large) was entered as a between-subjects factor. A problem inherent to this type of analysis is the possible existence of interdependencies between the levels of within-subjects factors. The resulting violation of the sphericity of the variance-covariance matrix invalidates the use of pooled error terms to test omnibus and partial within-subjects effects (O'Brien & Kaiser, 1985). In our analyses of main effects and interactions, this problem was dealt with by multiplying numerator and denominator degrees of freedom by the Huynh-Feldt estimator of sphericity, Θ_{HF}, whenever this statistic did not reach unity (Kirk, 1982, pp. 262). Partial effects, i.e. specific contrasts, were tested against the pooled error term for their parent effect whenever the circularity assumption for that effect was not violated. Tenability of the circularity assumption was evaluated by computing Mauchly's Criterium (Kirk, 1982, pp. 260). If the assumption appeared to be violated, specific contrasts were tested against their specific error terms, i.e. the variability of the contrast over subjects. These exact tests do not require the circularity assumption, but usually imply a loss of power. All partial effects involving the No Mask condition were tested against specific error terms.

Figure 2 plots the three dependent variables as a function of onset delay and mask size. Since the Mask Onset x Mask Size interaction did not reach reliability for any of the variables, we will first concentrate on the main effects. With respect to Mask Onset, no effects were found on fixation duration [F (3, 18) = 1.73, $p <$ .197, MS_e = 134.45, Θ_{HF} = 2.285] and saccadic amplitude [F (3, 18) < 1, MS_e = .0621, Θ_{HF} = 1.167]. Scene inspection time, however, mimicked the Rayner et al. (1981) findings and exhibited a Mask Onset effect [F (3, 18) = 7.7, $p <$ .0018, MS_e = .0117, Θ_{HF} = .973] which appeared to dissolve at onset delays between 45 and 75 ms. This asymptotic relation was confirmed by subjecting the data for the 4 onset delays to a Helmert transformation, i.e. a Dunn-Sidák test of a group of three contrasts comparing each onset level to the mean of the higher onset levels. Only the first of these contrasts proved to be reliable [tDS (18) = 2.998, $p <$ .05, MS_e = .0117]: Scene inspection time in the 15 ms condition was longer than the average of the other three conditions. Duncan comparisons of the onset levels revealed no irregularities in the asymptotic function: The 15 ms condition differed from the 45, 75, and 120 ms condition [$p <$ .05], which did not differ from each other. Finally, a Dunnett test of the differences between each of the onset conditions and the No Mask condition only showed a difference for the 15 ms condition [tD (6) = 3.42, $p <$ .05, MS_e = .0166] corroborating the conclusion that performance asymptoted to the base level when the mask was delayed around 45 ms.

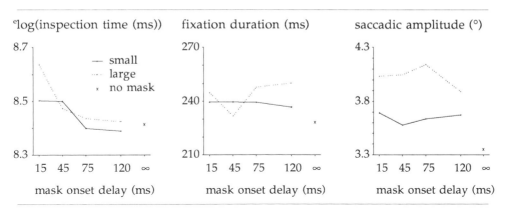

Figure 2: Effects of Mask Onset and Mask Size on inspection time, mean fixation duration, and mean saccadic amplitude. No Mask performance is plotted at the '∞' mark.

Mask Size only affected saccadic amplitude [F (1, 6) = 15.38, $p <$ .0078, MS_e = .153] with the larger mask producing longer saccades. In addition, both mask sizes increased saccadic amplitude relative to the No Mask condition [tD (6) = 4.474, $p <$ .01, MS_e = .0345 for the large mask, and tD (6) = 6.027, $p <$ .01, MS_e = .1011 for the small mask]. Inspection of the frequency distribution of saccadic amplitudes showed that this difference between the No Mask and the mask conditions was primarily due to a decrease in the frequency of small saccades (< 1.5°). This runs counter to the hypothesis that subjects could attempt to circumvent the mask by the strategic use of small eye movements which would turn the mask off. In addition, presence of the mask did not shift the mode of the frequency distributions to longer saccades, indicating that subjects did not actively attempt to steer the mask off the object of interest and then process it in peripheral vision. Saccade length simply seemed to be governed by where new information could be found, i.e. beyond the edge of the mask.

While effects on inspection time and saccadic amplitude are quite consistent with the Rayner et al. (1981) data, this does not seem to be the case for fixation durations which were unaffected by Mask Onset, Mask Size, and their interaction. Moreover, a Dunnett test of differences between the various onset levels and the No Mask condition showed that only an onset delay of 120 ms resulted in reliably longer fixation times [tD (6) = 4.055, $p < .05$, MS_e = 115.4], which is completely opposite to the Rayner et al. findings where the shortest onset delays produced the greatest differences with the No Mask condition. In view of these discrepancies and the somewhat erratic pattern of mean fixation durations (see Figure 2), we turned to a fuller description of fixation times. For each Mask Size x Mask Onset condition, all individual fixations were ordered in 20 ms bins. Frequencies in each bin were divided by the total number of fixations for the corresponding condition. Figure 3 plots the resulting distributions for both mask

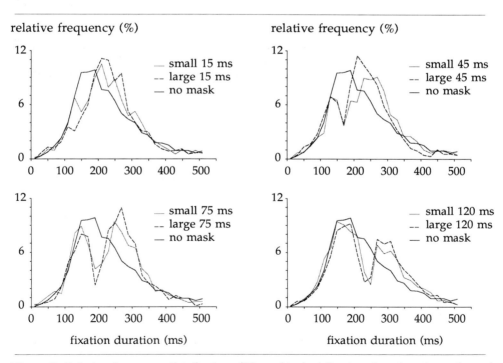

Figure 3: Relative frequency distributions (20 ms bins) of fixation durations for each mask condition compared to the No Mask distribution.

sizes at each onset level and compares them to the No Mask distribution. The original fixation data were used, without the correction for practice effects.

The most striking feature of these graphs is that all masking distributions, regardless of mask size and onset delay, exhibit a dip with a minimum at about 120 ms after mask onset. Hence, it seems reasonable to conclude that in the present experiment masking tended to increase the mean fixation duration about equally in all conditions, by sustaining a proportion of the fixations that would normally have been terminated in one specific time-interval.

Sofar, we have focussed on within-subjects effects without taking into account the between-subject manipulation of Mask Order. Neither saccadic amplitude nor fixation duration showed reliable effects involving Mask Order. However, for scene inspection time, both the Mask Order x Mask Size [F (1, 6) = 12.61, $p <$ .0120, MS_e = .0912] and the Mask Order x Mask Onset [F (3, 18) = 7.38, $p < .0022$, MS_e = .0117, Θ_{HF} = .9727] interactions were reliable. Inspection of simple main effects in the Mask Order x Mask Size interaction (plotted in Figure 4) revealed that Mask Size only affected subjects who first received the large mask [F (1, 6) = 8.57, $p < .0260$, MS_e = .0912], and not the subjects who first saw the small mask [F (1, 6) = 4.39, $p < .0811$, MS_e = .0912]. The most plausible explanation for this interaction is that a normal disadvantage (i.e. longer inspection times) of the large versus the small mask, is enhanced by the between-day practice effect in the large/small group and is counteracted by practice in the small/large group. In the Mask Order x Mask Onset interaction (Figure 4), onset delays reliably affected inspection times in the small/large group [F (3, 18) = 12.8, $p < .0001$, MS_e = .0117], and had no effect in the large/small group [F (3, 18) = 2.27, $p < .115$, MS_e = .0117]. In line with the above, this overall insensitivity of the large/small group to onset variations can be ascribed to the virtual disappearance of the effect of the small mask when measured on practiced subjects.

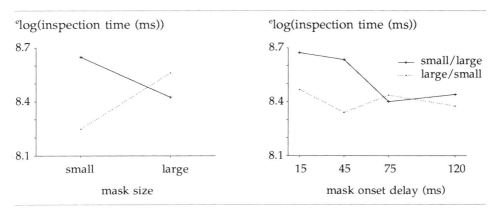

Figure 4: Mask Size x Order (left) and Mask Onset x Order (right) interactions on inspection time.

Discussion and conclusion

The main goal of our experiment was to use foveal masking in order to determine whether the fixations on objects in scenes possess a component which has been identified in word fixations in sentences, i.e. a short interval at the start of the fixation during which most foveal information is encoded from the image. Our results present mixed evidence. A global measure of information processing efficiency (total scene inspection time) revealed that the detrimental effect of foveal masking on performance did indeed dissolve if the mask appeared after about 45-75 ms. Clearly, this supports the notion that only the first part of an object fixation is necessary to actually encode that object. However, this effect on global processing time was not paralleled in the analysis of individual fixations: The appearance of a mask always tended to increase average fixation duration, even when the mask was presented late in the fixation. An inspection of the fixation duration frequency distributions showed that this average increase was in fact the result of the maintenance of a specific group of fixations: Relative to the control distribution, all experimental distributions exhibited a dip with a minimum at about 120 ms after mask onset, followed by a shoulder at longer fixation durations. This time-locked effect on fixation durations is not new. Very similar effects were found in reading research where brief text masking (Blanchard, McConkie, Zola, & Wolverton, 1984) and letter replacement during a fixation (McConkie, Underwood, Zola, & Wolverton, 1985) produced dips in the fixation duration distributions with a constant delay of about 90 ms. McConkie et al. (1985) and McConkie, Reddix, and Zola (1992) have interpreted this effect as an interruption of ongoing visual encoding, which only surfaces after a constant delay due to hard-wired transmission times in the neural system. Assuming that in the course of a fixation processing shifts from foveal to extrafoveal locations (Ishida & Ikeda, 1989), this interruption can perhaps best be described as the inhibiting of attention-shifts off the object in the early onset conditions, and the recapturing of shifted attention in the late onset conditions. Assuming that the eyes go where attention is allocated (Henderson, 1992b) this would result in sustained fixations at the current location. That the delay of this interruption effect is somewhat longer in the present experiment, can probably be ascribed to the fact that the examined distributions are not continuous, which allows bin-sized fluctuations in the estimate of the delay.

In view of the above, it seems reasonable to conclude that 45-75 ms of fixation time are generally sufficient to extract all the necessary information to identify an object. However, this does not mean that the system goes "foveally blind" after this period: The appearance of a foveal noise mask well after the foveal acquisition stage is sufficiently detectable to interrupt ongoing encoding. Judging by the absence of a negatively asymptoting relation between mask onset and fixation duration, decrease of stimulus quality is compensated for by an increase of the number of samples, i.e. the number of fixations, instead of longer fixation durations. In fact, this seems quite sensible in situations where the mask is maintained as long as the fixation is maintained. This raises the question why subjects in the Rayner et al. (1981) study did exhibit longer fixations at shorter onset delays. Our, speculative, explanation for this finding is that in reading, context and peripheral information can be used to cope with short mask onset delays. As the onset delays increase, the need for this compensatory construction of text becomes less and subjects can return to the normal, fast processing of the foveated text. In our study, however, scene context and simultaneously available peripheral information are unlikely to be of much help in deciding whether the fixated and masked object is real or not. Only a sufficient amount of object detail can conclusively answer that question, and multiple fixations are the only way to get such information.

Given these considerations, our findings indicate that two conditions need to be met before one can use foveal masking at various delays in the fixation in order to test the claim that fixation durations can measure ease of visual information encoding. First, one needs to ensure that longer fixation durations at shorter onset delays cannot be attributed to increased compensatory processing of periphery and context instead of to slower encoding as a result of interference from the mask. Second, presentation of the mask should hamper encoding, not shut it down completely. In the latter case, one risks to steer the visual system towards repeated fixations thus removing all effects on fixation durations. To meet this second condition, we plan to repeat our experiment and replace the high-density noise mask by stimulus manipulations that are better suited to affect the rate of object encoding. One possibility is to lower stimulus contrast (Loftus, Kaufman, Nishimoto, & Ruthruff, 1992): If this manipulation produces a negatively asymptoting relation between onset delay and fixation durations, then the latter can be concluded to truly reflect encoding time.

However, pending this validation of fixation durations, the data already have indicated that the foveal masking paradigm can be very useful in our exploration of the locus of scene-context effects on object perception. Specifically, scene inspection time appears to be sensitive to a premature interruption of object encoding. Hence, any manipulation of object-context relations that affects the ease of object encoding should result in a horizontal shift of the asymptote of the function that relates inspection time to mask onset delays: Facilitation of object encoding should advance the asymptote to shorter onset delays, inhibition of encoding should delay base level performance to longer onset times. Alternatively, if the contextual manipulations only shift the whole function upwards or downwards, scene context effects would have to be situated at higher levels in the object identification process (e.g. image-to-representation matching) or even at the level of object-in-scene integration. Either way, the foveal masking technique developed in reading research promises to play a role of equal significance in the domain of scene perception.

References

Antes, J. R., Singsaas, P. A., & Metzger, R. L. (1978). Components of pictorial informativeness. *Perceptual and Motor Skills, 47,* 459-464.

Antes, J. R., & Penland, J. G. (1981). Picture context effects on eye movement patterns. In D. F. Fisher, R. A. Monty, & J. W. Senders (Eds.), *Eye movements: Cognition and visual perception*, (pp. 157-170). Hillsdale, NJ: Erlbaum.

Biederman, I., Mezzanotte, R. J., & Rabinowitz, J. C. (1982). Scene perception: Detecting and judging object undergoing relational violations. *Cognitive Psychology, 14*, 143-177.

Blanchard, H. E., McConkie, G. W., Zola, D. & Wolverton, G. S. (1984). Time course of visual information utilization during fixations in reading. *Journal of Experimental Psychology: Human Perception and Performance, 10*, 75-89.

Boyce, S. J., & Pollatsek, A. (1992). An exploration of the effects of scene context on object identification. In K. Rayner (Ed.), *Eye movements and visual cognition: Scene perception and reading* (pp. 227-242). NY: Springer Verlag.

Breitmeyer, B. G., & Ganz, L. (1976). Implications of sustained and transient channels for theories of visual pattern masking, saccadic suppression, and information processing. *Psychological Review, 83*, 1-36.

Bush, L. K., Hess, U., & Wolford, G. (1993). Transformations for within-subject designs: A Monte Carlo investigation. *Psychological Bulletin, 113*, 566-579.

Crane, H. D., & Steele, C. M. (1985). Generation-V dual-Purkinje-image eyetracker. *Applied Optics, 24*, 527-537.

De Graef, P. (1992). Scene-context effects and models of real-world perception. In K. Rayner (Ed.), *Eye movements and visual cognition: Scene perception and reading* (pp. 243-259). NY: Springer Verlag.

De Graef, P., Christiaens, D., & d'Ydewalle, G. (1990). Perceptual effects of scene context on object identification. *Psychological Research, 52*, 317-329.

De Graef, P., De Troy, A., & d'Ydewalle, G. (1992). Local and global contextual constraints on the identification of objects in scenes. *Canadian Journal of Psychology, 46*, 489-508.

Friedman, A. (1979). Framing pictures: The role of knowledge in automatized encoding and memory for gist. *Journal of Experimental Psychology: General, 108*, 316-355.

Henderson, J. M. (1992a). Object identification in context: The visual processing of natural scenes. *Canadian Journal of Psychology, 46*, 319-341.

Henderson, J. M. (1992b). Visual attention and eye movement control during reading and picture viewing. In K. Rayner (Ed.), *Eye movements and visual cognition: Scene perception and reading* (pp. 260-283). NY: Springer Verlag.

Ishida, T., & Ikeda, M. (1989). Temporal properties of information extraction in reading studied by a text-mask replacement technique. *Journal of the Optical Society of America, 6*, 1624-1632.

Kirk, R. E. (1982). *Experimental design: Procedures for the behavioral sciences.* Monterey, CA: Brooks/Cole.

Loftus, G. R., Kaufman, L., Nishimoto, T., & Ruthruff, E. (1992). Effects of visual degradation on eye-fixation durations, perceptual processing, and long-term visual memory. In K. Rayner (Ed.), *Eye movements and visual cognition: Scene perception and reading* (pp. 203-226). NY: Springer Verlag.

Loftus, G. R., & Mackworth, N. H. (1978). Cognitive determinants of fixation location during picture viewing. *Journal of Experimental Psychology: Human Perception and Performance, 4*, 565-572.

McConkie, G. W., Underwood, N. R., Zola, D., & Wolverton, G. S. (1985). Some temporal characteristics of processing during reading. *Journal of Experimental Psychology: Human Perception and Performance, 11*, 168-186.

McConkie, G. W., Reddix, M. D., & Zola, D. (1992). Perception and cognition in reading: Where is the meeting point? In K. Rayner (Ed.), *Eye movements and visual cognition: Scene perception and reading* (pp. 293-303). NY: Springer Verlag.

Metzger, R. L., & Antes, J. R. (1983). The nature of processing early in picture perception. *Psychological Research, 45*, 267-274.

O'Brien, R. G., & Kaiser, M. K. (1985). MANOVA method for analyzing repeated measures designs: An extensive primer. *Psychological Bulletin, 97*, 316-333.

Rayner, K., Inhoff, A. W., Morrison, R. E., Slowiaczek, M. L., & Bertera, J. H. (1981). Masking of foveal and parafoveal vision during eye fixations in reading. *Journal of Experimental Psychology: Human Perception and Performance, 7*, 167-179.

Rayner, K., & Pollatsek, A. (1989). *The psychology of reading.* Englewood Cliffs, NJ: Prentice-Hall.

Rayner, K., & Pollatsek, A. (1992). Eye movements and scene perception. *Canadian Journal of Psychology, 46*, 342-376.

van Diepen, P. M. J., De Graef, P., & Van Rensbergen, J. (in press). On-line control of moving masks and windows on a complex background using the ATVista Videographics Adapter. *Behavior Research Methods, Instruments, & Computers.*

Van Rensbergen, J., & De Troy, A. (1993). *A reference guide for the Leuven dual-PC controlled Purkinje eyetracking system* (Psych. Rep. No. 145). Laboratory of Experimental Psychology, University of Leuven, Belgium.

Author notes

This research was supported by a Concerted Research Action from the University of Leuven and by agreement RFO/A1/04 of the Incentive Program for Fundamental Research in Artificial Intelligence. The work was conducted while PVD was a student at the Nijmegen Institute for Cognition and Information, University of Nijmegen, the Netherlands, and held an Erasmus scholarship at the Laboratory of Experimental Psychology, Leuven, Belgium. The authors are indebted to Andreas De Troy, Johan van Rensbergen, and Noël Bovens for their assistance in implementing the eyetracking infrastructure, and to Wim Fias, John Findlay and Karl Verfaillie for their comments on an earlier draft. Correspondence may be addressed to PVD or PDG, Laboratory of Experimental Psychology, University of Leuven, Tiensestraat 102, B-3000 Leuven, Belgium (e-mail: Paul.vanDiepen@psy.kuleuven.ac.be or Peter.DeGraef@psy.kuleuven.ac.be).

TRANSSACCADIC INTEGRATION OF BIOLOGICAL MOTION: OVERVIEW AND FURTHER EVIDENCE BASED ON REACTION TIMES

Karl Verfaillie, Andreas De Troy, and Johan Van Rensbergen

Laboratory of Experimental Psychology
University of Leuven, Leuven, Belgium

Abstract

In a transsaccadic integration paradigm, subjects had to detect saccade-contingent changes in a moving point-light walker. Verfaillie, De Troy, and Van Rensbergen (in press) provide a detailed description of the study. In the present article, we offer an overview and discuss converging evidence based on an analysis of the reaction times. In general, the reaction time to detect a display change decreased as the metric size of the change increased, whereas the time to decide that there was no change increased as the change became larger. First, changes in the relative positions of the limbs of the walker were detected fairly accurately. Moreover, the postsaccadic relative positions seemed to be anticipated across saccades. Second, the global image-plane position of the walker was not maintained accurately across saccades. Third, saccade-contingent changes in the walker's in-depth orientation were readily noticed. This suggests that transsaccadic object representations are position invariant but orientation dependent.

Keywords

information integration, saccadic eye movements, biological motion

Introduction

At present, it is not clear how information is integrated across saccadic eye movements. Verfaillie, De Troy, and Van Rensbergen (in press) report a number of experiments that focus on this problem of transsaccadic information integration. The experiments used a particular stimulus and experimental paradigm.

As far as the *stimulus* is concerned, subjects viewed a computer generated version of a human figure walking under biological motion conditions. In the perception of biological motion (Johansson, 1973; Verfaillie, 1993), the available visual material is confined to lights attached to the major joints of a moving actor. Despite the drastic impoverishment of the stimulus information in such a point-light walker, the visual apparatus organizes the moving dots in a rich percept of a human figure.

As far as the *experimental paradigm* is concerned, the detectability of several types of saccade-contingent changes in the point-light walker was investigated. In each trial, subjects were looking at a walker while their eye movements were monitored. At a specific moment, they made a saccade to a cued, new, location within the same figure. During the saccade, a display change occurred in a visual attribute of the walker on some of the trials, and subjects had to indicate whether they noticed that there had been a change. With this technique of saccade-contingent display changes,

Eye Movement Research/J.M. Findlay et al. (Editors)

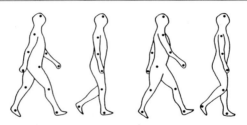

Fig. 1 - Four static sagittal views of a point-light walker.

the information available during the presaccadic fixation is different from the information in the postsaccadic fixation. Because of saccadic suppression, the actual occurrence of the display change itself is not perceptible, and detection that something has changed can only be based on the integration of presaccadic and postsaccadic information.

The study addressed two issues concerning transsaccadic integration. The two topics can be associated with the two components of a biological motion walker: the object in action (i.e., a human figure) and the action itself (i.e., human gait).

First, the nature of the *transsaccadic object representation* was investigated. If a particular object attribute is not inherent to the representation that survives a saccade, intrasaccadic transformations of the attribute should be hard to detect. Conversely, changes in characteristics that are incorporated into the transsaccadic representation should be easy to detect. For instance, suppose the brain represents an object in a particular orientation. The detection of saccade-contingent rotations (resulting in a change in the object's orientation) will be effortless, because the postsaccadic object activates a different orientation-dependent object representation than the presaccadic object. On the other hand, if the object representation is orientation invariant, the visual system's tolerance for intrasaccadic rotations will be high.

Second, *transsaccadic event course anticipation* was examined. When an observer is confronted with a moving walker, the motion event continues during the observer's saccade. Therefore, the walker's presaccadic state is different from the postsaccadic state, not only in the 2-D image but also in the 3-D world. The visual system might simply store the walker's state as it was before the saccade, thereby capitalizing on retrospective memory for the presaccadic information. Alternatively, the system might actively anticipate the future postsaccadic state of the walker.

We analyzed the percentage of detected saccade-contingent display changes as a function of the metric size of the change: How large must the change in a particular visual attribute be before it is detected? For more details on the theoretical underpinnings, the methodology, and the statistical analysis of the detection data, we refer to Verfaillie et al. (in press). The purpose of the present article is to explore whether the reaction time (RT) data reflect the same pattern as the detection data. We stress that the RT analysis is mainly exploratory. The experiments were set up to use detectability as the chief indicator of transsaccadic integration. For instance, the instructions stressed accuracy over speed. Nevertheless, the RT data might provide supporting evidence for the conclusions reached on the basis of the detection data.

Experiment 1: the articulatory motions of the limbs

In Experiment 1, the saccade-contingent change consisted of a change in the relative motions of the point-light walker's body parts (also known as articulatory motions), so that, at the beginning of the next fixation, the walker "reappeared" to the observer in a posture different from the posture reached at the end of the previous fixation. The observer had to indicate whether she/he noticed that the walker's posture had been changed. In addition to this condition in which the change took place during a saccade (the *intrasaccade* condition), we included a control condition in which the shift occurred during steady fixation (the *extrasaccade* condition).

Method

Eight subjects with normal uncorrected vision participated in the experiment.

The stimulus consisted of 11 dots corresponding to the head, the visible shoulder and hip, and the right and left elbow, wrist, knee, and ankle of a figure walking in the frontoparallel plane. Figure 1 shows a sample of static images. To clarify the perceived depth order of the limbs, the outline of a walker has been added. During the actual experiment, only the point-lights were visible. The translational component of motion was zero. Instead, the figure was presented as walking on a rolling treadmill. One step took 400 ms.

The point-light walkers were generated on a screen with a rapidly decaying P4 white phosphor. To reduce phosphor persistence visibility, stimuli were presented at a low level of luminance. Except for the dim light emitted by some control lamps outside of the subject's visual field, the experimental room was completely dark. Moreover, a reduction screen covered with black velvet cloth was placed in front of the screen. The subject was seated at a distance of 65 cm from the screen. Stimuli were viewed binocularly. Eye movements were monitored with a Generation-V dual-Purkinje-image eyetracker. Van Rensbergen and De Troy (1993) provide a description of the eyetracking configuration.

A trial in the *intrasaccade condition* consisted of the following events (Figure 2). After the presentation of a fixation point, a point-light walker appeared. Two seconds later, one of the point lights turned brighter (the head or the ankle light). The subject had to make a saccade to this cue (average saccadic amplitude was 1.94°). Once the tracker detected the cued saccade, the brightened light was converted into a normal light and the posture of the walker was changed. After the display change, the figure continued walking until the subject made a manual response.

For each presaccadic posture, there were seven possible postsaccadic postures (all presented with an equal probability), one of which was the posture which the figure naturally reaches because she/he continues walking during the observer's saccade (the "correct" postsaccadic posture). In the remaining six postures, the walker was shifted with 1/6, 1/3, or 1/2 of a step, either in the direction of articulatory motion (to a posture later in the step cycle) or opposite to it (to a posture earlier in the step cycle). Note that the shift sizes are defined relative to the 0 shift baseline and that this baseline corresponds to the correct *future* posture (therefore, even in the 0 shift case the postsaccadic posture differed from the presaccadic posture).

The course of a trial in the *extrasaccade condition* was identical, except that the change took place after a period of 100 ms in the fixation following the cued saccade.

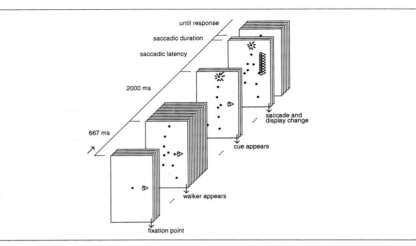

Fig. 2 - Schematic illustration of the time course of a trial in the intrasaccade condition (from Verfaillie, De Troy, & Van Rensbergen, in press, copyright by APA).

Summary of detection analysis

The percentage of trials in which the shift remained undetected was analyzed as a function of the metric size of the shift and the moment the shift took place (intrasaccadic vs. extrasaccadic). The average distributions are shown in the top panel of Figure 3. The 0 shift corresponds to the correct postsaccadic posture. Positive shifts represent shifts forward in the direction of articulatory motion. Negative shifts represent shifts to a posture earlier in the step cycle. The analysis of the detection data revealed the following effects.

First, in both the intrasaccade and the extrasaccade condition, the probability that a change remained undetected is determined by the metric size of the change: The larger the posture shift, the lower the percentage of undetected changes, as evidenced by the bell-shaped curves.

Second, the distribution is flatter in the intrasaccade condition than in the extrasaccade condition. This indicates that posture shifts were easier to detect when they occurred during steady fixation than when they occurred during the cued saccade. On the other hand, transsaccadic integration of the posture was at least partially successful.

Third, both in the intrasaccade and in the extrasaccade conditions, forward shifts (in the direction of articulatory motion) were harder to detect than backward shifts. This relates to the issue of transsaccadic event course anticipation. If the visual system anticipates the postsaccadic posture, the distribution will be symmetrical around the 0 shift (the correct postsaccadic posture). If, in contrast, the system simply retains the posture reached at the end of the presaccadic fixation, the distribution will be shifted backward: Shifts in the direction of the presaccadic posture (backward in relation to the 0 shift) will be detected least. The data do not support the latter prediction. Apparently, the visual apparatus anticipates the future event course. The forward skew of the distribution suggests that anticipatory processes even tend to

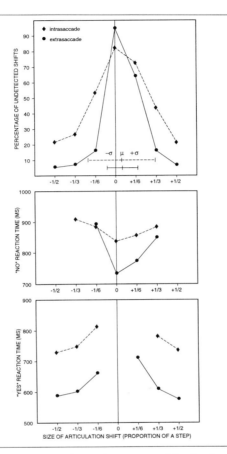

Fig. 3 - Data for Experiment 1. Diamonds connected by broken lines show data for the intrasaccade condition. Circles connected by solid lines show data for the extrasaccade condition. The top panel (reprinted from Verfaillie et al., in press, copyright by APA) depicts the percentage of undetected posture shifts, the middle panel the no RTs, and the bottom panel the yes RTs. Positive shifts on the abscissa represent posture shifts in the direction of articulatory motion. Negative shifts represent shifts opposite to the direction of motion.

overshoot the postsaccadic posture.

Reaction time analysis

It is necessary to make a distinction between the RT to detected changes ("yes" RTs) and the RT to undetected changes ("no" RTs). *Yes* RTs are predicted to decrease as the change becomes larger: When the difference between the postsaccadic posture as "expected" by the observer and the actually presented postsaccadic posture increases, it will be easier to respond that there was a change. Therefore, the distribution of yes RTs (going from the largest backward posture shift to the largest forward shift) is predicted to be bell shaped, similar in shape to the distribution of the percentage of undetected changes. The shortest RT will be found at the tails of the curve (corresponding to the largest changes), whereas the highest RT will be located near the central tendency. The prediction is in line with the general finding that the time for deciding that two patterns are different decreases as the patterns become more dissimilar (Finke, Freyd, & Shyi, 1986; Shepard & Cooper, 1982).

The opposite prediction is made for the *no* RTs. The time to decide that there was no perceptible change will increase as the actual change becomes larger. In other words, as the posture which the subject expects to occur after the saccade becomes more similar to the posture actually presented, the time to decide that the presented posture matches the expected posture will decrease. Therefore, the highest RT will be found at the tails of the curve and the lowest RT near the central tendency. This prediction follows from the general finding that response latency for matching patterns increases as the two patterns become less similar (Finke, Freyd, & Shyi, 1986; Shepard & Cooper, 1982).

However, a major problem is associated with the strategy of performing separate analyses on yes RTs and no RTs. As far as the yes RTs is concerned, there are few data points in the neighbourhood of the central tendency of the distribution (when detection accuracy varies with shift size, as is the case here). The postsaccadic posture associated with the central tendency corresponds to the postsaccadic posture that is actually expected by the subject. It is not surprising, then, that subjects almost never respond "Yes, there was a change" for a postsaccadic posture that matches the expected posture. Similarly, as far as the no RTs is concerned, there are few responses near the tails of the distribution. Again, this is not surprising, because the postsaccadic postures corresponding to the tails represent the largest posture shifts, eliciting fewest no responses.

As a consequence of this problem, the central region of the distribution had to be excluded from the analysis of the yes RTs. The number of data points was too small to guarantee reliable RT data (in most experiments, several subjects actually produced no yes responses at all for particular display change sizes). For analogous reasons, the tails of the curves cannot be included in the analysis of the no RTs.

Mean *no RTs* are shown in the middle panel of Figure 3. An analysis of variance (ANOVA) in the intrasaccade condition revealed a significant main effect of shift size, $F(4,28) = 3.02$, $MS_e = 33931.77$, $p < .04$. In the extrasaccade condition (to compare the detection of forward and backward shifts, we had to exclude two subjects, because they only gave yes responses to even the smallest backward shift), shift size also had a reliable effect, $F(3,15) = 3.67$, $MS_e = 106256.42$, $p < .04$. These findings parallel the conclusions of the detection analysis. First, in both conditions, the distribution has the shape of an inverted bell, as predicted. Second, variation in shift size led to larger RT differences in the extrasaccade condition than in the intrasaccade condition. In other words, both the detection distribution and the RT distribution are flatter in the intrasaccade condition than in the extrasaccade condition. We will come back to this point of convergence between detection and RT data in the discussion of Experiment 2. The third feature of the detection data, namely that the distribution is skewed to the right, is also present in the no RTs. Subjects are faster in responding no to forward shifts than to backward shifts. This converges on the conclusion that anticipatory processes tend to overshoot the postsaccadic posture. However, this trend is only modest in the intrasaccade condition.

The average *yes RTs* are depicted in the bottom panel of Figure 3. An ANOVA yielded a significant main effect of shift size, both in the intrasaccade condition, $F(4,28) = 4.21$, $MS_e = 55180.92$, $p < .01$, and in the extrasaccade condition, $F(4,28) = 11.36$, $MS_e = 50163.92$, $p < .001$. Yes RTs increased with decreasing shift size, as predicted.

Experiment 2: the global position of the walker

In Experiment 2, we focused on transsaccadic integration of the walker's global position in the image plane. There were two conditions in which the figure's position was changed during the cued saccade. In one condition, the figure was walking on a treadmill (so that the walker's global position remained constant before and after the intrasaccadic displacement). In a second condition, the walker translated forward in the frontoparallel plane (so that the walker's position changed continuously). A comparison of the detection of forward versus backward displacements in the latter condition may indicate whether the visual system anticipates the postsaccadic position. The two conditions with intrasaccadic changes were matched with two conditions with extrasaccadic displacements.

Method
Eight subjects took part in the experiment.

Four conditions resulted from the orthogonal manipulation of two variables. First, the point-light figure was presented either as moving on a treadmill or as translating across the screen. Second, the display change occurred during the cued saccade (intrasaccade conditions) or 100 ms after the end of the saccade (extrasaccade conditions).

Subjects had to detect a shift in the walker's position in the image plane. As in Experiment 1, shift size was varied parametrically. Pilot studies indicated that intrasaccadic position shifts were much harder to detect than extrasaccadic displacements. We therefore used a broader range of displacements for the former conditions than for the latter. In both extrasaccade conditions, the walker's position was shifted horizontally with -0.75°, -0.5°, -0.25°, 0°, +0.25°, +0.5°, or +0.75° of visual angle (the average amplitude of the cued saccade was 1.89° in the condition with a nontranslating walker and 3.88° in the condition with a translating walker). Positive shifts refer to displacements in the direction in which the figure was facing (and translating); negative shifts refer to displacements in the opposite direction. In both intrasaccade conditions, the walker was displaced with -1.5°, -1°, -0.5°, 0°, +0.5°, +1°, or +1.5°.

Summary of detection analysis
The top panel of Figure 4 shows the percentage of undetected position shifts as a function of shift size, for the condition with a nontranslating walker. The top panel of Figure 5 shows the corresponding data for the translating walker. Both figures include data for the intrasaccade and the extrasaccade condition.

First, in comparison with Experiment 1, detectability was much more dependent

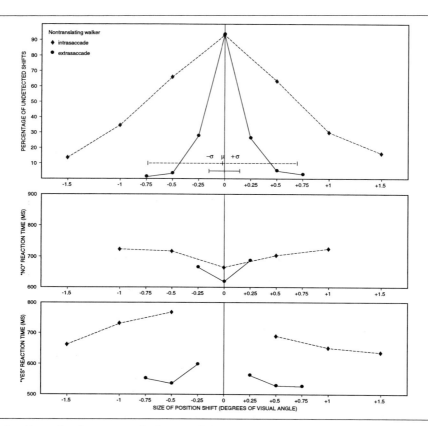

Fig. 4 - Data for the no-translation condition in Experiment 2. Diamonds connected by broken lines show data for the intrasaccade condition. Circles connected by solid lines show data for the extrasaccade condition. The top panel (reprinted from Verfaillie et al., in press, copyright by APA) depicts the percentage of undetected position shifts, the middle panel the average no RTs, and the bottom panel the yes RTs. Positive shifts on the abscissa represent shifts in the direction in which the walker was facing and negative shifts represent shifts in the opposite direction.

on whether the display change occurred during or after the cued saccade[1]. In the extrasaccade condition, a displacement of 0.25° was detected in about two thirds

---

[1] It is difficult to directly compare the detectability of intrasaccadic changes in the relative position of the limbs (Experiment 1) with the detectability of changes in the global position of the overall figure (Experiment 2). In an attempt to overcome this problem, the detectability of intrasaccadic changes is evaluated relative to the detectability of extrasaccadic changes. Therefore, to allow comparison between Experiments 1 and 2, Figures 3 and 4 (top panel) are matched to obtain a constant ratio of the standard deviation in the extrasaccade control condition to the height of the ordinate (we refer to Verfaillie et al., in press, for more details on the procedure of estimating the standard deviation).

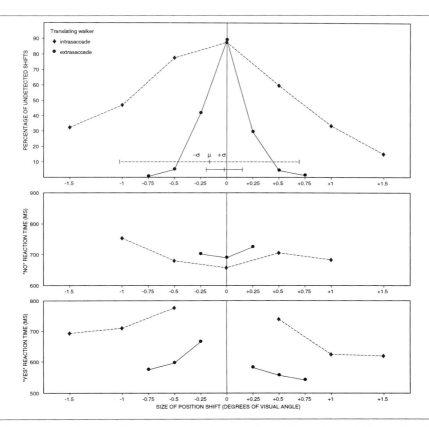

Fig. 5 - Data for the translation condition in Experiment 2. Positive shifts on the abscissa represent shifts in the direction in which the walker was translating and negative shifts represent shifts in the opposite direction.

of the trials, whereas the shift had to be more than four times as large to attain a comparable detection rate in the intrasaccade condition. The 50% threshold for detecting saccade-locked displacements amounted to about 0.75° of visual angle, which is more than one third of the average saccadic amplitude in the no-translation condition. The 50% threshold was even higher when the walker was translating and the shift occurred in the direction opposite to the translation direction (though the threshold was lower when expressed in relation to the saccade because saccadic amplitude was larger in this condition).

Second, the distributions are shifted to the left in the case of a translating walker (Figure 5). As far as the intrasaccade condition is concerned, this suggests that the walker's postsaccadic position was not anticipated. In fact, the mean of the distribution (Verfaillie et al., in press, provide more details on the estimation of the mean) is closer to the presaccadic position than to the correct postsaccadic position (the 0° displacement in Figure 5). Apparently, within the limits of overall poor integration, the position which survived the saccade best was the presaccadic position, suggesting that the visual apparatus capitalized on memory for the presaccadic position rather than on anticipation of the postsaccadic position. In the extrasaccade condition (with a translating walker), a similar, though less pronounced leftward skew was found: A backward displacement of -0.25° was slightly less detectable than a forward shift of +0.25°. Verfaillie et al. (in press) offer an explanation.

Reaction time analysis

The RT data are depicted in the middle (no RTs) and bottom panel (yes RTs) of Figures 4 (nontranslating walker) and 5 (translating walker).

For the analysis of the *no RTs* in the intrasaccade conditions, two subjects had to be excluded (in the translation or in the no-translation condition, they produced only yes responses to the -1° or to the +1° displacements; they even had few no responses to the -0.5° or the +0.5° displacements). An ANOVA with displacement size and the translation variable (translating vs. nontranslating walker) as independent variables yielded a reliable main effect of displacement size, $F(4,20) = 2.91$, $MS_e = 47530.81$, $p < .05$: No RTs decreased as the displacement became larger, giving rise to an inverted bell-shaped distribution. The fact that the walker was translating or not had no main effect, $F < 1$. Moreover, shift size did not interact significantly with the translation variable, $F(4,20) = 2.11$, $MS_e = 13219.47$, $p > .11$. The latter finding does not converge with the analysis of the detection data. A similar pattern emerged in the extrasaccade condition (again, the data of two subjects had to be discarded). Shift size had a reliable main effect, $F(2,10) = 7.21$, $MS_e = 14739.75$, $p < .02$. Neither the translation effect nor the interaction were significant, $F(1,5) = 1.69$, $MS_e = 131001.53$, $p > .25$, and $F < 1$, respectively.

As far as the *yes RTs* is concerned, the predicted bell-shaped curve is apparent in Figures 4 and 5 (bottom panel). For the intrasaccade condition, however, shift size only had a marginally significant effect, $F(5,35) = 2.17$, $MS_e = 182081.14$, $p < .08$ (other $Fs < 1$). The effect was reliable in the extrasaccade condition, $F(5,35) = 8.14$, $MS_e = 47439.04$, $p < .0001$ [the main effect of the translation factor approached significance, $F(1,7) = 4.60$, $MS_e = 150763.88$, $p < .07$, and the interaction effect was unreliable, $F(5,35) = 1.74$, $p > .15$].

The findings are partially in agreement with the detection data. First, the detection analysis revealed that intrasaccadic displacements were very hard to detect (in comparison to extrasaccadic displacements). In other words, in the intrasaccade condition, relatively large differences in displacement size only led to small difference in detectability. In the extrasaccade condition, small increases in displacement resulted in a relatively large decrease in the percentage of undetected displacements (until the floor of quasi perfect detection was reached). A similar effect showed up in the RT data. In the intrasaccade condition, relatively large increases in displacement only led to small increases in no RTs and small decreases in yes RTs, giving rise to relatively flat distributions. The relation between displacement size and yes RT did not even reach significance (though the effect was in the predicted direction). In the extrasaccade condition, displacement size had a more profound effect on RT: Even for relatively small increases in displacement, no RTs increased and yes RTs decreased, leading to sharper curves than in the intrasaccade condition.

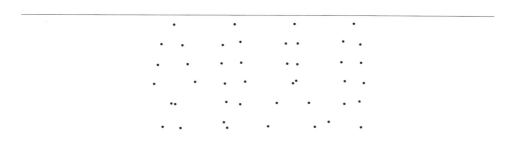

Fig. 6 - Four static 45° views of a walker, corresponding to the sagittal views shown in Figure 1. In contrast to Figure 1, the hip lights are absent and the point lights located behind the body are not occluded.

The second major finding of the detection analysis was that the distribution for a translating walker was skewed in a direction opposite to the direction of translational motion. For the intrasaccade condition, this suggested that the visual apparatus exploits memory for the walker's presaccadic position. However, such an effect was absent in the RT data (backward shifts indeed led to slightly longer yes RTs than forward shifts, but this was also the case for a nontranslating walker). In the extrasaccade condition, the RT distributions were more in agreement with the detection data, but the predicted interaction between displacement size and the translation variable was not significant.

Experiment 3: the in-depth orientation of the walker

Experiment 3 examined the detectability of changes in the point-light figure's orientation in depth. During the cued saccade, the walker's in-depth orientation was shifted over a particular angle of rotation around the figure's top-bottom axis. The orientation shift did not affect the walker's image-plane orientation; the figure always remained upright. Figure 6 depicts the four 45° views corresponding to the four sagittal views shown in Figure 1. Such a 45° in-depth orientation is an orientation between the sagittal orientation (in which the walker faces in a direction orthogonal to the observer's line of sight) and the frontal orientation (in which the walker faces towards the observer).

Method

Eight subjects participated in Experiment 3.

As in Experiment 2, there were four conditions, resulting from the orthogonal manipulation of two variables. First, the display change either occurred during the cued saccade (intrasaccade conditions) or 100 ms later (extrasaccade conditions). Second, either the walker's orientation remained constant before and after the display change (the no-rotation conditions) or the walker was rotating around his top-bottom axis during the entire trial (the rotation conditions). In the no-rotation conditions, the point-light figure was presented in a constant 45° orientation before the cued saccade (and the saccade-locked orientation change). In the rotation conditions, the upright walker was rotating with an angular velocity of 33.75°/s. To increase the comparability with the no-rotation condition, the saccade cue appeared just before the figure reached the 45° view, so that the intrasaccadic or extrasaccadic orientation shift took place when the walker approximately occupied a 45° orientation.

In all conditions, subjects had to detect a change in the upright walker's in-depth orientation. The orientation was shifted by a 3-D rotation around the figure's top-bottom axis over an angle of -33.75°, -22.5°, -11.25°, 0°, +11.25°, +22.5°, or +33.75°. Given a 45° starting orientation, the negative angles refer to rotations toward the 0° frontal view and the positive angles refer to rotations toward the 90° sagittal view. In the rotation conditions, the negative orientation shifts were opposite to the direction of rotation while the positive shifts occurred in the direction of rotation.

Summary of detection analysis

The percentage of undetected orientation shifts is depicted in the top panels of Figure 7 (left panel for a nonrotating walker and right panel for a rotating walker). To allow comparisons across experiments, the scale on the abscissa has been adapted so that Figure 7 (left top panel) matches Figures 3 and 4 as far as the ratio of the standard deviation in the extrasaccade control condition over the height of the ordinate is concerned.

First, though intrasaccadic shifts were again harder to detect than extrasaccadic shifts, the difference between the intrasaccade condition and the extrasaccade condition is much smaller than in Experiment 2: Transsaccadic integration of the in-depth orientation of the walker was remarkably accurate, especially in the condition with a nonrotating walker. In conjunction with Experiment 2, these findings suggest that the in-depth orientation in which an object appears is intrinsic to the object representation maintained across saccades, whereas the object's position is not.

The second main issue addressed in Experiment 3 was the question whether the visual system accurately anticipates the postsaccadic in-depth orientation of a rotating walker (right top panel of Figure 7). The analysis revealed that, for the intrasaccade condition, there was a large interindividual variation in the central tendency of the distribution (which is assumed to reflect the postsaccadic orientation anticipated by the observer). A more fine-grained analysis showed that the anticipated orientation appeared to be related to the walker's orientation at the moment of the orientation shift: Subjects showed a tendency to distort the anticipated object orientation in the direction of a canonical orientation (either the sagittal view or the frontal view). For instance, when the correct postsaccadic orientation consisted of an orientation between the 45° view and the sagittal view, the visual apparatus anticipated an orientation closer to the sagittal view than the correct postsaccadic view. When, on the other hand, the correct postsaccadic orientation was closer to the frontal view than to the sagittal view, the anticipated orientation was distorted in the direction of the frontal view.

Reaction time analysis

Mean *no* RTs (middle panels of Figure 7) in both the extrasaccade and the intrasaccade condition (one subject was not included in the latter analysis) show the predicted inverted bell-shaped curve, for a nonrotating as well as for a rotating walker. However, shift size only had a main effect in the extrasaccade condition, $F(2,14) = 5.11$, $MS_e = 66707.10$, $p < .03$ [the main effect of the rotation factor (rotating

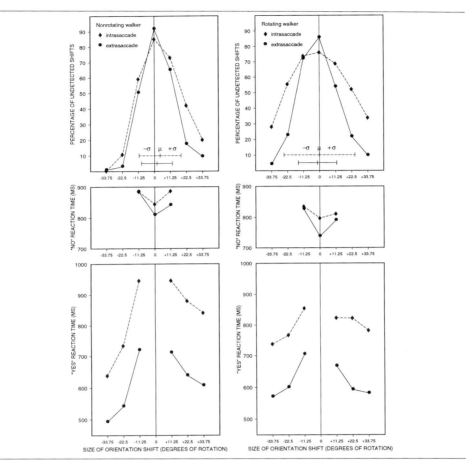

Fig. 7 - Data for the no-rotation (left panels) and the rotation condition (right panels) in Experiment 3. Diamonds connected by broken lines show data for the intrasaccade condition. Circles connected by solid lines show data for the extrasaccade condition. The top panels (reprinted from Verfaillie et al., in press, copyright by APA) depict the percentage of undetected orientation shifts, the middle panel the no RTs, and the bottom panel the yes RTs. Positive shifts on the abscissa represent shifts in the direction of the sagittal view (which is also the direction in which the walker was rotating in the rotation condition) and negative shifts represent shifts in the direction of the frontal view.

versus nonrotating walker) was not reliable, $F(1,7) = 1.03$, $p > .3$, nor was the interaction between shift size and the rotation factor, $F < 1$]. In the intrasaccade condition, the effect of shift size was not reliable, $F(2,12) = 1.72$, $MS_e = 66674.68$, $p > .2$ [the main effect of the rotation factor, $F(1,6) = 1.62$, $p > .25$, and the interaction effect were also unreliable, $F < 1$].

On the other hand, shift size had a reliable effect on the *yes RTs*, both in the intrasaccade and in the extrasaccade condition. Moreover, this effect was modulated by the rotation factor, as indicated by a significant interaction. In the intrasaccade condition (one subject was excluded), the main effect of shift size was significant, $F(5,30) = 3.80$, $MS_e = 300170.83$, $p < .01$: Yes RTs decreased as the orientation shift became larger. The rotation factor (nonrotating vs. rotating walker) had no reliable main effect, $F < 1$, but interacted with shift size, $F(5,30) = 3.58$, $MS_e = 88328.13$, $p < .02$: Increases in shift size led to larger decreases in yes RTs in the condition with a nonrotating walker than in the condition in which the walker rotated during the entire trial. This parallels the detection analysis.

A similar pattern emerged in the extrasaccade condition. Shift size had a reliable main effect, $F(5,35) = 6.19$, $MS_e = 146446.90$, $p < .0005$, and the interaction effect between shift size and the rotation factor was significant, $F(5,35) = 3.70$, $MS_e = 69510.02$, $p < .01$. The interaction has to do with the distribution in the no-rotation condition being slightly sharper and more skewed to the right than in the rotation condition. This was also observed in the detection analysis. Verfaillie et al. (in press) offer an explanation.

General discussion

The theoretical rationale for the present study was twofold. First, the nature of the *transsaccadic object representation* was examined. The low detection rate of changes in the image-plane position of the figure (Experiment 2) and the high detection of changes in the upright walker's in-depth orientation (Experiment 3) indicated that transsaccadic object representations are position invariant but orientation dependent. The second issue concerned *transsaccadic event course anticipation*. Subjects anticipated the postsaccadic relative positions of the walker's body parts (Experiment 1). In contrast, there was no anticipation of the postsaccadic global position of a translating figure; instead, subjects relied on memory of the figure's presaccadic position (Experiment 2). The anticipated in-depth orientation of a rotating walker seemed to be distorted in the direction of canonical views (Experiment 3).

In general, the RT analysis supported the conclusions reached on the basis of the detection data. First, as the metric size of the display change increased, the RT to undetected display changes increased (though the effect was not significant in the intrasaccade condition of Experiment 3), while the RT to detect a display change decreased (the effect was not significant in the intrasaccade condition of Experiment 2).

Second, when relatively large changes in display size led to relatively large variations in detection rate (i.e., when the distribution of undetected changes became sharper), there was also a larger variation in RTs (the distribution of yes and no RTs also became sharper). This was apparent in all three experiments, but it was especially manifest in Experiment 2, where the difference between the intrasaccade and the extrasaccade condition was most pronounced. Intrasaccadic displacements had to be relatively large before they resulted in an increase in detection rate, a decrease in yes RTs and an increase in no RTs. In the extrasaccade control condition, a comparatively small increase in displacement size led to relatively larger detection percentages, shorter yes RTs and longer no RTs.

Third, in the detection analysis, the central tendency of the distribution of undetected changes was interpreted as an indicator of anticipatory processes. Though the extrasaccade RT distributions in Experiments 1, 2, and 3 tended to have a comparable central tendency as the respective detection distributions, the parallel was mostly absent in the intrasaccade conditions. Variations in the central tendency of the detection distribution probably pertain to effects too subtle to show up in the RT data.

In summary, though the more subtle aspects of the detection data were not always manifest in the RT data, the RT analysis generally was in agreement with the detection analysis. However, the major obstacle in reaching firm conclusions on the basis of the RT data, is the fact that the number of data points for some display change sizes is too small (for the larger display changes in the case of the no RTs and for the smaller changes in the case of the yes RTs) to guarantee reliable RT data. Therefore, detectability remains the principal indicator of processes underlying transsaccadic integration.

Acknowledgements

Karl Verfaillie was supported by the Belgium Program on Interuniversity Poles of Attraction Convention Number 31, Andreas De Troy by the Incentive Program for Fundamental Research in Artificial Intelligence RF/A1/04, and Johan Van Rensbergen by GOA Contract 87/93-111. Andreas De Troy and Johan Van Rensbergen's contribution to this research project consisted of implementing the dual-Purkinje-image eyetracking infrastructure.

References

Finke, R. A., Freyd, J. J., & Shyi, G. C.-W. (1986). Implied velocity and acceleration induce transformations of visual memory. *Journal of Experimental Psychology: General, 115,* 175-188.

Johansson, G. (1973). Visual perception of biological motion and a model for its analysis. *Perception & Psychophysics, 14,* 201-211.

Shepard, R. N., & Cooper, L. A. (1982). *Mental images and their transformations.* Cambridge, MA: MIT Press.

Van Rensbergen, J., & De Troy, A. (1993). *A reference guide for the Leuven dual-PC controlled Purkinje eyetracking system* (Psychol. Rep. No. 145). Leuven: Katholieke Universiteit Leuven, Laboratory of Experimental Psychology.

Verfaillie, K. (1993). Orientation-dependent priming effects in the perception of biological motion. *Journal of Experimental Psychology: Human Perception and Performance, 19,* 992-1013.

Verfaillie, K., De Troy, A., & Van Rensbergen, J. (in press). Transsaccadic integration of biological motion. *Journal of Experimental Psychology: Learning, Memory, and Cognition.*

EYE MOVEMENTS AND LANGUAGE

A CHALLENGE TO CURRENT THEORIES OF EYE MOVEMENTS IN READING.

Françoise Vitu & J. Kevin O'Regan.

Laboratoire de Psychologie Expérimentale,
CNRS, EPHE, EHESS, Université René Descartes,
28 rue Serpente, 75006 Paris, FRANCE.

Abstract

The purpose of the present study was to precisely lay out and to test the predictions of two recent theories of eye movements in reading as concerns the within-word eye behavior: 'attention-type' theories where the main determinant of eye movements is the on-going visual and linguistic processing of the encountered words, and the 'Strategy-Tactics' theory where pre-determined oculomotor strategies are the main driving force of the eyes during reading. A comparison of the predictions made by both theories with the data available in the literature as concerns the probability of refixating words revealed that neither of the theories can account for the observed data and that an alternative theory is needed which takes into account both the influence of on-going processing and pre-determined oculomotor tactics as well as visuomotor constraints. Furthermore, an examination of the relative durations of fixations in the 1- and 2-fixations cases showed that the mechanism which is responsible for within-word refixations cannot be an oculomotor deadline as assumed in the 'attention-type' theory. Possible revisions of both theories are therefore proposed.

Keywords

Eye movements, Reading, Visual Attention, Linguistic Processing, Word Recognition, Visuomotor constraints, Oculomotor Strategies.

Introduction

During text reading, saccade sizes and fixation durations are quite variable: words are sometimes skipped and sometimes fixated; when they are fixated they may be fixated with one or several fixations; the positions and durations of these fixations are also quite variable. Up until now, a considerable amount of research has been devoted to understanding this variability. However, while it is clear at the present time that globally eye movements in reading result both from the on-going visual and linguistic processing of the encountered words and from visuo-motor constraints, it is still not clear how these two types of influences combine in order to give rise to the ocular behavior observed in the vicinity of each word.

It is only recently that some attempts have been made to answer this question. Particularly, two theories have been proposed: the 'attention-type' theories of Henderson & Ferreira (1990), derived from earlier proposals by Morrison (1984) and Rayner & Pollatsek (1989) where the main determinant of eye movements is on-going processing; and the 'Strategy-Tactics' theory proposed by O'Regan (1990; 1992; see also O'Regan & Lévy-Schoen, 1987) where the main determinants are pre-determined oculomotor strategies and tactics and visuo-motor constraints. These two theories,

Eye Movement Research/J.M. Findlay et al. (Editors)

whose purpose was initially to explain particular empirical data, can make precise predictions as concerns the probability of skipping words, the probability of refixating words and the positions and durations of each fixation. However, most of these predictions have never been clearly set out, and none have been directly tested.

The purpose of the present paper is therefore to examine the predictions made by both theories and to test these predictions on the basis of the data available in the literature. Here, we will examine only the question of the determinants of the probability of refixating words (or probability of making additional fixations after the initial one), and the durations of individual fixations.

Predictions

The 'Attention-type' theory.

In their 'attention-type' theory, Henderson & Ferreira (1990) (see also Morrison, 1984) assume that two mechanisms are responsible for eye movements during reading. The first mechanism is visual attention: saccades from one word to the next are assumed to be triggered by shifts of attention, and these shifts themselves are triggered in real time by ongoing processing of the encountered words. The second mechanism is an oculomotor deadline which prevents the eyes from staying too long at a single place without moving. This second mechanism gives rise to the occurence of saccades without a shift of attention.

More precisely, according to Henderson and Ferreira, the decision of refixating a word would be based on the efficiency of processing the fixated word. The sequence of events occuring before this decision would be as follows: when the eyes land on a word, attention is already focused on this word and processing of the fixated word is in progress. When processing of the fixated word terminates, attention shifts towards the next word in parafoveal vision and a saccade towards it starts being prepared. However, if processing of the fixated word takes too long, in particular, longer than the delay required by the oculomotor deadline, a saccade is programmed within the word since attention is still focused on this word.

In other words, according to Henderson & Ferreira (1990), a within-word refixating saccade should only be programmed in the cases where processing of the fixated word takes longer than a certain delay. This leads to two straightforward predictions (cf. Table 1): the first prediction is that the *probability of refixating* a word should depend on the efficiency of the processing of the fixated word and should therefore be affected by factors which influence visual and linguistic processing of words.

In particular, the probability of refixating a word should depend on the eyes' initial landing position in the word. This is because it is known that the location of the eyes in a word affects the amount of visual information which can be extracted from a word: the closer the eyes are to the middle of the word, the more letters of the word can be processed (Nazir et al., 1991). The probability of refixating a word should also depend on linguistic factors such as the word's frequency of occurence in the language and the word's predictability from the linguistic context since these have been shown to influence the time required to identify a word (i.e: Meyer et al., 1975; Balota & Chumbley, 1984; Fischler & Bloom, 1985; Paap et al., 1987; Monsell et al., 1989).

The second prediction made by the 'attention-type' theory concerns *fixation durations* and is a consequence of the fact that a word is refixated only in the cases where processing takes longer than the oculomotor deadline (cf. Tab. 1): the durations of the first fixations in the cases where several fixations occur on the word should be systematically equal to the oculomotor deadline. On the other hand, in the majority of cases where only a single fixation occurs on the word, the duration of this fixation should be shorter than the oculomotor deadline. There would be two exceptions to this scheme. A first exception would occur when a within-word refixation saccade is planned and processing of the fixated word manages to terminate during the refixation saccade programming period. In that case, the planned saccade would be re-oriented to the next word, and a single fixation whose duration is equal to the oculomotor deadline would occur. A second exception would occur when the word had been fully processed in parafoveal vision before being fixated, but not early enough to allow the eyes to skip this word. In this particular case, a single fixation would occur whose duration is very short and independent of processing. In summary, single fixation durations should be either shorter or equal to the duration of first fixations in the 2-fixation cases. Furthermore, the distribution of single fixation durations should be tri-modal, with one peak corresponding to 'oculomotor' single fixations, another peak corresponding to 'real' single fixations and a third peak corresponding to 'cancelled' refixations.

PREDICTIONS

	'Attention-type' theory (Henderson and Ferreira, 1990)	'Strategy-Tactics' theory (O'Regan, 1990; 1992)
REFIXATION PROBABILITY depends on:	-word's frequency -word's predictability -parafoveal preprocessing -initial fixation position.	initial fixation position
FIXATION DURATIONS:	Single < First of 2	Single > First of 2

Table 1: Predictions of the 'attention-type' and the 'Strategy-Tactics' theories.

The 'Strategy-Tactics' theory.

In the 'Strategy-Tactics' theory (O'Regan, 1990; 1992; see also O'Regan & Lévy-Schoen, 1987), the main idea is that although linguistic processing is occuring in parallel with eye movements, these are not determined in real time by on-going processing, but rather by pre-determined oculomotor strategies and tactics, and visuomotor constraints.

In particular, the decision of refixating a word is determined only by pre-determined oculomotor tactics based on the eyes' initial landing position in the word. The within-word tactics postulated by O'Regan are the following: if the eyes land near the middle of the word, a single fixation will occur on this word. On the contrary, if the eyes land at the beginning or the end of the word, a refixating saccade will be programmed. Note that whereas this tactic is coherent with the idea that processing is more efficient when the eyes are located at the middle of the word, it is actually not on-going processing which determines in real time the probability of refixating the word. Rather it is a pre-determined tactic based on the subject's prior experience that processing will be faster at this position.

Thus, if the eyes land at the middle of the word, a single fixation occurs on the word and the eyes stay at that location until processing of the word terminates. On the other hand, if the eyes do not land at the middle of the word, a second fixation occurs immmediately after the system has noticed that the eyes are inadequately located, and processing of the word terminates from the new eye position. In other words, while single fixation durations reflect the time required first to determine the position of the eyes, and then to finish processing, first fixation durations in the 2-fixation cases reflect only the time needed to determine the eyes' position: they do not depend on linguistic processing.

The above discussion of the Strategy-Tactics theory leads to two predictions. First, the *probability of refixating* a word should depend on the eyes' initial landing position in a word (as also predicted but for other reasons by the 'attention-type' theory) (cf. Tab. 1). However, contrary to the 'attention-type' theory, 'Strategy-Tactics' predicts no effect of linguistic factors on the refixation probability.

Second, as concerns *fixation durations*, contrary to the attentional theory, single fixation durations should be systematically longer than the duration of first fixations in the 2-fixation cases.

Empirical Data

We shall now confront the above predictions with the available empirical data. Let us consider first the case of *refixation probabilities*.

Both theories predict an effect of the initial fixation position on refixation probability, and indeed the existence of such an effect is now well established for the case of isolated word recognition (O'Regan & Lévy-Schoen, 1987; Holmes & O'Regan, 1987; Vitu et al., 1990) as well as for text reading (McConkie et al., 1989; Vitu et al., 1990): the probability of refixating a word is much smaller when the eyes initially land at the middle of the word than when they land at the beginning or the end of the word. This effect is illustrated in Figure 1a-b where the probability of refixating a

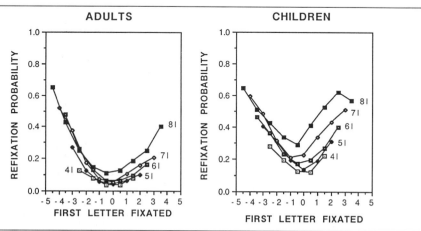

Fig. 1 - Probability of refixating words (of 4,5,6,7, and 8 letters) during text reading by adults (1a) and during the reading of half a book by 5th grade children (1b) as a function of the eyes' initial landing position in the words. The data in Figures 1a and 1b are taken respectively from studies done by McConkie et al (1989) and Vitu et al (in preparation).

word has been plotted as a function of the initial landing position in the word, for adults (Fig. 1a) and 5th-grade children (Fig. 1b) reading a text (Vitu et al., in preparation). We have chosen these two data corpuses since they are the largest existing ones.

As said above, both theories predicted this effect of the initial landing position on refixation probability. However, this was for different reasons. The 'attention-type' theory predicts it because it assumes that the probability of refixating a word is determined in real time by the efficiency of on-going visual and linguistic processing of the fixated word. On the other hand, 'Strategy-Tactics' predicts the effect of the initial landing position on the refixation probability because it assumes that the decision of refixating a word is based on predetermined oculomotor tactics which depend on how far the eyes are from the middle of the word.

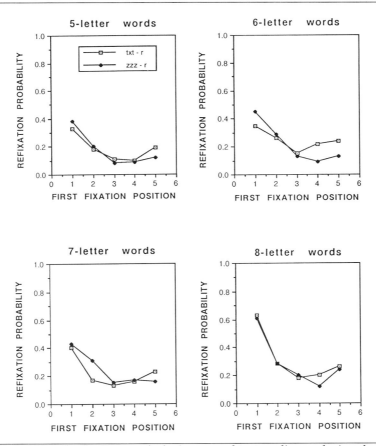

Fig. 2 - *Probability of refixating words during normal text reading or during the scanning of z-letter strings as a function of the initial landing position in the words or strings, for 5-,6-,7-, and 8-letter words (strings). These data come from a study by Vitu et al. (submitted).*

However, both theories should make different predictions as concerns the oculomotor behavior adopted in tasks which require no processing, such as the scanning of meaningless letter-strings. While the 'attention-type' theory would predict no effect of the initial fixation position on the probability of refixating meaningless letter strings, 'Strategy-Tactics' could predict such an effect. This was tested in a first experiment by Nazir (1991) and in a more recent experiment where subjects were presented either with normal texts or texts in which all the letters of the words were replaced by the letter 'z' (Vitu et al., submitted). In the 'z' letter string condition, subjects were asked to move their eyes as if they were reading. The results obtained in this experiment are presented in Figure 2a-d. They show that refixation probability depends strongly on the eyes' initial landing position both in words as well as in z-letter strings. These results argue in favor of the 'Strategy-Tactics' hypothesis that the decision of refixating words is based on predetermined oculomotor tactics. On the other hand, the results argue against the 'attention-type' theory, which explains the effect of the initial fixation position in terms of efficiency of on-going visual processing.

Another difference in predictions between the two theories concerns the effects of linguistic factors on refixation probability. Only the 'attention-type' theory predicts that a word's linguistic characteristics should affect refixation probability. Several data are available in the literature showing the effects on refixation probability of a word's frequency or a word's predictability from the linguistic context (Balota et al., 1985; Pollatsek et al., 1986; McConkie et al., 1989; Pynte et al., 1991; Vitu et al., in preparation). For example, refixation probability is smaller for high frequency words than for low frequency words (cf. Fig. 3 for illustration). Furthermore, as found by Vitu (1991), the effect of linguistic factors on refixation probability is interactive: for example the effect of word frequency on the refixation probability is only present in the cases where the word could be predicted by the linguistic context and/or preprocessed in parafoveal vision (cf. Fig. 4a-b).

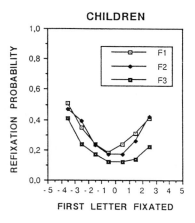

Fig. 3 - Probability of refixating 6-letter words for 5th grade children as a function of the initial landing position and the word's frequency of occurrence in the language.

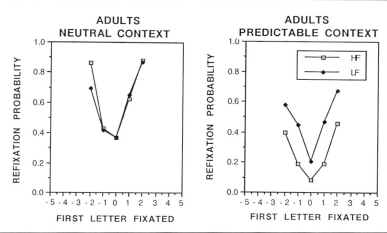

Fig. 4 - Probability of refixating 5-letter words as a function of the eyes' initial fixation position and the word's frequency. The presentation of the words was preceded by the presentation of a neutral (4a) or a predictable sentence (4b). These data come from a study by Vitu (1991).

These results showing an effect of linguistic factors on refixation probability are compatible with the 'attention-type' theory, but not with Strategy-Tactics': this latter theory would predict that the decision of refixating a word depends purely on pre-determined oculomotor tactics.

To conclude, together these results all show that the probability of refixating a word is determined both by on-going linguistic processing and by pre-determined oculomotor tactics. Thus, it appears that neither the 'attention-type' nor the 'Strategy-Tactics' theories account for the data obtained in the literature, and an alternative is needed.

As concerns now the case of *fixation durations*, the theories make totally opposite predictions. The 'attention-type' theory assumes that the durations of single fixations should be either shorter or equal to the first fixation durations in the cases where 2 fixations occur on the word. On the other hand, 'Strategy-Tactics' assumes that single fixation durations should be systematically longer than first fixation durations in the 2-fixation cases. Furthermore, while the 'attention-type' theory predicts that the distribution of single fixation durations should be tri-modal, 'Strategy-Tactics' predicts a unimodal distribution.

Most of the data available in the literature as concerns fixation durations do not allow us to distinguish between the two types of predictions since most of the studies have not distinguished single and two-fixation exploration tactics: that is, they have considered the durations of first fixations whatever the number of fixations that occurred on words. O'Regan & Lévy-Schoen (1987) did however do the appropriate analysis for the cases of isolated word recognition and found that single fixation durations are longer than the durations of first fixations in the 2-fixation cases. This has been confirmed for the reading of words embedded in sentences (Underwood et al., 1990) and words in texts (Vitu et al., submitted). We have also performed the necessary analysis on the two large corpuses of data collected while adults and 5th grade students were reading respectively a text and half a book (Vitu et al., in preparation). Only the results obtained for children are presented in Figure 5a-d since there were not enough data points for adults in the 2-fixation cases. As clearly appears, single fixation durations are systematically longer than the durations of first fixations in the 2-fixation cases. These results are in accordance with the predictions made by 'Strategy-Tactics' and incompatible with the 'attention-type' theory.

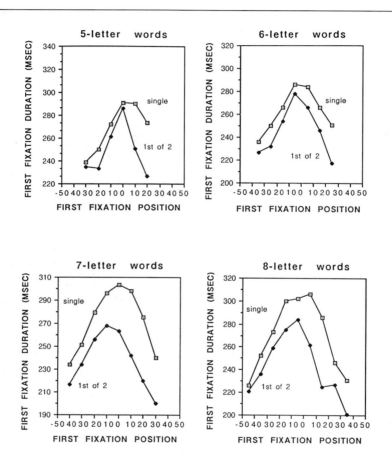

Fig. 5 - Durations of single fixations and first fixations in the 2-fixation cases as a function of the initial landing position in words of 5,6,7, and 8 letters for 5th grade children. These data come from a study by Vitu et al (in preparation).

Furthermore, if we look now at the distribution of single fixation durations (cf. Fig.6a-b), we can see that the distribution is unimodal suggesting that single fixations cannot be divided into three separate classes as assumed in the 'attention-type' theory.

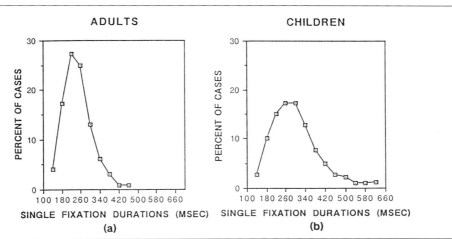

Fig. 6 - Distribution of single fixation durations on 7-letter words for adults (6a) and 5th grade children (6b). These data come from a study by Vitu et al (in preparation).

The prediction made by the 'attention-type' theory as concerns the difference between the durations of single fixations and first fixations when two occur results from the hypothesis that a refixation occurs only when processing of the fixated word has not terminated before the oculomotor deadline. The above results suggest therefore that such a mechanism cannot be operative during reading.

Furthermore, if we examine again the curves presented in Figure 5a-b, we can see that single fixation durations as well as first fixation durations in the 2-fixation cases are longer when the eyes start fixating the middle of the word than when they initially fixate other parts of the word. This effect for first fixations when two occur was previously found for the case of isolated word recognition (O'Regan & Lévy-Schoen, 1987), but the effect for single fixation durations has never been reported. The fact that the eyes stay longer when they are located at the middle of words cannot be explained in terms of processing since visual processing has been shown to be more efficient from this eye location. The effect may however result from the influence of visuo-motor constraints, but this remains to be tested.

An alternative theory should in any case also take into account this variation of fixation durations with the eyes' location.

Conclusion

The purpose of the present paper was to test, on the basis of the data available in the literature, the predictions made by two current theories of eye movements in reading as concerns the probability of refixating words and the relative durations of individual fixations. The data considered for this study clearly show that neither the 'attention-type' theory (Henderson & Ferreira, 1990) nor the 'Strategy-Tactics' theory (O'Regan, 1990; 1992) can correctly predict the variations of refixation probabilities and fixation durations observed during text reading. First, while the 'attention-type' theory cannot predict the effect of the eyes' initial fixation position on the probability of refixating meaningless letter strings, 'Strategy-Tactics' cannot predict the influence of a word's linguistic characteristics on its refixation probability. Second, the 'attention-type' theory with its notion of an oculomotor deadline cannot predict that single fixation durations are systematically longer than the durations of first fixations in the 2-fixation cases. Furthermore, neither theory can predict the variation of single fixation durations with the eyes' location.

These results suggest that a reliable theory of eye movements in reading should be less extreme with respect to the issues of on-going processing of the encountered words and pre-determined oculomotor tactics than the theories proposed up to now. However, before proposing such a theory, it would be necessary to determine empirically which of these two factors is the basic driving force of the eyes during reading. The fact that similar exploration tactics are observed during the reading of words as during the reading of z-letter strings (Vitu et al., submitted) and that refixation probabilities in words depend on linguistic factors, could be explained by two different hypotheses. Under one, pre-determined oculomotor tactics are the main determinants of eye movements in reading and they are modulated by on-going processing. Alternatively, the decision of refixating words during reading might mostly be based on on-going processing, and this decision might in particular cases be modulated by pre-determined oculomotor tactics. Thus, two solutions are available at the present time: either revising the 'Strategy-Tactics' theory or revising the 'attention-type' theory by introducing the possibility that the decision taken on the basis of the main mechanism be modulated respectively by on-going processing or by pre-determined oculomotor tactics. However, since fixation duration data are incompatible with the 'attention-type' theory's notion of an oculomotor deadline, the revision of this theory requires an additional step. This step would consist in finding a mechanism responsible for the decision of refixating words which is based on on-going processing, and which does not use the notion of an oculomotor deadline. In the next sections, we will therefore propose preliminary revisions of both theories.

For the 'attention-type' theory where processing is the main determinant of the decision of refixating words, we could propose a mechanism similar to the one assumed by Vitu (1991). When the eyes land on a word, processing of this word is in progress. After a certain delay, the system examines how much information has been processed on the fixated word. If the processing that has been accomplished is more than a certain threshold amount of processing, then processing of the fixated word continues from the same eye location. In the opposite case, a refixation saccade is programmed and processing continues from another eye location. The subject's reading strategy would determine the threshold amount of information which has to be processed before the critical moment for a single fixation to occur. For example, if subjects decide to read fast, they would choose a low value for the threshold and they would have a lower probability of refixating words. However, in particular cases where processing is very easy, or not successful or not useful (such as in the 'z-string' reading condition used by Vitu et al (submitted) where no processing is required), subjects would probably switch from this cognitive mode to an automatic scanning mode based on pre-determined oculomotor tactics.

As a revision for 'Strategy-Tactics', we could propose the following hypothesis. As assumed by O'Regan (1990; 1992), the decision to refixate a word would be determined first by pre-determined oculomotor tactics based on the eyes' initial fixation position. However, this decision could be changed as a function of the efficiency of on-going processing. Each time a decision is taken, the level of processing achieved at the moment of decision would be compared to a ceiling amount of processing which determines the non-necessity for refixating, and to a floor amount of processing which determines the necessity for refixating. In the particular cases where the result of this comparison process and the decision taken are in disagreement, the decision taken would be cancelled and changed. For example, in cases where the decision taken is to make only one fixation on the word (cases where the eyes land at the middle of the word), and the fixated word is taking particularly long to process, the final decision would be to make an additional fixation on the word.

References

Balota, D.A. & Chumbley, J.I. (1984). Are lexical decisions a good measure of lexical access? The role of word frequency in the neglected decision stage. J. Exp. Psychol. Human Percep. Performan. 10, 340-357.

Balota, D.A., Pollastek, A. & Rayner, K. (1985). The interaction of contextual constraints and parafoveal visual information in reading. Cognitive Psychol. 17, 364-390.

Fischler, I. & Bloom, P.A. (1985). Effects of constraint and validity of sentence context on lexical decisions. Memory and Cognition, 13, 128-139.

Henderson, J.M. & Ferreira, F. (1990). Effects of foveal processing difficulty on the perceptual span in reading: Implications for attention and eye movement control. J. Exp. Psychol. Learn. Mem. Cognit. 16(3), 417-429.

Holmes, V.M. & O'Regan, J.K. (1987). Decomposing french words. In Eye movements: From physiology to cognition, J.K. O'Regan & A. Lévy-Schoen Eds., North Holland, Amsterdam.

McConkie, G.W., Kerr, P.W., Reddix, M.D., Zola, D. & Jacobs, A.M. (1989). Eye movement control during reading: II. Frequency of refixating a word. Perception and Psychophysics 46, 245-253.

Meyer, D.E., Schvaneveldt, R.W. & Ruddy, M. (1975). Loci of contextual effects in visual word recognition. In Attention and Performance V, P.M.A. Rabbitt & S. Dornic Eds., Academic Press, New York.

Monsell, S., Doyle, M.C. & Haggard, P.N. (1989). Effects of frequency on visual word recognition tasks: where are they? J. Exp. Psychol. Gen. 118, 43-71.

Morrison, R.E. (1984). Manipulation of stimulus onset delay in reading: Evidence for parallel programming of saccades. J. Exp. Psychol. Human Percep. Performan. 10, 667-682.

Nazir, T. (1991). On the role of refixations in letter strings: The influence of oculomotor factors. Perception and Psychophysics 49(4), 373-389.

Nazir, T., O'Regan, J.K. & Jacobs, A.M. (1991). On words and their letters. Bulletin of the Psychonomic Society 29, 171-174.

O'Regan, J. K. (1990). Eye movements and reading. In Eye movements and their role in visual and cognitive processes, E. Kowler Ed., Elsevier.

O'Regan, J.K. (1992). Optimal viewing position in words and the strategy-tactics theory of eye movements in reading. In Eye movements and visual cognition, Scene Perception and Reading, K. Rayner Ed., Springer-Verlag, New York.

O'Regan, J.K., & Lévy-Schoen, A. (1987). Eye movement strategy and tactics in word recognition and reading. In The psychology of reading, Attention and Performance XII, M. Coltheart Ed., Erlbaum, Hillsdale, NJ, pp. 363-383.

Paap, K.R., McDonald, J.E., Schvaneveldt, R.W. & Noel, R.W. (1987). Frequency and pronounceability effects in visually presented naming and lexical decision tasks. In The psychology of reading, Attention and Performance XII, M. Coltheart Ed., Erlbaum, Hillsdale, NJ, pp. 221-243.

Pollatsek, A., Rayner, K. & Balota D.A. (1986). Inferences about eye movement control from the perceptual span in reading. Perception and Psychophysics 40, 123-130.

Pynte, J., Kennedy, A. & Murray, W.S. (1991). Within-word inspection strategies in continuous reading: Time course of perceptual, lexical, and contextual processes. J. Exp. Psychol. Human Percep. Performan. 17, 458-470.

Rayner, K. & Pollatsek, S. (1989). The psychology of Reading, Prentice-Hall, London.

Underwood, G., Clews, S. & Everatt, J. (1990). How do readers know where to look next? Local information distributions influence eye fixations. Q. J. Exp. Psychol. 42A, 39-65.

Vitu, F. (1991). The influence of parafoveal preprocessing and linguistic context on the optimal landing position effect. Perception and Psychophysics 50, 58-75.

Vitu, F., McConkie, G.W., Kerr, P.W. & O'Regan, J.K. The gaze duration on words. (In preparation).

Vitu, F., O'Regan, J.K., Inhoff, A.W. & Topolski, R. Mindless reading: Eye movement characteristics are similar in scanning strings and reading texts. (Paper submitted to Perception and Psychophysics).

Vitu, F., O'Regan J.K. & Mittau, M. (1990). Optimal landing position in reading isolated words and continuous text. Perception and Psychophysics 47, 583-600.

EFFECT OF LUMINANCE AND LINGUISTIC INFORMATION ON THE CENTER OF GRAVITY OF WORDS

Cécile Beauvillain & Karine Doré

Laboratoire de Psychologie Expérimentale, CNRS,
Université René Descartes, E.H.E.S.S.,
28 rue Serpente, Paris 75006, FRANCE

Abstract

Landing position in words was investigated as a function of the linguistic properties of words. The relative intensity of the word-initial or -final letters of 5-letter words was manipulated by keeping their sum constant. The data show that the orthographic structure of the word-initial letters modifies the global effect of luminance distribution on landing position. Whereas the eye is attracted by the word-initial or-ending high-contrasted letters in the case of redundant beginning words, a strong attenuation of the luminance distribution effect is observed for informative beginning words. It is shown that this effect of informativeness depends on the presence of orthographic irregular word-initial letters. The possible involvement of letter and word recognition mechanisms in early phases is discussed.

Keywords

Landing Position, Global Effect, Word Recognition, Orthographic structure, Reading.

Introduction

In reading, saccades are executed which bring the fovea to a new portion of the text where the information is integrated. The understanding of saccade programming is crucially important when elaborating a model of the construction of the internal representation of a linguistic stimulus. In the cases of saccades to a single word, the position of the target word on the retina provides a code for the position of the word in space and an eye rotation is executed to fixate the word. The structure of a linguistic stimulus offers a variety of potential target positions for the eye. For the selection of one position within the word, the preprocessing of parafoveal word information becomes an essential component of oculomotor behavior. The question addressed here concerns the nature of this parafoveal code and more specifically, how it can be modified by parafoveally available linguistic word information.

Work on saccade generation to simple visual targets has shown that saccade amplitude is generally determined in an automatic fashion by the parafoveal stimulation (Findlay, 1992; Becker, 1991). Word length is a low-level property that greatly influences the landing position. An inspection of landing positions within words shows a preference for fixating halfway between the beginning and the middle of word (O'Regan, 1979; Rayner, 1979). This location corresponds to the "center of gravity" of the word as used by Coren & Hoenig (1972) to describe the tendency of the eye to aim near the central position when a response is made to a configuration to which the eye was attracted. Interestingly, it has been shown that, when the one of two targets is made larger (Findlay, 1982), or more intense (Deubel, Wolf & Hauske, 1984), the saccade lands closer to that

Eye Movement Research/J.M. Findlay et al. (Editors)

target.   Findlay called this phenomenon the "global effect" and interpreted it in terms of the influence of the global characteristics of the amplitude of the saccade.   The explanation for these effects is perceptual and related to the processes involved in extracting the stimulus configuration. These findings may be accounted by the Becker & Jurgens' model (1979) that postulates that the saccadic computation system integrates stimulation over a large spatial and temporal window. Global factors influence the center of gravity calculation. In this calculation more weight is given to some elements as a function of their size, luminance and eccentricity. According to such a view, the deviation to the left of the geometric center of word should be a consequence of the fact that more weight is given to the nearer letters than to the more distant ones. In this type of functional relation, the eyes are attracted by the less eccentric letters that have been used to initiate the processing of a parafoveally presented word. Thus, the preferred viewing position in a word does not correspond to a position that coincides with the point where the eyes need to get further information about a word of which they have already processed the first letters. Rather, this preferred viewing position should be the consequence of the weight given to the first letters in the center of gravity calculation. Thus, if integration processes explain the global effect on the center of gravity, then it should be possible to modify the center of gravity of a word by increasing the intensity for some elements of the configuration. That is, it should be possible to modify the weight given to some letters of a target word to test precisely how these letters affect the landing position. The present experiment addresses this question by examining the effect of luminance distribution on the landing position in words. It is possible that the landing position within words only depends on "low-level" information extracted early in the parafovea. Thus, when manipulating letter intensity within words, the luminance distribution will deviate the center of gravity of words to the high-contrasted letters. It is also possible that the information derived from the parafoveally available word affects the computation of the saccade size. This hypothesis does not mean that the saccade programming has to contact lexical representations. Some linguistic stimulus characteristics, such as the orthographic regularity of the first letters, may affect saccade programming, as long as activation associated with such representations is available in the parafovea. If the acquisition of useful parafoveal word information affects the landing position, then the properties of a parafoveal sequence of letters should affect the positioning of the eyes on the target word. In experiments carried out in our laboratory, the stimulus consisted of an isolated target word. Consequently, the eccentricity of the target was precisely controlled as well as the moment of saccade triggering. The triggering of the saccade was delayed to improve the accuracy in the saccade amplitude computation by taking into account the complexity of the linguistic stimulus. In the experiments to be presented here, a 200msec saccade triggering delay was used in order to test precisely how the information extracted from a sequence of letters affects the landing position. During this delay, the word was presented $4^{0}30'$, that is 6 character spaces to the right of the fixation point before a saccade is initiated onto the word. If the saccade triggering is delayed beyond the spontaneous triggering, then more time is available for perceptual discrimination of the target word. This is consistent with a dual system that controls the metrics of saccades: a short-latency subsystem which has poor spatial resolution and lacks feature extracting capabilities, and a long-latency subsystem with good resolution and perceptual discrimination (Ottes, Van Gisbergen & Eggermont, 1985; Coeffe & O'Regan, 1987).

1. The effectiveness of saccade goal in the center of gravity of the target word

Our first experiment confirms that the concept of the center of gravity is also valid for various letter intensities. The intensity of the word-initial and -final letters was varied while keeping their sum constant. Five relative intensity conditions were used, the luminance of the initial letters relative to that of the whole word configuration being: 80%, 70%, 30%, 20% and a control condition of isoluminance (50%). Thus, in the 80% condition, the luminance of the three word-initial letters corresponded to 80% of the whole-word luminance whereas the two word-final letters represented 20% of the whole-word luminance. In the 20% condition, the luminance of the three word-final letters represented 80% of the whole-word luminance, whereas the two -initial letters corresponded to 20%. The levels of luminance were chosen such that the presence of the low-contrasted letters was nevertheless visible from the parafovea. (Detailed information about eye movement recording and data analysis are given elsewhere; Beauvillain, Doré & Baudouin, in preparation).

The experimental results obtained by 20 subjects reading orthographically regular 5-letter words, that is words composed of redundant word-initial and -final letters, is shown in Figure 1a. Whereas the eye lands at a position left of the geometric center of the 5-letter word in the isoluminance condition, it is attracted by the high-contrasted letters. For the different intensities, the data show a linear relation of landing position to relative letter intensity in such a way that the first orienting saccade lands relatively closer to the high-contrasted letters.

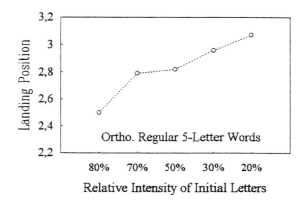

Fig. 1a - Mean landing position in character spaces on 5-letter words as a function of the relative intensity of the initial letters.

Figure 1b gives the means of the landing positions as a function of real saccade latencies, that is the true latencies without the imposed delay. This was done by partitioning the data on the basis of real saccade latencies. Clearly, the difference between the five relative intensity conditions occurs at almost all the points of the amplitude-latency distribution. Another interesting aspect of the distribution of landing positions is its time dependent nature. The mean landing positions vary with saccade latencies apparently converging toward a point slightly to the left of the center of words. This effect, present in isoluminance condition, is stronger when the high-contrasted letters are the word-initial letters. It suggests that if more time is given to the amplitude computation system, then more weight is given to the nearest letters. A similar pattern was observed by McConkie, Kerr, Reddix & Zola (1988) from saccades embedded in text reading. The authors reported that mean landing positions varied with prior fixation durations, converging toward a point left of the center of words. This was also found in Findlay's studies (1981; 1992) and seems in part to be a reflection of visual processing. The saccade endpoint that changes in time as word extraction proceeds is deviated toward the beginning of words. These data argue in favor of the notion of a center of gravity of the word weighted by intensity and eccentricity.

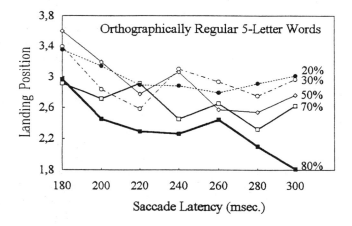

Fig. 1b - Landing position in character spaces on 5-letter words plotted as a function of the saccade latency for the different relative intensities of the initial letters.

2. The linguistic informativeness of word-initial letters influences the landing position

The saccade endpoint within words seems to be calculated mainly on the basis of low-level visual cues obtained in the parafovea, since such visual cues can be extracted rapidly. Thus, Vitu (1991) showed that the presence of weighting masked characters located above and below test-words influenced the eye's landing position within words. However, the saccade endpoint within words may be also influenced by feature-letter or letter analytic processes concerned with initial phases of the word recognition process. So, the letter sequence weight should depend critically on its informative content in a manner closely related to the lexical access process.

Parafoveal word information has been shown to influence the saccade size in reading situations (Rayner, Well, Pollatsek and Bertera, 1982; Morris, Rayner and Pollatsek, 1990; Inhoff, 1989). Such studies dealing with the parafoveal preview benefit effect show that masking letters of a parafoveally available word reduces the saccade size compared with no masking situations. The data are interpreted as showing that the eye is sent further into the word as some word information has been integrated in parafovea. However, when addressing the question of the critical information obtained from a whole-word preview, available studies have failed to find clear evidence about the source of word information that is integrated by the saccadic computation system. Initial letters of the word to the right of the word that is fixated might be identified and modify saccade computation as a function of the information derived from them. However, the available data do not allow the determination of the level of information of the initial letters that influences the saccade size.

The concern of the following experiment was to evaluate how the landing position within a word may be affected by the information derived from letters. If the letter integration process that is initiated in the parafovea is closely related to initial phases of the word recognition process, then the degree of informativeness of a sequence of letters should affect the positioning of the eyes on the target word. Moreover, the manipulation of the degree of informativeness of a letter sequence should give a better understanding of the effect of the acquisition of useful word information on the positioning of the eyes in a target word.

In Experiment 1, the effect of the weight of the initial and final letters was systematically tested, as these letters were clearly distinguished from the whole-word configuration. What happens if we vary the level of the informativeness of the word-initial and -final letters? In the following experiment, two types of 5-letter words were selected as a function of the lexical constraints imposed by the three initial and final letters. The "informative beginning/redundant ending words" begin with letters that are not shared by another word while their final letters are redundant. For instance, the French word CYGNE shares its initial trigram"CYG" with no word while it shares its final trigram "GNE" with many words. Such a word is informative by its initial letters that allow the word to be recognized. The "redundant beginning/informative ending words" end with trigrams that are not shared by another word while their initial letters are shared by many words. For instance, "CANIF" shares its final trigram "NIF" with no word while it shares its initial trigram "CAN" with many words. If saccades are affected by some pre-lexical or lexical information derived from word-initial or -final letters, then the level of informativeness of these letters should bias saccade computation.

The procedure was identical to that used in Experiment 1. However only three conditions of relative letter intensities were used: the 80%, and 20% condition and the   50% condition that corresponds to the control condition without contrast. 18 subjects were tested.

Figure 2a represents the mean landing positions as a function of the type of words and luminance conditions. Globally, the results replicate the linear relation of landing position to relative intensity observed in the previous experiment. Interestingly, the effect of luminance distribution interacts with the type of words, reflecting the fact that whereas, for redundant beginning/informative ending words, the center of gravity is deviated towards the high-contrasted letters, such an effect disappears for informative beginning/redundant ending words. More precisely, no deviation of the landing positions as a function of contrast condition is observed for the informative beginning/ redundant ending words. This attenuation of global effect of luminance when the initial letters are informative suggests that informativeness adds a weighting of its own in the center of gravity calculation that cancels the effect of luminance distribution. However, this attenuation is not observed when considering words that are informative by their final letters. For these words, the extraction of information about the final letters is not sufficient to modify the effect of luminance distribution.

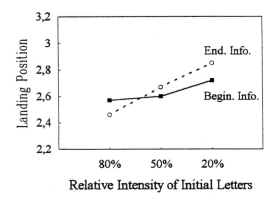

Fig. 2a - Mean landing position in character spaces on 5-letter words as a function of the relative intensity of the initial letters and the type of words. Words are informative by their first three letters ("Begin. Info.") or their last three letters ("End. Info.")

The Figure 2b shows that, for each type of words, the effect of luminance distribution on mean landing positions are observed at all points of the amplitude-latency distribution. For redundant beginning words (upper panel of Figure 2b), the configuration with 80% of relative luminance on the initial letters produces shorter saccades at all latencies whereas the deviation is not so clear when the final letters are more luminous. Clearly, the data for these words are similar to those observed in the preceding experiment (Figure 1b) when using orthographically regular words. Interestingly, a strong attenuation of the effect of luminance is observed for informative beginning/redundant ending words (lower panel of Figure 2b). For these words, the latency-amplitude distributions for the three luminance conditions tend to overlap. These data suggest that the informativeness of the beginning letters have biased the saccade computation.

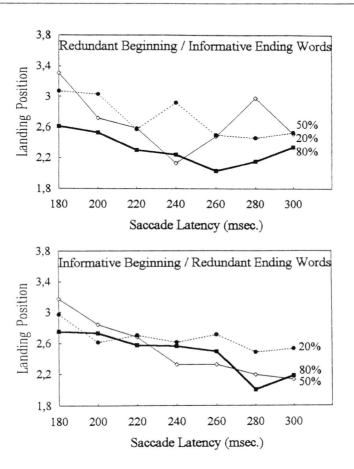

Fig. 2b - Landing position in character spaces on 5-letter words plotted as a function of the saccade latency for the different relative intensities. Upper panel concerns redundant beginning/informative ending words and lower panel concerns informative beginning/redundant ending words.

The results of this experiment demonstrate two points: first, the weight of the different letters of a target word depends on their eccentricity. When testing the informativeness location in words, strong effects of the initial letters are obtained. Second, the data suggest that the integration of the word-initial letters reflects letter extraction processes operating in the initial phases of word recognition. One possibility is that these letters have been recognized in parafoveal vision in order to activate lexical representations. According to such an hypothesis, identification of word-initial letters may yield sufficient lexical activation to affect the landing position as the informative value of these letters is unambiguous. However, it may be that the observed effects are the concern of intermediate coding between visual and lexical levels related to the orthographic encoding of the target word. Indeed, the informativeness of a letter sequence is correlated with the fact that such informative letters are unfrequent. For example, informative words are correlated with the presence of letter sequences such as "cyg" in "cygne", "ozo" in "ozone", that are very unfamiliar in French. Such unfrequent letter sequences may be particularly salient in that they generally differ markedly from the letter groupings used in a given language. Consequently, the weight of these informative letters could have modified the center of gravity calculation because they perceptually dominate in the word configuration.

To test whether or not the effects of informativeness on landing position are sensitive to the amount of orthographic irregularity of a letter sequence, we did an experiment which manipulated the degree of orthographic regularity of the three initial letters of a 5-letter words whereas the final letters were very frequent and redundant. Two types of informative beginning words were selected according to the orthographic regularity of the two initial bigrams: the mean frequency of use of the two first bigrams of irregular informative words such as "ozone" was very low (i.e. "oz"=1;"zo"=3), since the bigrams "oz" and "zo" are very unfrequent in French; the mean frequency of use of the two first bigrams of the regular informative words such "sosie" was high (i.e. "so"=27;"os"=21). For these two types of beginning informative words, the three initial letters were not shared by another word. A third type of redundant beginning word was included whose initial letters were shared by many 5-letter words; consequently they were orthographically regular by their initial letters (i.e. "salir" is composed of "sa"=47;"al"=61). The procedure and task were identical to that of the previous experiments. As this experiment tested the effect of the orthographic structure of the three initial letters we used only two contrast conditions: one condition of isoluminance and one condition where the relative intensity of the three initial letters was 80%.

If the effects of the informativeness of a letter sequence emerges as a correlate of the orthographic properties of the letter sequence in the language, then we expect an attenuation of global effects of luminance distribution only for orthographically irregular informative beginning words.

Figure 3 represents the mean landing positions obtained by 16 subjects for the three different types of words. The data clearly replicate the interaction observed in experiment 2 (Figure 2a) between the relative intensity and type of words when they differ in informativeness and orthographic regularity: the saccade endpoint is deviated to the most intense letters for redundant beginning words whereas no effect of luminance distribution is obtained for orthographically irregular informative words. Interestingly, the landing positions for orthographically regular informative words are located at a middle distance between the two other types of words. For these words, there is an effect of luminance distribution which is smaller than that observed for redundant words. The results suggest that the weight of informative word-initial letters is stronger when these letters are orthographically irregular than when they are regular.

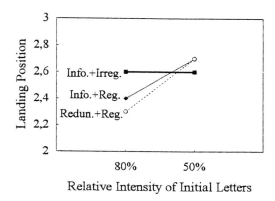

Fig. 3 - Mean landing position in character spaces on 5-letter words as a function of the relative intensity of the initial letters and the type of words determined by the first three letters. The first three letters are informative and orthographically irregular ("Info.+Irreg."), informative and orthographically regular ("Info.+Reg."), or redundant and orthographically regular ("Redun.+Reg.").

General Discussion

The data presented in this paper introduce a novel aspect in the investigation of saccade programming in words. When the eye saccades to a target word, processes related to early phases of word recognition can modify its tendency to saccade to a position determined by the global properties of the target word. When the word is composed of letter sequences which are very redundant in a given language because they are shared by many words, the visual word information is processed "globally" and the saccade depends on simple physical letter properties such as letter luminance. However, saccades may be affected by highly interesting properties of the linguistic stimulus, such as the informative value of the word-initial letters. These results suggest that the weight of a letter sequence at the beginning of word should be a function of its frequency of use in the language and its effectiveness to activate lexical representations. Indeed, orthographically irregular letter sequences are lexically unambiguous in that they give access to one unique lexical representation. Consequently they may modify the center of gravity calculation. The global effect observed in the case of redundant words can perhaps be regarded as a 'default option' operating when no other more elaborate form of processing is given by the subject to position his eye (Ottes & al., 1985). When signal processing related to letter discrimination and orthographic coding of a letter sequence can take place in the parafovea, then the weight of these highly informative letters is strong enough to affect the landing position. Interestingly, our data suggest that the salience of a letter sequence should be related to the effectiveness of an abstract coding related to lexical representations. However, much work has to be done in order to understand the properties that determine the salience of a letter sequence in the parafovea. Our work suggests that the salience of a letter sequence in saccade amplitude computation should be a direct function of its dissimilarity to the other letter sequences in a given language. Then the "weight" of an informative letter sequence in the calculation of the center of gravity seems to vary as a function of abstract orthographic structure -as long as lexical activation associated with such representations is available in the parafovea.

Our data suggest that the sensorimotor system can handle perfectly high-level visual processing related to letter discrimination processes and early phases of word recognition. This does not mean that voluntary influences direct the eyes to useful information (Underwood, Clews & Everatt's data, 1990; see Rayner & Morris, 1992 for a discussion). Obtaining letter information from a peripheral word preview affects the saccade amplitude computation in such a way that the saccade endpoint is modified by the letter information integrated in the parafovea. Our data suggest that the weight of the word-initial letters could depend on their informativeness correlated with the low frequency of use of orthographic letter sequences. This does not mean that the saccade computation system in reading has to contact the lexical representation. This may reflect an automatic operation of the recognition system that assigns more weight to a "salient" sequence of letters.

Conclusion

A considerable gap exists between simple models derived from the programming of saccades to isolated geometric targets and more complex linguistic information processing. In our experiments, the subject has to direct his eyes to a target word in which we systematically controlled the linguistic structure of word-initial and -final letters. In this situation, saccadic performance does not demand the recognition of the target word. Our work suggests that in some cases the saccade computation system has integrated some abstract orthographic properties of the target word at a level that allows the emergence of lexical influences.

References

Beauvillain, C., Doré, K. & Baudouin, V. Landing position in words. (in preparation).

Becker, W. & Jürgens, R. (1979). An analysis of the saccadic system by means on double step stimuli. Vision Research 19, 967-983.

Becker, W. (1991). Saccades. In Vision and Visual Dysfunction, Carpenter Ed., Macmillan Press, pp.95-137.

Coren, S. & Hoenig, P. (1972). Effect of non-target stimuli upon length of voluntary saccades. Perceptual and Motor Skills 34, 499-508.

Coëffé, C. & O'Regan, J.K. (1987). Reducing the influence of non-target stimuli on saccade accuracy: predictability and latency effects. Vision Research 27(2), 227-240.

Deubel, H., Wolf, W. & Hauske, G. (1984). The evaluation of the oculomotor error signal. In Theoritical and Applied Aspects of Eye Movement Research,. A.G. Gale, & F. Johnson Eds., Amsterdam: North Holland, pp. 55-62.

McConkie, G.W., Kerr, P.W., Reddix, M.D. & Zola, D. (1988). Eye movement control during reading: the location of initial eye fixations on words. Vision Research 27(2), 227-240.

Findlay, J.M. (1981). Local and Global Influences on Saccadic Eye Movements. In Eye movements: Cognition and Visual Perception D. F. Fisher, R.A. Monty, & J.W. Senders Eds., Hillsdale. NJ: Lawrence Erlbaum, pp.171-179.

Findlay, J.M. (1992) Programming of stimulus-elicited saccadic eye movements. In Eye Movements and Visual Cognition, K. Rayner Ed., Springer-Verlag New-York, pp.8-30. .

Findlay, J.M. (1982). Global visual processing for saccadic eye movements. Vision Research 22 1033-1045.

Inhoff, A.W. (1989). Parafoveal processing of words and saccade computation during eye fixations in reading. Journal of Experimental Psychology: Human Perception and Performance 15(3), 544-555.

Morris, R.K., Rayner, K., & Pollatsek, A. (1990). Eye movement guidance in reading: the role of parafoveal letter and space information. Journal of Experimental Psychology: Human Perception and Performance 16(2), 268-281.

O'Regan, J.K. (1979). Saccade size control in reading: evidence for the linguistic control hypothesis. Perception and Psychophysics 25, 501-509.

Ottes, S.P., Van Gisbergen, A.M. & Eggermont, J.J. (1985). Latency dependence of colour-based target vs nontarget discrimination by the saccadic system. Vision Research 25(6), 849-862.

Rayner, K. (1979). Eye guidance in reading: fixation location within words. Perception, 821- 30.

Rayner, K., Well, A.D., Pollatsek, A. & Bertera, J.H. (1982). The availability of useful information to the right of fixation during reading. Perception and Psychophysics 6, 537-550.

Rayner, K. & Morris, R.K. (1992). Eye movement control in reading: evidence against semantic preprocessing. Journal of Experimental Psychology: Human Perception and Performance 18(1), 163-172.

Underwood, G., Clews, S. & Everatt, J. (1990). How do readers know where to look next? Local information distributions influence eye movement. The Quaterly Journal of Experimental Psychology 42A(1), 39-65.

Vitu, F. (1991). The existence of a center of gravity effect during reading. Vision Research 31, 1289-1313.

PP-ATTACHMENT IN GERMAN: RESULTS FROM EYE MOVEMENT STUDIES[1]

Lars Konieczny, Barbara Hemforth, Christoph Scheepers, & Gerhard Strube

Institute of Computer Science and Social Research
Center for Cognitive Science
University of Freiburg, Germany

Abstract

In this paper we will present preliminary data on PP-attachment in German verb-second and verb-final structures. The results from our eye-movement studies will be shown to contradict the predictions of some well-known principles of human sentence comprehension. Parametrized Head Attachment will be introduced as a principle fully consistent with the data.

Keywords

Sentence processing, parsing principles, PP-attachment, verb-final structures

Introduction

Models of human sentence processing are usually claimed to be universally valid, independently of particular languages. A problem with this claim is that the empirical basis of such models is largely restricted to experiments on processing English. Only a couple of years ago, psycholinguistic research started to take different languages like Dutch (Frazier, 1987b), French (Mitchell, Cuetos, & Zagar, 1990), Spanish (Mitchell & Cuetos, 1991), German (Bader Lasser, in press; Hemforth, 1993; Konieczny, Hemforth, & Strube, 1991), or even Japanese (Mazuka, Itoh, Kiritani, Niwa, Ikejiru, & Naito, 1989; see Pritchett, 1992) into account. One of the main questions is, how well those models developed for processing English sentences fit to languages which vary in specific ways.

A syntactic feature of German, distinguishing it from English, is the placement of the verb at the end of the sentence surface-structure, as in perfect-tense sentences or subordinate clauses. As will be shown, these constructions present special problems to processing models such

---

[1] This research was supported by the German National Research Foundation (Deutsche Forschungsgemeinschaft, DFG, Str 301/4-1, "SOUL"). We want to thank Brona Collins and the reviewer for their helpful comments on an earlier draft of this manuscript.

as the Garden-Path theory (e.g. Frazier, 1987a) which are mainly based on the analysis of processing phenomena associated with structural ambiguities in English.

The Garden-Path Theory (Frazier & Rayner, 1982)

One of the most widely debated models of human sentence processing is the Garden-Path model (Frazier et al., 1982, Frazier, 1987a; 1991), which succeeded the well-known Sausage Machine (Frazier & Fodor, 1978; Fodor & Frazier,1980). The Garden-Path model is one of the main exemplars of a modular view of the language processing system in which an autonomous syntax module cannot be influenced by higher-level processes (semantics and pragmatics). The parser works incrementally word by word (left-to-right constraint) in a depth-first manner, with the result that, in cases of of structural ambiguities, only one analysis will be pursued (first analysis constraint). Measurable processing difficulties are caused by the initiation of reanalysis processes when the first analysis is unacceptable due to syntactic, semantic or discourse-pragmatic reasons.

In cases of structural ambiguties, two core principles of the model ensure that the most storage-efficient alternative will be chosen:

(p1)   *minimal attachment (MA)*
       Do not postulate any potentially unnecessary nodes.

(p2)   *late closure (LC)*
       Whenever possible, attach every new incoming item to the phrase currently being processed.

Frazier et al. (1982) found that in sentences such as (1), the [pp with binoculars] will preferably be attached to the verb-phrase (1a)[2]

(1)   *The spy saw the cop with binoculars.*
       a.The spy [VP saw [NP the cop] [PP with binoculars]]
       b.The spy [VP saw [NP [NP the cop] [PP with binoculars]]]

Taking a – certainly too – simple phrase structure grammar as a basis, it can be demonstrated by the syntactic alternatives for sentence (1) that this finding exactly meets the predictions of minimal attachment: When read from left to right, an attachment-conflict occurs on reading the article of the object-NP because it is possible either to attach this NP directly to the VP (1a) or to postulate a complex NP with an extra NP node (1b). However, because this extra NP node will possibly prove to be unnecessary in later processing, minimal attachment will hinder its construction in favor of a direct integration into the VP. Now, when at a later

---

[2] In some of our experiments on PP-attachment in German, we also found a slight preference for verb-attachment which was, however, strongly modified by verb-frame preferences (eg. Strube, Hemforth & Wrobel,1990).

stage the prepositional phrase is read, only integration of the PP into the VP (1a) can occur without reanalysis.

German verb-final constructions

For globally ambiguous verb-final sentences, minimal attachment predicts a preference for the verb-modifying reading, as it does for verb-second sentences. The alternative construction of a complex object-NP requires an extra preterminal node, independent of the verb placement.

In a couple of earlier experiments on reading German perfect-tense sentences and subordinate clauses we investigated whether or not the preferences predicted by minimal attachment can be found for these kinds of sentences too (Hemforth, Hölter, Konieczny, Scheepers, & Strube, 1991). In these experiments, word-by-word reading times (from the appearance of the word to the subject's key-press) were measured for sentences such as (2), which varied in the plausibility of the PP-attachment. World knowledge causes the instrumental reading in (2a), and the attributive reading in (2b) to be preferred. (2c) is ambiguous with respect to world knowledge.

(2)   a. *Ich habe gehört, daß Marion die Torte mit der praktischen Spritztülle verzierte.*
        I heard that Marion the cake with the icing funnel decorated.
        I heard that Marion decorated the cake with the icing funnel.

      b. *Ich habe gehört, daß Marion die Torte mit dem kräftigen Mokkageschmack verzierte.*
        I heard that Marion the cake with the mocca flavor decorated.
        I heard that Marion decorated the cake with the mocca flavor.

      c. *Ich habe gehört, daß Marion die Torte mit dem frischen Obst verzierte.*
        I heard that Marion the cake with the fresh fruit decorated.
        I heard that Marion decorated the cake with the fresh fruit.

The predictions of the Garden-Path Theory for these kinds of sentences depend on assumptions about the time course of thematic influences on sentences processing: If it is assumed that only the verb can hinder the attachment of a PP to the VP for semantic reasons, no differences in the processing of the PPs in (2a,b,c) should be determinable, because world-knowledge can only be fully employed after the subsequent verb has been read. But even if one accepts the incremental influence of thematic information on syntactic processing as in Frazier (1987a), it remains unclear if world-knowledge information about the prepositional object in (2) creates restrictions which are strong enough to force a reanalysis immediately after reading the PP. On the other hand, if the nouns of the PPs in sentences such as (2) can be judged as to whether or not the PP is a plausible (2a) or unplausible (2b) attribute of the direct object, certain processing difficulties are predictable within the PP itself: Minimal attachment (MA) hinders the construction of a complex object-NP in the first analysis of the sentence and thus the PP is integrated into the VP. When the PP-object is regarded as being an implausible attribute (2a), the analysis preferred by MA can be pursued. If, however, a PP-object is read which is a very plausible attribute (2b), then a reanalysis will occur at that position of the sentence, and the complex object-NP previously hindered by MA will be constructed. Therefore, the processing of prepositional objects should take longer in (2b) than in (2a).

None of these predictions could be confirmed empirically. In fact, the processing of an instrumental PP-object (2a) took significantly longer than the processing of a likely attributive PP-object (2b) or a semantically ambiguous PP-object (3c). In these experiments, processing difficulties therefore did not occur when world knowledge hinders a direct integration of the PP into the VP, but rather when it hinders the construction of a complex NP which, in sentences like (2a,b,c), is obviously preferred.

Parametrized Head Attachment

As an explanation of these results, Konieczny, Hemforth, and Strube (1991) offered the *head-attachment*-principle (p3a), which is incorporated in the three-part principle *parametrized head attachment* (p3; Konieczny, Scheepers, Hemforth, & Strube, 1994).

(p3)   *Parametrized Head Attachment (PHA)*

> a. *head attachment, HA (Konieczny, Hemforth & Strube, 1991)*
> If possible, attach a constituent g to a phrase with its lexical head already read.
>
> If further attachment possibilities exist for g, then
>
> a. *preferred role attachment*
> attach the constituent g to a phrase whose head provides a requested or expected theta- or place/time- role for g.
>
> If further attachment possibilities exist for g, then
>
> b. *recent head attachment*[3]
> attach the constituent g to the phrase whose lexical head was read most recently.

For verb-final constructions like (2a,b,c), *head attachment* (p3a) predicts that the PP or the NP, respectively, are preferentially attached to the left-adjacent noun-phrase. The lexical head of that noun-phrase, in contrast to that of the verbal phrase, has already been read, i.e., it is to be found to the left in the input string. If such an attachment later turns out to be implausible, a reanalysis will occur, resulting in additional processing load.

---

[3] Our principle of recent head attachment, like late closure does not account for the data from Spanish and French relative clause attachments in complex NPs (Mitchell, Cuetos & Zagar, 1990). Including new evidence from German, Hemforth, Konieczny, & Scheepers (forthcoming) will argue that these results are due to a "universal" mechanism for handling relative clauses.

Preferred role attachment (p3b), which is similar to Abney's (1989) *theta attachment principle* (but see Konieczny, Hemforth, Scheepers & Strube, 1994, for the differences) forces phrases like the PP in verb-second sentences to be attached to the head that can assign a (theta-) role to it. In most of the known cases its prediction coincides with that of *minimal attachment*, because it is usually the verb, not the object, which provides the theta-role for the constituent to be attached.

Finally, *recent head attachment* (p3c), a reformulation of *late closure* referring to the concept of heads, is proposed for cases where it is only permitted to attach to phrases whose heads are not distinguished with respect to their role-allocations, for example sentences like (3), where no role is given, or (4), where both noun and verb yield a *Beneficiary*-role. Here, an attachment to the phrase which has been read most recently can be observed.

(3) *Tom said Bill will think about it yesterday.*

(4) *Er organisierte die Hilfe für Rußland.*
     He organized the help for Russia.

Eye movement studies

The following sections describe some preliminary results of our recent eye-movement study which was designed to test the validity of *PHA*, the core principle of the *SOUL-Processor* (*Semantics-Oriented Unification-based Language Processor*; see e.g. Konieczny, Scheepers, Hemforth, & Strube, 1994) which has been developed in our research group. By collecting eye-movement data, we obtain results of higher ecological validity than by the reading time method: In particular, the subjects do not have to perform a secondary task, e.g. pressing a button, to read through the sentences. Most importantly, the eye-tracking method allows the measurement of regressive eye-movements which, in many cases, indicate quite directly the beginning of reanalysis processes (Just & Carpenter, 1987).

Method

Subjects: Twenty-four undergraduate students (native speakers of German) from the University of Freiburg were paid to participate in the study. All of them had normal, uncorrected vision and they were all naive concerning the purpose of the study. During an experimental session of 40 minutes, each of the subjects had to read 45 isolated sentences while her eye movements were monitored by a Dual Purkinje Eyetracker at the Institute of Computer Science and Social Research in Freiburg. 10 subjects had to be excluded from the analyses because of inaccuracies[4] or too many missing data. This is why the results have only preliminary status, since we still have to fill up the experimental design.

---

[4] Six subjects had to be deleted from the later analyses because they answered too many experimental questions incorrectly.

Materials and Design: The experimental sentences were manipulated according to a 2*2*2 within-subjects design (Table 1.) with the factors *verb placement* (final vs. initial), *lexical preference* (3-argument verb vs. 2-argument verb), and *world knowledge* (instrumental vs. attributive).[5] The order of the sentences was randomized. For each experimental condition, two sentences were presented, resulting in 16 target sentences per subject. These 16 sentences were taken from a pool of 24 sets of sentences different in content. The materials were rotated such that every sentence from every sentence set was presented to an equal number of subjects. One sentence had to be excluded because of too many missing data. Additionally, there were 29 filler sentences.

TABLE 1. Experimental design.

	verb-position: 2		verb-position: final	
	3-argument verb	2-arg. verb	3-argument verb	2-arg. verb
instrumental	5	7	9	11
attributive	6	8	10	12

(5) *Marion beobachtete das Pferd mit dem neuen Fernglas.*
     Marion watched the horse with the new binoculars.

(6) *Marion beobachtete das Pferd mit dem weißen Fleck.*
     Marion watched the horse with the white fleck.

(7) *Marion erblickte das Pferd mit dem neuen Fernglas.*
     Marion caught sight of the horse with the new binoculars.

(8) *Marion erblickte das Pferd mit dem weißen Fleck.*
     Marion caught sight of the horse with the white fleck.

(9) *Ich habe gehört, daß Marion das Pferd mit dem neuen Fernglas beobachtete.*
     I have heard, that Marion the horse with the new binoculars watched.

(10) *Ich habe gehört, daß Marion das Pferd mit dem weißen Fleck beobachtete.*
     I have heard, that Marion the horse with the white fleck watched.

---

[5] The materials were identical to those of the reading time experiments reported by Hemforth, Hoelter, Konieczny, Scheepers & Strube (1991).

(11) *Ich habe gehört, daß Marion das Pferd mit dem neuen Fernglas erblickte.*
      I have heard, that Marion the horse with the new binoculars caught sight of.

(12) *Ich habe gehört, daß Marion das Pferd mit dem weißen Fleck erblickte.*
      I have heard, that Marion the horse with the white fleck caught sight of.

Verb-2 sentences were presented in a single line on the computer screen, whereas verb-final sentences were presented in two lines, separated by two empty lines after the comma. This was done because verb-final sentences exceeded the maximum character space of the monitor.

*Procedure:*Prior to the experiment, the subject was fitted to a headrest to prevent head movements during reading. This was followed by a brief calibration procedure and a warming-up trial of five successive filler sentences. Then the experiment began which was built up of four trials. Each trial was initiated by a brief calibration procedure and contained 10 sentences - two filler sentences followed by eight randomly mixed target sentences or filler sentences, respectively. Before a sentence was presented, the subject had to fixate a cross-marking on the screen which indicated the start-position of the sentence-string. When the subject had finished reading the sentence, she pressed a button and the sentence disappeared from the screen. Each sentence of the experiment was followed by a simple yes/no-question which the subject was to answer by pressing one of two buttons (left-hand button: "yes", right-hand button: "no"). Subjects were told to read normally. The subjects whose data we included in our analyses answered with a high degree of accuracy (93%) which did not vary significantly across conditions.

*Apparatus:*The subjects' eye movements were monitored by a Generation 5.5 Dual Purkinje Eyetracker. Viewing was binocular, but eye movements were recorded only from the right eye. The eyetracker was connected to an AT 386 computer which controlled the stimulus-presentation and stored the output from the eyetracker for later data analysis. The time sampling rate for data collection was 1 KHz (one measure per millisecond). The sentences were presented on a 20-inch color monitor, beginning at the 6th column of the character matrix. The subject was seated 0.83 m from the face of the screen, so that 3 letters roughly equalled 1 degree of visual angle. To prevent disturbing light reflections, the monitor was covered by a tube, and the room was slightly darkened.[6]

Dependent Variables and Data Analyses: For our statistical analyses, the data were summarized for each word yielding three dependent variables, namely *first fixation duration, gaze duration*, and *total reading time*. First fixation duration refers to the time spent on fixating on a word for the first time during the normal left-to-right reading process without any refixations on the word.[7] Gaze duration refers to the amount of time that a reader looks at a word

---

[6.]The software we used is based on that originally developed by Charles Clifton and modified by Simon Garrod and his colleagues at Glasgow. We want to thank Simon very much for sharing his software with us.

for the first time before moving on to another word; it is the sum of all fixations on a word excluding any that result from regressions to the word. Finally, the total amount of time spent on a word including the time for re-reading was taken as the total reading time. In some cases the reader does not fixate a certain word at all, as is frequently observable for short, very frequent, or highly predictable words (e.g. Just & Carpenter, 1980; Rayner et al., 1989). These cases were excluded from data analyses. Consequently, the data reported in the following section represent *conditionalized fixation durations* (see Rayner et al., 1989).

Since we are primarily interested in the influence of sentential (syntactic and semantic) context on the processing of single words, we have to control for effects that occur only at the word level during lexical access. One very important factor which influences the processing of a word regardless of any context is the *word length*, defined by the number of character spaces of the word (Just & Carpenter, 1987). Therefore, most researchers in the field of reading either regard the word length as an additional covariate or eliminate its influence by computing a *reading time per character* measure. The latter computation, however, often implies an over-estimation of the word length. So we prefer to include word length as a covariate. Besides this, the word length is not the only source of influence on word-level processing. There are a number of other factors like *word frequency, concreteness* of a word etc. (see Just & Carpenter, 1987) which also should be taken into account. For that reason, if possible, we use the z-scored *lexical decision time* per word as a covariate in our statistical analyses which yields a more accurate estimate of word-level processing effects than word length alone.[8]

Hypotheses

Parametrized head attachment predicts different PP-attachment preferences, depending on the position of the verb. The critical constituent, at which a syntactic attachment preference at first can be detected is the PP itself, in particular the noun of the PP, because at this point sufficient world knowledge information becomes available for the thematic processor to determine the plausibility of the first analysis of the parser.

When the sentence is read from left to right, the head of the VP is still not available on the first reading of the PP in *verb-final* sentences. However, the head of the object-NP has already been read. Thus, rather than delaying the attachment to the end of the sentence, the PP should initially be attached to the object-NP (*head attachment*). If the thematic processor regards this preferred attachment implausible, as in the case of an instrumental PP-object, a *reanalysis* gets started, during which the alternative VP-modifying reading is generated. This initialization of a reanalysis should require some additional time for the processor. For that reason, the gaze duration on the noun of an instrumental PP should be reliably longer than on an attributive PP-object.

[7.]Only fixations of more than 10 msec were counted as fixations.

[8.]The lexical decision times were collected in an earlier experiment by Hemforth, 1993.

In *verb-2* sentences the lexical heads of both VP and object-NP have already been processed when the PP becomes available. Consequently, *preferred role attachment* causes the PP initially to be attached to the VP when the verb yields an (instrumental) argument role for it. If it comes to a clash between the preferred role attachment of the PP and world knowledge information, as is the case when the PP-object is an implausible instrument, a longer fixation duration should be observable which results from the initiation of a reanalysis. When the preferred reading of the verb does not require a prepositional argument, the PP is initially attached to the most recent head, i.e. the object noun. This causes the processor to initialize a reanalysis when an instrumental PP-object is read. Taken together, *PHA* predicts an interaction of *lexical preference* and *world knowledge* for verb-2 sentences with the pattern that the gaze duration on the PP-object should be longer if world knowledge contradicts either of the verb-dependent PP-attachment preferences.

Furthermore, an overall interaction of the factors *verb placement, lexical preference,* and *world knowledge* can be predicted which results from the differing PP-attachment preferences in the different conditions of verb placement.

Results

All reading time measures were submitted to a full factorial 2*2*2 analysis of variance for repeated measures including the factors *verb placement, lexical preference,* and *world knowledge.* For word-by-word analyses, the z-scored lexical decision times per word were included as an additional covariate. Note, that by analyzing *conditionalized fixation durations,* the degrees of freedom can vary over the conditions.

Parametrized head attachment. Table 2. shows the first fixation duration and gaze duration for the prepositional object by levels of verb placement and world knowledge. A significant interaction between verb placement and world knowledge resulted concerning the first fixation duration on the noun of the prepositional phrase. This interaction effect can be reduced to a significant world knowledge simple effect in verb-2 sentences (df=1,12; F1=9.25; p<.01; df=1,20; F2=4.30; p<.05): The first fixation on the PP-object takes longer if world knowledge demands an attributive interpretation. Comparing the world knowledge conditions in verb-final sentences revealed no reliable difference.

Regarding the gaze duration on the PP-object, the interaction of verb placement * world knowledge came up only as a tendency, and only by item-analyses. The linear contrasts for the two conditions of verb placement resulted in a marginal simple effect of world knowledge in verb-final sentences: The instrumental world knowledge condition showed a slightly longer gaze duration than the attributive world knowledge condition (df=1,12; F1=2.11; n.s;

df=1,20; F2=3.55; p<.08). No reliable difference between the world knowledge conditions was found in verb-2 sentences (F1, F2 < 1).

TABLE 2. First fixation duration and gaze duration (milliseconds) for the noun of the PP by levels of *verb placement* (final vs. second) and *world knowledge* (instrumental vs. attributive). The inferential statistics refer to the two-way interaction effects resulting from subject-analyses (F1) and item-analyses (F2), respectively.

	first fixation duration				
	verb-final	verb-2			
instrumental	219	248	df= 1,12	F1= 5.76	p< .04
attributive	202	353	(df= 1,21	F2= 4.11	p< .06)

	gaze duration				
	verb-final	verb-2			
instrumental	361	475	df= 1,12	F1= n.s.	
attributive	308	530	(df= 1,21	F2= 4.15	p< .06)

A significant three-way interaction of the factors verb placement, lexical preference, and world knowledge resulted for the gaze duration on the PP-object (see Table 3.). It refers to a significant interaction of lexical preference * world knowledge in verb-2 sentences (df=1,12;F1=7.50;p<.02; df=1,21; F2=4.71; p<.05). An examination of the different conditions in verb-2 sentences showed a reliable difference between the two world knowledge conditions in sentences with 3-argument verbs (df=1,12; F1=9.52; p<.01; df=1,21; F2=5.77; p<.03; all other contrasts: n.s.).

TABLE 3. Gaze duration (milliseconds) on the PP-object, related to *verb placement* (final vs. initial), *lexical preference* (3-arguments vs. 2-arguments), and *world knowledge* (instrumental vs. attributive). The inferential statistics refer to the three-way interaction.

	3-argument verb		2-argument verb	
	instrumental	attributive	instrumental	attributive
verb-final	360	342	361	274
verb-2	366	587	584	472

df=1,12; F1=5.73; p<.04; (df=1,21; F2=4.29; p<.05)

End-of-sentence effect. Table 4. shows the results of comparing fixation duration measures in the two conditions of *verb placement*. Whenever a constituent (the verb, the prepositional object, or the whole prepositional phrase) appeared at the end of the sentence[9], a significant increase of reading time resulted.

TABLE 4. Fixation duration measures (milliseconds, and milliseconds per character, respectively) by levels of *verb placement* for the verb, the PP-object, and the prepositional phrase. The inferential statistics refer to the respective main effect by subjects (F1) and by items (F2).

	verb placement				
	V-2	final	df1 (df2)	F1 (F2)	p1 (p2) <
verb					
first fixation duration	201	296	1,12 (1,21)	5.17 (6.94)	.05 (.02)
gaze duration	354	555	1,12 (1,21)	13.72 (37.47)	.01 (.001)
total reading time	604	799	1,12 (1,21)	3.32 (16.73)	.10 (.001)
noun of the PP					
first fixation duration	300	210	1,12 (1,21)	5.96 (11.52)	.04 (.01)
gaze duration	504	335	1,12 (1,21)	14.29 (32.78)	.01 (.001)
total reading time	751	581	1,12 (1,21)	4.43 (5.79)	.06 (.03)
region 4 (PP)					
first-pass RT/char	56	30	1,13 (1,22)	47.24 (66.54)	.001 (.001)
total RT/char	66	54	1,13 (1,22)	7.42 (4.33)	.02 (.05)

Regressive fixations. In what follows, some effects will be reported which could only be revealed for the *total reading time* measure on certain constituents. Since there are no corresponding effects in first fixation duration or gaze duration for these constituents, it can be concluded that they mainly result from re-reading the constituents.

A marginal three-way interaction effect of the factors *verb placement, lexical preference,* and *world knowledge* was found for the total reading time on the verb (Table 5.). However, examination of the different conditions revealed a significant two-way interaction between the factors *lexical preference* and *world knowledge* for sentences with *initial placement* of the verb (df=1,12; F1=4.53; p<.05; df=1,22; F2=5.98; p<.03): Significantly increased total reading times were found in conditions where *lexical preference* and *world knowledge* lead to diverging PP-attachment preferences in verb-2 sentences (*LP:3args/W:instr.* vs. *LP:3args/W:attr.:* df=1,12; F1=5.35; p<.04; df=1,11; F2=6.19; p<.03; *LP:2args/W:instr.* vs.

[9.]Note, that verb-2-placement implies final placement of the PP.

LP:2args/W:attr.: df=1,12; F1=3.89; p<.06; df=1,11; F2=1.13; n.s.; *LP:3args/W:instr.* vs. *LP:2args/W:instr.*: df=1,12; F1=2.96; p<.09; df=1,21; F2=3.70; p<.07; *LP:3args/W:attr.* vs. *LP:2args/W:attr.*: df=1,12; F1=4.37; p=.05; df=1,21; F2=3.10; p<.10). No reliable difference in total reading time was found for verb-final sentences.

TABLE 5. Total reading time (milliseconds) for the verb related to *verb placement* (final vs. second), *lexical preference* (3-arguments vs. 2-arguments), and *world knowledge* (instrumental vs. attributive). The inferential statistics refer to the three-way interaction.

	3-argument verb		2-argument verb	
	instrumental	attributive	instrumental	attributive
verb-final	778	823	791	802
verb-2	515	693	657	549

df=1,12; F1=1.99; p<.15; (df=1,21; F2=3.48; p<.08)

Another effect which reached significance by both subject-analyses (F1) and item-analyses (F2) was the two-way interaction between *lexical preference* and *world knowledge* regarding the total reading time per character for the object-NP (region 3) as is shown in Table 6. Linear contrasts showed that this interaction refers to the lower total reading time in the condition *lexical preference: 3-arguments, world knowledge: instrumental* compared with all other conditions (*LP:3args/W:instr.* vs. *LP:2args/W:instr.*: df=1,13; F1=8.76; p<.01; df=1,22; F2=8.49; p<.01, *LP:3args/W:instr.* vs. *LP:3args/W:attr.*: df=1,13; F1=7.51; p<.02; df=1,22; F2=6.56; p<.03, all other contrasts: F1, F2<1).

TABLE 6. Total reading time (milliseconds per character) for region 3 (object-NP) related to levels of *lexical preference* (3-argument verb, 2-argument verb) and *world knowledge* (instrumental, attributive). The inferential statistics refer to the two-way interaction.

	3-argument verb	2-argument verb			
instrumental	43	58	df = 1,13	F1 = 9.21	p<.01
attributive	59	53	(df = 1,22	F2 = 8.58	p<.01)

A similar interaction effect has been found when only the head of the object-NP was considered (Table 7.). The total reading time for the object noun was reliably shorter in the condition *lexical preference: 3-arguments, world knowledge: instrumental* than in the other conditions (*LP:3args/W:instr.* vs. *LP:2args/W:instr.*: df=1,10; F1=9.63; p<.02; df=1,18; F2=4.12; p<.05; *LP:3args/W:instr.* vs. *LP:3args/W:attr.*: df=1,10; F1=3.04; p<.09; df=1,9; F2=3.78; p<.09; all other contrasts: F1, F2 n.s.).

TABLE 7. Total reading time (milliseconds) for the direct object noun related to levels of *lexical preference* (3 arguments, 2 arguments) and *world knowledge* (instrumental, attributive). The inferential statistics refer to the two-way interaction.

	3-argument verb	2-argument verb			
instrumental	330	486	df = 1,10	F1 = 6.14	p<.04
attributive	504	405	(df = 1,19	F2 = 4.65	p<.05)

Discussion

Parametrized head attachment. Although the reported data have only a preliminary status yet, their general pattern seems to confirm the predictions of *PHA*. The most important measure to be considered here is the gaze duration on the noun of the PP since the first fixation duration reflects only a part of the processes which become relevant on that word with respect to the attachment preferences.[10] Regarding the gaze duration on the PP-object, a reliable overall interaction of the factors *verb placement, lexical preference,* and *world knowledge* was found which had been predicted by different preference principles in different verb placement conditions. For verb-2 sentences, a significant interaction of *lexical preference* and *world knowledge* was found, as was predicted by *preferred role attachment*: If the verb yielded an instrumental argument role for the PP, the gaze duration was significantly longer for an attributive PP-object than for an instrumental PP-object. In verb-final sentences, the gaze duration on the PP-object was slightly shorter in the attributive world knowledge condition than in the instrumental world knowledge condition, as it was predicted by *head attachment*. However, this effect did not reach significance in the reduced experimental design which we have already tested.

End-of-sentence effect. By comparing the fixation durations for constituents in different positions of the sentence revealed significantly higher fixation durations for constituents which were placed at the end of the sentence. This effect is consistent with the frequently reported sentence wrap-up effect which refers to processes during which a final interpretation of the sentence is computed after it has been read (see Just & Carpenter, 1980).

Regressive fixations. It has been found that in sentences where lexical preference and world knowledge conflict with each other regarding the attachment of the ambiguous PP, the total reading times on the heads of the attachment-alternatives, i.e. the object noun and the verb, tend to be higher. This is consistent with the prior findings of an interaction of lexical preference and world knowledge at the end of the sentence which can be referred to as a *discordance effect.* With the results from our eye-movement study, we now have some evidence that this effect refers to a selective reanalysis process in which the lexical heads of the attachment-alternatives are considered for a second time to come to a final interpretation. When the verb is placed at the end of the sentence, such selective re-readings only occur at the noun of the object-NP.

However, the total reading time effects on certain constituents reveal only a rough picture of the underlying reanalysis-processes, since they contain no information about the position from which the re-reading was started. In word-by-word presentation experiments the interaction of *lexical preference* and *world knowledge* was observed at the end of the sentence, indicating that a reanalysis takes place at this point where all relevant information of the two

---

[10.]Some researchers appear to use the first fixation duration on a word only as a measure for lexical access (e.g. Inhoff & Rayner, 1986).

factors becomes available. By using the word-by-word presentation method, the effects which result from a reanalysis must be locally limited to the point where the reanalysis starts, because of the impossibility of regression to prior passages. In the eye-movement data, we find selective regressions to the heads of the VP and object-NP, respectively. It might be the case that the effects in both types of studies refer to the very same processes which start at the end of the sentence. Thus, for a more accurate comparison of the eye-tracking data with the data from the word-by-word presentation method we need a measure that captures the whole amount of time spent for a reanalysis that begins at a certain position in the sentence. This is not only true for the *discordance effect*, but also for all other expected reanalysis-effects, especially those which start at the noun of the PP.

For future analyses, we will therefore consider a *first regression path duration* measure which is the duration of all regressive fixations beginning at a word (or region) until the word (or region) is fixated again.

Concluding remarks

In contrast to principles such as *minimal attachment* and *late closure/low attachment*, *parametrized head attachment* does not appeal to structures underlying the sentence but to the lexical information of heads alone and their ordering on the surface of the sentence. *Parametrized head attachment* is thus independent of presumptions about syntactic structures.

While the "deeper meaning" of principles such as *minimal attachment* and *late closure/low attachment* is to guarantee the least possible memory load by constructing structures as shallow as possible, *head attachment* achieves an improvement by allowing earlier semantic integration because it suggests attachments which are dependent on the presence of lexical heads. It is this aspect of economical processing, i.e., of achieving the earliest interpretation possible, which realizes the central idea behind the model presented here, namely a *semantics oriented syntax processing*. Note, however, that *parametrized head attachment* is supposed to guide the initial analysis without referring to semantic or pragmatic information. Thus, *semantics-oriented* processing must be distinguished from *semantics-driven* accounts. *Parametrized head attachment* works merely at the syntactic level in order to provide a semantic interpretation as fast as possible, therefore being *semantics-oriented*.

References

Abney, S. (1989). A computational model of human parsing. Journal of Psycholinguistic Research, 18.1, 129-144.

Bader. M. & Lasser, I. (in press). German Verb-Final Clauses and Sentence Processing: Evidence for Immediate Attachment. In C. Clifton, L. Frazier & K. Rayner (Eds.), *Perspectives in Sentence Processing*.

Fodor, J. D., & Frazier, L. (1980). Is the HSPM an ATN? *Cognition, 8*, 417-459.

Frazier, L. (1987a). Sentence processing: A tutorial review. In M. Coltheart (Ed.), *The psychology of reading* (pp. 559-586). Hove/London/Hillsdale: Lawrence Erlbaum.

Frazier, L. (1987b). Syntactic processing: evidence from Dutch. *Natural Language & Linguistic Theory, 5*, 519-559.

Frazier, L. (1991). Exploring the architecture of the language-processing system. In G. Altmann (Ed.), *Cognitive Models of Speech Processing* (pp. 409-433). Cambridge, Mass.: MIT Press

Frazier, L., & Fodor, J. D. (1978). The sausage machine: a two stage parsing model. *Cognition, 6*, 291-325.

Frazier, L., & Rayner, K. (1982). Making and correcting errors during sentence comprehension: eye movements in the analysis of structurally ambiguous sentences. *Cognitive Psychology, 14*, 178-210.

Hemforth, B. (1993). *Kognitives Parsing: Repräsentation und Verarbeitung sprachlichen Wissens.* Sankt Augustin: Infix.

Hemforth, B., Hölter, M., Konieczny, L., Scheepers, C., & Strube, G. (1991). Kognitive Modellierung und empirische Analyse von Prozessen der Satzverarbeitung. *4. Zwischenbericht im DFG-Schwerpunktprogramm Kognitive Linguistik.* Ruhr-Universität Bochum.

Hemforth, B., Konieczny, L., & Scheepers, C. (forthcoming). On the universality of parsing principles.(will appear in *IIG-Berichte: Reanalysis and Repair II).*

Inhoff, A. W., & Rayner, K. (1986). Parafoveal word processing during eye fixations in reading: Effects of word frequency. *Perception and Psychophysics, 40*, 431-439.

Just, M. A., & Carpenter, P. A. (1980). A theory of reading: From eye fixations to comprehension. *Psychological Review, 87,* 329 - 354.

Just, M. A., & Carpenter, P. A. (1987). *The psychology of reading and language comprehension*. Boston: Allyn & Bacon.

Konieczny, L., Hemforth, B., & Strube, G. (1991). Psychologisch fundierte Prinzipien der Satzverarbeitung jenseits von Minimal Attachment. *Kognitionswissenschaft, 2.*

Konieczny, L., Scheepers, C., Hemforth, B., & Strube, G. (1994). Semantikorientierte Syntaxverarbeitung. In S. Felix, C. Habel, & G. Rickheit (Eds.), *Kognitive Linguistik: Repräsentation und Prozesse* (pp. 129-158). Opladen: Westdeutscher Verlag.

Mazuka, R., Itoh, K., Kiritani, S., Niwa, S., Ikejiru, K. & Naito, K. (1989). Processing of Japanese garden path, center-embedded, and multiply-left-embedded sentences. *Annual Bulletin of the Research Institute of Logopedics and Phoniatrics, 21.*

Mitchell, D. C. & Cuetos, F. (1991). The origins of parsing strategies. In C. Smith (Ed.), *Current Issues in Natural Language Processing.* Austin, Texas: Center for Cognitive Science, University of Texas.

Mitchell, D. C., Cuetos, F. & Zagar, D. (1990). Reading in different languages: Is there a universal mechanism for parsing sentences? In D. Balota, G.B. Flores d'Arcais & K. Rayner (Eds.), *Comprehension Processes in Reading* (pp. 285-302). Hillsdale. NJ: Erlbaum.

Pritchett, B. (1992). *Grammatical competence and parsing performance.* Chicago: University of Chicago Press.

Rayner, K., Sereno S. C., Morris, R. K., Schmauder, A. R., & Clifton, C. (1989). Eye movements and on-line language comprehension processes. *Language and Cognitive Processes, 4,* 21-50.

Strube, G., Hemforth, B., & Wrobel, H. (1990). Resolution of Structural Ambiguities in Sentence Comprehension: On-line Analysis of Syntactic, Lexical, and Semantic Effects. *The 12th Annual Conference of the Cognitive Science Society.* Hillsdale, NJ: Erlbaum, pp. 558-565.

INDIVIDUAL EYE MOVEMENT PATTERNS IN WORD RECOGNITION: PERCEPTUAL AND LINGUISTIC FACTORS

Ralph Radach, Jo Krummenacher, Dieter Heller & Jörg Hofmeister

Institute of Psychology, Technical University of Aachen,
Jaegerstrasse 17, 52056 Aachen, Germany

Abstract

The present paper deals with a central question of research on eye movement control in reading and word recognition: the determination of initial fixation positions within words. Four subjects were asked to perform a semantic decision task on words presented at different eccentricities left vs. right of a central fixation point. An individual analysis of primary saccade landing site distributions and refixation frequency curves suggests that subjects may use different visual processing strategies in solving the task. The redundancy of word beginnings as determined by the frequency of word forms with the same initial letter trigram had a significant effect on response times but no effect on primary saccade landing sites. This result provides further evidence against a role of parafoveal preprocessing on saccade amplitudes and landing sites in word recognition.

Keywords

Word recognition, eye movement control, preferred landing position, parafoveal preprocessing.

Introduction

One of the central issues of current research on eye movement control in reading is what factors determine the positions of initial fixations within words. A basic finding on this subject is that fixations tend to cluster at locations left of the center of words (Dunn-Rankin 1978). This phenomenon, generally referred to as the "preferred landing position" (Rayner 1979) was confirmed in a number of reading studies (Kliegl, Olson & Davidson 1983, Inhoff 1989, Vitu, O'Regan & Mittau 1990, Kennedy & Murray 1991, Radach & Kempe 1993).

McConkie, Kerr, Reddix & Zola (1988) have shown that in text reading initial fixation positions depend strongly on where the saccade into the critical word was launched. The position initially fixated within a word is shifted to the left by about 0.5 letter for each increment of launch distance. They relate this finding to a general "saccadic range error" (Kapoula 1985) that is made responsible for the apparent deviation between mean landing sites and a "functional target location" assumed to be located at the word center. This interpretation was challenged by Vitu (1991b) who in a word recognition experiment with words presented at different eccentricities to the right of a central fixation point found no systematic effect of target eccentricity on landing sites of primary saccades into the word.

The functional base of the preferred landing position is assumed to be related to a second phenomenon, the "optimal viewing position" (O'Regan 1990). When words are presented for short durations, recognition performance is maximal when the viewing position is slightly left of the word center (Nazir, Heller & Sußmann 1992). When presentation is self-paced, word recognition time is minimal when a position close to the word center is intially fixated, mostly because in this case refixations on the same word are least likely. O'Regan (1990) asserts that during reading the eye is deviated from landing at the optimal viewing position by a cortically weighted "center of gravity effect" or "global effect" of the visual configuration formed by the upcoming text (Findlay 1981, 82, but see also Findlay, Brogan & Wenban-Smith 1993 for a modified view on the "global effect"). This position was further elaborated by Vitu (1991a) to the notion that the eyes tend to land at the cortically weighted center of gravity within a critical peripheral region that includes about seven letters from the beginning of a target word.

Rayner & Morris (1992) favour a different account for the preferred landing position. On the basis of a large body of evidence on the utilization of peripheral letter information during reading they suggest that "readers may often move their eyes to a position in a word that coincides with the point at which they need to get further information about the word given that they have already processed the first few letters of the word on the prior fixation". For such a mechanism to work there does not need to exist on-line feedback about the results of parafoveal processing. Alternatively, a visual strategy could be developed to send the eyes to a location where the acquisition of letter information can be expected to be most effective.

If target words of medium length (i.e. seven letters) are presented eccentrically to the left vs. right of a central fixation point and subjects are asked to identify the words, on the basis of the above discussion two predictions can be made:

1. One could assume that when programming the primary saccade the word should in any case be treated as a low level (low spatial frequency) target. Under the assumption that the "optimal viewing position" is generally located near the word center, and saccades tend to undershoot this position, landing site distributions should be symmetric: If in the case of a right hand presentation saccades would land at letter position 2 to 3, they should land at letter positions 5 to 6 in left hand presentation. (If the optimal position is assumed to be slightly left of the word center, a symmetric undershoot would result in the distribution for the words presented to the left hand to be shifted a bit to the left). This hypothesis of "symmetric" landing position distributions will be further referred to as the "low level hypothesis".

2. A second possible hypothesis would be that over a certain period of time a strategy is being adopted to bring the eye to a location where a maximum of lexical information can be extracted irrespective of where the critical word is presented. In this case the eye tend to land close to a position optimal for solving the cognitive task required in the experiment and landing site distributions for left hand and right hand presentation can be expected to be very similar. This hypothesis will be referred to as the "processing adaptation hypothesis".

Our experiment was carried out primarily to test these general hypotheses. A second objective of the experiment is closely related to the current discussion about what role linguistic factors may play in determining saccade amplitudes and initial fixation positions during reading. Despite considerable debate on details of eye movement control models most researchers in the area seem to agree that the position to be fixated next is generally selected on the basis of low-level parafoveal information. Initial landing positions are believed to be determined by this basic visual information, perhaps with some mediation by oculomotor constraints (McConkie, Kerr, Reddix & Zola 1988, Rayner & Morris 1992, O'Regan, Vitu, Radach & Kerr, in press).

Underwood and his co-workers recently summarized a differing position as follows: "The reader's eyes are said to move to the next fixation according to the state of the comprehension calculation and according to syntactic and semantic variations in the text ahead of the eyes...The strongest version of this hypothesis suggests that the eyes are attracted to linguistically informative parts of the text following parafoveal preprocessing, and this hypothesis continues to find empirical support" (Underwood & Everatt 1992, p. 127).

The empirical support for this "parafoveal guidance hypothesis" comes from studies where the "distribution of information" within words was determined by dictionary counts or using a word-completion test. In these experiments subjects read words with informative vs. redundant ending and/or beginning either in isolation or embedded in sentences. When the landing positions of initial saccades were analysed, the typical finding was that the eyes go further into words with "redundant" beginnings. In a recent review Hyona (1993) lists several such studies in which an effect of "informativeness" on the initial landing position was found.

With respect to the functional basis for these results there seem to be different views. Hyona (this volume) modified the original interpretation of a parafoveal "semantic preprocessing" effect in favour of a more moderate view including the idea that the observed effects may be due to the "orthographic saliency" of word beginnings. These ideas as well as the original notion of "lexical informativeness" were put to a test in our experiment by studying the effects of frequency and redundancy of word initial letters on saccade amplitude.

Methodology

Four unpaid subjects with normal vision participated in the experiment. They read 7-letter nouns from a CRT-screen and indicated the grammatical gender of the word by pressing a response button. To prevent possible effects of word shape target words were displayed in capital letters. Each trial began with the presentation of a central fixation mark consisting of random letters changing at a rate of 3 per second. When the subject fixated these letters, he or she pushed a button and a fixation cross appeared for a period of about 500 ms. The target word was displayed until a second button press indicating the response terminated the trial.

Target words were presented at five positions: peripheral left and right, parafoveal left and right, and central. In central presentation trials, the location of letter four of the seven-letter target word corresponded to the location of the fixation mark. In the eccentric conditions word beginnings were 11 vs. 19 letters left and 4 vs. 12 letters right of the fixation mark. These positions will be further referred to as "peripheral left", "parafoveal left", "parafoveal right" and "peripheral right". The width of one letter corresponded to 0.33 deg of visual angle.

X

BERATER VORTRAG SCHELLE ZYNIKER PAZIFIK

Figure 1. Illustration of the spatial arrangement of fixation cross and target words positions in the experiment. Presentation positions are separated by one empty space. Only one target word at a certain position was presented during a single trial. The fixation cross has been vertically displaced in the figure to enable its position to be seen more clearly.

Stimuli were displayed on a Sony GDM 2040 21 inch black Trinitron monitor as white text on dark background. An AMTech infrared pupil reflection eye tracker sampling at 200 Hz was used to collect eye movement data during the experiment. As determined in an independent series of experiments, the recording system has a spatial resolution (reliability of fixation position in absolute coordinates) of about 0.25 deg. Saccades were identified using a simple velocity threshold algorithm as described by Lemij and Collewijn (1989). The velocity threshold for including a data point into a saccade was set to 15 deg per sec., and the minimum saccade amplitude was 0.25 deg.

A sample of masculine and feminine nouns was selected using the German CELEX-corpus provided by the Max-Planck-Institute of Psycholinguistics in Nijmegen. Word frequency was restricted to a range between 1 and 100 per million. For each word, we determined the frequency of the word initial letter trigram (trigram frequency) as well as the number of word forms beginning with the same trigram (trigram redundancy). Words were assigned to the final list of 500 nouns such as to minimize the word frequency difference between items of 4 ranges of trigram redundancy that corresponded to the quartiles of the trigram redundancy distribution. The mean word frequency for the four trigram redundancy bands ranged from 11 to 19 per million and mean trigram redundancy values within the bands were 14, 37, 79 and 270. The correlation between word frequency and trigram frequency was 0.06. Each subject completed 20 blocks with 100 trials each, displayed at fixed random presentation positions.

Results

Landing sites of primary saccades

Included in the analysis of data were all trials that did not contain blinks and in which the primary saccade did not miss the target word by more than one letter position. There were 98.1 %, 96.2 %, 98.2 % and 98.6 % of such valid trials in subjects 1 to 4, respectively. Figure 2 presents individual primary saccade landing site distributions for all 4 eccentricities and table 1 gives the corresponding means and standard deviations. The distributions are all close to gaussian in shape and their relatively small standard deviations suggest the existence of a narrow saccade target area within the critical word.

subject	peripheral left	parafoveal left	parafoveal right	peripheral right
1	4.18 (1.07)	3.70 (0.69)	2.52 (0.65)	1.49 (0.88)
2	5.12 (1.08)	4.59 (0.88)	2.93 (0.81)	2.74 (0.94)
3	2.75 (1.24)	2.76 (0.85)	2.81 (0.75)	2.81 (1.16)
4	3.14 (1.04)	2.18 (0.72)	2.99 (0.69)	2.36 (0.91)

Table1. Means and standard deviations of landing sites of primary saccades into target words presented at different positions.

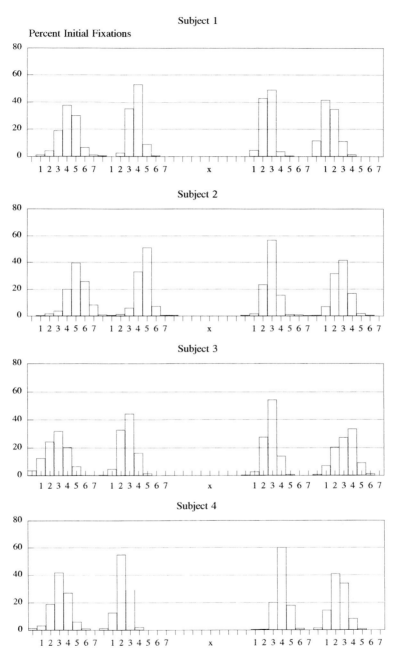

Figure 2. *Individual landing side distributions of primary saccades for all 4 subjects. The distributions are arranged in accordance with presentation positions and numbers on the abscissa correspond to letter positions within target words.*

To allow for an easier assessment of the data with respect to the hypotheses stated above, the means of all individual landing site distributions and mean landing position averaged across subjects are presented in Figure 3.

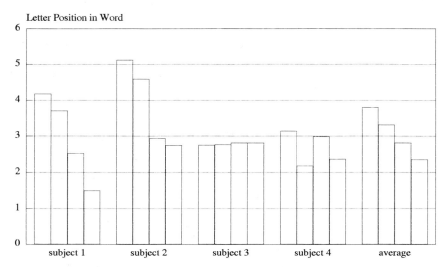

Figure 3. Mean landing positions of primary saccades into words presented at different positions. For each subject, the means are arranged in the following order: peripheral left, parafoveal left, parafoveal left, parafoveal right and peripheral right.

When only the averaged values are taken into account, the data seem to indicate that the "low level hypothesis" nicely accounts for the data. A straightforward interpretation of the observed pattern of results would be that primary saccades undershoot a position at letter 3 by a certain amplitude depending on the distance between fixation mark and target word. An ANOVA of averaged mean landing sites indicates a significant main effect of presentation position [F (1,3) = 450.4, p<0.01] with all contrasts between the respective means also being highly significant.

When looking at the individual data, a different picture emerges. The data of subject 1 and subject 2 are as predicted by the "low level" - hypothesis. Their intial landing sites appear "symmetric" and seem to undershoot a target position slightly left of the word center. Nonparametric Kruskal-Wallis ANOVAs show significant effects of presentation position on landing sites (chi² = 951.4, p<0.01 for subject 1 and chi² = 865.4, p<0.01 for subject 2). As indicated by pairwise T-tests, in both subjects there are as well significant differences between mean landing sites for left and right peripheral vs. parafoveal presentation.

In subject 3 however, there are no significant differences in landing sites between conditions (chi² = 1.75, n.s.). The largest pairwise difference is 0.06 letters between peripheral left and peripheral right presentation (t = 0.70, n. s.). Wherever target words are presented, the mean landing site of the primary saccade is always very close to the third letter position. This result can be accounted for by the "processing adaptation hypothesis".

In subject 4 the general effect of position on primary saccade landing site is again significant (chi^2 = 282.4, p<0.01). Compared to subject 1 and 2 however, there is one marked difference: for parafoveal left presentation, landing sites are considerably shifted to the right. The size of this effect is quite remarkable, since mean landing site for parafoveal left presentation is located even right of landing site for peripheral right presentation (t = 3.03, p<0.01). All other pairwise differences are also significant in this subject.

In order to show that primary saccade landing sites are indeed relevant for solving the semantic decision task required in the experiment, probabilities of refixating the target word have been plotted as a function of landing sites. Figure 4 presents individual refixation probability curves for all four presentation conditions.

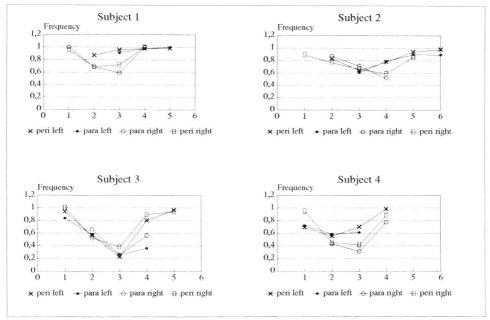

Figure 4. Relative frequency of refixating the target words as a function of primary saccade landing site for peripheral left and peripheral right presentation.

Except for the left hand presentation conditions in subject 1, where a ceiling effect is apparent, the curves are u-shaped with a clear minimum between letter position 2 and 4. In two presentation conditions for subject 2 there are only rudimentary curves because observations for the rest of the curve are missing in the respective landing site distributions. When the minima of the complete refixation curves are assumed to be an estimate of the desired saccade target position and compared to mean landing sites, an interesting observation can be made. In subjects 1,2, and 4 primary saccades clearly tend to fall short of this position, whereas in subject 3 the mean of the landing site distribution and the minumum of the refixation curve are essentially identical.

Influence of word beginning "informativeness" on response times and saccades

The notion of "lexical informativeness" of word beginnings has been operationalized in the present study by selecting target words with very different levels of their initial letter trigram redundancy as determined by the number of word forms with the same inital letter trigram. This measure is in good accordance with the definition of word beginning informativeness as proposed by Underwood, Clews & Everatt (1990). Table 2 presents the results of an analysis of semantic decision times for words in ranges 1 to 4 of the trigram redundancy distribution formed by the items of our word sample.

Subject	Quartile 1	Quartile 2	Quartile 3	Quartile 4	Chi² (K-W)	T-test (1 - 4)
1	644 (113)	671 (115)	673 (103)	690 (112)	9.52 (p<0.05)	2.77 (p<0.01)
2	699 (132)	746 (132)	764 (123)	773 (122)	21.4 (p<0.01)	3.88 (p<0.01)
3	596 (85)	663 (98)	656 (109)	657 (120)	26.15 (p<0.01)	4.04 (p<0.01)
4	648 (106)	660 (87)	670 (99)	655 (92)	4.73 (n. s.)	0.48 (n. s.)

Table 2. Mean semantic decision times as a function of intial trigram redundancy quartiles as well as results of nonparametric Kruskal-Wallis ANOVAs and T-tests for quartile 1 vs. quartile 4. Data for parafoveal right presentation.

The data summarized in table 2 indicate that in 3 out of our 4 subjects there is a large and significant effect of word beginning redundancy on semantic decision times. The size of this effect ranges from 46 ms to 61 ms. In subject 4 there is a similar tendency, but the effects fail to reach significance. It should be noted however, that in central presentation (where the initial fixation is always at letter position 4 within the target word) there were significant effects for all 4 subjects, including a difference of 26 ms between trigram frequency quartiles 1 and 4 for subject 4.

From these data it can be concluded that the variation of word beginning redundancy in the present experiment had a considerable effect on solving the semantic decision task. The fact that responses on target words with redundant beginnings like VORGABE or SCHELLE take longer than those on words like ZYNIKER or PAZIFIK cannot be attributed to the (relatively small) difference in word frequency, because more frequent words are generally recognized faster. The oberserved effect is likely to be due a cost associated with selecting within a greater neighborhood of activated lexical entries (see i.e. Paap, Newsome, McDonald & Schvaneveldt 1982) when making the grammatical gender decision.

Table 3 presents results of an analysis of initial word trigram redundancy on primary saccade length in degrees of visual angle. As suggested by Inhoff (1989), saccade amplitude was chosen as the dependent variable here in order to allow small saccade size changes to be noticed that might not have been picked up in the letter position metric. For this analysis, saccade amplitudes were sampled in units of 0.01 deg in order to take advantage of the higher precision of our recording apparatus with respect to measuring relative rather than absolute position.

Subject	Quartile 1	Quartile 2	Quartile 3	Quartile 4	Chi² (K-W)	T-test (1-4)
1	2.11 (0.20)	2.13 (0.19)	2.15 (0.18)	2.14 (0.20)	2.73 (n. s.)	1.23 (n. s.)
2	2.37 (0.34)	2.33 (0.24)	2.28 (0.25)	2.31 (0.26)	7.43 (n. s.)	1.25 (n. s.)
3	2.35 (0.22)	2.27 (0.21)	2.30 (0.24)	2.25 (0.22)	7.70 (n. s.)	3.18 (p<0.01)
4	2.39 (0.22)	2.37 (0.21)	2.36 (0.18)	2.34 (0.22)	2.35 (n. s.)	1.79 (n. s.)

Table 3. Mean amplitudes of primary saccades into target words as a function of initial trigram redundancy quartiles as well as results of nonparametric Kruskal-Wallis ANOVAs and T-tests for quartile 1 vs. trigram quartile 4. Data for parafoveal right presentation.

There is no significant main effect of initial letter trigram redundancy on mean saccade amplitude in any of the subjects. There appear to be small differences in subject 3 between saccade amplitudes for quartiles 1 and 4 of the trigram redundancy distribution. But these differences are actually opposite to the hypothesis that saccades go further into words with more redundant beginnings.

To test a second variant of the "parafoveal guidance hypothesis" recently proposed by Hyona (this volume), we partitioned our target word sample post hoc into 4 ranges of "orthographic saliency" according to the quartiles of the initial letter trigram frequency distribution. Over these frequency ranges, mean trigram frequency increased from 146 to 7712 (a factor of more than 1 to 50). Table 4 shows the results of the respective analysis of individual primary saccade amplitudes.

Subject	Quartile 1	Quartile 2	Quartile 3	Quartile 4	Chi² (K-W)	T-test (1-4)
1	2.09 (0.29)	2.14 (0.19)	2.15 (0.19)	2.14 (0.19)	4.78 (n. s.)	1.75 (n. s.)
2	2.34 (0.34)	2.34 (0.27)	2.29 (0.22)	2.32 (0.27)	2.84 (n. s.)	0.08 (n. s.)
3	2.34 (0.24)	2.27 (0.21)	2.30 (0.21)	2.26 (0.23)	7.01 (n. s.)	2.50 (p<0.05)
4	2.41 (0.23)	2.34 (0.19)	2.38 (0.19)	2.34 (0.21)	6.49 (n. s.)	2.37 (p<0.05)

Table 4. Mean amplitudes of primary saccades into target words as a function of initial trigram frequency quartiles as well as results of nonparametric Kruskal-Wallis ANOVAs and T-tests for quartile 1 vs. trigram quartile 4. Data for parafoveal right presentation.

The results summarized in table 4 are very similar to the data discussed above. There is no main effect of initial letter trigram frequency on primary saccade amplitude. The observed small differences in subject 3 and 4 are again not in accordance with the hypothesis. This similarity of results is not very surprising given the fact that initial letter trigram frequency ("orthographic saliency") and number of words with the same initial trigram ("word beginning informativeness") are highly correlated ($r = 0.92$). Taken together, the data provide no support for either version of the "parafoveal guidance hypothesis".

General discussion

Up to now intial fixation positions in word recognition and reading have been studied in most cases by averaging across subjects. Only recently have there been attempts to analyse individual data and to relate "preferred" and "optimal" viewing positions in single subjects (Kerr & McConkie 1992, Radach & Kempe 1993). In an analysis of a large body of reading data we have recently identified several factors that determine or mediate initial fixation positions. Among these are the surrounding configuration of words as well as the fixation pattern on the prior word when saccade launch site is held constant. Interestingly, there is also a large influence of the position of the critical word on a line of text, providing further evidence for the role of the general spatial layout of text on eye movement parameters (Heller 1982). In general, there are marked interindividual differences in preferred landing positions, whereas fixation probability curves appear relatively stable, with minima slightly left or right of the word center (Radach & Kempe 1993).

The results of the present study show that considerable interindividual differences can be found in a very simple word recognition task as well. They further suggest that subjects may use different visual information processing strategies to solve such a task: In two subjects the obtained results can be explained by a "low-level" strategy. The word is taken as a low-level visual target and the landing sites of primary saccades into the word tend to undershoot the target location as estimated by refixation probability curves. These results correspond to the idea of a "center of gravity" effect of the present visual configuration deviating the saccade from the desired target (Findlay 1981, 82, O'Regan 1990).

The results of subject 4 show a similar pattern, except for a significant leftward shift of mean landing site in the left parafoveal presentation condition. This may seem to suggest an interpretation in terms of a range effect (Kapoula 1985), with saccades to parafoveal targets overshooting and saccades to peripheral targets undershooting the target. But this idea is not supported by a comparison of mean landing sites and refixation curves, where for the parafoveal presentation conditions there is no indication of landing site overshoot.

Interestingly, in one subject mean landing positions for all presentation conditions were almost identical. Moreover, initial fixation positions clustered exactly at the minimum of the refixation curves suggesting a close proximity between preferred and optimal viewing position. We interpret this as evidence for an alternative (and possibly more effective) visual processing strategy in this subject. Solving the task required in the experiment clearly profits from sending the eye directly to a location where a maximum of information can be extracted. Obviously the subject was able to optimally adjust the amplitude of the primary saccade for this purpose. The possible objection, that this result does not necessarily reflect a strategic adjustment but may be partly due to visuomotor factors (exceptionally small saccadic undershoot in this subject) needs to be addressed in a follow-up experiment.

Our finding of significant differences in primary saccade landing sites between parafoveal and peripheral presentation in 3 of 4 subjects contradicts results of Vitu (1991b). At least for paradigms like the one used here, the present results casts doubt on her idea of a "critical zone" within which initial fixation positions are determined by a cortically weighted center of gravity effect (almost) irrespective of target eccentricity (Vitu 1991a).

On the other hand, our results are in good accordance with the literature on saccades to eccentric targets in general. As an example, de Bie, van den Brink & van Sonderen (1987) reported a considerable increase in primary saccade undershoot to simple stimuli presented at

2, 4 and 8 deg. eccentricity to the left vs. right of a central fixation point. Heller (1982) obtained similar results for return sweeps in text with increasing line length. We agree with Vitu (1991b) that the linear relation between launch distance and landing sites of saccades in reading observed by McConkie et. al. (1988) as well as Radach & Kempe (1993) should not exclusively be interpreted as a saccadic range effect. The general relation between target distance and saccade amplitude however, as reported in many studies with different stimulus material, should be present in recognition tasks with eccentric words as well.

With respect to a possible role of semantic, lexical or orthographic parafoveal preprocess- ing in determining saccade landing positions the results of the present study are quite clear. There is no evidence for such a role, especially since in our experiment the full variability of word beginning redundancy and "orthographic saliency" present in German content words has been used. There are now a number of studies that fail to support the "parafoveal guidance hypothesis", including an exact replication of the Underwood, Clews & Everatt (1990) key experiment (Rayner & Morris 1992). There are also several studies involving eye movement contingent image changes showing that in conditions where a parafoveal preview is prevented or is meaningless, saccade landing positions are not altered (Kerr & McConkie 1992, Vitu 1991b, but see also Inhoff 1989, for a slightly different view). Taken all these results together, it seems unlikely that the "parafoveal guidance hypothesis" can account for the spatial distribution of initial fixation positions during word recognition and normal reading, at least in languages like German or English.

Literature

de Bie, J., van den Brink, G. & van Sonderen, J.F. (1987). The systematic undershoot of saccades: A localization or an oculomotor phenomenon? In J.K. O'Regan & A. Lévy-Schoen (Eds.), Eye movements: From Physiology to Cognition. Amsterdam: Elsevier (North-Holland), 85-94.

Dunn-Rankin, P. (1978). The visual characteristics of words. Scientific American 238, 1, 122-130.

Findlay, J.M. (1981). Local and global influences on saccadic eye movements. In D.F. Fisher, R.A. Monty, & J.W. Senders (Eds.) Eye movements: Cognition and Visual Perception Hillsdale, NJ: Erlbaum, 171-179.

Findlay, J.M. (1982). Global processing for saccadic eye movements. Vision Research 22, 1033-1045.

Findlay, J. M., Brogan, D. & Wenban-Smith, G. (1993). The spatial signal for saccadic eye movements emphasizes visual boundaries. Perception & Psychophysics 53, 633-641.

Heller, D. (1982). Eye movements in reading. In R. Groner & P. Fraisse (Eds.), Cognition and Eye Movements. Berlin: Deutscher Verlag der Wissenschaften, 139-154.

Hyona, J. (1993). Eye movements during reading and discourse processing. Psychological Re- search Reports. University of Turku.

Inhoff, A.W. (1989). Parafoveal processing of words and saccade computation during eye fixations in reading. Journal of Experimental Psychology: Human Perception and Perfor- mance 15, 544-555.

Kapoula, Z. (1985). Evidence for a range effect in the saccadic system. Vision Research 25, 1155-1157.

Kennedy, A. & Murray, W. (1991). The effect of flicker on eye movement control. Quarterly Journal of Experimental Psychology, 43a (1), 79-99.

Kerr, P. & McConkie, G.W. (1992). Towards a mathematical description of eye movement behavior in reading. Paper held at the symposium "Eye movements in reading: a confrontation of current theories". XXV Int. Congress of Psychology. Brussels, 19-24 July 1992.

Kliegl, R., Olson, R.K. & Davidson, B.J. (1983). On problems of unconfounding perceptual and language processes. In K. Rayner (Ed.), Eye Movements in Reading. Perceptual and Language Processes. New York: Academic Press, 333-343.

Lemij, H.G. & Collewijn, H. (1989). Differences in accuracy of human saccades between stationary and jumping targets. Vision Research 29, 12, 1737-1748.

McConkie, G.W., Kerr, P.W., Reddix, M.D. & Zola, D. (1988). Eye movement control during reading: I. The location of initial eye fixation on words. Vision Research 28, 1107-1118.

McConkie, G.W., Kerr, P.W., Reddix, M.D., Zola, D. & Jacobs, A.M. (1989). Eye movement control during reading: II. Frequency of refixating a word. Perception & Psychophysics 46, 245-253.

Nazir, T.A., Heller, D. & Sußmann, C. (1992). Letter visibility and word recognition: The optimal viewing position in printed words. Perception & Psychophysics 52 (3), 315-328.

O'Regan, J.K. (1990). Eye movements and reading. In E. Kowler (Ed.). Eye Movements and Their Role in Visual and Cognitive Processes. Amsterdam: Elsevier (North Holland).

O'Regan, J.K., Vitu, F. Radach, R. & Kerr, P. (in press). Effects of local processing and oculomotor factors in eye movement guidance in reading. In Ygge, J. & Lennerstrand, G., Eye movements in reading. New York: Pergamon Press.

Paap, K.R., Newsome, S.L., McDonald, J.E. & Schvaneveldt, R.W. (1982). An activation-verification model for letter and word recognition. Psychological Review, 89, 573-594

Radach, R. & Kempe, V. (1993). An individual analysis of initial fixation positions in reading. In d'Ydewalle, G. & Van Rensbergen, J. (Eds.) Perception and Cognition. Amsterdam: Elsevier (North Holland).

Rayner, K. (1979). Eye guidance in reading: Fixation locations within words. Perception 8, 21-30.

Rayner, K. & Morris, R. (1992). Eye movement control in reading: Evidence against semantic preprocessing. Journal of Experimental Psychology: Human Perception and Performance 18, 163-172.

Underwood, G. & Everatt, J. (1992). The role of eye movements in reading: Some limitations of the eye-mind assumption. In Chekaluk, E. & K.R. Llewellyn: The role of eye movements in perceptual processes. Amsterdam: Elsevier (North Holland).

Underwood, G., Clews, S. & Everatt, J. (1990). How do readers know where to look next? Local information distributions influence eye fixations. Quarterly Journal of Experimental Psychology 42a, 39-65.

Vitu, F. (1991a). The existence of a center of gravity effect during reading. Vision Research 31, 1289-1313.

Vitu, F. (1991b). Against the existence of a range effect during reading. Vision Research 31, 2009-2015.

Vitu, F., O'Regan J.K. & Mittau, M. (1990). Optimal landing position in reading isolated words and continuous texts. Perception & Psychophysics 47, 583-600.

LEXICAL INFLUENCES ON PARSING STRATEGIES: EVIDENCE FROM EYE MOVEMENTS

C. Frenck-Mestre and J. Pynte

Département de Psychologie, Université de Provence - CNRS (URA 182)
29 avenue Robert Schuman, 13621 Aix-en-Provence, France

Abstract

The present eye-movement study aimed at testing the hypothesis, forwarded by advocates of an exclusively syntactically-guided parser, that the parser initially blindly assigns syntactic roles without regard to lexical information (minimal attachment principle). The analysis of mean first pass reading times and of the mean number of fixations clearly showed that processing of the critical (noun phrase) region of sentences differed in difficulty depending upon information provided by the verb. This result is thus in line with those obtained in previous studies which have shown that structural preferences associated with a given verb and/or subcategorization information in general, can influence parsing processes We will argue, on the basis of conjoint evidence from interzone regressions and first pass reading times, that our results reflect initial parsing strategies, and, as such, that extra-syntactic factors can influence the parser's decisions, while not actually superseding syntactic analysis.

Introduction

In the last twenty years, numerous authors have chosen to record subjects' eye movements during the reading of text as a means of revealing on-line syntactic parsing decisions (Altman, Garnham & Dennis, 1992; Britt, Perfetti, Garrod & Rayner, 1992; Ferreira & Clifton, 1986; Ferreira & Henderson, 1990; Frazier and Rayner, 1982; Kennedy & Murray, 1984; Kennedy, Murray, Jennings & Reid, 1987; Pynte & Kennedy, 1992; Rayner, Carlson & Frazier, 1983; Rayner & Frazier, 1987, to name just some). Underlying the popularity of this relatively non-invasive measure is the assumption that the grammatical processing of any given part of the sentence can be precisely time-locked to the fixation occurring at that position in the sentence (first formulated by Just & Carpenter (1980) as the immediacy assumption, coupled with the eye-mind assumption). Given this, it has been suggested that the recording of eye movements offers the sensitive measure that is needed to decide between models of syntactic processing, in contrast to various other techniques which may encompass processing decisions at several different levels of analysis (cf. Ferreira & Henderson, 1990; Mitchell, 1987; 1989).

A case in point is the debate over whether the human syntactic parser is completely autonomous (Ferreira & Henderson, op.cit.; Frazier, 1987) or on the contrary, guided by lexical information (Ford, Bresnan & Kaplan, 1983; Tanenhaus, Carlson & Trueswell, 1989) when assigning initial structures to grammatical (although perhaps ambiguous) strings of words. The issue of debate, as brought out in extensive reviews (Frazier, op.cit.; Mitchell, op.cit), is not if but exactly when detailed lexical information

Eye Movement Research/J.M. Findlay et al. (Editors)

can be used by the parser[1] . The results of several studies which support the hypothesis that the parser is initially guided in its decisions by lexical information (or subcategorization information furnished by the verb) have been dismissed by some authors on the grounds that the results could be explained as well, if not better, in terms of re-analysis strategies (Ferreira & Henderson, op.cit.; Mitchell, op.cit.). This argument has been levied in particular against studies using the self-paced reading technique (whether word-by-word (Stowe, 1988) or using larger segments (Holmes, 1987; Mitchell & Holmes, 1985)) as well as those employing techniques such as an on-line lexical decision during auditory sentence presentation (Nicol & Osterhout, 1988; cited in Mitchell, 1989).

To date, relatively few eye movement studies have addressed the issue of whether or not lexical information (i.e. verb subcategorization frames) plays a guiding role in the syntactic analysis of sentences. In one study (Kennedy et al., 1989), the structural bias of verbs that could take various types of complements (for example, the verb believe allows both a clausal complement and a NP complement, however subjective ratings show the former to be the preferred structure) was included as one of several factors thought to influence the parsing process. The results of that study failed to find an effect of verb bias on either of the two dependent variables (first pass reading times and regressions) employed to reveal parsing strategies. In another study, which was principally intended to test for the effect of verb bias (Ferreira & Henderson, op cit.), this factor again failed to significantly affect initial (first pass) parsing strategies when the dependent measure was the first fixation on the first word that disambiguated the sentence structure. That is, for sentences such as 1, below, subjects did not spend less time reading the second (embedded) verb of the sentence when the first verb was

(1) Bill (hoped/wrote) Jill arrived safely today.

one that rarely took a direct object (such as hope) than when the first verb often took a direct object (such as write). Hence the conclusion that verb subcategorization information failed to influence initial parsing, since it should have blocked the direct object analysis of the NP for the first class of verbs and thus eliminated structural ambiguity before the processing of the embedded verb. In the latter study, the same materials were also presented in a word-by-word self-paced reading task. Again, the results failed to reveal an effect of verb bias on the mean reading times of the disambiguating verb. A facilitating effect of verb bias was found, however, at the word immediately following the disambiguating verb. Ferreira and Henderson argued from these results that results from previous self-paced reading studies showing an effect of verb bias reflected re-analysis strategies, not initial parsing decisions, given that in those studies the disambiguating region was several words long.

In both of the eye movement studies cited above, it was concluded that readers do not use verb information to guide the initial parsing of temporarily ambiguous sentences. However, this conclusion has recently been questioned. On the one hand, Altman et al. (op cit.) cast some doubt on the conclusiveness of Ferreira & Henderson's results (i.e. the absence of an effect of lexical bias on syntactic ambiguity resolution) on the grounds that these authors did not examine reading times in the absence of

---

[1] *Note, the same issue is at stake concerning the role of referential context; cf . Altman et al. (1992) and Ferreira and Clifton (1986) for recent discussions*

regressions whereas Altman et al. have found that this regression-contingent analysis can in fact significantly modify results. More convincingly, perhaps, Altman et al. also provided some evidence, in their eye- movement study which examined the effect of referential context on initial parsing decisions, that subcategorization information can prevent readers from making errors when assigning syntactic structures. That is, readers did not experience difficulty in interpreting sentences (preceded by a referential context) such as 2, below,

(2) He (told/invited) the woman that he was confident about to return in a fortnight.

when the first verb was one that could not take a that-complement clause (such as invite, ask, beg, etc.) as opposed to when the first verb of the sentence could take such a complement (e.g. tell, assure, etc.). In other words, readers could apparently use the information contained in the verb in order to initially treat the noun phrase following the critical (first) verb as the head of a complex NP, rather than as a simple NP followed by a that-complement clause. This result is quite the contrary to what would be predicted by advocates of a syntactic parser that is initially blind to non- syntactic information (Ferreira & Henderson, op.cit.; Frazier, op.cit.; Mitchell, op.cit.).

At present, thus, we do not appear to have a truly conclusive account of affairs as concerns the role of verb information in parsing. The experiment we will present here was another attempt to address this question, through the study of subjects' eye-movements during the reading of single sentences. Our study aimed in particular at testing the hypothesis forwarded by advocates of an exclusively syntactically-guided parser, that the parser blindly assigns the role of object to a noun phrase following the verb without regard to the properties of the verb (minimal attachment principle). Support for this hypothesis has been found in studies which contrasted two classes of verbs: intransitive verbs which cannot take a direct object complement versus optionally transitive verbs which can either take a direct object complement or be used intransitively (Ferreira & Henderson, op.cit.; Mitchell, 1987). Results of these studies show that in the case of both types of verbs, subjects apparently initially assigned the role of object to the NP following the verb, as attested by the fact that no processing advantage was found for intransitive verbs as compared to "mixed" verbs at the beginning of the zone which disambiguated the role of the NP. It can be noted, however, that most intransitive verbs accept other types of complements and, importantly, in many cases intransitive verbs can accept indirect object complements (as was in fact the case of some of the verbs employed in the studies just cited). As such, it is not clear that all intransitive verbs should initially impede the parser from assigning the role of object to a noun phrase following the verb, since in some cases the NP could in fact be the (indirect) object . This is illustrated in 3, below:

3) Because the boy always lied...

In the above example, the noun phrase which would normally follow the intransitive verb could equally as plausibly be an indirect object (as in "to his mother..") as the head of a sentential complement (as in "his mother.."). This type of "pseudo- intransitive" verb might thus be expected to initially engage the parser in the same strategy as optionally transitive verbs, if one ignores the precise syntactic construction required (i.e. the presence of a preposition) and considers only thematic roles.

The present experiment contrasted two classes of verbs: intransitive verbs that accept (and often take) indirect object complements, versus "purely" intransitive verbs that do not accept such complements[2].   Our aim was to show that there is a difference between these two classes of verbs, as a test of the hypothesis that (indirect object) complement verbs would be initially interpreted in the same manner as optionally transitive verbs (i.e. the tendency to initially assign the syntactic function of object to the immediately following NP), whereas non-complement intransitive verbs (those that do not accept object complements) should not be subject to such an interpretation. By contrasting these two types of verbs, the two models of syntactic parsing so far discussed (and the principle of minimal attachment) could be tested. According to the hypothesis of an autonomous parser which is blind to lexical information, both of these verbs should be treated in the same manner. That is, in sentences such as 4, below, the noun phrase following the critical verb should initially be assigned the role of object whatever be the verb. According to the lexical-guidance hypothesis, however, these two classes of verbs should lead to differential processing. For the object-complement verbs, the above NVN = SVO strategy (Bever, 1970; Frazier, 1987) should be applied, whereas it should be blocked by grammatical information associated to the verb in the case of "purely" intransitive verbs.

4. Whenever the baby (smiled/squirmed) his mother turned around.

These predictions can readily be tested by examining the patterns of subjects' eye-movements during the reading of such sentences. In line with the lexical-guidance hypothesis, either first pass reading times of the noun phrase following the verb should be longer and/or more regressions out of the NP region should occur in the case of complement-verbs than in the case of non-complement verbs, thus reflecting the need to revise an initial erroneous analysis in the case of the former class of verbs. If the parser is initially blind to lexical information, however, than there should not be evidence of greater processing difficulty for one class of verbs over the other.

Methods

Subjects
A total of 24 unpaid students of psychology at the University of Provence participated in the experiment which lasted approximately 30 minutes. All had normal vision and were naive with regard to the purpose of the study.

Materials and design
Twenty pairs of experimental sentences were constructed. All experimental sentences contained a main clause preceded by a subordinate clause. The two sentences in a pair were identical except for the verb in the subordinate clause which was one of two types: either a verb that accepted an indirect object complement, or one that did not accept either a direct or an indirect object complement (Bescherel, 1966). All subordinate verbs were employed intransitively in the experimental sentences,

---

[2] *It can be noted that, in French (the language in which the study was conducted), there are in fact a good number of intransitive verbs which are readily used in constructions with an indirect object complement.*

however, the first set of verbs could also accept an indirect object (i.e. preceded by a short (one or two letter) preposition), which was not the case of the second set of verbs (see examples 1 and 2, below). For purposes of analysis, the sentences were divided into four regions: the initial region up to the first (subordinate) verb; the subordinate verb; the following subject noun phrase (which ranged in length from 2 to 4 words); the region beginning at the verb of the main clause up to the penultimate word of the sentence. Examples of the two types of sentences in a pair and the regions of analysis (marked by slashes) are shown below:

1. Just before George / spoke / his ageing mother / came into the / room.
2. Just before George / died / his ageing mother / came into the / room.

The two sets of verbs were matched as closely as possible for length and frequency however, the complement verbs were in fact slightly more frequent than were the non-complement verbs (Tresor de la langue francaise (1971) frequency counts). The experimental materials are provided in the Appendix. Included with the experimental sentences were 40 fillers of varying structures and comparable in length to the experimental sentences. The 20 pairs of experimental sentences were split into two counterbalanced lists such that subjects saw only one sentence of each pair, thus ten complement verb sentences and ten non-complement verb sentences. Eight of the fillers were used for initial practice on the task. The remaining 32 fillers and 20 experimental sentences of each list were presented in a different random order for each subject. Each list was seen by a group of 12 subjects.

Apparatus and procedure

Subjects' eye movements were recorded using an infrared limbus-tracking device mounted on adjustable eyeglass frames. Horizontal signals from the right eye were sampled every 5 ms using a 12-bit A/D device interfaced to an Opus 386 computer. Subjects viewed the display binocularly while seated 60 cm from the screen, with their heads restrained with a bite-bar and adjustable head and chin rests. The apparatus was calibrated every four sentences using an array of five single digits across the display screen. Overall accuracy of the system was approximately 1 character. Sentences were presented individually on a high-resolution display monitor linked to the Opus 386. The sentences were displayed on a single line in normal upper and lower case. Sentence presentation was contingent on the subject's fixating a fixation point located at the far left of the screen. Subjects read the sentences and indicated whether or not they were sensible by pressing one of two buttons marked "oui" and "non" located under the subject's right and left hand. All 20 experimental sentences called for a positive response; of the 32 fillers, 6 called for a positive and 26 called for a negative response.

Results and discussion

The data were first analysed in terms of first pass reading times per region, number of first pass fixations per region, and the probability of making a regressive saccade both to and from a region. First pass reading time per region was defined, in line with previous work (Kennedy et al, 1989; Pynte & Kennedy, 1992), as the sum of all first-pass fixations (left-to-right saccades) within a specified region, plus any within-zone regressions up to and including the final fixation (here, defined as the rightmost fixation within the final region of analysis). Note that we did not measure reading time in terms of average fixation per character or per word, given that the sentences were identical with the exception of the verb in the subordinate clause, and the two sets of verbs were matched for length, and as closely as possible for frequency. Subsequent analyses were also performed on the data obtained specifically at the critical region of interest (the noun phrase). Namely, mean reading times in the absence of regressions, mean first fixation durations at the noun phrase, and mean number of regressions launched from the beginning of the noun phrase region.

First pass reading times

The mean first pass reading times for each region appear in Table 1. Analysis of variance of the data revealed a tendency for reading times to be overall longer for the complement than for the non-complement verb sentences [F1 (1,22) = 3.94, p<0.06; F2 (1,18) = 2.59, p<.12]. Verb type interacted with Region [F1 (3,66) = 8.91, p<.001; F2 (3,54) = 4.67, p<.005], showing that reading times were longer in the third region (i.e. the NP immediately following the verb of the subordinate clause) of sentences containing complement verbs than in the same region of these sentences when they contained a non-complement verb. A Newman-Keuls analysis of the interaction effect confirmed that reading times differed between the two types of sentences (as defined by verb type) only at the critical noun phrase.

Number of fixations

The mean number of first pass fixations per region are shown in parentheses in Table 1. Analysis of variance of the data revealed the same pattern of results as that observed for first pass reading times. Globally, subjects made slightly more fixations in the complement verb than in the non-complement verb sentences [F1 (1,22) = 13.13, p<.001; F2 (1,18) = 3.69,p<.07].

	Sentence Region			
Verb Type	Initial	Verb	NP	Final
Complement	935(4.3)	516(2.2)	913(4.1)	538(2.4)
Non-Complement	931(4.5)	550(2.1)	759(3.3)	518(2.4)

Table 1. Mean First Pass reading times (ms) and mean number of fixations (in parentheses) per region for each sentence type (as defined by verb type).

This was true, however, only at the third region of sentences, as revealed by an interaction between Verb type and Region [F1 (3,66) = 8.61, p<.001; F2 (3,54) = 6.95, p<.001], and as confirmed by a Newman-Keuls analysis of the interaction effect.

Interzone Regressions

Table 2 shows, for each verb type, the mean percentage of regressive saccades that subjects made both to and from each of the four regions of analysis, that is, the landing and launch sites of interzone regressions. Analysis of variance of the data revealed that subjects made significantly more interzone regressions in the case of complement verb than in the case of non-complement verb sentences [F1 (1,22) = 8.05, p<.01; F2 (1,18) = 5.83, p<.03] The analysis of the launch sites of these regressions revealed, as with reading times, that the main effect of Verb type was modified by an interaction with Region. Specifically, subjects launched more interzone regressions from the third region of analysis (i.e. from the NP following the subordinate verb) in sentences containing complement verbs than in the non-complement verb sentences [F1 (2,44) = 3.41, p<.05; F2 (2,36) = 3.35, p<.05]. A Newman-Keuls analysis of the interaction effect confirmed that the two types of sentences differed only at the critical noun phrase. The analysis of the landing sites of interzone regressions also revealed an interaction between Verb type and Region [F1 (2,44) = 3.20, p<.05; F2 (2,36) = 3.28, p<.05]. Subjects made significantly more interzone regressions to the verb of the subordinate clause in the case of complement verb than in the case of non-complement verb sentences, whereas there were no significant differences between the two sentences at the other regions of analysis (as confirmed by a Newman-Keuls analysis of the interaction). Otherwise stated, subjects made more regressions from the noun phrase and to the subordinate verb as a function of verb type.

	Sentence Region			
	Initial	Verb	Noun phrase	Final
	launch/land	launch/land	launch/land	launch/land
Verb Type				
Complement	** / 14.6	12.5 / 29.6	30.0 / 7.1	8.8/ **
Non-Complement	** / 10.8	9.6 / 17.9	18.3 / 6.7	7.5/ **

Table 2. Percentage of interzone first pass regressions to (landing site) and from (launch site) the four regions of analysis of sentences, as a function of verb type (Comp. = Complement, Non-Comp. = Non- complement). Inspection of the critical noun phrase:

First pass reading times in the absence of regressions

The results cited above show a robust effect of verb type at the critical noun phrase, on first pass reading times and on the mean number of first pass fixations. It is possible that the longer reading times and the greater number of fixations observed at the NP for complement-verb sentences as opposed to non-complement- verb sentences are linked to the probability of making a regression, either out of the NP or within the NP region itself. This is because our first pass measure included all left-to-right going saccades (including those occurring after a regression out (to the left) of a region but before any fixations to the right of the region), as well as any within-zone regressions. Given that subjects made significantly more regressions out of the NP in the case of complement-verb sentences, reading times may have been elevated by any left-to-right going fixations that occurred in the NP following such a regression. In order to examine this question, we performed two subsequent analyses on the data obtained at the critical NP. First, all trials in which an interzone regression (i.e. from the NP to a previous region of the sentence) had occurred were excluded. The analyses performed on the remaining data revealed significant effects of verb type at the NP, both as concerns mean reading times [$F_1(1,22) = 12.89$, $p<.01$; $F_2(1,18) = 7.42$, $p<.01$] and mean number of fixations [$F_1(1,22) = 15.64$, $p<.001$; $F_2(1,18) = 9.69$, $p<.01$]. Subjects spent more time reading the noun phrase, and made a greater number of fixations in this region when the NP was preceded by a complement-verb than by a non-complement verb (878 vs. 711 ms, and 3.8 vs. 3.1 fixations, respectively). From this analysis, we can conclude that the longer reading times we observed were not linked to the probability of making a regressive saccade out of the NP. Furthermore a second analysis, performed on the data excluding all trails in which a within-zone regression had occurred, also showed longer mean reading times and a greater mean number of fixations for the NP when it followed a complement-verb than a non-complement verb (838 vs. 682 ms, respectively: $F_1(1,22) = 18.51$, $p<.001$; $F_2(1,18) = 5.38$, $p<.03$; and 3.5 vs. 2.9 fixations, respectively: $F_1(1,22) = 18.28$, $p<.001$; $F_2(1,18) = 6.42$, $p<.02$). Thus, it can be stated that the effects we found on first pass reading times and on first pass fixations at the NP were quite independent of the probability of making a regression, either out (to the left) of or within the NP region.

First fixation durations and interzone regressions

We conducted analyses specifically on the region of the noun phrase in order to determine whether the effects that we observed occurred early in processing. Instead of considering the mean first pass reading times of the entire NP, we examined the duration of the first fixation within a delimited space of seven character spaces following the verb of the subordinate clause[3]. We also re-examined the data for interzone regressions launched from the critical NP region, limiting the size of this region to seven character spaces into the noun phrase (roughly the size of a single word in French). The analysis of mean first fixation durations (which excluded trials on which the region was not fixated, or the fixation durations were less than 100 ms), did not reveal a significant effect of verb type [F1 and F2 both <1]. In fact,   the

---

[3] *The first word of the noun phrase was, in almost all cases, a definite artic le. Given that subjects have a tendency to skip short words, we did not delimit the region of analysis to the first word of the NP, but took the first fixation within the NP region up to 7 character spaces.*

difference between the two means was only 8 ms (242 and 250 ms, for the non-complement and complement-verb conditions, respectively). In contrast, the analysis of interzone regressions (launched from the initial region of the NP) did reveal at least a strong tendency for verb type to affect processing (12.5% and 19.2% regressions, for the non-complement and complement-verb conditions, respectively [$F1(1,22) = 4.00$, $p<.06$; $F2 (1,18) = 4.27$, $p<.05$]).

General Discussion

The analysis of mean reading times of the entire noun phrase, and of the mean number of fixations within this region, clearly show that processing of the noun phrase was more difficult when the verb preceding it was one that readily took an indirect object complement than when the verb did not accept such a complement. This indicates that subjects were obliged to re-analyse the NP in the case of complement-verbs, due to unfulfilled expectations concerning its role in the sentence. This result is thus in line with those obtained in previous studies which have shown that structural preferences associated with a given verb and/or subcategorization information in general, can influence parsing processes (Ford, Bresnan & Kaplan, 1982; Holmes, 1987; Mitchell & Holmes, 1985; Tanenhaus et al., 1989).

The pertinent question, however, is not whether but exactly when this type of information exerts its influence. It has been argued that lexical information does not guide the parser, but acts at a later stage, in the re-analysis process (Ferreira & Henderson, 1990; Mitchell, 1989). Whether or not our results reflect initial parsing decisions or only re-analysis strategies may be open to question. As reported in the introduction, Ferreira & Henderson (op.cit) failed to find significant effects of verb bias on initial parsing decisions, although this information did tend to affect the ease of re-analysis of misparsed sentences. In their study, the critical region of analysis (the disambiguating region) consisted of a single word, and the dependent measure they employed was the mean duration of the first fixation within that region. In the present study, the critical region ranged in length from two to four words. Ferreira & Henderson's data show that the effect of verb bias became apparent only at the "post-disambiguating" region, that is, one to two words following the disambiguating word. Otherwise stated, it could be claimed that the size of our critical region was large enough to include both initial parsing decisions and re-analysis processes (which, in fact, is the argument that Ferreira & Henderson put forward to explain the effect of subcategorization information upon parsing of structurally ambiguous sentences, found in previous studies).

The subsequent analyses that we carried out on the first fixation durations specifically in the critical (noun phrase) region did not in fact reveal a significant effect of verb type. However, verb type did have a strong tendency to affect the probability of launching a regression from the initial region of the NP back to the verb. As such, it cannot be said that the effects we observed on the entire noun phrase were absent at the earliest processing of this region, i.e. that they were the result solely of re-analysis processes.

Taken as a whole, the results of the present experiment pose a rather provocative question. Namely, when eye movement recording is adopted as the measurement technique, just what measure should be taken as an indication of initial parsing decisions? Rayner et al. (1989) have suggested that a range of measures (first fixation duration, gaze duration, second pass reading times, probability of making a regression out of a region) should be taken into account before concluding that a given effect is the product of an initial analysis process rather than reflecting re- analysis processes. According to Ferreira and Henderson's (op.cit.) argument, however, it would seem that any processing beyond that observed for the first fixation in a designated region should be considered as a reflection of re-analysis, rather than of initial decisions.

This, in itself, would appear to be a rather strong interpretation of the immediacy assumption forwarded by Just & Carpenter (1980). It is also in apparent contradiction with the suggestions that first fixation durations are largely independent of lexical (let alone syntactic) influences (O'Regan, Levy-Schoen, Pynte & Brugalliere, 1984; but see Inhoff, 1984), and that the processing of a given word can "spill over" onto subsequent fixations (Rayner & Pollatsek, 1987). For all of these reasons, it would seem reasonable not to regard processing beyond that of the first fixation in a designated region as being totally independent of initial syntactic processing decisions. We will argue, on the basis of conjoint evidence from interzone regressions and first pass reading times, that our results reflect initial parsing strategies and not only re-analysis. As such, the claim is made that structural preferences can influence the parser. It is not necessary, however, to adopt a strong view of the lexical- guidance hypothesis. In line with suggestions made by Tanenhaus et al. (1989), we would argue that extra-syntactic factors can influence the parser's decisions, while not actually superseding syntactic analysis. Moreover, as suggested by Altman et al. (1992), we believe that there may in fact be little empirical value to the distinction between processes linked to syntactic processing and sentence interpretation when applied to the resolution of local syntactic ambiguities.

References

Altman, G.T., Garnham, A. & Dennis, Y. (1992). Avoiding the garden path: Eye movements in context. J. of Mem. and Lang., 31, 685-712.

Bescherel, Les frères. (1966). L'art de conjuguer. Hatier: Paris.

Bever, T.G. (1970). The cognitive basis for linguistic structures. In J.P. Hayes (Ed.) Cognition and the development of language. New York: John Wiley

Britt, M.A., Perfetti, C.A., Garrod, S. & Rayner, K. (1992). Parsing in discourse: Context effects and their limits. J. of Mem. and Lang., 31, 293-314.

Ferreira, F & Clifton, C. (1986). The independence of syntactic processing. J. of Mem. and Lang., 25, 348-368.

Ferreira, F. & Henderson, J.M. (1990). Use of verb information in syntactic processing: Evidence from eye movements and word-by-word self-paced reading. J. Exp. Psych.: Learn, Mem and Cog., 16, 555-568.

Ford, M., Bresnan J.W. & Kaplan, R.M. (1982). A competence based theory of syntactic closure. In J.W. Bresnan (Ed). The mental representation of grammatical relations. Cambridge, Mass.: MIT Press.

Frazier, L. (1987). Sentence processing: A tutorial review. In M. Coltheart (Ed). Attention and performance XII. Hillsdale, N.J., Lawrence Erlbaum Associates Ltd.

Frazier, L. & Rayner, K. (1982). Making and correcting errors during sentence comprehension: Eye movements in the analysis of structurally ambiguous sentences.
Cog. Psych., 14, 178-210.

Holmes, V.M. (1987). Syntactic parsing: In search of the garden path. In M. Coltheart
(Ed).Attention and performance XII. Hillsdale, N.J., Lawrence Erlbaum
Associates Ltd.

Inhoff, A.W. (1984). Two stages of word processing during eye fixations in the reading
of prose. J. of Verb. Learn. and Verb. Beh. 23, 612-624.

Just, M.A. & Carpenter, P.A. (1980). A theory of reading: From eye fixations to
comprehension. Psych. Review, 87, 329-354.

Kennedy, A. & Murray, W.S. (1984). Inspection times for words in syntactically
ambiguous sentences under three presentation conditions. J. Exp. Psych: HPP,
10, 833-849.

Kennedy, A., Murray, W.S., Jennings, F. & Reid, C. (1989). Parsing complements:
Comments on the generality principle of minimal attachment. Lang. and Cog.
Proc., 4, 51-76.

Mitchell, D.C. (1987). Lexical guidance in human parsing: Locus and processing
characteristics. In M. Coltheart (Ed).Attention and performance XII. Hillsdale,
N.J., Lawrence Erlbaum Associates Ltd.

Mitchell, D.C. (1989). Verb-guidance and other lexical effects in parsing. Lang. and
Cog. Proc., 4, 123-154.

Mitchell, D.C. & Holmes, V.M. (1985). The role of specific information about the verb
in parsing sentences with local structural ambiguity. J. of Mem. and Lang., 24,
542-559.

O'Regan, K., Lèvy-Schoen, A., Pynte, J. & Brugallière, B. (1984). Convenient fixation
location within isolated words of different length and structure. J. Exp. Psych:
HPP, 10, 250-257.

Pynte, J. & Kennedy, A. (1992). Referential context and within- word refixations:
Evidence for weak interaction. In G. d'Ydewalle & J. Van Rensbergen (Eds.),
Perception and Cognition: Advances in eye movement research. Amsterdam,
North Holland, 1992.

Rayner, K., Carlson, M. & Frazier, L. (1983). The interaction of syntax and semantics
during sentence processing: Eye movements in the analysis of semantically
biased sentences. J. of Verb. Learn. and Verb. Beh.,

Rayner, K. & Frazier, L.(1987) Parsing temporarily ambiguous complements. Quart. J.
of Exp. Psych., 39A, 657-673.

Rayner, K. & Pollatsek, A. (1987). Eye movements in reading: A tutorial review. In M.
Coltheart (Ed). Attention and performance XII. Hillsdale, N.J., Lawrence
Erlbaum Associates Ltd.

Rayner, K., Sereno, S.C., Morris, R.K., Schmauder, A.R. & Clifton, C. (1989). Eye
movements and on-line language comprehension processes. Lang. and Cog.
Proc. 4, 21-50.

Stowe, L.(1988). Thematic structures and sentence comprehension. In G. Carlson & M.
Tanenhaus (Eds.), Linguistic structure and language processing. Dordecht:
Klower Academic.

Tanenhaus, Carlson & Trueswell, (1989). The role of thematic structures in
interpretation and parsing. Lang. and Cog. Proc., 4, 211-234.

Trésor de la langue française (1971). Dictionnaire des fréquences. Paris: Klincksieck.

APPENDIX

1. Chaque fois que le bébé souriait (gigotait) la petite fille se retournait pour le voir.

2. Du fait que l'eau de la montagne jaillisait (pétillait) la source était devenue célèbre.

3. Pendant que le professeur démissionnait (divaguait) l'équipe pédagogique cherchait une solution.

4. Pendant que l'adolescent téléphonait (ralait) ses parents adoptifs préparaient le départ.

5. Alors que cette coutume survivait (prosperait) certains changements ébranlaient la société.

6. Au moment où le viellard mourait (agonisait) l'épidemié de grippe faisait des ravages.

7. Du fait que la villageoise mentait (radotait) tout le monde cherchait a l'éviter.

8. Lorsque le garcon désobéissait (chomait) sa pauvre grand-mère était toute triste.

9. Lorsque le train disparut (dérailla) la galerie d'exposition fut fermée définitivement.

10. Du fait que le monarque déplaisait (faiblissait) la population locale souhaitait son départ.

11. Pendant que la concierge médisait (ironisait) la voisine d'à côté traversa la cour.

12. Alors que le convoi arrivait (bifurquait) la gare de triage fut soudainement bombardée.

13. Tandis que la fille résistait (accourait) son jeune amant faisait d'ardentes déclarations.

14. Chaque fois que l'enfant tombait (pedalait) sa vieille bicyclette faisait du bruit.

15. Chaque fois que le délinquant sortait (recidivait) la gendarmerie nationale était avertie.

16. Du fait que le jus de raisin dégoulinait (fermentait) la cuve en aluminium débordait un peu.

17. Pendant que le surveillant réagissait (enrageait) la cohue des élèves s'attenuait lentement.

18. Au moment où le militant adherait (ergotait) l'aile gauche du parti votait les crédits.

19. Même si cette bacterie convenait (proliferait) la recherche medicale n'avançait pas vite.

20. Lorsque l'écolier parlait (bouffonnait) ses camarades de classe riaient aux éclats.

EFFECTS OF A WORD'S MORPHOLOGICAL COMPLEXITY ON READERS' EYE FIXATION PATTERNS

Jukka Hyönä, Matti Laine and [2]Jussi Niemi
Academy of Finland and University of Turku, Finland
[2]University of Joensuu, Finland

Abstract

In the present study morphological processing during word recognition was examined by employing the lexical decision and naming tasks in combination with eye movement registration. Three types of nouns with different morphological complexity were used: monomorphemic base-form nouns, and two types of multimorphemic nouns - derived and inflected. Subjects were asked to look at each word as long as they needed to be able to name it aloud (Expt. 1) or make a lexical decision (Expt. 2) to it. The results showed that inflected words attracted longer first and second fixations than derived or base-form words. On the other hand, the fixation frequency was lower for the inflected words than for the other wordtypes. The morphological effect was not found to interact with the experimental task. The study lends support to the view that some morphological parsing takes place during lexical access. Moreover, the results are in accordance with the SAID (Stem Allomorph/Inflectional Decomposition) model proposed by Laine, Niemi, Koivuselkä-Sallinen, Ahlsén and Hyönä (in press), according to which a speaker of Finnish represents and accesses inflected words in a morphologically decomposed form, while derived words are accessed as single entities.

Keywords

fixation duration, fixation frequency, morphological processing, word recognition, naming, lexical decision, Finnish.

Introduction

The role of morphology in word recognition has recently gained increased attention. Some models claim that morphologically complex forms (i.e., all non-monomorphemic words) are accessed by decomposing the word form into its root (or stem) and affix(es) (e.g., Taft & Forster, 1975). In contrast to decomposition theories, models have been put forth that assign negligible significance to morphology in word recognition. For example, according to the Full Listing model proposed by Butterworth (1983), all word forms are stored as single entities and thus no morphological parsing is needed for lexical access. The third set of models, the so-called hybrid models, combines features of the decomposition and full listing models. For instance, according to the Morphological Race model (Frauenfelder & Schreuder, 1992), words can be accessed either as single entities (via the direct route) or via morphological decomposition (via the parsing route). The surface frequency of a multimorphemic word is the most important factor in determining the winner of the race, high frequency words being usually recognized by the direct route. In addition, the phonological and semantic transparency is assumed to influence the outcome of the race for medium and low frequency words.

Eye Movement Research/J.M. Findlay et al. (Editors)

The available evidence does not conclusively favor any of the above mentioned model types (see Henderson, 1985). One reason for the controversial results may be due to the structure of the languages studied. In languages with restricted morphology (e.g., English), the morphological structure of words may not occupy a central role in the recognition process. On the other hand, in a morphologically rich language such as Finnish, which is studied here, the morphological structure of words presumably plays a much more significant role during word identification, as the number of possible word forms the speaker must be able to handle is enormous. It has been estimated that in Finnish there are over 2 000 possible noun forms; the number of possible verb forms exceeds 10 000; all this even without the effect of derivation and highly productive compounding (Karlsson, 1983).

Comprehensive studies of morphological processing in Finnish have been lacking. However, recently Laine and Niemi took the first steps towards unravelling the effects of a word's morphological structure on lexical access in Finnish (Laine, Niemi, Koivuselkä-Sallinen & Hyönä, in press; Laine, Niemi, Koivuselkä-Sallinen, Ahlsén & Hyönä, in press; Niemi, Laine & Tuominen, in press). These data suggest that inflected forms of nouns would be harder to process than monomorphemic base-form words, and also harder than multimorphemic derived words. This finding is taken to suggest that only inflected forms, not all polymorphemic words (such as derived words), would be decomposed during lexical access. Evidence supporting this view comes from both normal subjects as well as from single-case studies of aphasic patients. On the basis of this and other findings Laine and Niemi have proposed the Stem Allomorph/Inflectional Decomposition (SAID) model (Laine et al., in press; Niemi et al., in press), according to which a speaker of Finnish represents and accesses inflected words in a morphologically decomposed form (with the exclusion of most frequent inflected forms), while derived (and base-form) words are accessed as single entities.

Lexical decision is the most commonly used paradigm in studies of morphological processing. In lexical decision, the subject has to decide as fast as possible whether or not a given word form exists in the language studied. Lexical decision is argued to consist of two consecutive processing phases - the lexical access phase and the post-access checking and integration phase. As a consequence, by only using the lexical decision task one may not be able to distinguish which of the two processes the word's morphological complexity is influencing one's results. To avoid the somewhat complex nature of the lexical decision task, some investigators have favored the naming task over lexical decision. The naming task is presumed to tap only lexical access and not to involve any post-access integration. Consequently, if one finds an effect of morphological complexity in lexical decision but not in naming, one may ascribe the effect to be primarily post-lexical in nature.

In the present study, these two most commonly used experimental paradigms were used in combination with eye movement registration. The study was conducted to further test the SAID model by using durations of eye fixations on the target words as the critical measure to study morphological processing. In two experiments, subjects were presented words with different morphological complexity in isolation of any linguistic context. In Experiment 1, they were asked to look at each word as long as they needed to be able to name it aloud, whereafter they were to look away from the word and pronounce the word. In Experiment 2, subjects were required to make a lexical decision to each target item, half of which were real words of varying morphological complexity, the other half being pronounceable pseudowords. Subjects responded by gazing at either the word 'Yes' or the word 'No' presented simultaneously to the right of the target. In other words, the procedure was different from the traditional lexical decision task, where subjects give their responses manually by pushing a button.

Durations of individual fixations as well as their summed total (i.e., gaze duration) were used as indices of the ease of processing. If the word's morphological complexity is to influence the earlier stages of processing, one should find an effect already in the first fixation duration on the word. If a later stage is relatively more affected, the duration of the second fixation or perhaps only the gaze duration would be sensitive to morphological complexity. Moreover, if primarily post-lexical in nature, the morphological effect should be observable in lexical decision, but not so much in naming.

In the following, a pooled analysis is reported of the naming and lexical-decision experiments.

Method

Subjects. Thirty university students took part in the naming experiment and twenty-four in the lexical decision experiment, as a part of their course requirement. No subject participated in both experiments.

Apparatus. Eye movements were recorded using an Applied Science Laboratories Model 1994 eye-tracker. This monitoring system is video-based and makes use of pupil and corneal reflections. Its accuracy has been estimated as 0.65 degrees horizontally and 0.35 degrees vertically (Muller, Cavegn, d'Ydewalle & Groner, 1993). Eye positions were sampled every 20 msec using an IBM compatible microcomputer. A chin rest and a head restraint were used to restrict possible head movements. Fixations less than 100 msec were excluded from the analysis.

Materials. Three types of nouns with different morphological complexity were used: monomorphemic *base-form* nouns, and two types of multimorphemic nouns - *derived* and *inflected*. Base-form nouns were words without any type of morphological affix; all derived words had a highly productive deverbal agentive suffix -jA corresponding to the English -er (e.g., juoksi+ja 'runner'); all inflected words comprised the inessive case ending -ssA, which corresponds to the preposition 'in' in English (e.g., auto+ssa 'in the car'). There were 20 of each type of words. The words were matched for length and frequency, both for lemma and surface frequency. The average lemma frequencies were 20.9 (sd=9.8), 21.0 (sd=8.2), and 22.1 (sd=10.0), for the base-form, derived, and inflected words, respectively. The respective surface frequencies were 19.0 (sd=14.3), 21.9 (sd=22.4), and 15.1 (sd=13.1). The length of the target words ranged between 6 and 11 letters. In the lexical decision experiment, real words were mixed with 60 pronounceable length-matched pseudowords.

Procedure. Prior to the experiment, the eye-tracker was calibrated. Each experimental trial was initiated by presenting a fixation point on the computer screen. Subjects were asked to gaze at the fixation point, while the experimenter checked the accuracy of the calibration. Immediately after the calibration check the word appeared on the right side of the initial fixation point. In the naming experiment, subjects were asked to look at the word as long as they needed to be able to name it aloud, after which they were to look at a dot on the right side of the target word and pronounce the word. In the lexical decision experiment, subjects were asked to make a lexical decision to each lexical item that were presented on the computer screen. They were asked to gaze at the target as long as they needed to decide upon its lexical status. If it was a real word, they were to fixate at the word Kyllä 'yes' presented to the right of and slightly above the target; in the case of a pseudowords, they were to fixate at the word Ei 'no' presented to the right of and slightly below the target.

Results

Repeated measures ANOVAs were performed on the data using Wordtype (base-form, derived, inflected) as the within-subject variable and Experiment (naming vs. lexical decision) as the between-subject variable. All analyses were carried out both by subjects (F') and by items (F''). The first and second fixation durations are reported first for the complete set of data, followed by a separate analysis of the one-fixation and the two-fixation cases. The analysis of the one fixation cases is based on data from 28 subjects in the naming experiment and from 22 subjects in the lexical decision experiment. One subject had to be discarded from the analysis of the two fixation cases due to missing data points for each cell. The means for all eye movement parameters are presented in Table 1.

In naming, 33 % of words were read with a single fixation, 53 % with two fixations, and 14 % with more than two fixations. In lexical decision, the respective percentages were 36 %, 51 %, and 13 %.

In lexical decision, subjects gave a wrong response only in 0.9 % of the trials, most errors being made on pseudowords. Using an eye response for lexical decision, we were also able to record possible hesitations in responding by computing the probability of first giving a wrong answer followed by a correction. The overall frequency of hesitation was 4.7 %, 57 % of the hesitations occurring on pseudowords. Of the real words that produced an initial hesitation, 49 % were inflected words, 35 % base-form words, and 16 % derived words.

First fixation duration
All data. The duration of the first fixation seemed to be affected by Wordtype, but only in the subject analysis ($F'(2,104)=4.33$, $p<.025$; $F''(2,57)=1.48$, $p>.1$). Pairwise contrasts revealed that inflected words received longer first fixations than base-form ($F'(1,52)=4.45$, $p<.04$) or derived words ($F'(1,52)=6.83$, $p=.01$). The main effect of Experiment was clearly significant ($F'(1,52)=8.00$, $p<.01$; $F''(1,57)=83.67$, $p<.0001$). On the average, the lexical decision task produced 39 msec longer first fixations than naming. The Wordtype x Experiment interaction was far from significant (both F' and F'' < 1).

	Lexical decision			Naming		
	B	D	I	B	D	I
First fixation duration						
All data	321	319	338	285	282	295
One fixation cases	423	420	438	400	414	394
Two fixation cases	289	297	294	258	249	255
Second fixation duration						
All data	218	216	239	215	234	234
Two fixation cases	221	216	239	216	234	236
Fixation frequency	1.83	1.84	1.74	1.83	1.87	1.75
Gaze duration	507	506	514	463	485	471

Table 1. *Eye fixation parameters in the lexical decision and naming experiment as a function of Wordtype (B = Base-form noun, D = Derived noun, I = Inflected noun).*

One fixation cases. For the cases where there was only one fixation on the word, the duration of this single fixation did not seem to be affected by the word's morphological structure (both F' and F" < 1). The main effect of Experiment did not reach significance by subjects (F' < 1) but did so by items (F"(1,57)=53.02, p<.0001). There was a tendency for lexical decision to produce longer single fixations than naming. The Wordtype x Experiment interaction was non-significant (F'(2,96)=1.51, p>.1; F" < 1).

Two fixation cases. When there were exactly two fixations on the word, the first fixation was longer in lexical decision than in naming (F'(1,51)=5.09, p<.05; F"(1,57)=22.15, p<.0001). The main effect of Wordtype and the Wordtype x Experiment interaction were both non-significant (both F' and F" < 1).

Second fixation duration

All data. In the analysis of all the cases when there was a second fixation on the word, the main effect of Wordtype reached significance (F'(2,104)=5.16, p<.01; F"(2,57)=3.24, p<.05). The inflected words attracted significantly longer second fixations than base-form words (F'(1,52)=10.47, p<.01; t(57)=2.69, p<.01 by items) and marginally longer than derived words (F'(1,52)=2.86, p<.1; t(57)=1.08, p>.1 by items). The main effect of Experiment (both F' and F" < 1) and the Wordtype x Experiment interaction (F'(2,104)=2.24, p>.1; F" < 1) were both non-significant.

Two fixation cases. For the cases with exactly two fixations, there was a significant main effect of Wordtype (F'(2,102)=4.45, p=.01; F"(2,57)=2.37, p=.1). Pairwise contrasts revealed that inflected nouns received significantly longer second fixations than base-form nouns (F'(1,51)=8.85, p<.01; t(57)=2.15, p<.05 by items) and marginally longer second fixations than derived nouns (F'(1,51)=3.39, p<.1; t(57)=1.37, p>.1). The main effect of Experiment failed to reach significance (F' < 1; F"(1,57)=2.97, p<.1), and so did the Wordtype x Experiment interaction (F'(2,102)=2.02, p>.1; F" < 1).

Fixation frequency

With the number of fixations, the main effect of Wordtype proved significant by subjects (F'(2,104)=12.14, p<.0001), but not by items (F''(2,57)=1.64, p>.1). The subjects tended to make less fixations on inflected words compared to base-form (F'(1,52)=12.04, p=.001) or derived words (F'(1,52)=21.00, p<.0001), although this trend did not generalize across the items used. The main effect of Experiment (both F' and F'' < 1) and the Wordtype x Experiment interaction (both F' and F'' < 1) were both non-significant.

Gaze duration

For the gaze duration, all fixations landing on the word before moving away from it were summed to yield the total fixation time measure. With gaze duration, the main effect of Wordtype was not significant (F'(2,104)=2.03, p>.1; F'' < 1). The main effect of Experiment was far from significant by subjects (F' < 1), but was clearly significant by items (F''(1,57)=29.44, p<.0001). The Wordtype x Experiment did not reach significance (F'(2,104)=2.55, p<.1; F''(2,57)=1.13, p>.1).

Discussion

The pooled analysis of the naming and lexical decision experiments showed that the duration of the initial fixation on the target word was influenced by the word's morphological structure. Inflected words attracted greater first fixation durations than derived or monomorphemic base-form words. An analogous finding was observed for the duration of the second fixation on the target. The morphological effect was not found to interact with the experimental task. This lends support to the view that some morphological parsing is taking place already during lexical access. If the effect had been primarily post-lexical in nature, it should have been more pronounced in lexical decision than in naming.

The effect of morphological complexity seemed to be restricted to inflected words. There was no indication in the data suggesting that derived words would need an extra processing effort like morphological decomposition. This is in accordance both with Laine and Niemi's earlier studies, which demonstrated a morphological complexity effect only for inflected words, as well as with their Stem Allomorph/Inflectional Decomposition (SAID) model. Using the same stimulus material as in the present study and a standard lexical decision task, Niemi, Laine & Koivuselkä-Sallinen (1991) found that inflected nouns produced slightly longer response times (658 msec) than base-form (635 msec) or derived nouns (629 msec). According to the SAID model, a speaker of Finnish represents and accesses most inflected words in a morphologically decomposed form, while derived words are accessed as single entities. The evidence rests on the assumption that the extra processing time observed for inflected words is spent in decomposing the word into its morphological constituents.

While inflected nouns produced longer individual fixation durations, they were at the same time read with fewer fixations than the two other wordtypes. This finding is quite interesting. It may reflect the fact that the stem of the inflected words (i.e., auto of auto+ssa) was, on the average, 3 letters shorter than that of the base-form or the derived words, where the test items were the free-standing stems. Consequently, if the stem is used to access the internal lexicon, as suggested by Taft and Forster (1975) (see also Jarvella, Job, Sandström & Schreuder, 1987, for a similar argument in Italian), there would be less need to make a refixation towards the end of the word. In other words, being of shorter length, the stem of an inflected word can more easily be perceived with one fixation, which in turn would trigger the lexical access and lessen the need to make another fixation on the word. Across the two experiments, 38.4 % of the inflected words were read with one fixation, the occurrence of the one fixation cases being less frequent for the other wordtypes; the percentages were 33.0 and 31.7 for the base-form and the derived words, respectively. The increased fixation durations for inflected words may in turn reflect the extra processing needed for the identification of the inflection after the lexical lookup (cf. Jarvella et al, 1987).

The fact that an analogous morphological effect was obtained both for the first and the second fixation durations argues against all models that posit a qualitative distinction between the first and second fixation on a word (cf. Inhoff, 1984; O'Regan, 1992). The present data are more consistent with the view that cognitive processes taking place during word recognition are spilled over successive fixations (see also Hyönä & Jarvella, 1993). The processing initiated during the first fixation may be continued during the second fixation. Consequently, gaze duration (i.e., the total fixation time) is sometimes found to reflect the on-line word processing more reliably than individual fixation durations. It is surprising that in the present study gaze duration was not affected by morphological structure. This insensitivity seems to be due to the multiple fixation cases (i.e., cases with more than two fixations) not behaving in accordance with the one and two fixations cases.

When standardly used, lexical decision is found to be a clearly slower mental process than word naming (e.g., Jarvella et al., 1987). In the present study, however, the total processing times, measured as gaze duration, did not differ markedly between the two tasks. This may be due to the fact that in the present study subjects were not required to give a manual response. When compared to Niemi et al.'s (1991) standard lexical decision times, the gazes in the lexical decision experiment were, on the average, 132 msec faster. Responding by eyes seems to be very natural as our subjects felt comfortable with both tasks from the very beginning.

In lexical decision, the eye response method also yields a measure of initial hestitation in giving an answer. We found that about 5 % of the target items were initially given a wrong response followed by a correction. Inflected words tended to produce more hesitations than other words. All in all, employing an eye response in conducting lexical decision and naming experiments proved very promising, and the method certainly merits further experimenting (see also Stampe & Reingold, this volume).

Acknowledgements

This study was financially supported by the Academy of Finland. We are grateful to Heli Hujanen for her help in the data analysis.

References

Butterworth, B. (1983). Lexical representation. In Language Production, Vol. 2, B. Butterworth Ed., Academic Press, New York, pp. 257-294.

Frauenfelder, U.H. & Schreuder, R. (1992). Constraining psycholinguistic models of morphological processing and representation: The role of productivity. In Yearbook of Morphology 1991, G. Booij & J. van Marle Eds., Kluwer, Dordrecht, pp. 165-183.

Henderson, L. (1985). Toward a psychology of morphemes. In Progress in the Psychology of Language, A.W. Ellis Ed., Erlbaum, London, pp. 15-72.

Hyönä, J. & Jarvella, R.J. (1993). Time course of context effects during reading: An eye fixation analysis. In Perception and Cognition: Advances in Eye Movement Research, G. d'Ydewalle & J. Van Rensbergen Eds., North-Holland, Amsterdam. pp 239-249.

Inhoff, A.W. (1984). Two stages of word processing during eye fixations in the reading of prose. J. Verbal Learn. and Verbal Beh., 23, 612-624.

Jarvella, R.J., Job, R., Sandström, G. & Schreuder, R. (1987). Morphological constraints on word recognition. In Language Perception and Production: Relations between Speaking, Listening, Reading, and Writing, A. Allport, D. MacKay, W. Prinz, & E. Schreerer Eds., Academic Press, London, pp. 245-262.

Karlsson, F. (1983). Suomen kielen äänne- ja muotorakenne ['Finnish Phonology and Morphology']. WSOY: Juva.

Laine, M., Niemi, J., Koivuselkä-Sallinen, P., Ahlsén, E. & Hyönä, J. (in press). A neurolinguistic analysis of morphological deficits in a Finnish-Swedish bilingual aphasic. Clin. Ling. & Phon.

Laine, M., Niemi, J., Koivuselkä-Sallinen, P. & Hyönä, J. (in press). Morphological processing of polymorphemic nouns in a highly inflecting language. Cogn. Neuropsych.

Muller, P.U., Cavegn, D., d'Ydewalle, G. & Groner, R. (1993). A comparison of a new limbus tracker, corneal reflection technique, Purkinje image eye tracking and electro-oculography. In Perception and Cognition: Advances in Eye Movement Research, G. d'Ydewalle & J. Van Rensbergen Eds., North-Holland, Amsterdam, pp. 393-401.

Niemi, J., Laine, M. & Koivuselkä-Sallinen, P. (1991). Recognition of Finnish polymorphemic words: Effects of morphological complexity and inflection vs.
derivation. In Papers from the 18th Finnish Conference of Linguistics, J. Niemi Ed., Studies in Languages 24, University of Joensuu, pp. 114-132.

Niemi, J., Laine, M. & Tuominen, J. (in press). Cognitive morphology in Finnish: Foundations of a new model. Lang. & Cogn. Proc.

O'Regan, J.K. (1992). Optimal viewing position in words and the strategy-tactics theory of eye movements in reading. In Eye Movements and Visual Cognition: Scene Perception and Reading, K. Rayner & H.A. Whitaker Eds., Erlbaum, Hillsdale, NJ, pp. 363-383.

Taft, M. & Forster, K.I. (1975). Lexical storage and retrieval of prefixed words. J. Verbal Learn. and Verbal Beh., 14, 638-647.

EFFECTIVE VISUAL FIELD SIZE NECESSARY FOR PROOFREADING DURING JAPANESE TEXT EDITING

Naoyuki Osaka[a], Mariko Osaka[b] and Hitoshi Tsuji[c]

[a] Department of Psychology, Faculty of Letters, Kyoto University, Kyoto 606, Japan

[b] Department of Psychology, Osaka University of Foreign Studies, Mino, Osaka 562, Japan

[c] Educational Center for Information Processing, Kyoto University, Kyoto 606, Japan

Abstract

Effective visual field size and reading time necessary for visual search was measured during proofreading of Japanese text. Proofreading performance in detecting erroneously printed characters in both hirakana and kanji was measured using a variable moving window, through which the subject could read the areas of the text. The moving window based on the current eye position, was generated in real time on a computer-controlled CRT screen, determined by an eye movement recording system (600 points/sec). Movement of the window in which text was presented was dynamically controlled. Subjects were asked to detect transposed characters (Mozer, 1983) both in hirakana and kanji (words were composed of 2 to 4 characters). Detection of erroneously printed words was found to be relatively easier in kanji than hirakana. The effective visual field size tended to change depending on the type of script. Furthermore, as the window size increased, the proofreading rate increased significantly.

Individual differences between effective visual field size and reading span score were also investigated. The subjects were divided into high and low score groups using reading span tests, which can measure working memory efficiency during Japanese text reading (Osaka & Osaka,1992). The results suggest that effective field size tended to increase for subjects with higher scores and in general, reading performance was better for subjects with higher scores. The results are discussed in terms of working memory efficiency and field size.

Keywords

reading, proofreading, effective visual field, Japanese text, peripheral vision

Introduction

Many studies have investigated the process of proofreading. Although results of proofreading studies have important implications for reading, the two tasks appear quite different (Rayner & Pollatsek, 1989). Proofreading asks the subject to detect typographical errors, misspellings, and omitted words, while reading asks the subject to understand the text. Thus, proofreading is likely to depend on visual search.

Eye Movement Research/J.M. Findlay et al. (Editors)

Performance of proofreading during text processing is likely to depend on several factors such as text font, text/background color, letter/word density, text difficulty, orthographic-, logographic-, and semantic-similarity as well as the reader's lexical knowledge. Furthermore, it was found that proofreading accuracy was significantly worse on a CRT than on paper when using the same method of signaling errors in each case (Wilkinson & Robinshaw,1987).

For normal reading, the size of the effective visual field during a single eye fixation has been found to be relatively small (McConkie & Rayner, 1975; O'Regan, 1979; Osaka, 1987, 1992; Osaka & Oda, 1991). This small area contains the foveal and parafoveal regions of the retina.   Visual acuity declines sharply as the distance from the fovea increases, with the difficulty in identifying a word or character increasing as the distance from the fovea increases. Thus, saccadic eye movement plays an important role in bringing words into the foveal region for detailed analysis as we read. However, the size of the effective visual field size necessary for proofreading and visual search has not been studied. The question is whether a misspelled word can be detected in the parafoveal preview region or in the foveal decision area only.

Using sophisticated computer-controlled experiments, McConkie and Rayner (1975) found that parafoveal vision plays an important role during normal reading. To estimate the size of the effective visual field during text reading, they used a moving-window technique to control the amount of information available to the reader during each eye fixation. The reader's eye movements were monitored by a computer and information concerning current eye position was fed into a computer program which controlled the cathode-ray tube (CRT) from which the subject was reading. Changes were made on the basis of the position of the subject's fixation. McConkie and Rayner (1975) used a passage of mutilated text which was initially presented on the CRT with each character in the original text replaced by the character x, then wherever the subject fixated, a region around the current fixation point reverted to readable text. This window region moved in synchrony with the subject's eye movements so that wherever the subject fixated, normal text was exposed. However, everywhere outside the window region mutilated text remained. By changing the size of the window, the size of the effective visual field for proofreading could be estimated.

This study used a method similar to that of McConkie and Rayner (1975) to estimate the extent to which subjects can obtain sufficient information outside the foveal region for proofreading. As McConkie and Rayner (1975) did, we changed window size to determine whether parafoveal information used in proofreading tends to be obtained further to the right of the fixation point during Japanese text reading.

Method

Apparatus

Eye movements were recorded with an eye-mark recording system (NAC Corporation, EMR 600), in which eye positions could be recorded at a rate of 600 points/sec using a corneal reflected infrared light. The current eye position data were utilized for movement of the window. Current raw positional data from EMR600's digital output were fed into the computer so that linear position data could be calculated by simultaneous equations. The calculated linear position data were used to control movement of the current window. The computations were performed and resulting moving window position was adjusted using a PC system with numerical co-processor (Epson Corporation, PC386).

The moving-window technique involves highly sophisticated real-time feedback and provides definitive information about the size of the effective visual field. It uses on-line recording of the subject's eye positions to generate text on a computer-controlled display, depending on where the subject is fixating (Osaka & Oda, 1994). Characters of the text are replaced with gray-mask patches, except in an experimenter-defined window region around the subject's current point of fixation. Wherever the subject focuses, the original text is observed through the window, while outside this window the text is obliterated. When the subject moves his or her fixation point, a new region of text appears, and the previous region is obliterated.

Using the moving-window technique with various window sizes, we estimated the visual field size necessary to proofread effectively.

Figure 1 shows the block diagram of the moving-window generator system. Five different windows sizes, i.e., 1, 2, 4, 6, and 8 character windows, were used for each text.

Text for proofreading

Twenty different sets of written Japanese text, each of which contained about 200 (25 characters x 8 lines) characters, were used. They consisted of normal Japanese text, composed of both kanji and hirakana. One character subtended a visual angle of about 0.5 deg. The passages were selected from modern texts. The kanji contribution factor (Osaka, 1989) of the composite texts was about 30%. Each set contained four misspelled words (two each kanji and hirakana). The misspelled words were selected from content words to avoid the function-word-dependent missing effect, i.e., the tendency to miss characters in short function words (Haber & Schindler, 1981; Healy, 1980; Healy & Drewnowski, 1983). The misspelled words were selected from both kanji (50%) and hirakana (50%). Misspelled words were

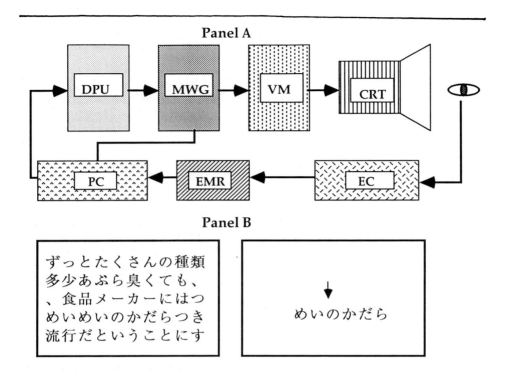

Figure 1. Panel A: Moving window generator system: EMR600 NAC eye movement recording system; EC, eye camera; EMR, eye mark recorder; DPU, data process unit; VM, video memory; MWG, moving window generator; CRT, cathode ray tube; PC, personal computer system (Epson PC386). Panel B: Text screen (left; partial view) and the text observed through the moving window (6-character window)(right). An arrow indicates an experimenter-defined fixation position (not actually visible).

created as follows: We interchanged (migrated) the position of the given character with the position of another character in a 3- to 5-character-word for hirakana or kanji. The order they appeared in the text was randomized.

Procedure

After a calibration period for saccade and fixation measurements, each subject was asked to read the text from left to right (horizontal reading) silently, as it appeared on the screen. Training sessions were performed for 8, 4, and 1 character windows before the experimental session.

When the subject detected a misspelling, he or she was asked to report it immediately. Proofreading time was measured for each window condition. Each subject received 4 sessions for each of the 5 window conditions, in random order. Eye movements were recorded from either the left or right eye of the subject.

After each session, the subject was asked to answer 3 questions regarding the text contents to determine whether he/she understood the text.

Subjects

Seventeen undergraduate students with normal vision were the subjects.

Reading span test

The Japanese reading span test (RST) (Osaka & Osaka,1992) was used to measure working memory efficiency during text reading (Daneman & Carpenter,1980; Just & Carpenter,1992). Two groups were selected of high (7 subjects with RST score of 4.0 - 4.5) and low (7 subjects with RST score of 2.0 - 2.5) reading ability.

Results and discussion

Proofreading vs window size

As Figure 2 shows, proofreading time decreased as window size increased up to the 4 character window. An ANOVA was performed on reading time data. The main effect of window size was found to be statistically significant, $F(4,64)= 27.04$, $p<.0001$.

Figure 3 shows the correct answers (%) on the comprehension test after each session. The ANOVA showed no significant differences among window sizes. Figure 4 shows the percentage of missing error of both hirakana and kanji as a function of moving window size. A two-way [5 (window sizes) x 2(target type; hirakana and kanji] analysis of variance (ANOVA) revealed a significant main effect of target type $F(1,16)=24.60$, $p<.0001$. As Figure 4 indicates, hirakana errors appear to be more frequently missed than kanji errors. Window size had significant effect on error detection $F(4,64)=3.10$, $p<.05$, and the interaction between the target type and window size was also found to be significant $F(4,64)=2.99$, $p<.05$. Close inspection of differences between the two types of targets for each window size, showed a tendency for the errors to increase as window size increased up to a 4 character window for hirakana but not for kanji. This suggests that the detection of missing characters depends on critical window size, i.e., a 4 character window: Kanji performance remained stable in all window sizes, while hirakana migration errors tended to

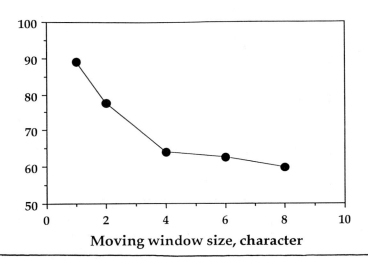

Figure 2. Proofreading time as a function of moving window size.

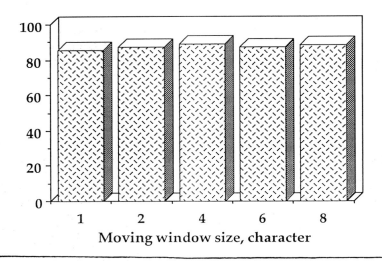

Figure 3. Correct answers (%) on the comprehension test as a
function of window size

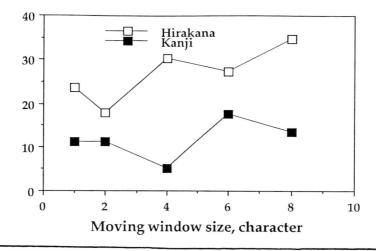

Figure 4. Percentage of missing error as a function of window size.
Parameter is type of erroneously printed word (hirakana and kanji)

escape detection in windows larger than 4 characters.

As Figure 4 shows, the percentage of hirakana missed was larger than that for kanji. This indicates that an illusory word conjunction or letter migration (Mozer, 1983; Treisman & Souther, 1986) occurred only for alphabet-like hirakana and not for logograph-like kanji.

Reading span test

Figure 5 shows reading time as a function of window size of two RST groups. Reading time is longer for subjects with low RST. However, there was no significant difference in reading time between subjects with high or low RST scores. Figure 6 and 7 each shows missing character (%) of hirakana and kanji, respectively. The parameter is the RST score. Three-way ANOVA revealed that the differences due to type were significant, $F(1,12)=1.5$, $p<.001$. However the main effect of RST and the interaction term (type x window) was not significant. Further analysis (two-way ANOVA; RST x window for each type) indicated that, as Figure 7 shows, the significant effects of RST was detected on missing character of kanji, $F(1,12)=5.65$, $p<.05$. However, as for detecting missing hirakana characters, there was no difference between high RST and low RST groups. Thus, kanji appears to be affected by working memory while hirakana is not. Interestingly, we could find distinctive working-memory-dependent differences between kanji and hirakana; which possibly causes aphasia in the Japanese patient. i.e., so called *Gogi* aphasia (Sasanuma,1975).

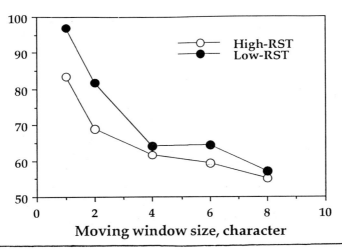

Figure 5. Reading time as a function of window size. Parameter is RST score

One study suggested that more information is being processed per fixation during the search task than during reading, and therefore visual search rate was found to be faster than reading rate (Spragins, Lefton, & Fisher, 1976). However, the present study suggests that this is not true: The usual effective visual field size was found to range from 4 to 6 characters during Japanese text reading (Osaka, 1989, 1990,1992; Osaka & Oda, 1991). As Figure 2 indicates, proofreading time decreased as window size increased, while missing characters increased as window size increased (Figure 4).

The error detection performance for narrow window sizes (from 1 to 2) indicates the reader has to read the text using character-by-character tactics so that he or she could maintain good error detection performance in the text area near the target during proofreading in Japanese. We changed the window size to determine whether parafoveal information used in proofreading tends to be obtained further to the right of the fixation point in terms of Japanese text reading. However, the results showed that preview information, obtained from the parafovea, did not improve the proofreading performance. Thus, the size of the effective visual field for proofreading appears to be relatively small (less than 4 character-window estimated from Figure 4) as compared with normal Japanese text reading.

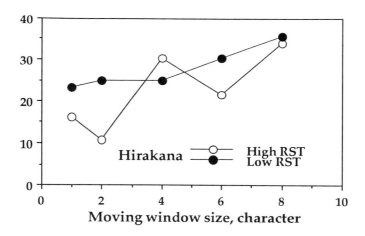

Figure 6. *Missing hirakana character (%) as a function of window size.*
Parameter is RST score

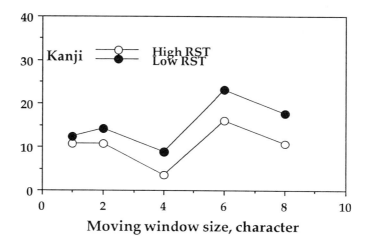

Figure 7. *Missing kanji character (%) as a function of window size.*
Parameter is RST score

References

Daneman, M., & Carpenter, P.A. (1980). Individual differences in working memory and reading. *Journal of Verbal Learning & Verbal Behavior*, 19, 450-466.

Haber, R.N., & Schindler, R.M. (1981) Error in proofreading: Evidence of syntactic control of letter processing? *Journal of Experimental Psychology: Human Perception & Performance*, 7, 573-579.

Healy, A.F. (1980) Proofreading errors on the word the: New evidence on reading units. *Journal of Experimental Psychology: Human Perception & Performance*, 6, 45-57.

Healy, A.F.,& Drewnowski, A. (1983) Investigating the boundaries of reading units: Letter detection in misspelled words. *Journal of Experimental Psychology: Human Perception & Performance*, 9, 413-426.

Just, M.A., & Carpenter, P.A. (1992). A capacity theory of comprehension; Individual differences in working memory. *Psychological Review*, 99, 122-149.

McConkie, G.W., Rayner, K. (1975). The span of the effective stimulus during a fixation in reading. *Perception & Psychophysics*, 17, 578-586.

Mozer, M.C. (1983). Letter migration in word perception. *Journal of Experimental Psychology: human Perception & Performance,* 9, 531-546.

O'Regan, K. (1979). Saccade size control in reading: Evidence for the linguistic control hypothesis. *Perception & Psychophysics*, 25, 501-509.

Osaka, N. (1987). Effect of peripheral visual field size upon eye movements during Japanese text processing. In J.K.O'Regan & A.Levy-Schoen (Eds.), *Eye movements: From physiology to cognition* (pp.421-429). Amsterdam: North Holland.

Osaka, N. (1989). Eye fixation and saccade during kana and kanji text reading: Comparison of English and Japanese text processing. *Bulletin of the Psychonomic Society*, 27, 548-550.

Osaka, N. (1990). Spread of visual attention during fixation while reading Japanese text. In R.Groner, G.D'Ydewalle & R.Parham (Eds.),*From eye to mind* (pp.167-178). Amsterdam: North Holland.

Osaka, N. (1992). Size of saccade and fixation duration of eye movements during reading: Psychophysics of Japanese text processing. *Journal of the Optical Society of America A*, 9, 5-13.

Osaka, N., & Oda, K. (1991). Effective visual field size necessary for vertical reading during Japanese text processing. *Bulletin of the Psychonomic Society*, 29, 345-347.

Osaka, N., & Oda, K. (1994). Moving window generator for reading experiments. *Behavior Research Methods, Instrumentation & Computers.* 26, 49-53.

Osaka, M., & Osaka, N. (1992). Language-independent working memory as measured by Japanese and English reading span tests. *Bulletin of the Psychonomic Society*, 30, 287-289.

Rayner, K.,& Pollatsek, A. (1989). *The psychology of reading*. New Jersey: Prentice-Hall.

Sasanuma, S., & Monoi, H. (1974). The syndrome of *Gogi* (word meaning) aphasia. *Neurology*, 25, 627-632.

Spragins, A.B.,Lefton, L.A.,& Fisher, D.F. (1976) Eye movements while reading and searching spatially transformed text: A developmental examination. *Memory & Cognition*, 4, 36-42.

Treisman, A.,& Souther, J. (1986) Illusory words: The roles of attention and of top-down constraints in conjoining letters to form words. *Journal of Experimental Psychology: Human Perception & Performance,* 12, 3-17.

Wilkinson, R.T.,& Robinshaw, H.M. (1987) proof-reading: VDU and paper text compared for speed, accuracy and fatigue. *Behaviour & Information Technology*, 6, 125-133.

Acknowledgement

This work was supported in part by grant from the Japan Society for the Promotion of Science under the *International Project on Comparative Reading Behavior* and by grants *Kansei information processing projects* (representative Prof. S.Tsuji; modelling section represetntative M.Nagao), #02801014, #03801010, and #03401003 from the Ministry of Education, Japan. Thanks are due Koichi Oda for help in making moving-window software for previous (EMR-V) version.

DISPLAYS AND APPLICATIONS

SELECTION BY LOOKING: A NOVEL COMPUTER INTERFACE AND ITS APPLICATION TO PSYCHOLOGICAL RESEARCH

Dave M. Stampe and Eyal M. Reingold

Department of Psychology
University of Toronto, Toronto, Canada

Abstract

Real time monitoring of a subject's gaze position on a computer display of response options may form an important element of future computer interfaces. Response by gaze position can also be a useful tool in psychological research, for example in a visual search task.

Experimental tasks reported include visual search, and typing from an alphabetic menu. A lexical decision study revealed gaze response RT to be much more powerful than button press RT and new phenomena including self-correction were observed. Methods of improving response reliability are introduced, including drift correction by dynamic recentering, gaze aggregation, and automated selection.

Keywords

Eye-movements, eye-tracking, computer-interface, control, handicapped, lexical-decision, typing, disabled, visual search

Introduction

Psychological research using eye movements has focused on the study of natural tasks such as reading or problem solving. Typically, gaze position is simply recorded for later analysis, and is incidental to the experimental task. One exception is the gaze-contingent display paradigm, where the displayed stimuli are changed rapidly in response to eye movements in order to present different images to the foveal and peripheral regions of the subject's visual field (e.g. McConkie and Rayner, 1976). In this paper, we explore the active use of gaze in performance of an experimental task: selection by looking. We investigate implementation and psychophysical issues, and demonstrate tasks which illustrate the unique potential of this paradigm.

In a typical task, the subject's gaze position is monitored while viewing displays such as those in Figure 1. The subject registers responses or commands by directing his gaze to targets on the screen and holding his gaze until the response is registered. The gaze-response system's computer processes the eye tracker output in real time to compute gaze position on the screen and detect response events, then modifies the image displayed to the subject in response to the gaze input.

The use of control by eye-movement as an aid to handicapped persons is not new (e.g. Laefsky and Roemer, 1978), but such systems have usually been developed with little research into psychophysical or cognitive factors.

Eye Movement Research/J.M. Findlay et al. (Editors)

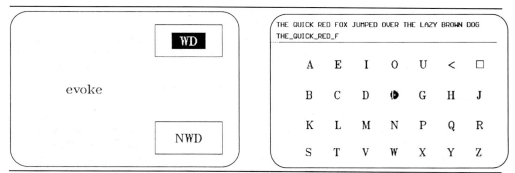

Figure 1. Gaze control displays from two experimental tasks. Left: the lexical decision task display. Sufficient gaze duration on the "WD" or "NWD" response areas registers an appropriate response, and feedback is given by highlighting the selected response. Right: the eye-typing task. Fixating a character will result in it being typed in the upper part of the screen, and feedback is given by a round dot.

Recently, eye movements have been proposed as computer interface devices for normal users (e.g. Jacob, 1991), but despite high expectations, the advantages and limitations of this interface modality have yet to be determined. There are many tasks in which gaze control would be advantageous, for example as an input device when hands are otherwise occupied (Charlier, Sourdille, Behague, and Buquet, 1991).

In this paper, we will discuss some of the advantages of gaze position response, then summarize the practical aspects of implementing gaze response systems. Dynamic recentering will be introduced as a technique to correct for eye tracker drift to prevent response errors. Results from several experimental tasks will be presented that explore implementation issues, demonstrate the methods discussed, and compare response by gaze to more traditional button press responses.

Advantages of Gaze Response in Research

Gaze response is well suited to typical eye movement research tasks such as visual search within pictures or stimulus arrays. Once the search target is found all the subject need do is to continue fixating the target until the response is registered. This is a highly intuitive response method, requiring no training and resulting in fast reaction times. Targets can be selected even when embedded within pictures or dense arrays of distractors. Gaze response acts as a pointer, providing two-dimensional input similar to devices such as touch screens, light pens or computer mice. However, these physical input devices require gross motor movements, which may introduce noise into the eye movement record. Verbal responses can be used to implement complex responses paradigms, but are not practical with eye tracking systems that require bite bars or chin rests. Gaze response not only does not exhibit these problems, but actually enhances the eye movement record.

In eye movement research, gaze response can help to disambiguate the temporal relationship between eye movements and responses during task performance.

Asynchrony between eye movements and manual or vocal responses is often seen, as eye movements are free to continue with task execution after the motor program for the response is initiated. With gaze response, the temporal relationship between eye movements and responses is clear, since both cannot be performed at once. Task-related fixations (e.g. searching) and response fixations are easily disambiguated during analysis, allowing the eye movement record of sequential (multiple step) tasks to be clearly divided at the gaze responses.

The use of gaze response in eye movement research is likely to reveal new aspects of cognitive processes. It is important to compare gaze response to more classical response methods such as button presses: such a comparison will be reported in this paper for a lexical decision task. The combination of intuitive operation, spatial selection capability and potential for new findings make gaze position response a powerful tool for psychological research.

Implementation Issues

For gaze response to be practical, especially for psychological research, the implementation is critical. The method chosen for detection of gaze responses in the eye movement data must be carefully chosen to prevent unintended responses. In the following discussion we will evaluate the effects of eye tracker performance on response accuracy, and discuss methods for correcting eye tracker drift.

Detecting Gaze Responses

The most natural technique for gaze response is simply to hold gaze on the response area for a critical time period. Subjectively, the subject simply concentrates on a target until the selection response occurs. Most subjects can use this technique immediately, and need little or no practice to perform well. The duration of gaze needed to select a target, also called dwell time (Jacob, 1991), must be short enough to be comfortable for the subject, yet long enough to prevent unintentional triggering. Fixations with durations greater than 500 msec are often seen during cognitive integration phases of difficult tasks, and could be mistaken by the computer for response events. Pilot studies indicated that a dwell time of 1000 msec makes such false selections unlikely, and 700 msec or less works well for simple tasks.

System responses to gaze input must be quick, correct and predictable to encourage linking of gaze and response by the user. It is disconcerting if the target next to the intended selection is selected because the eye tracking system has miscalculated the gaze position due to eye tracker drift. Subjects become frustrated if the gaze time required to select a response is unpredictable or if selection does not occur at all due to system instabilities. Careful layout of response targets, use of high quality eye tracking systems, drift correction, and reliable response detection methods can prevent these problems.

Ideally, gaze on a response target would be detected as a single long fixation. Initial investigations indicated that gaze durations longer than 800 msec gaze are often broken by blinks or corrective saccades. Single fixations are an unreliable measure of gaze duration: it is necessary to aggregate several nearby fixations into a gaze period, for example by cluster analysis (Kundel, Nodine, and Krupinski, 1989). Fixations may also be grouped within a region surrounding each response target. Responses are registered when the sum of the duration of all fixations within the cluster or region exceeds the dwell time threshold, and the average position of gaze may be computed.

Eye Tracker Drift Correction

It is important for any eye monitoring system to have good resolution, accuracy, and stability. Almost all eye tracking systems exhibit drift over time, with computed gaze position gradually moving away from the subject's true gaze location. Severe drift can be caused by head movement, or motion of eye cameras relative to the eye, but even systems that control these may require periodic drift correction.

While a complete system recalibration will correct drift, it is much more efficient to measure the drift directly and compensate for it. This is performed by displaying a fixation target to the subject, then measuring the deviation of computed gaze from the target position. This process of recentering may be done between each trial or block of trials, and dramatically improves stability (Stampe, 1993).

Dynamic Recentering

A novel drift correction technique was developed which is unique to gaze position control and is invisible to the subject. Assuming that the average gaze position during target selection falls at the center of the response target, we can compute drift as the mean offset between the target and gaze position at each selection. Small variations in gaze position on targets will be averaged out over several selections. This technique dynamically performs the recentering operation to correct system drift at each gaze response event, and can be combined with normal recentering to correct for larger drifts. Drift usually accumulates slowly, and the mean error from several fixations of targets will track it closely. Sudden increases in drift will be corrected over several selections.

Dynamic recentering is implemented by a low-pass filter which tracks the drift component of target fixation error while ignoring small random differences in target fixation. A step-by step description of the dynamic recentering algorithm is:

1) Subtract the estimated drift from the uncorrected gaze position to compute the corrected gaze position. The drift estimate is initially set to zero after recentering or calibration, or is carried over from step 3 of the previous selection's correction.

2) Subtract the location of the true target center from the corrected gaze position to compute the residual fixation error.

3) Add a fraction (1/4 to 1/6) of the residual fixation error to the estimate of drift. This will reduce the error in the estimate of drift, eventually eliminating the fixation error. Random variations in target fixation will average out over time.

Experimental Tasks

Three experimental tasks were performed to validate the gaze response system and to explore important aspects of the gaze control paradigm. The first task measured fixation accuracy and selection error rates, and evaluated the effectiveness of dynamic recentering in correcting drift. The second task demonstrated the efficacy of dynamic recentering in a typing by eye paradigm. The third task contrasted button press and gaze response methods in a simple lexical decision task.

General Method for Tasks

Experimental tasks were implemented using a prototype eye tracking system developed by SR Research Ltd. This system uses a headband-mounted video camera and a proprietary image processing card to measure the subject's pupil position 60 times per second. Resolution of the system is very good (0.005° or 15 seconds of arc), with extremely low noise. A second camera on the headband views LED targets attached to the display monitor to compensate for head position, correcting gaze position to within 0.5° of visual angle over ±20° of head rotation, and allows head motion within a 100 cm cube.

Task displays were presented in black on a white background on a 21" VGA monitor located 75 cm in front of the subject, with a field of view of 30° horizontally and 24° vertically. A second VGA monitor was used by the experimenter to perform calibrations and monitor subject gaze in real time during experiments. Gaze position accuracies of better than 0.5° on all parts of the screen were routinely obtained.

Twelve subjects, five male and seven female with an average age of 25 years were run on all tasks in a single 60-minute session. Tasks were run in the same order for all subjects. All subjects had normal or corrected to normal vision, four with eyeglasses and three with contact lenses. A system calibration was performed before each task, and repeated if needed to meet a 0.5° accuracy criterion (Stampe, 1993). The experimenter monitored gaze position during trials to ensure continuing accuracy.

TASK 1: Array Search and Selection

The first task explored the effect of target layout and dynamic recentering in tasks where arrays of targets representing multiple response alternatives are used. The distance between targets affects the likelihood that an eye tracking or fixation error will cause a target adjacent to the intended one to be selected in error. One dimensional (line) and two dimensional (grid) arrays were investigated, and visual density of the target array was also manipulated.

Method
Subjects were required to indicate a search target "T" hidden in an array of "O" distractors by gaze. The search target was highly salient, keeping trials short and

minimizing non-selection errors. All characters in the search array were selectable by gaze, with selection of distractor characters counted as selection errors.

Three search array configurations were used in the task: a square grid, a horizontal line, and a vertical line. Line arrays contained 6 characters spaced by 3° or 8 characters spaced by 2°. The small character spacing was designed to increase error rates, which were known from pilot studies to be vanishingly small for target spacings of 4° or greater. Grids consisted of a square array of 6 by 6 characters spaced by 3° or 8 by 8 characters spaced by 2°. All characters were 0.6° in size.

A total of 124 trials were presented, consisting of 32 8x8 grids, 36 6x6 grids, 16 each of horizontal and vertical lines arrays spaced by 2°, and 12 each of horizontal and vertical lines arrays spaced by 3°. Complete sampling of all target position and arrays required 496 trials, which were divided between 4 subjects to reduce task length, for a total of 124 trials per subject.

Trials were presented in a random sequence in four blocks of 31 trials each, separated by recentering screens to correct eye tracking system drift. The first 8 trials for each subject were discarded as practice. Search arrays were displayed until a target was selected by a gaze duration of 1000 msec, and the screen blanked for 200 msec between trials. Gaze position was aggregated by cluster analysis with a diameter of 1.5°, with position recorded for use in analysis. Dynamic recentering was simulated during analysis.

Results and Discussion

Two error measures were computed for each selection: fixation error and selection error. The calculation of fixation error was determined by the target array type. For horizontal line arrays, the horizontal distance from the search target to the gaze location was used as the fixation error, the vertical distance for vertical line arrays, and the largest of either the horizontal or vertical distance for grid arrays. Selection errors were considered to have occurred if the fixation error exceeded half of the intertarget spacing.

Gaze position was corrected by the dynamic recentering method, allowing both corrected and uncorrected target fixation error magnitudes (target position minus gaze position) to be measured. Dynamic recentering was found to be effective with mean corrected fixation errors of 0.38° versus 0.51° before correction: $t(11) = -4.72$, $p < .001$. More importantly, dynamic recentering was highly effective in reducing the frequency of selection errors from 6.6% before correction to 2.4% after correction: $t(11) = 2.71$, $p < 0.05$. The relationship between the magnitude of fixation error and probability of selection error (gaze falling on response targets neighboring the intended target) is best illustrated by a plot of the fixation error distribution before and after dynamic recentering was applied (Figure 2).

A fixation error larger than half of the distance between response targets places the gaze position closer to one of the search distractor targets than the search target itself, producing a selection error. For example a fixation error greater than 1.5° will cause selection of an adjacent target in an array of targets spaced by 3°.

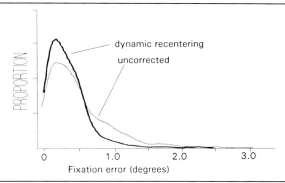

PROPORTION

dynamic recentering

uncorrected

0   1.0   2.0   3.0
Fixation error (degrees)

Figure 2. *Distribution of fixation errors by magnitude in the search task before and after correction by dynamic recentering. Note the long error tail in the uncorrected error distribution caused by eye tracker drift. The incidence of fixation errors larger than 0.6° is greatly reduced by recentering, decreasing the incidence of selection errors.*

Figure 2 also reveals a long tail caused by eye tracker drift, which results in significant numbers of selection errors at target spacings of up to 3°. For example, at 2° target spacing, fixation errors greater than 1° produce selection errors, and occur in 9.5% of trials before dynamic recentering, and 2.8% of trials after correction.

Three dependent measures (fixation error, selection error, and number of fixations) were analyzed after dynamic recentering was applied, using a 3 by 2 analysis of variance. Array type and target spacing were evaluated as within subject factors. Means and $\underline{F}$ values are summarized in Table 1.

The horizontal line arrays showed significantly smaller fixation errors than vertical and grid array, attributable to the largely vertical component of drift for this eye tracking system, a characteristic shared by most eye trackers we have tested. This is also show by comparing horizontal and vertical errors collapsed across array types, the mean vertical error of 0.38° was greater than mean horizontal error of 0.23°: $\underline{t}$ = -3.73, $\underline{p} < .01$. Note that no selection errors occurred for horizontal one-dimensional target arrays at all: thus it should be possible to place response targets closer together horizontally than vertically when designing gaze response screens.

Comparisons of selection error rates of 2.8% for 2° target spacing and 1.9% for 3° target spacing were not significant (Table 1) as the rarity of selection errors inflated the variance. The means do indicate that errors decrease with large target spacings, and pilot studies had indicated that errors with target spacings of 4° or greater are very rare.

The number of fixations per trial was greater for 2° target spacing, indicating that the task became more difficult as visual density increased. This implies that well-spaced target arrays will decrease search time and improve task performance.

Variable	Level	Fixation Error	Selection Errors	Fixations
Array Type	Horizontal	0.24°	0.0%	3.55
	Vertical	0.36°	2.5%	3.51
	Grid	0.44°	3.3%	3.72
		$F(2,22) = 8.03*$	No variance	$F(2,22) = 2.70$, n.s.
Spacing	2°	0.36°	2.8%	3.68
	3°	0.39°	1.9%	3.59
		$F(1,11) = 2.88$, n.s.	$F(1,11) = 1.72$, n.s.	$F(1,11) = 10.56*$

* $p<.01$

Note: Two-way interactions were not significant.

Table 1: Summary of Main Effects in Task 1

TASK 2: Typing by Eye

Typing by eye is one of the most common applications of gaze-controlled aids for the handicapped. It is unfortunate that little investigation has been done into the cognitive aspects and efficiency of this paradigm, perhaps because of the emphasis on the implementation of such systems rather than on research with them.

Method

This task was set up to evaluate the subjects' impressions and types of error made during performance. The screen layout, as shown in Figure 1, used a 7 by 4 grid of 1.2° characters spaced by 4° horizontally and vertically. The top of the screen contained a line of text to be typed and a space for display of the typed output.

To type, subjects fixated a desired letter for 750 msec, with all fixations within a 4° region centered on each target counted towards the gaze period. Dynamic recentering was applied at each selection to correct for system drift. Selection feedback was given by placing a round highlight spot on the letter for 300 msec. If the subject continued to fixate the character, it was typed repeatedly. Typing ended when a button was pressed by the subject.

Each subject performed three typing trials. The subject first typed random input (usually their name), then two test sentences of 48 and 44 characters each. Characters typed and use of the backspace function were recorded for later analysis.

Results and Discussion

In this preliminary investigation, only simple statistics and subjective impressions were collected. Subjects enjoyed the task, but found it slow compared to manual typing. The gaze selection time of 750 msec subjectively seemed limiting, but in reality selection time was only 40% of the 1870 msec average time required to type each character, with the remaining 60% of the time spent searching for the next character in the typing array.

Errors were classified by counting backspaces and examining the typed output: transcription errors included missed characters or spelling mistakes (4 instances in 1400 characters typed, 0.29%). Selection errors were scored if a spelling mistake involved a letter adjacent to the correct letter on the selection grid, and occurred 5 times out of 1400 characters typed (0.36%). Error rates compare well to the 1.3% reported for a 54-character typing screen (Spaepen and Wouteers, 1989). All selection errors involved selection of a target above or below the intended character, analogous to the vertical selection errors seen in Task 1.

It is apparent that typing by eye is much slower than manual typing, with most time spent searching for the character to be typed. With much practice, search time may be minimized and the dwell time may be reduced further. Research has shown that typing by touch screen can be as fast as 500 msec per character (25 words/minute), and by mouse at 700 msec (17 words/minute) (Andrew, 1991). Typing by eye can probably be as fast once character positions in the typing array are memorized.

TASK 3: Lexical Decision Comparison

To demonstrate the potential of selection by looking in psychological research, this task compared reaction times for gaze response to a more typical button press response method. The experimental paradigm used was a simple lexical decision task, classifying five-letter strings as words or nonwords. Both reaction time and response accuracy measures were compared for button and gaze response methods in the analysis.

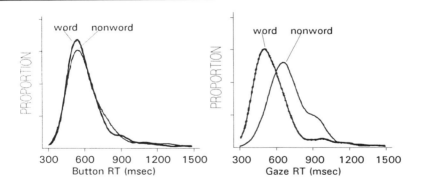

Figure 3. Smoothed histograms of word/nonword reaction times in the gaze response and button response conditions of the lexical decision task. The gaze response condition shows much greater sensitivity to the word/nonword manipulation, and a secondary peak at 900 msec that may be associated with self-correction phenomena.

Condition	Button Response		Gaze Response		
	Accuracy	RT (msec)	Accuracy	RT (msec)	Time on Stimulus (msec)
Overall	97.9%	621	99.4%	660	476
Word	97.5%	603	99.4%	572	423
Nonword	98.3%	639	99.4%	750	531

Table 2: Response Accuracy and Reaction Time (RT) Results in Task 3

Method

In the gaze response condition the subject classified the stimulus by fixating the word or nonword response areas on the screen (Figure 1). The response areas were also displayed in the button response condition to indicate word/nonword response assignments to the upper and lower buttons on a three switch button box. In both conditions the third button was used to start each trial.

Gaze and button response conditions were blocked and their order counterbalanced across subjects. The first 12 trials in each block served as practice, with 60 experimental trials following. The stimuli were selected randomly from 144 common five-letter words and 144 nonwords created by randomizing the letters of the word set. The strings were displayed in characters of 1.2° in size.

The subject initiated each trial by pressing a button. A fixation point was displayed for 300 msec, followed by the word/nonword string. The subject responded by fixating one of the response areas or by pressing a response button, depending on the block type. In the gaze response condition, selection was triggered by 850 msec of gaze within each response area. The selected response was then highlighted for 300 msec, and dynamic recentering was automatically applied at each selection to correct for system drift.

Results and Discussion

In the gaze response condition, response reaction time was measured from stimulus onset to when gaze last entered a response area: the start of the gaze period that resulted in the response. Because gaze must leave the stimulus to register a response, a second temporal measure is available: duration of gaze on the stimulus. This provides a secondary measure of processing required to make the decision. In the button response condition, the subject's gaze remained on the stimulus throughout the trial and the only meaningful measures were button-press reaction time and response accuracy.

Table 2 summarizes button and gaze reaction time measures. The overall mean RT for gaze response is not significantly different from overall button RT: $F(1,11) = 1.55$, $p = .24$. When reaction times for word and nonword stimuli are compared, the interaction between word/nonword and gaze/button conditions indicates that gaze response RT is much more sensitive to the word/nonword stimulus dimension than button RT: $F(1,11) = 21.27$, $p < .001$. Smoothed histograms of word/nonword RTs are shown in Figure 3 for gaze and button conditions. These clearly show the greater sensitivity of the gaze response paradigm to the word/nonword stimulus dimension, and display the positively skewed RT distributions typical of this task. Similarly, gaze duration on the stimulus in the gaze response condition is also shorter for words versus nonwords, $t(1,11) = 6.64$, $p < .001$, and is also more sensitive to the word/nonword manipulation than button RT: $F(1,11) = 5.47$, $p < .05$.

The proportion of correct responses in the gaze and button response conditions are summarized in Table 2. Fewer errors occur with gaze response, $F(1,11) = 10.34$, $p < .01$, as the long gaze dwell needed for target selection allowed the subject to correct their choice. Analysis of fixations preceding target selection shows that such corrections occurred in 11.5% of trials. Typical fixation time on the first response target before correction was 100 msec: such transient cognitive events can only be observed using the gaze response paradigm.

General Discussion

In general, the results from the experimental tasks suggest that gaze response is intuitive and reliable enough to be practical in many psychological research and computer interface applications. All subjects performed a wide variety of control tasks without need for any training, and were enthusiastic about the natural quality of selection by looking. These positive subjective impressions were supported by the speed and accuracy scores for the tasks.

Important to the success of the paradigm was the ability to precisely place gaze on response targets and to hold the gaze for long enough to trigger the response. Although natural gaze is often broken by blinks or refixations, the aggregation of gaze by cluster or region resulted in reliable selections and predictable gaze times. Subjects had no difficulty with dwell times requiring gaze periods as long as 1000 msec. This is in marked contrast to reported difficulty with dwell times over 700 msec by Jacob (1991), who used only single fixation as a measure of gaze duration.

Psychophysical limits on accuracy of gaze placement were not large enough to be a problem in response selection. The main source of selection errors appeared to be the result of occasional drifts in the eye tracking system. Such drift could be corrected by the use of dynamic recentering, which in Task 1 reduced selection errors by 64%. Selection errors were also reduced by increasing spacing between response targets: error rates were 2.8% for the 2° target spacing, 1.9% for 3° target spacing, and were 0.4% for the 4° target spacing used in the typing task. No selection errors occurred with horizontal arrays of targets even for 2° target spacing, as most drift in the eye tracking system used for the experiment was in the vertical direction.

If eye trackers with low resolution or with rapid drifting such as that caused by head movements are used, response targets must be widely separated. For example, the screen layout used in the lexical decision task used targets spaced by 11°, and would probably work reliably with most eye trackers. Dynamic recentering works best in systems where drift changes slowly over several responses. Sudden changes in drift will require several responses to repair. If the drift is large enough to cause target selection errors, correction will be toward the center of the erroneously selected response target and can result in shifting of responses in the target array. A conventional recentering using a single target will correct the offset.

The typing task was representative of gaze control computer interfaces. Tasks requiring reliable selection between many response targets, require the use of high-quality eye tracking devices with good accuracy and low drift. User comfort is important if gaze control is to be accepted by computer users, requiring headband-mounted or desktop eye trackers that do not constrain head motion. Computer control systems must be able to be set up and used by the user without assistance.

Gaze response shows great promise in psychological research. A commonly used lexical decision task was utilized to compare gaze and button response methods. The gaze response RT proved much more sensitive to the word/nonword manipulation than the button response RT. The new phenomena of rapid self-correction was revealed and the new measure of gaze time on the stimulus was made possible by use of gaze response. Investigation of the effects of the gaze response method in other research paradigms is likely to show similar advantages.

Acknowledgments

This research was supported by NSERC grant OGP0105451 to E. Reingold. We thank Elizabeth Bosman for her helpful and constructive comments on a preliminary version of this paper.

References

Andrew, S. (1991). Improving touchscreen keyboards: Design issues and a comparison with other devices. *Interacting with Computers, 3(3)*, 253-269.

Charlier, J., Sourdille, P., Behague, M., & Buquet, C. (1991). Eye-Controlled Microscopes for Surgical Applications. *Developments in Ophthalmology, (22)*, 154-158.

Jacob, R.J.K. (1991). The use of eye movements in human-computer interaction techniques: What you look at is what you get. *ACM Transactions on Information Systems, 9(3)*, 152-169.

Kundel, H.L., Nodine, C.F., & Krupinski, E.A. (1989). Searching for lung nodules: Visual dwell indicates locations of false-positive and false-negative decisions. *Investigative Radiology, 24*, 472-478.

Laefsky, I.A. & Roemer, R.A. (1978). A real-time control system for CAI and prosthesis. *Behavioral Research Methods & Instrumentation, 10(2)*, 182-185.

McConkie, G.W. & Rayner, K. (1976). Identifying the span of the effective stimulus in reading: Literature review and theories of reading. In H. Singer and R. Ruddel (Eds.), *Theoretical models and processes of reading*. Newark, Del.: International Reading Association.

Spaepen, A.J. & Wouters, M. (1989). Using an eye-mark recorder for alternative communication. In A.M. Tjoa, H. Reiterer, and R. Wagner (Eds.), *Computers for Handicapped Persons* (pp. 475-478). Vienna: R. Oldenbourg.

Stampe, D.M. (1993). Heuristic filtering and reliable calibration methods for video-based pupil-tracking systems. *Behavioral Research Methods, Instruments, & Computers, 25(2)*, 137-142.

EYE MOVEMENT RECORDINGS TO STUDY DETERMINANTS OF IMAGE QUALITY IN NEW DISPLAY TECHNOLOGY

Gerhard Deffner
Digital Imaging Venture Projects, Texas Instruments,
P.O. Box 655474, MS 429, Dallas, TX 75265, U.S.A.

Abstract

The paper describes and illustrates a paradigm for in-depth study of perceived quality of visually displayed images. The development of new, all-digital display technology offers unique possibilities for fine tuning the match between display devices and the human perceptual system to achieve superior image quality. Engineering efforts need to be prioritized, however, according to which image characteristics are the most important for the perception of quality. Detailed studies have frequently run into problems because subjective data have not been very reliable. The present approach uses eye-movement data recorded while subjects view and compare images on different displays. These recordings provide quantitative data, indicating which parts of the test images received how much of a viewer's attention. In addition, these recordings are played back to provide cues for subsequent retrospective verbalizations, providing more data about a subject's thought processes when judging perceived image quality.

Keywords:

Visual displays, perception of image quality, eye-movement recordings, cued retrospective verbalization

Introduction

Background

The present study is conducted to support ongoing research in product development at Texas Instruments. Digital Micromirror Device (DMD) by Texas Instruments is a new technology incorporating a fully-digital design for visual displays (Sampsell, 1993). A DMD chip can be described as a conventional memory chip with extremely small mirrors above each cell which may be tilted +/- 10 degrees, depending on the content of the underlying memory cell. In combination with a light source, projection optics, and screen, the DMD chip may be used to build projection display systems. There is no need to scan successive pixels on the screen, and modulation of brightness is achieved via pulse width modulation, i.e., very rapid on/off switching. Three primary colour lights, red, green, and blue, within each pixel are projected onto an identical location, and the desired grey level

Eye Movement Research/J.M. Findlay et al. (Editors)

or colour is achieved by controlling the length of time a particular colour of light is reflected onto a given pixel.

This technology has great potential for overcoming limitations of cathode ray tubes (CRT) and producing a new generation of high quality, high definition displays. It is possible to control critical image parameters completely because the DMD technology allows development of fully digital displays. Therefore, artifacts, such as flicker, poor colour balance, off-center loss of resolution and poor convergence, can be avoided altogether. Although full control and complete optimization of displays are possible, it may not be cost effective because of high demand it places on processing and memory resources. For this reason, it is important to determine the priority of image characteristics in their contribution to overall image quality. It is advantageous to understand what determines perceived image quality in order to prioritize engineering efforts and allocate manpower to a limited number of high priority image characteristics. Thus, examination of critical image characteristics such as contrast ratio, shadow detail, highlight detail, colour saturation, colour balance, geometric patterns, motion artifacts, etc. is the first step to optimize perceived image quality.

Examination of Image Quality Evaluation Processes: A New Approach

Ordinary consumers perform image quality evaluations of displays when they shop for a new television (TV) set. Typically, consumers conduct a series of paired comparisons, rapidly narrowing their choices down to a few sets. Once consumers have chosen a TV set, they are not likely to compare two similar sets side by side again or to perform stringent evaluations of image quality.

Previously, a study was conducted to simulate image quality evaluations by ordinary consumers in a showroom. A series of short video clips was shown on two displays, placed side by side. Subjects were asked to compare image quality of the two displays as if they were trying to make a purchase decision. They were instructed to state their preferences and express the magnitude of perceived differences of image quality in terms of expected differences in price. The results indicated that the task, though conducted in a laboratory, was realistic enough for subjects to spend considerable time and effort as if they were actually making a major purchase. Unfortunately, outcome data such as final preferences, time to reach decisions, and perceived differences, did not shed any light on cognitive processes of display image quality evaluation. Indeed, when asked to explain the reason for their preferences, subjects provided an assortment of statements unrelated to specific image characteristics; for example, subjects stated that images on one display were "crisper, " "more realistic," "easier on my eyes," "more vivid," and "smoother" than images on the other. Clearly, people are not able to express how they have reached a particular preference decision regarding image quality intelligibly. Their statements by themselves are too vague and idiosyncratic to be subjected to any analysis.

Understanding complex cognitive processes of image quality evaluation should supply a wealth of information about critical image characteristics in displays;

therefore, a method to study such processes in addition to the decision outcomes would be invaluable. Eye tracking recording serves as an ideal method to study such complex processes. It provides an opportunity for a detailed study of image quality evaluation, resulting in 'process' data indicating changes of attention over time and sequences of gazes during comparisons.

Data on cognitive processes have a greater importance than mere outcome data in some research disciplines such as human problem solving. Researchers in the area of human problem solving have made rigorous attempts to refine techniques to obtain process data. One of the techniques these researchers utilized successfully was eye-movement recording.

However, display image quality was incidental to the tasks in these studies; image quality per se, that is, apart from the content, became an issue only when poor quality of display led to degradation of performance. Even applied studies which utilized eye movement recordings have been content-bound. For example, studies on advertising efficiency (Treistman & Gregg, 1979), X-ray image examinations (Carmody, Nodine, & Kundel, 1981), search for discrepancies between nearly identical pictures (Fisher, Karsh, Breitenbach & Barnette, 1983), embedded target search (Nodine, Carmody, & Herman, 1979), and face perception (Janik, Wellens, Goldberg, & Dell'Osso, 1978) were all concerned with the contents of display images which were explicitly related to a well-defined primary task. The display media were assumed to be of sufficient fidelity, and display quality was not the object of study. The study of legibility of text displayed on screen (Kolers, Duchnicky, & Ferguson, 1981) is another example of content-bound studies, although there was indirect reference to screen quality. Only a recent study by Jorna and Snyder (1991) came close to a study of image quality. They studied display quality of still images as an important determinant of reading speeds; however, they did not include an analysis of underlying perceptual and cognitive processes.

The focus on perceived image quality in the study of eye movements appears to be a departure from traditional research paradigms. Evaluation of image quality, especially of naturalistic scenes employed in the present study, requires a complex multiple-stage processing of image materials. First, one must perceive and recognize the image. Recognition often plays an important role in the evaluation of image quality of naturalistic scenes. People have clear memory representations of certain image features, for example skin tones and colours of foods. Thus, they are likely to be more critical about the fidelity of their appearance. Second, one must abstract the content and identify critical features for evaluative comparisons. At this stage, special eye movement behaviour is likely to occur which is not directed at content; rather some eye movements may represent efforts to find and evaluate critical image features that support the task of perceptual judgment. Finally, one must assess image quality and come to a decision. Therefore, image evaluation examined in the present study requires complex multiple-stage processes.

The other successful technique used in studying cognitive processes is verbal protocols (Ericsson & Simon, 1993). There has been much debate over whether concurrent verbalization would potentially alter thought processes. Deffner (1984,

1989) found little support for such reactivity of concurrent verbalization procedures in experiments using simple tasks. Bowers and Snyder (1990) found little difference when comparing concurrent verbalization with retrospective verbalization. However, Bowers and Snyder recommend the use of retrospective verbal protocols when subjects are engaged in complex tasks. Furthermore, the possibility of concurrent verbalization inducing reactivity should not be ignored when tasks have visual and automatic components. In such cases, retrospective verbalization may be better suited.

Earlier research at the University of Hamburg combined verbal protocols with eye-movement technology to curtail the necessities to remind subjects to "keep talking" or to prompt them with questions. Deffner, Koebe, Preuss, and Völkl (1986) studied how students debugged computer programs. Initially, subjects were asked to talk about how they had found and corrected a bug in a program after the task had been completed. The subjects' answers were brief and provided little information regarding how they had found the error. Next, investigators recorded eye-movements during reading of the program text, and replayed the recordings as cues for retrospective verbalization. When confronted with this detailed trace of their prior task performance, many subjects realized they had spent a considerable amount of time reading lines not related to the error; this inititated more detailed verbalization and thus provided much richer and more detailed data. Retrospective verbalization, when coupled with eye tracking, is a powerful method which allows a deeper understanding of such cognitive processes.

An Example Experiment

If data from eye-tracking recording and retrospective verbalization converge, the resultant experimental paradigm provides a powerful technique to study image quality evaluation. Eye-movement technology has matured to the point where it may be used routinely. However, it generates large amounts of data to be integrated into a detailed analysis of the processes. The following is a description of an emerging approach. The illustrated example is taken from a pilot study, still in progress, which challenges the assumption that expert's evaluation of display quality may be reliably used to predict consumer's evaluation. Image evaluation performances of experts and non-experts are compared by recording their eye movements during image evaluation and by collecting their retrospective verbalization about decision processes. The hypothesis is that display experts will focus on critical features sooner and more exclusively while disregarding overall image content more frequently than non-experts.

Method
Subjects

Optical, electrical, and system engineers currently involved in the development of display-related products are recruited for expert subjects, and administrative personnel for non-experts. Only subjects with 20/20 uncorrected visual acuity and normal colour vision are included in this study.

Material

Fourteen still images were selected for this study. Twelve images are single-image still pictures while the remaining two are panels of nine images. Prior to the selection of the images, image characteristics considered critical in image evaluation were identified. These critical features include, colour saturation, colour brightness, white highlight, black-level retention, colour fidelity, geometric patterns, picture clarity, shadow and highlight details, contrast, texture, and text. The single-image still pictures were selected to contain only two or three of the features described above. Moreover, the critical features are dominant parts of the pictures and separated from one another. The individual pictures in 9-image panels each contain only one dominant feature representing a single critical image characteristic. Figure 1 shows one of the single-image still pictures used in the experiment. This image contains two primary critical image characteristics, colour fidelity of skin tones and geometric patterns of the blind. Secondary image characteristics within this picture include highlight details of the blind where light is coming through and contrast and edge definition between the blouse and jacket.

Apparatus

An Applied Science 1998 system upgraded to the current 4250 system is utilized to collect eye-tracking data. Images to be evaluated are played by a laser disk player which outputs to two 35" colour televisions placed side by side. Brand names of the TV sets are masked.

Figure 1. Example of a single-image still picture.

Procedure

Experiments are conducted as individual sessions. Upon arrival, subjects receive a brief description of the purpose of the study and the eye-tracking device. They are tested for visual acuity and colour deficiencies. Subjects are then seated in front of the TV sets with their heads stabilized by a set of pillows fixed against the back of the chair. After the eye-tracking device is calibrated, subjects are instructed to view a series of still images presented on the two TV sets and to express their preference regarding the quality of the images. Subjects' eye movements are recorded on video tape and computer disks.

After all images are shown, a replay of the original recording with super-imposed crosshairs, indicating subjects' points of gazes, is shown. Subjects are instructed to stop the replay to voice their thoughts as they occurred during original viewing. Their comments are recorded digitally. Following retrospective verbalization, subjects are debriefed and dismissed from the experiment.

Analysis and Results

The analyses of eye-movement and retrospective-verbalization data are discussed below, using data obtained from one expert and one non-expert to illustrate analytical methodologies.

First, raw fixation data are corrected using correction factors derived from a subject's calibration matrix. In addition, fluctuations of fixations within the equipment's measurement errors (i.e., +/- 2 degrees visual angle) are aggregated to *gazes* in order to exclude micro-saccades from analyses. Second, discrete clusters of gazes are identified which involve a single critical image characteristic. One method utilized to specify discrete areas of gazes is data-driven; a k-means cluster analysis is used to select a cluster solution which optimizes within and between cluster error terms (Hartigan, 1975), and assigns cluster membership as a new identifier. Another method employed in this study is a content-driven approach where the researcher specifies areas around critical features, such as a face, hands, a sharp edge, and white highlight, and assigns cluster identifiers. All subsequent analyses are conducted using cluster areas specified by the content-driven method. Clusters obtained by cluster analyses are not suitable for this purpose because such clusters differ considerably across subjects. Furthermore, clusters can vary between identical images on two displays for a single subject. Such inconsistencies within and across subjects make further analyses incoherent. Therefore, the outcome of content free, mathematical cluster analyses is only used to confirm the validity of critical areas specified by the researcher. Figures 2 and 3 show examples of data-driven cluster specification, expert and non-expert respectively. This picture contains three critical image features; colour fidelity of egg and bacon, edge definition of the plate, and background shadow. The center of each ellipse indicates the center of the cluster, and horizontal and vertical distances from the center to the edge of ellipse represent one standard deviation in those directions. The durations of total gazes falling within the clusters are shown below the images. As seen in Figures 2 and 3, image features examined by the expert and non-expert subject are considerably different.   While the expert compared edges and background shadow, the non-expert appeared to check colours of food extensively.

Both aggregate and sequential data analyses use gaze cluster areas specified by the content-driven method. Descriptive statistics, such as frequency and total duration of gazes falling in each area are computed; such statistics can be subjected to inferential testing when sufficient data are collected. Figures 4 and 5 show examples of content-driven cluster classification of expert's and non-expert's gazes.

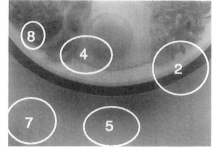

Display A   Display B

Cluster 1: t = 3.9 sec.	Custer 2: t = 2.8 sec.	Cluster 3: t = 2.5 sec.	Cluster 4: t = 4.0 sec.
Cluster 5: t = 3.5 sec.	Cluster 6: t = 2.6 sec.	Cluster 7: t = 0.5 sec.	Cluster 8: t = 0.9 sec.
Cluster 9: t = 1.7 sec.	Cluster 10: t = 2.9 sec.		

Figure 2. An example of data-driven cluster specification using the expert's data.

Display A   Display B

Cluster 1: t = 7.3 sec.	Cluster 2: t = 2.2 sec.	Cluster 3: t = 1.7 sec.	Cluster 4: t = 2.9 sec.
Cluster 5: t = 0.8 sec.	Cluster 6: t = 1.7 sec.	Cluster 7: t = 1.6 sec.	Cluster 8: t = 0.6 sec.
Cluster 9: t = 2.8 sec.			

Figure 3. An example of data-driven cluster specification using the non-expert's data.

The total durations of gazes falling in cluster areas are listed below the images. These analyses also show that the expert spent some time studying plate edges and background shadow whereas the non-expert compared food colours exclusively. Results of content-driven and data-driven analyses were consistent for these two cases.

Display A

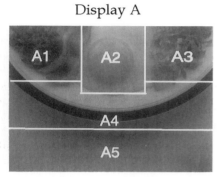

Display B

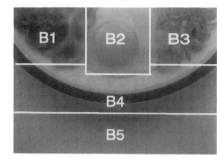

Cluster A1: t = 2.6 sec.	Cluster A2: t = 2.2 sec.	Cluster A3: t = 1.8 sec.	Cluster A4: t = 0.2 sec.
Cluster A5: t = 1.2 sec.	Cluster B1: t = 1.0 sec.	Cluster B 2: t = 1.0 sec.	Cluster B3: t = 0.9 sec.
Cluster B4: t = 0.8 sec.	Cluster B5: t = 2.3 sec.		

Figure 4. An example of content-driven cluster specification using the expert's data.

Display A

Display B

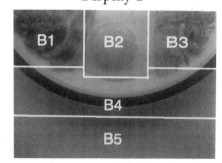

Cluster A1: t = 2.2 sec.	Cluster A2: t = 2.9 sec.	Cluster A3: t = 0.4 sec.	Cluster A4: t = 0.0 sec.
Cluster A5: t = 0.0 sec.	Cluster B1: t = 5.3 sec.	Cluster B 2: t = 5.5 sec.	Cluster B3: t = 0.6 sec.
Cluster B4: t = 0.2 sec.	Cluster B5: t = 0.4 sec.		

Figure 5. An example of content-driven cluster specification using the non-expert's data.

Next, simple pattern recognition is employed to identify eye movements aimed at finding and evaluating critical features of image quality. Rhenius and Heydemann (1984) and Deffner (1985) have successfully demonstrated that one could identify groups of gazes directed at a specific task by deriving a gaze sequence from task analyses and scanning the whole gaze sequence for occurrences of similar gaze sequence patterns. The detection of sequential cluster patterns provides an alternative to pattern recognition and is a content free approach independent of task analyses (Rhenius & Locher, 1992; Deffner, 1986). The initial set of gaze patterns selected for the present study include sequences of gazes between two critical image clusters within a display, between the same critical image clusters on two displays, and between different critical image clusters on two displays. In order to make certain that gaze shifts from cluster to cluster represent underlying cognitive processes of image quality evaluation and not meaningless saccades or artificial fluctuations across cluster borders, fixation sequences shorter than 100 ms are excluded from the analysis. Table 1 shows examples of gaze shift sequences of the expert and non-expert viewing the egg-and-bacon image. The areas to which the expert and non-expert directed their gazes varied somewhat; however, the type and number of gaze shifts did not differ much between the expert and non-expert.

Retrospective verbalization data describing how subjects reached their preference decisions are transcribed. Two raters, blind to subjects' expertise and performance, then classify their statements into two categories: content independent and content dependent. Content independent statements describe image quality without reference to image content whereas content dependent statements evaluate image quality contingent upon content. An example of content independent statements is "the edges are sharper in Display A than in B," and an example of content dependent statements is "a muted light seems to improve this image because it makes the little boy's face look soft." Statements are further categorized by critical image characteristics they represent. Prescribed categories of critical image characteristics include brightness, edge definition, detail, colour fidelity, etc. After statements have been categorized, the raters compare their results. Any disagreements between the raters are resolved by discussion, and inter-rater reliability is computed. The resultant data may be analyzed statistically and examined for coherence with results of eye-movement data analysis. Table 2 shows sample statements of the expert and non-expert subjects describing how they have made their preference decisions while viewing the egg-and-bacon image on the two displays. Table 3 shows sample results of a content analysis on the statements shown in Table 2. In general, the expert subject made content independent comments almost exclusively while all statements made by the non-expert subject were content dependent. Moreover, the expert focussed on highlight and shadow details and edge definition whereas the non-expert based her decisions solely on brightness and colour fidelity of the images. The non-expert made vague comments frequently; the expert made no unclear comments. Retrospective verbalization of these subjects agreed with their eye movements closely.

| EXPERT | | NON-EXPERT | |
Display A	Display B	Display A	Display B
	egg => plate edge	egg => bacon	
	plate edge => egg	bacon ===> egg	
	egg => potato		egg => potato
	potato => backgrnd.		potato => egg
background <===* background		egg <===* egg	
background => egg		egg ===>* egg	
egg => potato			egg => plate edge
potato => egg			plate edge => backgrd.
egg ===> plate edge			background => bacon
	plate edge => egg		bacon => background
	egg => bacon		background => egg
	bacon => egg	egg <===* egg	
potato <=== egg		egg ===>* egg	
potato => bacon			egg => bacon
bacon => egg			bacon => egg
egg => bacon			egg => bacon
bacon ===>* bacon			egg => potato
bacon <===* bacon		potato <===* potato	
bacon => potato		potato ===> bacon	
potato ===> bacon		egg <=== potato	
	bacon => potato	egg => potato	
	potato => egg	potato => egg	
potato <=== egg		egg ===>* egg	
potato => plate edge			egg => bacon
plate edge => egg		bacon <===* bacon	
egg => background		bacon ===>* bacon	
background ===>* background			bacon => egg
background <===* background			
background ===>* background			

Number of gaze shifts between two
  clusters within a single display = 19
Number of gaze shifts between areas of identical
  critical image feature on two displays = 6
Number of gaze shifts between two different
  critical features on two displays = 3

Number of gaze shifts between two
  clusters within a single display = 16
Number of gaze shifts between areas of identical
  critical image feature on two displays = 8
Number of gaze shifts between two different
  critical features on two displays = 3

* Gaze shifts between areas of an identical critical image feature on two displays.

Table 1. An example of gaze shift sequences from the expert and the non-expert viewer.

Expert: "This image here was really, to my mind, close. I didn't have a good reason for selecting one over the other. I was trying to look at the edge breaks between the plate and the table. I couldn't tell a lot of difference between (displays) A and B..."

Non-expert: "I chose (display) B on this because the brighter the better in a picture like this. It made me hungry too. Food looks more appealing in the clearer one."

Table 2. Examples of retrospective verbalizations from the expert and the non-expert viewer.

Subject	Content Dependent	Content Independent	Critical Image Characteristics
Expert		"...edge breaks between the plate and the table."	edge definition
Non-expert	"...brighter... in a picture like this... Food looks more appealing in the clearer one."		brightness colour fidelity?

Table 3. An example of content-analysis of retrospective verbalization: expert and non-expert

Discussion

The methodology presented above is not yet an established procedure to study determinants of perceived image quality. Rather, this paper proposes a new approach in which several theoretical and technical techniques have been integrated. A paired comparison of displays is studied by means of eye movement recording and retrospective verbalization; pattern recognition and sequential clusters detection provide frameworks to examine the process of image quality evaluation in a meaningful fashion. Inclusion of non-expert subjects and utilization of naturalistic scenes in lieu of test patterns are also significant modifications to more traditional image evaluations; the author believes that human factors evaluation of display image quality should include non-expert consumers viewing naturalistic images because this combination approximates consumer purchase behaviour. Data collection has been facilitated by on-line recording, indexing, and transcription of verbal protocols. The experiment described above is still in progress; it has been reported here to illustrate a new methodology developed to study the process of image quality evaluation of displays. This will serve as a framework for further refinement of this methodology, and the development of a new paradigm is expected to clarify the mystery surrounding the process of human perception of image quality.

References

Bowers, V.A. & Snyder, H.L. (1990). Concurrent versus retrospective verbal protocol for comparing window usability. Proceedings of the Human Factors Society 34th Annual Meeting, 1270-1274.

Carmody, D.P., Nodine, C.F. & Kundel, H.L. (1981). Finding lung nodules with and without comparative visual scanning. Perception and Psychophysics, 29(6), 594-598.

Deffner, G. (1984). Lautes Denken - Untersuchung zur Qualität eines Datenerhebungsverfahrens [Thinking aloud - Study of the quality of a research instrument]. Frankfurt: Lang.

Deffner, G. (1985). Identification of solution strategies on the basis of eye-movement data. In R. Groner, G.W. McConkie, & C. Menz (Eds.), Eye movements and human information processing, 309-322. Amsterdam: North-Holland.

Deffner, G. (1986). Das Erkennen von Gruppierungen um Center in sequentiellen Nominaldaten. [Identification of groupings around centers in sequential nominal data]. Psychologische Beiträge, 28, 375-383.

Deffner, G. (1989). Interaktion zwischen Lautem Denken, Bearbeitungsstrategien und Aufgabenmerkmalen? Eine experimentelle Prüfung des Modells von Ericsson und Simon. [Interaction of thinking aloud, solution strategies and task characteristics? An experimental test of the Ericsson and Simon model]. Sprache und Kognition, 9, 98-111.

Deffner, G., Koebe, S., Preuß, A., & Völkl, T. (March, 1986). Verstehen von Fehlermeldungen bei der Arbeit mit SPSS [Understanding error messages while working with SPSS]. Presented at: 28 Tagung experimentell arbeitender Psychologen in Saarbrücken.

Ericsson, K. A., & Simon, H. A. (1993). Protocol analysis (Revised edition). Cambridge: MIT Press.

Fisher, D.F., Karsh, R., Breitenbach, F., & Barnette, B.D. (1983). Eye movements and picture recognition: Contribution or embellishment. In R. Groner, C. Menz, D.F. Fisher & R.A. Monty (Eds.), Eye movements and psychological functions: International Views. Hillsdale, N.J.: Erlbaum.

Hartigan, J.A. (1975). Clustering algorithms. New York: Wiley & Sons.

Janik, S.W., Wellens, A.R., Goldberg, M.L., & Dell'Osso, L.F. (1978). Eyes as the center of focus in the visual examination of human faces. Perceptual and Motor Skills, 47, 857-858.

Jorna, G.C. & Snyder, H.L. (1991). Image quality determines differences in reading performance and perceived image quality with CRT and hard-copy display. Human Factors, 33(4), 459-469.

Kolers, P.A., Duchnicky, R.L., & Ferguson, D.C. (1981). Eye movement measurement of readability of CRT displays. Human Factors, 23(5), 517-527.

Nodine, C.F., Carmody, D.P., & Herman, E. (1979). Eye movements during visual search for artistically embedded targets. Bulletin of the Psychonomic Society, 13(6), 371-374.

Rhenius, D. & Heydemann, M. (1984). Lautes Denken beim Bearbeiten von RAVEN Aufgaben [Thinking aloud while solving RAVEN matrix problems]. Zeitschrift für experimentelle und angewandte Psychologie, 31(2), 308-327.

Rhenius, D. & Locher, J. (1992). Auswertungsalgorithmus für Folgen von Augenbewegungen während eines Problemlöseprozessess [Algorithms for sequential analysis of eye movements during problem solving]. Zeitschrift für experimentelle und angewandte Psychologie, 39(4), 646-661.

Sampsell, J. (1993). An overview of the Digital Micromirror Device (DMD) and its application to projection displays. Proceedings of the 1993 SID International Symposium, 1012-1015.

Treistman, J. & Gregg, J.P. (1979). Visual, verbal, and sales responses to print ads. Journal of Advertising Research, 19, 41-47.

EYE-GAZE DETERMINATION OF USER INTENT AT THE COMPUTER INTERFACE

Joseph H. Goldberg[a] and Jack C. Schryver[b]

[a]Department of Industrial Engineering, The Pennsylvania State University, 207 Hammond Building, University Park, PA 16802 USA

[b]Cognitive Systems & Human Factors Group, Engineering Physics & Mathematics Division, Oak Ridge National Laboratory, Oak Ridge, TN 37831-6360 USA

Abstract

Determination of user intent at the computer interface through eye-gaze monitoring can significantly aid applications for the disabled, as well as telerobotics and process control interfaces. Whereas current eye-gaze control applications are limited to object selection and x/y gazepoint tracking, a methodology was developed here to discriminate a more Abstract interface operation: zooming-in or out. This methodology first collects samples of eye-gaze location looking at controlled stimuli, at 30 Hz, just prior to a user's decision to zoom. The sample is broken into data frames, or temporal snapshots. Within a data frame, all spatial samples are connected into a minimum spanning tree, then clustered, according to user defined parameters. Each cluster is mapped to one in the prior data frame, and statistics are computed from each cluster. These characteristics include cluster size, position, and pupil size. A multiple discriminant analysis uses these statistics both within and between data frames to formulate optimal rules for assigning the observations into zoom-in, zoom-out, or no zoom conditions. The statistical procedure effectively generates heuristics for future assignments, based upon these variables. Future work will enhance the accuracy and precision of the modeling technique, and will empirically test users in controlled experiments.

Keywords

Eye-Gaze, Eye Movements, Computer Control, Cluster Analysis, Discriminant Analysis, User Intent, Gaze-Contingent Control User Intent and Eye-Gaze

User Intent and Eye-Gaze

User intent discrimination is an essential feature of truly adaptive user interfaces. When implemented correctly, intent discrimination generates interface transparency, i.e., the user does not notice the workings of the interface and is not encumbered with the need to input control actions in correct syntax. Application software that infers user intent does not rely on traditional input "devices." Rather, it uses all available information to discover what the user is trying to accomplish at the interface. Eye-gaze is a rich source of information regarding user intent, as evidenced by almost 20 years of research demonstrating the relationship between cognitive processing and eye-gaze position (Just and Carpenter, 1976).

Computer Interface Control

Control of the computer interface from a user's eye-gaze location is an application of user intent discrimination. Prior reports of eye-gaze control have been limited to using eye-gaze location as an analog to concrete mouse operations such as cursor control, object selection, or drag and drop. Applications have been developed for word processing (Gips, et al., 1993; Frey, et al., 1990), selecting menu items (Hutchinson, et al.,

Eye Movement Research/J.M. Findlay et al. (Editors)

1989), information disclosure in fictional worlds (Starker and Bolt, 1990) and object selection and movement in a tactical decision-making simulation (Jacob, 1991).

Mouse substitution may not be effective for eye-gaze control of more abstract operations such as interface zooming. Such operations are likely driven by more covert visual attention or cognitive processes. Uncovering overt patterns among otherwise covert operations represents a challenge to the eye-gaze investigator. An interface control heuristic formed from these overt patterns is essentially a marker of user intent, if gathered prior to the execution of a desired operation.

Automated discrimination of user intent for some mundane tasks can potentially free up human operator resources for more important decision making. While user intent analysis can occur at many levels (e.g., intent determination from physical positions or motions, or from speech or language patterns), analysis of eye-gaze patterns were chosen here for several reasons:

- Eye-gaze can be non-invasively monitored while using a computer.
- Based upon clinical psychology techniques, eye-gaze may possibly provide significant information on user intent.
- A small literature on using eye-gaze for computer interface control exists.
- Eye-gaze location, a spatial mechanism, should have a natural compatibility with control of spatial devices, such as cameras or robots.
- If natural eye-gaze tendencies are used, eye-gaze based control should prove relatively automatic and attention-free.

Spatial Clusters of Attention

Spatial clusters of eye-gaze locations signal important loci of visual attention. Fig. 1 shows a scanpath with spatial clusters circled. The clusters may be formed by a serial connected set of fixations (lower right-hand cluster in Fig. 1), or may be formed from a series of refixations within a spatial area (upper cluster in Fig. 1). The formation of spatial clusters from time-separated fixations implies that cluster formation methods must inherently neglect scanpath analysis methodologies. Cluster formation methods from eye-gaze have been presented in prior papers (Pillalamarri, et al., 1993; Latimer, 1988; Belofsky and Lyon, 1988; Scinto and Barnette, 1986). Thus, independent consideration of eye-gaze location on a display is insufficient to understand visual attention loci, without an accompanying cluster analysis. Interpretation of user intent, if based upon visual attention patterns, should primarily consider the location and characteristics (e.g., size, shape) of eye movement clusters.

Analysis of clusters of eye movement or visual attention locations may be static or quite dynamic. In the latter case, clusters may expand or contract, and possibly spawn additional clusters as visual attention becomes more distributed over an area. If clusters are composed of at least 3-4 fixations, then the time represented by a cluster should be at least one second.   Analysis of several clusters should thus require several seconds. A clustering scheme based solely on fixation locations will be quite slow and nonresponsive. The present approach sidesteps this difficulty by neglecting the concept of fixations altogether. Instead, locations of eye-gaze point-of-regard are sampled at the maximum rate of the tracking system; successive samples may or may not be within the same fixation. Eye movements that occur within a fixation (e.g., microsaccades, ticks) could form multiple clusters, given input clustering criteria described below, if an adaptive clustering technique is used. Such clusters are treated and characterized just as clusters formed between fixations. At a 30 Hz eye movement sampling rate, up to 30 observations per second are provided by this methodology, greatly aiding the power of the cluster approach.

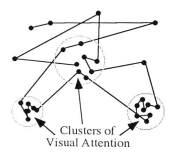

Clusters of Visual Attention

Fig. 1. Spatial clusters of eye-gaze location, formed independently of serial scanpath.

This paper describes an empirical methodology for construction of on-line algorithms which perform intent discrimination from eye-gaze characteristics. Artificial results are provided to illustrate the use of the methodology to construct decision rules for zooming in and out, or not zooming, at the computer interface.

Methodology

The general procedure for user intent discrimination consisted of 6 sequential steps. While currently conducted off-line, the procedure has large potential for real-time analysis. The steps, described further below, were: (1) Eye-gaze sampling, (2) Data reduction of eye-gaze samples, (3) Spatial clustering of eye-gaze samples, (4) Characterization of spatial clusters, (5) Multiple discriminant analysis across spatial and temporal cluster characteristics, and (6) Visualization of significant variable relationships.

Eye-Gaze Tracking Apparatus

An LC Technologies eye tracking system collected serial records at 30 Hz of eye-gaze screen location, pupil diameter and cumulative time. The system camera was mounted beneath a Sun Sparcstation 2 monitor, and pointed toward the user's right eye, which was at a distance of 50 cm. The eyetracking output data was transferred, via a host computer, to the Sun workstation for subsequent analysis. The system provided accurate records of serial eye-gaze location, based upon prior reference calibration. Using a chin rest to stabilize the head of the user, the average angular bias error was less than 0.5, with about 4 cm of head motion tolerance in the horizontal and vertical frontal planes. The system worked equally well with eyeglasses or contact lenses.

Eye-Gaze Sampling

The example data reported here were collected during an experiment requiring users to view a display and make a decision whether to zoom-in or out to gather further information. The user performed the zoom, when necessary, by pressing one of two buttons on a mouse. The user's task was a same/different judgement between an initially memorized test stimulus, and a presented comparison stimulus probe. On each trial, eye-gaze samples were collected following the probe stimulus object presentation, and ended with the user's zoom response. Thus, the serial sample of eye-gaze locations preceded the motor zoom response. An example trial is provided in Fig. 2, showing relevant events. Fig. 2A shows a sample trial on which a zoom-in was required, while Fig. 2B shows a zoom-out trial. Two responses were actually required for each trial: an initial zoom response on the mouse, then a keyboard response of "s" (same) or "d" (different). Though this paper is a report on methodology, note that several different stimuli, randomly permuted, were actually used.

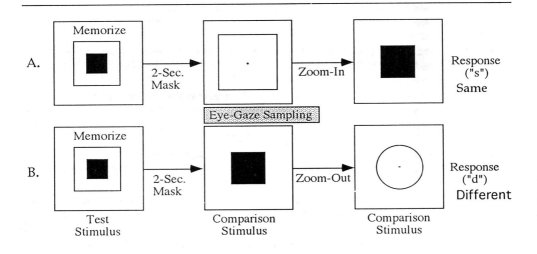

Fig. 2. Example trial events, shown within screen borders. Approximate eye-gaze sampling interval is also indicated; all sampling preceded actual zoom responses by the user. A. Zoom-in trial, showing stimulus border as the comparison stimulus. B. Zoom-out trial, showing stimulus interior as the comparison stimulus.

On a typical trial, the decision to zoom-in, zoom-out, or not to zoom required just under one second; about 20-30 30 Hz serial eye-gaze locations were collected during this interval. An experiment of 100 trials thus recorded about 3000 records for subsequent, off-line analysis.

Initial Data Reduction
Data reduction and analysis was conducted off-line for this project, although real-time methods are feasible extensions. Each of the serial eye-gaze locations (hereafter referred to as data samples) was represented by an X/ Y location pair and a time stamp. The eye-gaze analyst controlled several aspects of data reduction via user inputs to the analysis software.

Data Frames. The data framing process was initially defined via user inputs. The user split up each trial into successive data frames, with defined overlap, by inputting the number of samples within each frame. All samples within each frame were, in essence, considered simultaneous. A frame size of one sample simply led to analysis of individual eye-gaze samples. A frame size as large as the number of collected samples (or larger) within a trial created one data frame for that trial. For analyses below, 2 to 4 data frames with 15 samples were formed, with overlaps of 5 samples per frame. Improved resolution was obtained by smaller frames, and improved stability or accuracy in spatial locations was obtained by larger frames.

Minimum Spanning Trees. All samples within a data frame were connected into a shared minimum spanning tree (MST) data representation. The data treatment here differs substantially from the fixation and scanpath analyses by others, which attempt to group successive samples into fixation locations via temporal and location-based heuristics. As these heuristics are rather arbitrary and not necessary for the present work, the actual sample locations were used.

An additional advantage of the MST representation over scanpath analysis concerns the identification of attentional foci. Scanpaths often loop back on themselves, visiting former locations. Identification of attentional foci must consider these refixations before dividing the screen into separate areas. The MST representation inherently considers refixations as additional samples, given a large enough sampling frame. By iteratively varying the size of data frames, optimal, converging evidence for number of attentional foci may be found. Note that all data within a data frame is considered to be simultaneous; the MST thus loses all sequential information within a data frame. Smaller data frames, with fewer observations, regain this temporal information, however, for subsequent analyses.

Minimum spanning trees were formed by Prim's algorithm (Camerini, et al., 1988), due to its rapid and efficient operation. Given a spatial array of eye-gaze locations, only one such MST can be formed. The MST has the beneficial property of no closed circuits, so it can be rapidly searched. In addition, it defines the minimum distance network that interconnects all locations or nodes. Starting from an arbitrary eye-gaze location, the minimum distance location was connected, creating the first edge of the MST. The next shortest distance to either of the two nodes of the tree was next connected. These connections continued until no more unconnected eye-gaze samples were present. Fig. 3 presents a set of eye-gaze samples interpreted in two different ways. Fig. 3A shows a classic temporal scanpath interpretation (assuming each sample is a fixation location, for present purposes of illustration). Fig. 3B shows the same data, interpreted as an MST. Note the lack of completed circuits in the latter representation, which enables cluster definition.

Spatial Cluster Formation

Spatial MSTs were clustered to define the user's actual attentional foci on the computer interface. If a relatively long time is spent observing a particular screen area, more samples are taken within this area, and the resulting MST will be fairly tightly clustered. The clusters were computed following input of several controlling parameters. As opposed to eye-gaze clustering techniques requiring significant user interaction (Pillalamarri, et al., 1993), the MST representation allowed rapid, statistically-based clustering from the user defined parameters.

Using a depth-first search (Gibbons, 1984), the mean and standard deviation (SD) of internodal (between eye-gaze location samples) distances were computed, in screen pixels. Analyst-selectable inputs for defining clusters included the required branching depth (BD), distance ratio (DR), and distance SDs (DSDs). Due to the lack of circuits in the MST, two clusters were separated by only one internodal edge (i.e., one branch connecting two eye-gaze location samples). Thus, clusters could be separated if the distance between these nodes was greater than that of the locally surrounding nodes.

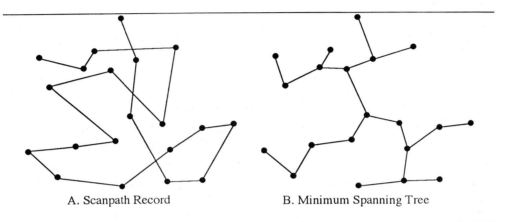

A. Scanpath Record   B. Minimum Spanning Tree

Fig. 3. Comparison of data representations for identical eye-gaze samples. A. Scanpath interpretation, temporally linking adjacent samples. B. Minimum spanning tree representation, easing cluster separation.

The clustering process considered each edge in the MST. At a given edge, the actual branching depth had to meet or exceed the user-input BD at each of its end nodes, to insure that cut edges separating clusters were not too close to the cluster edges. An edge not meeting this requirement at both nodes was not further considered as a cut edge. (Note, however, that BD=O effectively disabled this check.) Given a successfully passed check of branching depth for an edge, edge lengths (in pixels) were collected from each of its defining nodes outward, to a depth of BD. The BD essentially defined the local edge environment with which a potential cut edge was compared. The mean and standard deviation edge lengths were computed from this collection. The considered edge was marked as a cut edge if two criteria were satisfied:

$$\text{Mark as cut edge if:} \begin{cases} \dfrac{\text{Edge Length}}{\text{Edge Mean}} > \text{ER, (Edge Ratio) and} \\[2mm] \text{Edge Length} > \text{Edge Length Mean} + \text{DSD (Edge SD)} \end{cases}$$

Values in the range of 2-4 for ER and DSD generally provided intuitive spatial clusters. Larger input values forced clusters to be separated by greater distances, and were thus more conservative. Increasing ER relative to DSD forced greater emphasis on mean distance as the separation criteria, whereas increasing DSD relative to ER forced more emphasis on variance in edge lengths.

Characterization of Spatial Clusters

Each cluster of eye-gaze samples was characterized for further analysis. For the present study, nine variables were computed for each cluster; many more are possible, for specific purposes. The variable symbols, units, and descriptions are presented in Table 1. The software also allowed all clusters within a data frame to be pooled into one representative cluster. In this case, the variables for each cluster were averaged, weighted by the number of nodes contained within each cluster. The frame is then represented by only one cluster, expressing the central cluster tendency on that frame.

Table 1. Statistics generated from each cluster

Symbol	Units	Brief Description
N	--	Number of nodes within cluster
M_x	pixels	Mean horizontal spatial cluster location
M_y	pixels	Mean vertical spatial cluster location
M_e	pixels	Mean of cluster edges
SD_e	pixels	Standard deviation of cluster edge length
M_d	pixels	Mean distance from nodes to cluster x/y mean
SD_d	pixels	SD distance from nodes to cluster x/y mean
M_p	mm	Mean pupil diameter across cluster samples
SD_p	mm	SD pupil diameter across cluster samples

Cluster Mapping Between Frames

Because all eye-gaze samples within a data frame are considered to be simultaneous in time, additional characterization was necessary to discover temporal changes between data frames. For example, a large cluster on the first frame may split into two or more clusters on subsequent frames. Each of these may further split, or may reconsolidate as the user's attentional focus dynamically changes. Tracking of these clusters between frames requires a mapping procedure, whereby each cluster within a frame is mapped to its precursor cluster in the immediately preceding frame.

The cluster mapping procedure implemented here used a constrained closest-distance algorithm. Between frames, clusters whose spatial means were closest were mapped to one another. The assignments considered all possible mappings, to provide an optimally closest mapped set. Note that additional constraints could be implemented, such as mapping clusters of approximately similar size or numbers of nodes. The direction of cluster movement could also be modeled to extrapolate the location of the mapped cluster on the next fame.

The mapped clusters provided a second set of parameters to discriminate the zoom-in/zoom-out/no zoom conditions. The same basic statistics that were presented in Table 1 were computed, except that computations were made between successive frames. For example, the mapped M described the mean difference in horizontal distance between each cluster in a frame, and its mapped predecessor in the preceding frame. These provided a glimpse into the dynamic characteristics of the eye-gaze clusters, in addition to the previously computed static characteristics. Note that additional sets of variable mappings could have been constructed for frames lagged by more than one. These may describe longer-term changes in the computer variables.

Multiple Discriminant Analysis

Given the static and dynamic cluster characteristics described above, a multiple discriminant analysis (MDA) procedure attempted to locate the best set of characteristics that would optimally separate the zoom-in, zoom-out, and no zoom conditions. In essence, the discriminant function minimized within-condition variance, while maximizing between-condition variance for a set of variables. The procedure computed either one or two linear discriminant functions to separate the groups; additional discriminatory power could potentially be obtained from nonlinear functions. The specific MDA used here was adapted from public domain code (Murtagh and Heck, 1987) and allowed the user to select 2-5 different model dependent variables.

The present methodology used the MDA as a classification technique. As the zoom condition classification was known by the software a priori, the input data was a training set for supervised MDA classification. A confusion matrix was generated to evaluate the success of a particular model. Significance of a particular classification was computed from a test statistic developed by Press (1972). The statistic tests equal probability of assignment to each of the three groups as its null hypothesis. For the three zoom conditions, or three classification groups here, the

statistic was $Q = \dfrac{(N - 3n)^2}{2N} \sim \chi_1^2$, where N and n are, respectively, the total number of clusters, and the number of clusters correctly classified. A significant statistic indicated that the MDA classification was significantly better than chance, which was equal probability of assignment across the three zoom conditions.

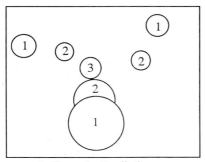
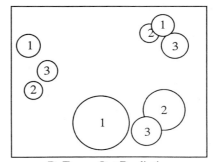

A. Zoom-In Prediction   B. Zoom-Out Prediction

Fig. 4. Two possible zoom condition predictions, based upon changes in eye-gaze cluster size and position over time. Clusters are shown on a screen interface, and are all prior to actual zoom-in or zoom-out. Data frame sequential number is indicated at the center of each cluster. A. Prediction for zooming-in, showing converging cluster locations over three data frames. B. Prediction for zooming-out showing cluster divergence.

Example. To provide a concrete example of the MDA procedure, consider a two variable model. For simplicity, also consider only two assignment conditions: zoom-in or zoom-out. (Thus, we ignore the choice not to zoom.) Also assume that, when signaling an intent to zoom-in, the eye-gaze spatial cluster sizes become smaller and tend towards one location on the screen. Assume the reverse occurs when signaling intention to zoom-out: clusters increase, and tend towards many screen locations. Fig.4 illustrates these two predictions, by showing clusters over three data frames.

An MDA model here might include M_x and M_d the mean horizontal cluster screen position, and the mean distance, within each cluster, from each node to its spatial x/y center. In this example, small values of M_d and mid-screen values of M_x indicate zoom-in, whereas larger values of M_d , with any value of M_x indicate zoom-out. This data pair is plotted from each cluster and zoom condition. Fig. 5A graphically shows these hypothetical data as lightly shaded ellipses, whose means are indicated by a solid circle. Two discriminant functions are formed that simultaneously maximize the data variance between these two groups and minimize the data variance within each group. The second discriminant function is used only when its eigenvalue can explain significantly more variance in the data. The projections from each of the zoom condition means onto each of the discriminant functions are computed. Fig. 5B plots these condition means and projections in discriminant function space.

The zoom heuristic is now formed by computing the perpendicularly bisecting function between these group means. After remapping back to parameter space, it will take the form:

$$\text{Zoom-In if:} \quad \begin{cases} \beta_1\left(M_{d_1}\right) + \delta_1\left(M_{x_1}\right) < \theta_1, \text{ and} \\ \beta_2\left(M_{d_2}\right) + \delta_2\left(M_{x_2}\right) > \theta_2 \end{cases}$$

Otherwise Zoom-Out.

The formulated zoom heuristic may now be used to classify new observations. It may remain static, or could be updated or recalibrated to suit varying users, conditions, or software needs.

Generalization of Variable Relationships

Zoom condition heuristics from the MDA are specific to an individual and a set of interface conditions. Broader generalization of these heuristics requires improved data and variable visualization procedures, both within and between individuals.

Pooled Data. The MDA can potentially be computed across subjects, by simply pooling cluster data into one file. Robust, user-independent phenomena will stand out, but some within-user characteristics will be lost. As an example, pooled-subject results showed that the most significant MDA assignments are found after a few data frames have been collected. That is, at least one second of eye-gaze sampling may be required before reliable zoom discrimination may occur. The model significance dramatically dropped for appreciably shorter or longer sampling intervals.

Composite Variable Plots. Superimposing several subjects' derived zoom regions on composite plots of variable pairs can provide valuable intuition on user-independent trends. Fig. 6 demonstrates this, using the same two example variables, for a

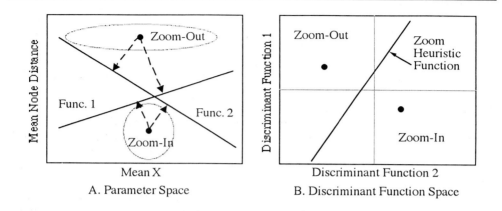

A. Parameter Space B. Discriminant Function Space

Fig.5. Illustration of two dimensional MDA-based heuristic development. A. In parameter space for selected variable pair, projections of means from each zoom condition are mapped onto each of two discriminant functions. Actual data, represented by broader ellipses, may be highly spread. Zoom condition means are represented in discriminant function space. The zoom heuristic, which must be remapped to parameter space, perpendicularly bisects the group means in this space.

hypothetical set of users' data. There may be extensive overlap between zoom regions, but central tendencies are still quite apparent.

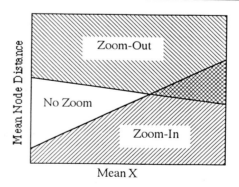

Fig. 6. Example composite heuristic plot across several hypothetical users. Each area represents a zoom heuristic for a variable pair.

Discussion

A methodology was presented to assess and discriminate user zoom intent, at the computer interface, from eye-gaze cluster characteristics. The technique is empirical and adaptive. It may use either static or temporally dynamic characteristics. The methodology may also be generalized well beyond analysis of eye-gaze characteristics.

Other Interface Operations
Besides determining whether to zoom-in or out, this technique may be applied to several other computer interface operations. Object selection, rotation, dragging, and inversion are three spatial operations that could potentially be controlled by eye-gaze.

Other Classification Techniques
An MDA was used here to discriminate among the zoom conditions, but other techniques are also available. A neural net is one important example. Consisting of at least three layers of 'neurons', the input layer must contain as many neurons as parameters of the variables of interest. Here, 18 variables were defined; 9 defined characteristics within frames, and 9 between frames. The output layer in this zoom example would always contain 3 neurons; zoom-in, zoom-out, or no zoom. Once the neural net is set up, it need only be trained on a fixed set of zoom trials, for a particular user under fixed conditions.

Acknowledgment

This work was sponsored by the Office of Technology Support Programs, U.S. Department of Energy, under contract DE-AC05-84OR21400 with Martin Marietta Energy Systems, Inc., and U.S. DOE and Oak Ridge Associated Universities, under Contract DE-AC05-76OR00033.

References

Belofsky, M.S., and Lyon, D.R. (1988). Modeling eye movement sequences using conceptual clustering techniques. Air Force Human Resources Laboratory Technical Report AFHRL-TR-8S-16, Air Force Systems Command, Brooks Air Force Base, Texas.

Camerini, P.M., Galbiati, G., and Maffioli, F. (1988). Algorithms for finding optimum trees: Description, use, and evaluation. Ann. Op. Res., 13, 265-397.

Frey, L. A., White, K. P., and Hutchinson, T. E. (1990). Eye-gaze word processing. IEEE Trans. Sys., Man, and Cyber., 20(4), 944-950.

Gibbons, A. (1984). Algorithmic Graph Theory. Cambridge, England: Cambridge University Press.

Gips, J., Olivieri, P., and Tecce, J. (1993). Direct control of the computer through electrodes placed around the eyes. pp. 630-635 in Proc. 5th Int. Conf. Hum.-Computer Inter., Amsterdam: Elsevier.

Hutchinson, T. E., White, K. P., Martin, W. N., Reichert, K. C., and Frey, L. A. (1989). Human-computer interaction using eye-gaze input, IEEE Trans. Sys., Man, and Cyber., 19(6), 1527-1534.

Jacob, R. J. K. (1991). The use of eye movements in human- computer interaction techniques: What you look at is what you get. ACM Trans. Info. Sys., 9(3), 152-169.

Just, M. A. and Carpenter, P. A. (1976). Eye fixations and cognitive processes. Cog. Psych., 8, pp. 441-480.

Latimer, C.R. (1988). Eye-movement data: Cumulative fixation time and cluster analysis. Beh. Res. Meth., Instr., & Comp., 20(5): 437-470.

Murtagh, F., and Heck, A. (1987). Multivariate Data Analysis. Boston, MA: Kluwer and Dordrecht.

Pillalamarri, R.S., Barnette, B.D., Birkmire, D., and Karsh, R. (1993). Cluster: A program for the identification of eye-fixation-cluster characteristics. Beh. Res. Meth., Instr., & Comp., 25(1): 9-15.

Press, S.J. (1972). Applied Multivariate Analysis. New York: Holt, Rinehart and Winston, Inc.

Scinto, L.F., and Barnette, B.D. (1986). An algorithm for determining clusters, pairs or singletons in eye-movement scan-path records. Beh, Res. Meth, Instr., & Comp., 18(1): 41-44.

Starker, I., and Bolt, R. A. (1990). A gaze-responsive self-disclosing display, CHI '90 Proceedings: Empowering People, Seattle, WA: Association for Computing Machinery, pp. 3-9.

KNOWLEDGE ENGINEERING IN THE DOMAIN OF VISUAL BEHAVIOUR OF PILOTS

Axel Schulte and Reiner Onken

Universität der Bundeswehr München, Fakultät für Luft- und Raumfahrttechnik
Institut für Systemdynamik und Flugmechanik
Werner-Heisenberg-Weg 39, D-85577 Neubiberg, Germany

Abstract

Low level flight training carried out in flight simulators with computer generated image visual displays can lead to visual habit patterns (Comstock et al., 1987; Dixon et al., 1990) of the pilots, which may cause disorientation in real world flight (Haber, 1987). This study is an attempt to contribute to the improvement of visual simulation systems by means of knowledge about visual skills in different situations of flight. This knowledge can be taken account of in the design process of computer image generation so that there are elements of the visual scene, which pilots use as source of visual information by fixating them in special situations in real world flight (Kleiss, 1990; Kennedy et al., 1988). Situation in this context means the complete set of input information to the pilot at one point of time including the visual input of the out-of-window scene and the aircraft displays, the speech input from the copilot, the pilot's knowledge about the mission plan and the current task and other inputs like mechanical cues. The pilot responds on each situation with certain actions including aircraft operations and eye movements.

The eye movements have to be measured in a suitable experimental environment. The Universität der Bundeswehr München is carrying out a study for the development of such an environment. In order to investigate under the most realistic situations, the experiments should be performed in real flight. However, because of principal operational restrictions, this is not possible. Therefore, a video replay of a real flight's visual scene is used in a research flight simulator. The paper presents the experimental concept, using the NAC-EMR 600 eye movement measuring system on pilots engaged in visual tasks in connection with the video replay. The data analysis is described. The variety of possible situations is represented by a number of feature vectors, allowing a numerical classification. The fixated scene elements can thereby be related to the corresponding situation classes. The visual behaviour is defined by the most significant correlations of fixated scene element classes and certain situation classes.

Keywords

Pilot; low-level flight; visual behaviour; classification; eye movements; flight simulator.

Introduction

The visual scene is the most important source of information to the pilot in low level flight training (Johnson et al., 1987). The information needed depends strongly upon the current task. For example, in terrain following flight pilots need cues about the structure of the terrain contour. During visual navigation pilots search for particular scene elements which are related to this task. So, one reason for flight training to be conducted is to improve the pilots' performance of selecting and finding the suitable

Eye Movement Research/J.M. Findlay et al. (Editors)
1995 Elsevier Science B.V.

visual scene elements for each task. (Barfield et al., 1987; Grunwald et al., 1988; Hoh, 1985; Kellogg et al., 1988; Kleiss et al., 1988; Lintern et al., 1987; Lintern et al., 1989; Martin et al., 1983).

Nowadays low level flight training has to be more and more transferred into flight simulators. To become more than a device for aircraft procedure training simulators have to be equipped with a visual system, usually of computer generated image type (CGI). However, the image generator cannot display a picture of the real world in it's diverse variety because of performance limits. So, the right visual terrain features have to be selected and integrated in a visual database. But which kinds of visual cues and scene objects are really needed and looked for by pilots as sources of information? Selecting the wrong cues can lead to training of wrong visual habit patterns concerning real world flight (Kruk et al., 1983). This study tries to give advice to the designers of visual databases by investigating the visual behaviour of pilots in certain situations of low level flight. Therefore, the eye movements, especially the objects fixated by pilots, will be observed.

Concept

The aim of the investigation is to describe the visual behaviour of pilots. Only the input and the output of the pilot are considered in order to define a concept (Figure 1).

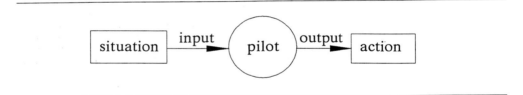

Fig. 1 - Pilot as input/output system

The input is the situation which continuously changes over time. In low level flight training the situation consists of:

1. visual stimuli from the visual scene and the displays,

2. mechanical stimuli from the aircraft movement and the control stick force,

3. verbal speech input from the copilot and

4. pilot's knowledge about the mission plan and the current task.

The pilot responds to each situation with certain actions as output. The actions are:

1. eye movements,

2. aircraft operations and

3. verbal speech output.

The hypothesis of the investigation is that the action depends on the corresponding situation. In different situations different actions can be observed. In similar situations the actions of a pilot are similar. Different pilots also show certain similarities in identical situations. To prove the hypothesis and to make statements about pilot's visual behaviour pilots have to be put into several situations of low level flight training and their actions have to be observed. The experimental design is described in the next section. In the consecutive analysis most significant correlations are found by use of suitable representations of the situations and actions related to each other. The approach is described in more detail in one of the following sections.

Experimental design

Most favorably the measurement of eye movements should take place in the aircraft during flight. This is not possible because of principal operational restrictions in the con-text of high performance aircrafts. Therefore, an available video replay (Kleiss, 1990) of a real flight's visual scene is used to put pilots in situations as close as possible to real flight situations. The experimental design is described in two parts, first the experimental procedure and second the experimental environment.

Experimental procedure

Figure 2 shows the flow diagram of the steps to be followed during the course of the experiment.

Two aspects have to be considered. On the one hand, there are experimental steps concerning the situation input, on the other hand there are experimental steps of recording the pilot's actions. In chronological order the steps are:

Briefing: During the briefing the pilot is made familiar with the mission plan of the replay mission. The succession of the tasks is explained. The pilot has the opportunity to build up a mental representation of how and by which landmarks to navigate and to comply with the mission goals. Therefore, he uses a topographical map.

Calibration: The calibration of the eye movement measuring system has to be done immediately before the flight replay starts. Thus, the measurement of the pilot's fixations in the visual scene is possible.

Flight replay with measurement: During the flight replay the pilot is confronted with the visual stimuli of a real low level flight. The out-of-window view was recorded by video. The aircraft displays are reconstructed by visual animation. A control stick moves according to it's recorded movements during the real flight. This gives the pilot mechanical clues for the following aircraft movements before they become visible in the displays or the visual scene. A copilot gives the pilot verbal information concerning tasks such as visual navigation at waypoints during the 45 minute run of flight replay. Thereby, the pilot's attention is linked to the flight task, in particular, the navigation task. In this complex situation the pilot's actions are measured. The pilot's eye movements in the visual scene are the most interesting output data. Additionally, the pilot's operations on certain cockpit controls are recorded.

Debriefing: After each run of the flight replay the crew is interviewed about the success of the mission tasks and the visual behaviour. The crew has the possibility to give comments on the experiment.

Experimental environment

The experimental environment (Figure 3) is the so-called flight replay system (Schulte et al., 1992). It includes technical components to synchronously provide the variety of stimuli to the pilot and to perform the calibration and the measurement. The out-of-window visual stimuli are given by a virtual image visual display in a field of view of 135 x 30 driven by a video system. This display guarantees a highly realistic reproduction of the visual scene. The pilot accommodates his eyes to infinity like in real flight, due to the collimation optics.

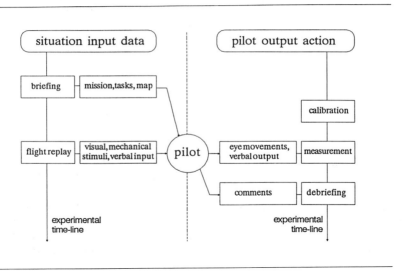

Fig. 2 - *Experimental procedure*

A computer generated head-up-display (HUD) is superimposed to the video picture. A head-down-display (HDD) monitor simulates the major aircraft displays. The control stick is driven by a hydraulic system. A so-called pilot's-hand-controler is installed in order to enable the pilot to operate on certain subsystems. The copilot has additional displays informing him about the replay flight status to verbally give the pilot navigational advice.

The NAC EMR-600 is used as eye movement measuring system . For calibration a matrix of LEDs is positioned in the collimation optics in order to achieve the identical optical conditions to the pilot's eyes in calibration and flight replay. The viewing distance is a critical parameter in use of the EMR-600. The pilot's head-centered view and his eye movements provided by the head unit are recorded as video images with superimposed reticles for both eyes on a video recorder.

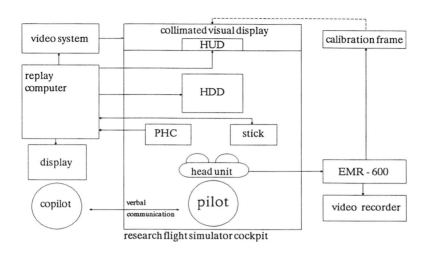

Fig. 3 - Experimental environment

Analysis

The experiments with educated pilots provide a considerable amount of data for analysis. So, the flight replay with it's large number of situations has to be correlated with the recording of pilots' output actions.

The aim of the analysis is to distinguish between a few stereotypes of situations, the so-called situation classes, and to determine the visual behaviour of pilots for each of the classes. The visual behaviour is described by the distribution of typical combinations of output actions, the action classes, in situations belonging to one situation class. This distribution will be made up for each situation class.

In order to classify the situations and actions relevant features have to be selected. The features are organized in a feature vector. All situations considered and the respective actions are represented as a list of feature vector samples over time. (McDermott et al., 1987; Rosenking et al., 1990).

After the process of classification each single situation feature vector belongs to oneunique situation class and each action feature vector belongs to one unique action class. By observing the occurrences of situation and action classes over time it is now possible to distribute the action classes over the situation classes. Significant differences between the distributions of action classes in different situation classes are called the visual behaviour. This theoretical approach will be applied to the visual behaviour of pilots during low level flight training in the following sections.

Preliminary evaluations

The first flight-replay experiments following the experimental design described above have been conducted. In order to prove the task assignment of the experimental pilot and the realism of the sensory inputs by the replay system a subjective rating was conducted (+2: positive, 0: neutral, -2: negative assessment). Table 1 shows the rating of the first check pilot.

Rating	+2	+1	0	-1	-2
Importance of out–the–window view		100			
Disturbance by perspective errors	100				
Visual recognition of objects		75	25		
Disturbance by image geometric discontinuities	25	75			
Importance of head-up-display	100				
Completeness HUD symbology	100				
Importance of head-down-display	100				
Completeness of HDD instruments	100				
HDD instrument lay out	25	75			
Importance of stick movement	100				
Stick feel pitch	50	50			
Stick feel roll		75	25		
Operation of pilots handcontroller				100	
Importance of copilot		100			
Information from copilot			100		
Importance of briefing	50	50			
Briefing contents	100				
Overall Task assignment	100				

Tab. 1 - Pilot's acceptance of flight replay system and experiment in % of entries

This shows that there is a very good acceptance of the reproduction of aircraft specific systems in the replay system. Due to the perfect task assignment it is ensured that pilots recall the associated visual routines in the flight replay experiments. A closer investigation of the objective data will be conducted after the experimental phase.

The methods of analysis were already applied on some material available of low level flight training conducted in a flight simulator with advanced computer generated imagery. The visual display system was equipped with an eye-tracking facility. The recorded data were:

- a video recording of the visual CGI scene slaved to the pilot's eye movements,

- a cockpit voice recording,

- numerical flight and crew action data and

- a crew interview.

The material was analysed under different aspects concerning the selection of features for situation and action description:

aspect	situation features	action features
1.	visual scene: surface coverage, objects, aircraft attitude	fixated objects, object classes,
2.	verbal information from the copilot, pilot's knowledge about map	fixated objects, object geographic position, object distance to ownship
3.	visual scene: terrain structure, attitude and position	aircraft fixated object classes

The following sections briefly summarize the results of the aspects. The results give a first few hints to improve the design of databases for computer generated low-level flight simulator imagery.

Fixated objects under certain surface coverage conditions

In this evaluation study 941 situations were represented by use of eighteen features. The features were flight mechanical parameters, the texture covering the surface and the occurrence of special point- and line-features in the computer generated image. The task was to fly a,s low as possible without any altitude above ground information. No additional navigation activities were required of the pilots.

The data were converted into metric data and classified by a, heuristic cluster analysis algorithm (Späth, 1975). Four clusters with more than 30 members and another six clusters with 20 and more members could be established. The largest clusters are described as follows:

S_1: straight and level flight between 300 and 700 ft altitude over mostly wood texture covered terrain, very few objects except some roads. (members 163)

S_2: flight over field texture covered terrain with wood texture in the background, many single trees, few other objects. (members 128)

S_3: flight over wood texture covered terrain with field texture in the background, only objects some roads and single trees. (members 77)

S_4: climbing flight between 150 and 200 feet over field texture covered terrain with wood texture in the middle- and background, all kinds of objects. (members 44)

During the considered situations 278 visual fixations were observed. The only features for action description were the object class of the fixated object and the duration of the fixation. The objects were classified into:

A_1 = TEX: texture (e.g. wood, town, fields)

A_2 = LOB: linear objects (e.g. road, railway line, woodside)

A_3 = POB: planar objects (e.g. lake, field)

A_4= VXO: vertical extended objects (e.g. tower, powerline)

A_5 = SOB: single objects (e.g. tree, house, bridge)

A_6 = CPO: corner points (e.g. road bend, railway intersection)

Table 2 shows the percentage of fixations in each class and the mean duration T of fixation in the upper part. The distributions of fixated object classes in the considered situation classes are shown in the lower part.

	TEX	LOB	POB	VXO	SOB	CPO
%	3.96	40.65	1.80	12.23	33.81	7.55
T	440ms	876ms	440ms	580ms	662ms	722ms
S_1	7%	58%	—	10%	15%	10%
S_2	—	23%	4%	19%	50%	4%
S_3	8%	67%	—	—	25%	—
S_4	—	84%	—	8%	8%	—

Tab. 2 - Fixation distribution under different surface coverage condicions

The most obvious result is that the pilots' choice of fixations in the computer generated out-of-window scene is led by the existing scene elements which are relatively poor even with a high performance image generator. Linear and single objects are the favourite fixations. Vertical extended objects are fixated whenever available (Martin et al., 1983). Approaching the edge of a wood this scene element seems to be of most interest to the pilot (S4).

Fixations of geographical landmarks for navigation
The situation in this study was characterized by a low scale navigation task in order to find a special target point in the visual scene during low-level flight.

Before the flight the pilots had the opportunity to study the relevant map and to learn some so-called lead-in features to the target. During flight the copilots recalled these features. So, the situation is described by the presence or absence of memorized scene elements extracted from a cockpit voice recording during the flight and a crew interview after the flight. 25 items were called out during 208 seconds of four investigated flights. In the interviews the crews named 26 led-in features. The objects were identified and marked in a topographical map. On the output side 211 fixations of 123 objects were measured and also marked in the map. This approach led to some interesting results:

- 65% of the memorized scene elements were fixated.

- 40% of the fixated objects were fixated more than one time.

- The distance of the fixated objects and the ownship position is between 1 and 5 km.

- The plot of the viewing distance (*dist*) over time (Figure 4) shows a sawtooth function. The pilots' gaze stays in the area of a memorized scene element while approaching (= distance decreases according to speed). Then the gaze jumps to the next area of interest without doing many fixations in between. Local and global eye movement patterns could be distinguished (*map*). The spatial density of fixations (*dens*) along the Right track shows distinct maxima.

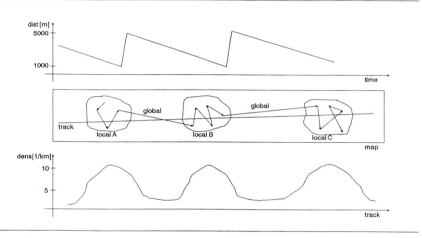

Fig. 4 - Geometric distribution of fixated landmarks for navigation (schematic)

Fixations of objects for terrain contour recognition

The investigation under the 3rd aspect is very similar to the 1st aspect. The flight task was to fly terrain masking that means to fly in valleys of hilly terrain. The situations were classified by the momentary manoeuvring task:

S_1 = SVF: straight valley following flight

S_2 = TVF: turning valley following flight

S_3 = VIC: valley interception

S_4 = RCC: ridgeline crossing, climb

S_5 = RCH: ridgeline crossing, horizontal

S_6 = RCD: ridgeline crossing, dive

S_7 = STF: straight terrain following

S_8 = TTF: turning terrain following

The relevant actions were fixations of different classes of visual scene elements. 816 fixations were found and distributed over the following object classes:

A_1 = LOB: linear objects (e.g. road, railway line, woodside)

A_2 = POB: planar objects (e.g. lake, field)

A_3 = VXO: vertical extended objects (e.g. tower, powerline)

A_4 = CPO: corner points (e.g. road bend, railway line intersection)

A_5 = SOB: single objects (e.g. tree, house, bridge)

A_6 = LCF: linear terrain contour features (e.g. hill crest, ridgeline)

A_7 = PCF: planar terrain contour features (e.g. hillside)

Table 3 shows the overall distribution of the fixations and the distributions of fixated object classes in the different situations.

	LOB [%]	POB [%]	VXO [%]	CPO [%]	SOB [%]	LCF [%]	PCF [%]
$\sum$	44.2	12.6	4.2	9.9	5.8	19.4	4.9
SVF	45	11	5	6	6	14	13
TVF	46	12	4	12	6	14	6
VIC	41	12	15	—	2	24	6
RCC	25	10	8	10	2	40	5
RCH	38	10	4	3	14	30	1
RCD	66	14	4	1	5	9	1
STF	44	22	4	9	4	13	4
TTF	50	8	1	20	4	16	1

Tab. 3 - Fixation distribution over maneuvering tasks

The most striking result of this approach is the low frequency of single object (SOB) fixations in comparison with the situations of the 1st aspect. Single objects contain no information about the terrain contour. Comparing different situations during terrain masking the table shows a frequency peak of linear terrain contour features (LCF) while intercepting a valley (VIC) and climbing up a ridgeline (RCC, RCH). Linear objects (LOB) like roads and rivers which are assumed to run along the valley bottom mostly are used to lead along the valleys. The left and right hillsides (PCF) are the most important obstacles and therefore fixated more often during straight valley following (SVF).

Summary

The aim of the investigation was to determine the visual behaviour of pilots during manual low level flight. Visual behaviour was defined as the relation between certain fixated stereotype objects in the visual scene and the corresponding situation. The video replay of a typical low-level flight mission was used to create a sequence of so-called situations. Pilots were confronted with the replay. During the replay the pilot's eye movements were measured. The pilot's subjective assessments of the experimental environment concerning the degree of realism of the flight replay were positive.

Some first results concerning the visual behaviour could be achieved by applying the methods of analysis on simulator flight training data. The visual stimuli from a computer generated visual image are very poor in comparison to real-world scenes. So, pilots often had not the opportunity to choose between different types of scene elements. Pilots preferred single objects (e.g. house) and linear objects (e.g. road) opposed to textured surfaces. There are typical classes of fixated objects for the different tasks. During terrain masking linear objects were preferred, due to the main task to recognize the terrain contour. During terrain following the major task is altitude control. Then single objects were fixated more frequently.

References

Barfield, W.; Rosenberg, C.; Kraft, C. (1989): The effects of visual cues to realism and perceived impact point during final approach. Proceedings of the Human Factors Society 33rd Annual Meeting.

Comstock, J.R.; Harris, R.L.; Coates, G.D.; Kirby, R.H. (1987): Time-locked time-histories: A new way examining eye-movement data. Proceedings of the Fourth Inter-national Symposium on Aviation Psychology. Jensen, R.S. (ed).

Dixon, K.W.; Krueger, G.M.; Rojas,V.A.; Martin, E.L. (1990): Visual behavior in the F-15 simulator for air-to-air combat. NASA Technical Report, Issue 16.

Goldin, S.E.; Thorndyke, P.W. (1982): Simulating Navigation for Spatial Knowledge Acquisition. Human Factors, 24(4), 457-471.

Grunwald, A.J.; Ellis, S.R. (1988): Spatial orientation by familiarity cues. Training, Human Decision Making and Control. J.Patrick, K.D.Duncan (eds), Elsevier Science Publishers B.V., North-Holland.

Haber, R.N. (1987): Why Low-Flying Fighter Planes Crash: Perceptual and Attentional Factors in Collisions with the Ground. Human Factors, 29(9), 519-532.

Hoh, R. (1985): Investigation of outside visual cues required for low level speed and hover. AIAA Paper 85-1808-CP, Snowmass, Col.

Johnson, W.W.; Bennett, C.T.; Tsang, P.S.; Phatak, A.V. (1987): The visual control of simulated altitude. Proceedings of the Fourth International Symposium on Aviation Psychology. Jensen, R.S. (ed).

Kellogg, R.S.; Hubbard, D.C.; Sieverding, M.J. (1989): Field-of-view variations and stripe-texturing effects on assault landing performance in the C-130 weapon system trainer. Air Force Human Resources Laboratory, Final Technical Report.

Kennedy, R.S.; Berbaum, K.S.; Collyer, S.C.; May, J.G.; Dunlap, W.P. (1988): Spatial Requirements for Visual Simulation of Aircraft at Real-World Distances. Human Factors, 30(2), 153-161.

Kleiss, J.A.; Curry, D.G.; Hubbard, D.C. (1988): Effect of three dimensional object type and density in simulated low-level flight. Proceedings of the Human Factors Society 32nd Annual Meeting, Anaheim, CA, 1299-1303.

Kleiss, J.A. (1990): Terrain visual cue analysis for simulating low-level flight: a multidimensional scaling approach. IMAGE V Conference, Phoenix, Arizona,.

Kruk, R.; Regan, D.; Beverley, K.I.; Longridge, T. (1983): Flying Performance on the Advanced Simulator for Pilot Training and Laboratory Tests of Vision. Human Factors, 25(4), 457-466.

Lintern, G.; Thomley-Yates, K.E.; Nelson, B.E.; Roscoe, S.N. (1987): Content, Variety, and Augmentation of Simulated Visual Scenes for Teaching Air-to-Ground Attack. Human Factors, 29(1), 45-59.

Lintern, G.; Sheppard, D.J.; Parker, D.L.; Yates, K.E.; Nolan, M.D. (1989): Simulator Design and Instructional Features for Air-to-Ground Attack: A Transfer Study. Human Factors, 31(1), 87-99.

McDermott, D.; Gelsey, A. (1987): Terrain Analysis for Tactical Situation Assessment. Proceedings 1987 Workshop Spatial Reasoning and Multi-Sensor Fusion, Morgan Kaufmann, Los Altos, CA.

Martin, E.L.; Rinalducci, E.J. (1983): Low-level flight simulation vertical cues. AFHRL Technical Report 83-17, Operations Training division, Air Force Human Resources Laboratory, Williams AFB.

Rosenking, J.P.; Hayslip, I.C.; Eilbert, J. (1990): Traditional and automated approaches for acquiring expert knowledge. SPIE Vol. 1923 Applications of Artificial Intelligence VIII.

Schulte, A.; Prevot, T. (1992): Visuelle Routinen von Piloten bei der Durchführung von Flugmissionen. Interner Bericht, UniBwM/LRT/WE 13/IB/92-4.

Späth, H. (1975): Cluster-Analyse-Algorithmen zur Objektklassifizierung und Datenreduktion. Oldenbourg Verlag München Wien.

DISTRIBUTION OF VISUAL ATTENTION REVEALED BY EYE MOVEMENTS AND NEURAL NETWORKS

P. Krappmann*

Interdisciplinary Graduate College, University of Göttingen,Germany

Abstract

The present study investigates the relationship between the distribution of visual attention and the information content of selected points of interest in a complex problem-solving situation by eye tracking technology and neural networks.

In a first experiment subjects were asked to direct a computer simulated factory. All information necessary for problem management was presented on a projection screen in front of the subject. Four out of twenty information items could be influenced. While the subject was deciding on the economic measures, eye movements were recorded. The results indicated that successful subjects use a more selective information gathering strategy than unsuccessful subjects.

In a second experiment the relevance of each item for making decisions was assessed. For this purpose neural networks were trained by error back-propagation. Information, presented in the problem-solving situation before, now provided the input values. The decisions of the subjects determined the four output units. After training the networks, an index for every input unit was formed. These indices, which stand for the contribution of information items on decision making, were correlated with relative fixation frequencies and fixation duration of information items. High correlation coefficients were found for the relationship between the index of items and fixation frequency.

The results suggest that the amount of attention, as measured by the number of gaze frequencies, reflects the relative importance of information items. It was also found that problem-solvers used only simple logical rules when deciding on the economic measures.

Keywords

eye movements; visual attention; information gathering; problem-solving; neural networks

Introduction

In cognitive psychology, visual search is usually conceived of as a process in which the visual information uptake does not only depend on the presented stimuli but rather on the cognitive structure of the perceiving person. The rationale behind this position of information processing theory is that subjects build up representations of the external environment which aid them by guiding attention to stimuli with relevant content of information and facilitating encoding of that information into existing knowledge to handle a given task. Identifying the relationship between information gathering strategies and cognitive structures is often accomplished by eye movement recordings (see Rayner, 1978). Eye movement studies in human information processing research have proceeded with the intention to demonstrate a correlation between eye fixation parameters and cognitive

---

* Present address: Brain Research Institute, University of Bremen, POB 330440, 28334 Bremen, Germany.

Eye Movement Research/J.M. Findlay et al. (Editors)

processing. Since the retina does not have a uniform distribution of spatial resolution, eye movements are supposed to bring regions of special interest into the region of highest spatial resolution, the fovea. So, visual search data reveals important aspects of the attentional and cognitive processes of subjects as they engage in activities such as picture viewing, reading or performing cognitive tasks. Differences in eye movement patterns between individuals engaged in these activities implies differences in attention which reflects differences in cognitive functioning (Koga & Groner, 1989).

In the case of problem-solving eye fixation measurements emerged to be a useful technique for analysing mental operations. Previous studies of eye movement behaviour during problem-solving have already shown that the individual performance level correlates with the information gathering strategy of problem-solvers (Lüer et al., 1986; Putz-Osterloh, 1979). Differences in the internal representation of the so-called problem space resulted in different individual scanning patterns. So the act of looking appears to be closely related to the internal information processing.

Direct evidence that the subject's cognitive structure influences the perception of information has been indicated by various experiments (e.g. Antes, 1974; Mackworth & Morandi, 1967). In these studies the relevance of stimuli for performing a given task has been assessed by subjective judgement. The assessed degree of informativeness of a stimulus has relied on judgements of experts or subjects after performing the task. However, the achieved ratings do not account for the cognitive structure of the perceiving subject at the moment of information gathering. Also methods like "thinking aloud" are not suitable since they influence the structure of cognitive processes (Lass et al., 1991). So, if we want to define the informativeness of a stimulus for a subject in an unbiased way, the cognitive structure of the perceiving subject has to be taken into account at the moment of information gathering. But there is no study which has analysed the relationship between the information gathering strategy and the subjective informativeness of the parts of a scene on the basis of the above mentioned principle.

The present study was performed to accomplish this analysis. In the first experiment it was hypothesised that differences in the cognitive structures of successful and unsuccessful problem-solvers would be reflected by differences in their information gathering strategies. This assumption was tested by examining eye movement behaviour of persons dealing with a complex problem-solving situation. In the second experiment it was hypothesised that the visual search pattern would depend on the subjective relevance of stimuli for performing a given task. In order to verify this, the degree of the behavioural significance of the stimuli was analysed on the basis of the individual decision behaviour.

Experiment I

In the first experiment eye movements were recorded while subjects controlled a given computer-simulated system as problem situation. Changes of eye movement behaviour in a complex problem-solving situation have already been reported by Lüer and co-workers (1985, 1986). The results indicated that in the first trials the information gathering strategy of successful subjects could be discriminated from unsuccessful with respect to fixation frequency.

Characterisation of scanning patterns of successes and failures has also been made during observational concept learning (Itoh & Fujita, 1982). Correlation analysis of the sequential change of fixation points revealed a less restricted movement in the group of successful performers. Unsuccessful subjects tended to fixate more often on unimportant points at the cost of free movement.

The purpose of this experiment was to find out whether differences in the representation of problem space of successful and unsuccessful problem-solvers reflected in differences in their eye movement behaviour. The problem-solving situation in which subjects have to achieve a solution for an unsolved problem was realized by a computer simulation program.

Methods

Task
  Subjects were asked to play the role of a manager, who had to direct an electric bulb
producing factory. The factory and the economic situation were simulated by a modified
version of the computer program *OPEX* consisting of 20 economic measures. The given goal
was to increase the factory's capital resources. By choosing values for the four so-called
manipulation items the factory's profit could be influenced.

Information Presentation
  The presented economic measures consisted of manipulation items, feedback items,
internal and external items. Four manipulation items could be changed directly by the subject,
like the selling price. Four feedback items contained information about the success of the
manipulations, like the profit. Eight internal items contained information about the state of the
factory, like the raw material inventory. And four external items included information about
the economic environment, like the level of prices. All subjects started with the same values
of the variables as shown in figure 1.

selling price 6.40	quantity of production 400,000	raw material requirement 650,000	marketing expenditure 300,000
raw material inventory 1,002,296	finished products inventory 3,834	debts due 6,553,115	capital resources 3,638,430
profit -14,570	competitors' profit -14,570	seasonal trend 90 %	seasonal forecasting 95 %
sales volume 447,170	saleable volume 447,170	market share 25 %	capacity per manshift 424,379
unit cost price 2.98	level of prices 6.40	grossprofit on sales 2,861,888	capital expenditure 2,876,458

*Figure 1: Starting values of the variables which represent the simulated electric bulb factor
and the economic situation.*

  The items were projected by a LCD panel in a 5 x 4 matrix format onto a screen in front of
the subject. On the screen the matrix subtended a visual angle of about 30 deg horizontally
and 24 deg vertically. The distance between adjacent items was 5 degree of visual angle.

Procedure
  At the start of the experiment subjects were instructed with detailed knowledge about the
economic measures and the simulated situation. After the instruction and the calibration of the
subject's eye movements the experimental session began. The session consisted of five
subsequent trials.
  In each trial the subject had to decide on the four manipulation items. Once the subject had
given his last decision orally to the experimenter, the recording was stopped and the
experimenter entered the decisions into the computer. The resulting changes of economic
measures were visible for the subject in the subsequent trial.

Statistical Distribution of Fixation Duration

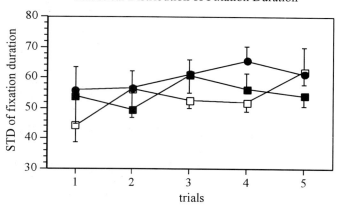

Statistical Distribution of Fixation Frequency

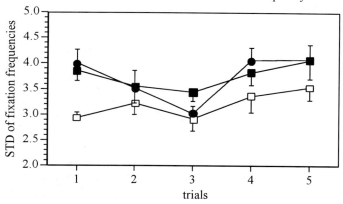

Figure 2: To examine the hypothesis that very successful (■), moderately successful (●) and unsuccessful (□) problem-solver differ in their distribution of visual attention, the mean standard deviation of fixation frequency and fixation duration (mean ± standard error of mean) was calculated for each group. The standard deviation is an index of redundancy.

Subjects

33 undergraduate students of economics and psychology participated as subjects in this experiment. After the performance of the experiment all subjects were divided into three groups according to the achieved capital resources: there was an unsuccessful group (n = 9), a moderately successful group (n = 12) and a very successful group (n = 12). Successful and unsuccessful problem-solvers were discriminated with respect to the achieved capital resource after five trials.

Apparatus and eye fixation parameters

Eye movement recordings were performed during the whole decision process using a DEBIC 84 system. The position of the centre of eye fixation was calculated every 20 ms. The

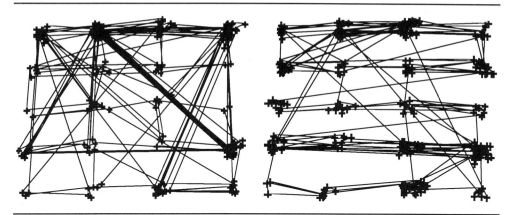

Figure 3: Scanpath of a successful (left) and an unsuccessful problem solver (right) viewing the presented matrix consisting of 20 variables. The presented eye movement patterns were recorded during the first trial.

fixation of an item termed *item fixation* was defined as a sequence of subsequent eye fixations within the region of the item and a minimum duration of 70 ms. The item region was defined as a circle with a radius of 2 deg around the midpoint of the cell.

To examine the hypothesis that successful and unsuccessful problem-solvers differ in their distribution of attention, the standard deviation of item fixation frequency and fixation duration was calculated for each subject as an index of redundancy (Attneave, 1965). The idea was that, if all items were viewed equally often and for the same length of time, the standard deviation should be zero or almost zero. The larger the standard deviation was, the stronger some items were regarded and others disregarded.

The distribution of item fixations of successful and unsuccessful subjects was compared by means of the standard deviation of relative item fixation frequency and duration. Differences in the redundancy index for successful and unsuccessful subjects as well as in all subsequent statistical tests were tested using the analysis of variance (ANOVA). The Scheffe F-test was used to isolate the source of any effect significant at an alpha level of .05. In figure 2 the mean of redundancy and the standard error of mean is presented.

Results and Discussion

Figure 2 shows that the distribution of fixation duration is very similar in all the groups. Groups do not differ remarkably in any of the five trials. The distribution of item fixations, however, showed more redundancy in the groups of successful problem-solvers. This difference is particularly high on the first trial ($F_{2,30} = 6.138$; $p < .01$). That means that the successful problem-solvers concentrate on some items whereas in the group of unsuccessful problem-solvers the item fixations are much more equally distributed. The fixation patterns of moderately and very successful problem-solver differ significantly from fixation patterns of unsuccessful subjects. Differences in overall statistical distribution of fixations between successful (suc.) and unsuccessful (unsuc.) problem-solver emerge in manipulation items (frequency$_{suc.}$ > frequency$_{unsuc.}$), feedback items (frequency$_{suc.}$ < frequency$_{unsuc.}$) and external items (frequency$_{suc.}$ < frequency$_{unsuc.}$).

Typical eye movement patterns of an unsuccessful and a successful problem-solver are given in figure 3. As the figure shows the pattern of an unsuccessful subject is unstructured.

However, the scanpath of a successful problem-solver shows many different relationships between manipulation and information items. For example, there are many saccades between the item "quantity of production" and the item "capacity per manshift". Both measures have a direct economically relevant relationship. This example indicates that the differences of eye movement patterns reflect the individual differences of cognitive functioning.

A further indication for the existence of a direct relationship between eye movement patterns and decision behaviour is given in table 1. After carrying out the experiment the decision behaviour of the subjects was analysed it became obvious that unsuccessful subjects had not bought enough raw material for the electric bulb production. Thus unsuccessful problem-solver were not able to produce as much as they had decided before and, therefore, they lost a lot of money. As the amount of raw material was essential for the decision making on the item "raw material requirement" the poor attention paid to the item "raw material inventory" as revealed by fixation frequency seems to mirror this inattention and the wrong decision behaviour of unsuccessful subjects ($F_{2,30} = 4.99$; $p < .02$), respectively. The relative importance of information for decision making obviously influenced the number of item fixation frequencies.

group	decision behaviour			distribution of visual attention	
	decided quantity of production	realized quantity of production		relat. fixation frequency	fixation duration
unsuccessful	m = 631,010 sem = 49,346	m = 514,590 sem = 38,555	t = 2.98 p < .02	m = 6.24 sem = 0.783	m = 346.37 sem = 43.7
moderately successful	m = 697,680 sem = 28,945	m = 672,530 sem = 21,089	t = 1.80 p > .05	m = 10.22 sem = 0.566	m = 361.72 sem = 39.1
very successful	m = 617,720 sem = 25,379	m = 614,400 sem = 26,107	t = 1.48 p > .1	m = 8.93 sem = 1.108	m = 384.07 sem = 38.9
				$F_{2,30} = 4.99$ p < .02	$F_{2,30} = 1.44$ p > .2

Table 1: Intragroup comparison between decided and realized quantity of production (mean ± standard error of mean) and intergroup comparison between the parameters of visual attention paid to the item "raw material inventory" (mean ± standard error of mean).

The results demonstrate that the groups of successful problem-solvers distinguished between the presented items much more than the group of unsuccessful problem-solvers, did. In both groups of successful problem-solvers some items were fixated very often, while other items were hardly noticed. The individual knowledge of relations between visual objects seems to be reflected by the scanpaths of the subjects.

Experiment II

Eye fixations are not distributed randomly over a picture, but rather are allocated to a relatively small portion of the scene (Yarbus, 1967). One idea of cognitive psychology is, that the gaze is attracted to informative areas of the scene. The informativeness of a stimulus is usually defined in terms of ratings by independent observers or experts. However, the informativeness depends on characteristics of the perceiving subject.

The purpose of the second experiment was, therefore, to analyse the informativeness of the presented items by determining their behavioural significance for decision making. To define the degree of informativeness on the basis of the individual decision behaviour without scaling the relative importance of items by subjective expert rating, neural networks were trained.

In a two-layer network, with input and output units, simple linear relations can be modelled (Minsky & Papert, 1969). Non-linear functions can only be represented by a network with at least one hidden layer (Hinton, 1989). Both architectures were used here to simulate the performance of a neural network on the same problem to that used with humans.

Methods

Values of information and decision items recorded in the first experiment now served as training patterns for the networks. Since there were four decision items, four network architectures were constructed. The input layer of the networks was composed of 23 units: representing the 20 information items and 3 out of 4 decision items (fig. 4). Each network was trained to respond to a single decision dimension by presenting all information values of the matrix as well as the values of the remaining decisions measures as input values. The procedure of network learning was performed for each experimental group separately.

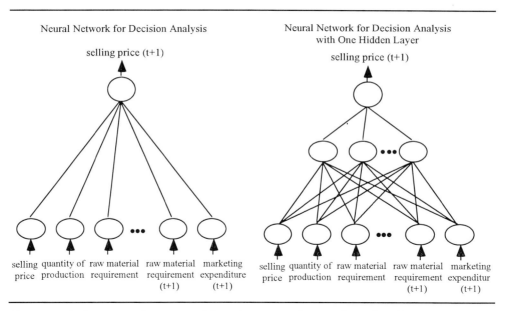

Figure 4: Architecture of a two- and three-layer neural network used in the simulations. 23 input units each send connections to the output unit or to the 12 hidden units.

The networks were trained by a standard back-propagation algorithm to modify the strength of relationship between two units as a function of the discrepancy between the observed and desired activation levels of the output units as described elsewhere (Rumelhart et al., 1986). The learning rate was 0.1 and the momentum coefficient was 0.9.

After training the networks, an index for every input unit was formed by summing the absolute value of all weights leaving the input unit. These indices, which stand for the contribution of items on decision making, were correlated with relative item fixation frequency and item fixation duration. The degree of relationship is indicated by the product moment correlation coefficient.

Results and Discussion

As shown in table 2 high correlation coefficients were found for the relationship between the index of an item and its relative fixation frequency. The density of visual fixations on presented items corresponds to the contribution of items on decision making. Behaviourally significant details within visual scenes were selected more often than irrelevant stimuli. No such correlation was found for the relationship between the index of an item and its relative fixation duration. The relevance of a stimulus for performing a given task has no influence on the fixation duration of this stimulus.

training	unsuccessful		mod. successful		very successful	
cycles	frequency	duration	frequency	duration	frequency	duration
	2-layer neural network					
25	0.5547 *	0.1360	0.3760	0.2334	0.5244 *	0.0616
50	0.5329 *	0.1943	0.5082	0.3664	0.6942 **	0.3235
100	0.4329	0.2726	0.5120	0.3535	0.7239 **	0.4115
200	0.3613	0.3934	0.5695 *	0.3039	0.7351 **	0.4072
500	0.2844	0.4383	0.5479 *	0.2277	0.8015 **	0.3785
1000	0.1963	0.3765	0.4905	0.1647	0.7940 **	0.3917
	3-layer neural network					
100	0.5326 *	-0.0491	0.1804	0.0451	0.4762	-0.1550
200	0.4809	0.1389	0.5089	0.3100	0.7873 **	-0.0398
500	0.4178	0.4297	0.5905 *	0.2146	0.8070 **	0.2895
1000	0.3770	0.3169	0.5601 *	0.1357	0.8084 **	0.3034

* $p < 0.01$　　** $p < 0.001$

Table 2: Correlation coefficients between relative item fixation parameters and indexes of items. These indexes stand for the contribution of items on decision making in a neural network simulation.

However, a comparison of the coefficients also shows that if enough training cycles were accomplished the correlation indexes are higher in the groups of successful problem-solvers. This difference is in close accordance with different degrees of redundancy of item fixations in the groups of successful and unsuccessful subjects. Since the weights are very similar at the beginning of the network training and the absolute value of all weights leaving the input units change insignificantly after a small number of training cycles, high correlation coefficients were found between these weights and the relative equally distributed frequencies of item fixations in the group of unsuccessful subjects. With increasing training cycles, however, the weights of input items differ more and more. Since the density of item fixation frequency varied much more in the groups of successful problem-solver, the correlation coefficients increase as a function of training sequence length in these groups. This effect of training sequence length on the degree of correlation coefficient seems to reflect differences in the internal representation of the so-called problem space.

Figure 5 shows the error rates for two-layer network's performance on the training patterns of all three groups. The congruence between the observed and desired activation levels of the output units increased with the number of training cycles. But the limes of error rates is much better after training a network with the patterns of the successful groups. One can conclude from this that the decision behaviour of successful problem-solvers is much more homogeneous.

The similarity of results yielded with two- and three-layer networks suggests that all subjects realized only simple linear relations between information and decisions as it was found that simple two-layer networks learned the correct output for the learning pattern in a similar way to three-layer networks. In a two-layer network, with an input- and an output-layer, only simple logical relations can be modelled. More complex problems involving non-linear functions can be represented by a network with at least one hidden layer. But there is no evidence that problem-solver make use of such rules. Although back-propagation is unfeasible neurophysiologically the algorithm is an interesting tool for studying the mechanism of cognitive processes as shown here.

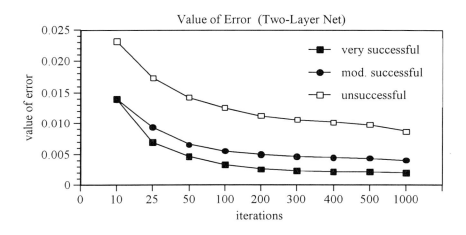

Figure 5: Error rates decreased with the number of learning cycles; in the figure the averaged error rates are demonstrated for the two-layer network.

General discussion

Research on the cognitive processes involved in visual tasks profits from measuring eye movements of subjects while they are engaged in these tasks. For this reason the eye movement behaviour of persons dealing with a complex problem-solving situation was examined. The hypothesis was that differences in the cognitive structures of successful and unsuccessful problem-solvers would be reflected by differences in their information gathering strategies. In the present study it was directly shown that the cognitive structure of a perceiving person influences the informativeness of the parts of a scene and, therefore, the acquisition of information from a stimulus pattern

As it can be seen from figure 2, the groups of successful problem-solvers differentiated between the presented items much more than the group of unsuccessful problem-solvers did. Differences in the density of item fixations mirror the individual differences of cognitive functioning. Successful problem-solvers concentrate on some items whereas in the groups of unsuccessful problem-solvers the fixation frequency per item is much more equally distributed. In contrast to item fixation frequency the distribution of item fixation duration is very similar in both groups. These results and interpretations are in close agreement with very different experiments (e.g. Bard & Fleury, 1981; Kundel & La Follette, 1972; Witruk, 1982).

An example of the close relationship between eye movement patterns and cognitive functioning is also given by a study of Koga and Groner (1989). The authors demonstrated a change of cognitive functioning by typical eye movement patterns before and after learning characters. In their study non-Japanese subjects were asked to learn the relationship between characters and meanings. In the pre-learning phase, the fixations were not organized. After the learning phase, the scanpaths changed dramatically: the fixations were concentrated on the more informative parts of the character. That means, that the change of scanpaths represented a change of cognitive functioning.

This relationship between eye movement behaviour and cognitive functioning can also be seen in the presented experiments. In figure 3 the eye movement pattern of an unsuccessful problem-solver is demonstrated. As it can be seen, the scanpath is unstructured. Just as little as this scanpath of the unsuccessful problem-solver shows many different relationships between manipulation and information items, novices are able to select task-relevant information (Dee-Lucas & Larkin, 1988). In both mentioned cases the information uptake is not controlled by a adequate internal representation of the problem space. The visual attention of successful subjects like that of experts is directed to selected parts of a scene. Those parts are most relevant for performing a given task. In the present experiment, for example, there are many saccades between related items. Here the difference of eye movement patterns reflects the difference of cognitive functioning, too.

In addition, neural networks are a useful tool for studying visual attention and complex problem-solving. By comparing the performance of humans and networks it was found that problem-solver used simple logical rules. There was no evidence for using complex relations. Similar results are known from investigations about the ability of humans to solve a variety of pattern classification problems (Thorpe et al., 1989). The comparison between the performance of a neural network and that of human subjects showed that humans do not make use of back-propagation.

The results of experiment II also provide strong evidence that the relative importance of information, as measured by neural network representation of information-decision relations, influences the distribution of visual attention, as measured by the number of eye fixation frequencies. Comparable findings were reported by Antes and co-workers (1985). They examined attentional process of subjects while the subjects were engaged in map reading. In this task the distribution of eye fixations was apparently to map regions providing most information for successful performance.

From the reported results it can be concluded that the cognitive structure of a perceiving subject influences the distribution of eye fixation frequency. Since the density of eye fixations on presented items corresponds to the contribution of items on decision making it can be supposed that the informativeness of visual patterns depends importantly on the cognitive structure of the perceiving subject. Indications are also found that subjects realized only simple linear relations between information and decisions.

Acknowledgements

This study was supported by the Volkswagen Stiftung and the Land Niedersachsen. The author is grateful to G. Lüer for initiating and supporting the project and to J. Biethahn for placing the program OPEX at his disposal. The author would also like to thank J.M. Findlay for his helpful comments on the manuscript.

References

Antes, J.R. (1974). The time course of picture viewing. Journal of Experimental Psychology 103, 62-70.

Antes, J.R., Chang, K.-T. & Lenzen, T. (1985). Eye movements in map reading. In R. Groner, G.W. McConkie & C. Menz (eds.), Eye movements and human information processing. Amsterdam: North-Holland.

Attneave, F. (1965). Informationstheorie in der Psychologie. Bern: Huber.

Bard, C. & Fleury, M. (1981). Considering eye movement as a predictor of attainment. In I.M. Cockerill & W.W. MacGillivary (eds.), Vision and sport. Cheltenham: Thornes.

Dee-Lucas, D., Just, M.A., Carpenter, P.A. & Daneman, M. (1982). What eye fixations tell us about the time course of text integration. In R. Groner & P. Fraisse (eds.), Cognition and eye movements. Amsterdam: North-Holland.

Hinton, G. E. (1989). Connectionst learning procedures. AI 40, 185-234.

Itoh, H. & Fujita, K. (1982). An analysis of eye movements during observational concept learning: characterization of individual scanning patterns of succeses and failures. In R. Groner & P. Fraisse (eds.), Cognition and eye movements. Amsterdam: North-Holland.

Koga, K. & Groner, R. (1989). Intercultural experiments as a research tool in the study of cognitive skill acquisition: Japanese character rcognition and eye movements in non-Japanese subjects. In H. Mandl & J.R. Levin (eds.), Knowledge acquisition from text and pictures. Amsterdam: North-Holland.

Kundel, H.L. & La Follette, P.S. (1972). Visual search patterns and experience with radiological images. Radiology 103, 523-528.

Lass, U., Klettke, W., Lüer, G. & Ruhlender, P. (1991). Does thinking aloud influence the structure of cognitive processes? In R. Schmid & D. Zambarbieri (eds.), Oculomotor control and cognitive processes: normal and pathological aspects. Amsterdam: North-Holland.

Lüer, G., Hübner, R. & Lass, U. (1985). Sequences of eye-movements in a problem solving situation. In R. Groner, G.W. McConkie & C. Menz (eds.), Eye movements and human information processing. Amsterdam: North-Holland.

Lüer, G., Lass, U., Ulrich, M. & Schroiff, H.W. (1986). Changes in eye movement behavior in complex problem solving. In F. Klix & H. Hagendorf (eds.), Human memory and cognitive capabilities. Amsterdam: North-Holland.

Mackworth, N.H. & Morandi, A. (1967). The gaze selects informative details within pictures. Perception and Psychophysics 2, 547-552.

Minsky, M. & Papert, S. (1969). Perceptrons. Cambridge: MIT Press.

Putz-Osterloh, W. (1979). Blickbewegungsparameter als Indikatoren für die Güte der Informationsverarbeitung bei Raven-Aufgaben. In H. Ueckert & D. Rhenius (eds.), Komplexe menschliche Informationsverarbeitung. Bern: Huber.

Rayner, K. (1978). Eye movements in reading and information processing. Psychological Bulletin 3, 618-660.

Rumelhart, D.E., Hinton, G.E. & Williams, R.J. (1986). Learning internal representations by error propagation. In D.E. Rumelhart & J.L. McClelland (eds.), Parallel distributed processing. Explorations in the microstructure of cognition. Cambridge: MIT Press.

Thorpe, S.J., O'Reagan, K. & Pouget, A. (1989). Humans fail on XOR pattern classification problems. In L. Personnaz & G. Dreyfus (eds.), Neural networks: from models to applications. Paris: I.D.S.E.T.

Witruk, E. (1982). Eye movements as a process indicator of interindividual differences in cognitive information processing. In R. Groner & P. Fraisse (eds.), Cognition and eye movements. Amsterdam: North-Holland.

Yarbus, A.L. (1967). Eye movements and vision. New York: Plenum.

GAZE CONTROL IN BASKETBALL FOUL SHOOTING

J. N. Vickers
Neuromotor Psychology Laboratory
University of Calgary, Calgary, Alberta. Canada.

Abstract

Gaze of 8 expert (E mean 75%) and 8 near-expert (NE mean 42%) basketball shooters was assessed as they performed 10 accurate and 10 inaccurate foul shots to a regulation basket wearing an eye movement helmet. The E and NE shooters did not differ in the temporal phasing of the movement but did in mean frequency, duration, location and onset of gaze. Six E shooters fixated the hoop and were significantly different from all others in that they maintained fixation (mean 1789 ms) from early in the preparation, through the preshot and into the initial 103 ms of the shot phase. During the remainder of the shot phase (mean 385 ms) they appeared to suppress the intake of new visual information by not initiating a fixation at this time, by letting their gaze shift with their body movement and/or by blinking. This style contrasted with the NE who fixated the hoop for shorter durations and higher frequencies but maintained fixation on the hoop longer into the shot phase suggesting simultaneous planning and control of the action. Four subjects fixated the backboard and initiated a fixation late in the preparation for a duration under 1 s. In the discussion, a sequence of gaze behaviors is presented that appears to underlie skill development and accuracy in foul shooting.

Keywords

gaze, eye movements, aiming, vision, basketball.

Introduction

In basketball foul shooting, a player is provided a "free" shot as a result of a foul or infraction by the other team. The shooter stands behind the foul line, which is located 15 feet (4.6 m) from the basket and shoots a ball 9.7 in (24.7 cm) in diameter into a hoop 18 in (45.7 cm) in diameter. The hoop is located directly in front of the shooter, at a height of 10 ft (3.0 m) from the floor. The rules of basketball state a player must initiate the foul shot within five seconds of being handed the ball by the official. The biomechanical techniques used in foul shooting vary, with players developing techniques that include dribbles of the ball, spins of the ball, full jump, modified jump or set shot biomechanics (Knudson, 1993; Wooden, 1988). Typically a basketball player orients his or her eyes toward the basket at some time during the preparation of the shot and attempts to maintain a steady gaze prior to and during the shooting action. Expert shooters orient their gaze toward the hoop sooner and for longer periods of time than novices (Ripoll, Bard, & Paillard, 1986), however no research has determined the location of gaze during the phases of the shooting action, or the role gaze control plays in either skill development or accuracy.

Eye Movement Research/J.M. Findlay et al. (Editors)

Gaze control in foul shooting

Schmid & Zambarbieri (1991, p. 229) define gaze as "the absolute position of the eyes in space and depends on both eye position in orbit and head position in space." According to their model, sensory information is received from the environment, with a voluntary generating of commands that direct the gaze according to the goals and intentions of the subject. At the lower levels, the head, body, and oculomotor control systems act as one closely coupled system (Zangemeister & Stark, 1982; Guitton & Volle, 1987). In basketball foul shooting, performers develop a fluid shooting motion, one in which the eyes, head, arms and body work together as one coordinated unit. When a subject orients his or her gaze to a target, such as the hoop in basketball, the most common movement of the eyes is a sequence of events in which the eyes move prior to the head (Zangemeister & Stark, 1982; Guitton & Volle, 1987; Gauthier et al, 1991; Schmid & Zambarbieri, 1991). Gauthier et al (1991) explain that the head follows because of its greater inertia, with the eyes localizing the target first and visual discrimination beginning immediately and maintained steadily on the target even though the head is moving. The movement of eyes and head to the target is normally smooth, with the processing of information occurring as soon as the eyes stabilize on the target. Gaze control in basketball foul shooting is therefore defined as the manner is which a performer moves the head and eyes to take in available information while preparing and executing the shot. Gaze control in the study reported here was specified by the percentage of trials in which a gaze was used, as well as the mean frequency and duration of gaze across the temporal phases of the movement.

Research in eye movements and gaze control currently supports two views of the role of visual search in physical performance. Abernethy & Russell (1987), Abernethy (1990), Goulet, Bard & Fleury (1989), Helsen & Pauwels (1990) have found that when expert and novice athletes watch video representations of sport problems and are asked to select a best answer from available choices, their eye movements are not an indicator of the decisions made. Abernethy (1990, p. 63) has concluded that differences between expert and novice performers is not due to "an inappropriate search strategy but rather an inability to make full use of the information available from fixated display features." This conclusion is not supported by research where the visual behavior of subjects is monitored as they perform physical skills (Ripoll, Papin, Guezennec, Verdy, & Philip, 1985; Ripoll, Bard & Paillard, 1986; Vickers, 1992; Helsen & Pauwels, 1993). In pistol shooting, Ripoll et al (1985) found that expert and near-experts shooters differ in how they sight and track the gun to the target. In basketball shooting, Ripoll et al (1986) found that E shooters orient their gaze to the basket earlier and maintain head stability for longer durations than novices. In golf putting, Vickers (1992) found that low handicap golfers[1] hold their gaze fixated on the ball through the backswing, foreswing, and contact phases of the stroke, a behavior that contributed to greater accuracy. Lesser skilled golfers did not hold their gaze fixated, but shifted their gaze at intervals under one second taking in available information at a set rate

---

[1] A low handicap signifies a higher level of play.

irrespective of its form or potential function. It would appear that in research where subjects are required to physically perform skills, significant differences . in gaze have been found between skilled and lesser skilled performers. These differences may be centered in the multiple demands of controlling not only the aiming system but also the gaze and accompanying input of environmental information.

In the current study expert (E) and near-expert (NE) athletes performed foul shots while their gaze was monitored simultaneously. An expert (E) was defined as one who made over 75% of their foul shots over a full season of play. A near-expert (NE) competed in the same environment but was unable to shoot above 60%. Within this highly skilled group of athletes it has remained a mystery why some excel in foul shooting, while others struggle given similar years of practice, training and competition. Reasons solely attributable to physical ability are difficult to rationalize since athletes at the upper levels of basketball tend to be very similar in cardiac and aerobic function (Hartley, 1992; Astrand, 1992), anaerobic power (McArdle, Katch & Katch, 1991), muscular strength and power (Komi, 1992), and body composition (Titel & Wutscherk, 1992; Wilmore & Costill, 1988). Because of the equivalence of these factors, it is important to look for other contributing reasons, such as the control of gaze while shooting. The E foul shooters were expected to use significantly fewer gaze of longer mean duration, while the NE would shift their gaze at higher frequencies and lower durations, irrespective of type of gaze behavior, location of gaze, or phase of the movement. The E shooters were also expected to hold their gaze fixated on a critical target location across two or more phases of the shooting movement, a form of gaze independence in which the eyes remain fixated on a critical location while the aiming movements are made.

Method

Subjects

Sixteen highly skilled female basketball players were invited to be the subjects. Thirteen played for the women's intercollegiate team that finished second in Canada in 1991, with five subjects playing at the national and Olympic level. Eight were classified as E shooters, based on their foul shooting percentages over a season of regular and playoff competition of 75%; eight were classified as NE shooters, based on their mean percentage of 42%. Two of the NE subjects were national/Olympic team members, illustrating the paradox of foul shooting where it is not unusual to find accomplished players having great difficulty with this shot which is taken unguarded with a clear view of the basket. The groups did not differ in age, E mean 20.8 years and NE 21.1 years, nor did they differ in the number of years they had played competitive basketball, E mean 9.9 years and NE 9.8 years.[2]

---

[2] Years in competition was defined as the number of years the subject had been on a basketball team that practiced a minimum of three times per week.

Figure 1. A split frame of video data showing (left) the scene from the external camera of the subject shooting and (right) from the scene camera on the EVM helmet. Phases durations of the shot were derived from the left and the subject's gaze from the right. The black cursor indicated the location of the shooter's gaze every 33.3 msec.

Collecting gaze and movement phase data

Gaze data was collected using the mobile helmet based ASL 3100H Eye View monitor shown in Figure 1. The ASL 3100H is a monocular corneal reflection system that measures eye-line-of-gaze with respect to the helmet. The helmet has a 30 metre cord attached to the waist, interfaced to the main computer, thus permitting the subject near normal mobility. Miniaturized optics, an illuminator, solid state sensor, relay lens, and visor are mounted on the helmet, with a total weight 700 g. The eye is illuminated by a near infrared light source, beamed coaxially with the optical axis of a solid state camera (pupil camera). The result is a backlighted bright, rather than dark, pupil image. Both paths (illuminator and camera) are reflected from the visor, which is coated to be transparent in the visible spectrum but reflective to the near infrared. The system measures eye line of gaze using a miniature scene camera mounted on the helmet, thus the measurable field of view is virtually unlimited. In order to avoid parallax, the scene camera video taped the reflection from the outside of the mirror, thus providing a view that is nearly aligned with the view from the eye being monitored. A cursor is superimposed on the scene camera to show precise point of gaze. The measurement update rate for video is 30 hz. The system has an accuracy of ±1 degree visual angle, and precision of 1 degree.

In order to determine the phases of the shooting movement simultaneous with the subject's gaze, an external video camera was synchronised with the eye monitor, thus creating the split frame of video data shown in Figure 1. The left portion of the frame shows a subject performing the foul shot and the right side shows the shooter's gaze indicated by the black cursor. Location of gaze and phase of the movement were therefore monitored simultaneously every 33.3ms. Superimposing the external image of the shooter was achieved using a Panasonic Special Effects Generator, Model WJ 4600a. During data collection, the video data as shown in Figure 1 was on-line, thus permitting the constant monitoring of the video for accuracy and quality of scenes and eye data.

Data was collected in the gymnasium at a regulation basket where the subjects played in competition. Each subject took approximately 20 warm-up shots, was fitted with the helmet and calibrated. The subjects then took consecutive shots until they made 10 hits and 10 misses, a research goal they were unaware of. All subjects recorded their 10th hit very quickly, but a considerable number of trials were needed before the 10th miss was recorded. The greatest number of shots taken by a subject before their 10th miss was 108, the least 23. Practice, calibration, and data collection took an average of 40 minutes.

Video coding procedures

The video data was coded in an editing suite equipped with frame by frame shuttle control and counter. Ten misses were coded and 10 hits randomly selected but sequentially coded for each subject, resulting in a total data set of 320 shots (160 hits, 160 misses). Count and duration of gaze was coded by phase of the shot, gaze behavior, gaze location, and accuracy (hit or miss). Coding reliability followed procedures developed earlier (Vickers, 1988; Vickers, 1992) in which randomly selected trials were recoded by a second independent coder. Code-recode reliability's for phase was r = 1.0; gaze behavior, r = .97, and gaze location, r = .90.

Movement phases

Four phases of the foul shot were derived from the sagittal view of the subjects provided by the external camera as shown in Figure 1 (left). Duration of phase (msec) was determined from a count of video frames by 33.3 ms. The preparation phase was a constant 60 frames or 1998 msec prior to the ball moving into the shooting motion. The preshot phase began with the frame showing the initial drop of the ball, a downward movement put on the ball prior to the upward shooting action. The shot phase began with the frame showing the first upward motion of the ball and until the ball was observed to leave the fingertips of the player. The flight phase began with the frame showing the ball leaving the players fingertips and recorded gaze while the ball travelled to the hoop. The flight phase ended when the ball contacted with the hoop, hit the backboard or the subject ceased to view.

Gaze location

Gaze location was defined as the subject's gaze to a location within the targeting scene. Six gaze locations were coded: 1) the ball, 2) hands, 3) floor, 4) the hoop, 5) the backboard and 6) out of range. Out of range was coded when the subject's gaze moved outside of the backboard or the visor of the helmet.

Gaze behavior

In defining the gaze behavior of the subjects, no attempt was made to separate eye from head and body motion. Instead, a gaze behavior was defined as a fixated gaze on a location in the targeting environment or a shift in gaze from one environmental location to another. Since a shift in gaze is normally initiated by eye movements (Zangemeister & Stark, 1982; Guitton & Volle, 1987; Schmid & Zambarbieri, 1991; Gauthier et al, 1991; Ron & Berthoz, 1991), four gaze behaviors were defined using minimum duration parameters derived from the eye movement literature for fixations, saccades, tracking movements and blinks (Optican, 1985; Fischer & Weber, 1993; Carl & Gellman, 1987; Millodot, 1986). Minimum gaze duration of a fixation has varied from 80 to 150 msec in the literature, with the lower durations found in situations where subjects performed highly practiced tasks (Optican, 1985; Carl & Gellman, 1987). Since the subjects in this experiment were highly skilled, minimum fixation duration was set at 99.9 msec. (3 or more frames) and defined as the stabilization of the gaze on a location. A saccade was coded when a shift in gaze was observed from one location to another, with a minimum duration of 133.2 msec. (4 frames). A tracking or smooth pursuit eye movement was coded when the subject's gaze tracked a moving object for a minimum of 3 frames or 99.9 msec. A blink was coded when the gaze cursor disappeared due to the subject's eye lid dosing and occluding the optics on the system for a minimum duration of 3 frames or 99.9 msec. Finally, a shift in gaze due to the body movement was coded when the gaze was observed to move due to the shooter's flexing and extending their knees during the shot. During this time the gaze was unlike that observed during a fixation, saccade, tracking or blink. This last category permitted an assessment of eye and head stability during the shooting action.

Results

Experimental accuracy

The E shooters made 10 hits in fewer attempts (E mean 12 shots, SD 1.7) than the NE (mean 14 shots, SD 2.2). They were also more successful in prolonging their 10th miss (mean 55 shots, SD 25) than the NE (mean 44 shots, SD 17). Experimental shooting percentage was 79% for the E shooters and 74% for the NE. The E shooters percentages were quite similar to their competitive levels while the NE shooters performed at a higher level. Possible reasons for the better performance of the NE shooters may have been the repetitive nature of the shooting action in the experiment and the absence of game pressure. Overall, the E shooters were more accurate given the available opportunities.

Mean duration of the shot and phases of the shot

Repeated measures ANOVA of skill and accuracy found no significant differences in mean duration of the foul shot. The E shooters averaged 3597 msec per shot and the NE 3396 msec; mean duration of hits and misses was also similar, 3483 msec for hits and 3386 msec for misses. There were also no significant differences in the duration of the phases of the shot due to skill level or accuracy. Figure 2 (bars) shows the E shooters taking 1712 msec in the preparation phase and the NE 1533 msec. Mean duration of the preshot was 391 msec for the E and 402 for the NE; mean duration of the shot phase was E 488 msec and NE mean 454 msec. By combining preshot and shot durations, mean movement time (MT) for the E shooters was 879 msec and for the NE 855 msec. Flight time of the ball to the hoop was near identical, E mean 1006 msec and NE mean 1008 msec. These results show that the E and NE groups were quite similar in the temporal organization of the foul shooting action.

Mean frequency of gaze during the shot and phases of the shot

The groups were also similar in the frequency of gaze during the shot (all phases combined). The E shooters averaged 7.3 gaze per shot and the NE 7.4; mean frequency of gaze on hits and misses was also similar at 7.1. However, analysis of frequency of gaze by phase of the shot (hits and misses combined) found significant differences due to phase (F (3, 468) = 43.56, p <.0001) and skill by phase (F (3, 468) = 2.85, p <.04) as shown in Figure 2 (points). Mean frequency of gaze of the E shooters was lower during the preparation and preshot phases but higher during the shot and flight phases. Post hoc comparison of simple means showed the E shooters initiated significantly fewer gaze in the preparation phase (E mean 2.5 and NE mean 3.3). The lower frequency of gaze of the E shooters was expected in the preparation and preshot phases, however the higher frequency in the critical shot phase was not expected. As the shot was delivered the NE shooters initiated 1.0 gaze while the E initiated 1.5. In summary, we see the temporal phasing of the movement to be very similar for both E and NE shooters, however the frequency of gaze within phases of the movements differed significantly.

Gaze behaviors while shooting.

Figure 3 presents the percent of trials in which a gaze to a location was initiated by the E and NE shooters.[3] Within group differences were minimal on both hits and misses so these were combined. Four gaze behaviors were observed in the preparation phase[4] (fixation on the hoop, fixation on the backboard, gaze shift due to body movement, blink), one in the preshot (fixation on the hoop, fixation on the backboard), two in the shot phase (blink, gaze shift due to body movement) and three during the flight (fixation on hoop, blink and tracking the ball).

---

[3] Gaze behaviors used in a very low percentage of trials (< 5% of trials by both groups) were omitted.

[4] Figure 3 Shows the phase in which a gaze behavior was initiated, and does not reflect the gaze extending across phases of the shot.

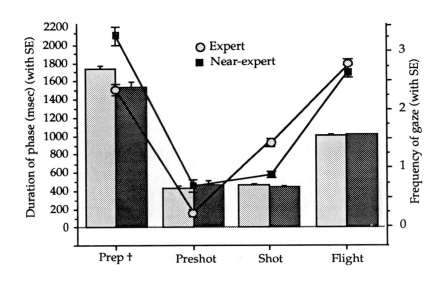

Figure 2. Mean duration of phases of the foul shot (bars) and mean frequency of gaze (points) for E and NE shooters. The interaction of skill by phase was significant for frequency, †means sig different for frequency.

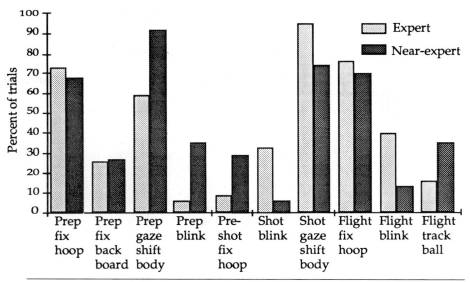

Figure 3. Percentage of trials in which a gaze was initiated by E and NE shooters during the preparation, preshot, shot and flight phases (hits and misses combined).

During the preparation phase, the E shooters fixated the hoop in 73% of trials and the NE in 68%. Fixations on the backboard were also very similar for both groups, the E in 25% of trials and NE in 26%. In the preshot phase, the NE initiated a fixation on the hoop in 27% of trials, while the E shooters did this in only 9% of shots. Also apparent in Figure 3 is the difference in head and gaze stability of the two groups. During the preparation phase, the E shooter's shifted their gaze due to body movement in 59% of trials and the NE shooters in 92%. However these percentages were reversed in the shot phase, the E shooters shifted their gaze as a function of body movement in 95% of trials and the NE in 73%.

Figure 3 is also informative in indicating what the foul shooters did not do. In the all phases (except flight) the shooters did not fixate or track the ball or their hands. They also did not saccade from the ball or their hands to the hoop, nor did they saccade from one location to another on the target. During the shot phase, the subjects rarely initiated a fixation on the target.

Unexpected was the high percentage of trials in which the shooters blinked, the E in 78% of shots (all phases combined) and the NE in 54%. Other studies (Ripoll, Papin, Guezennec, Verdy, & Philip, 1985; Ripoll, Bard & Paillard, 1986; Helsen & Pauwels, 1992; Vickers, 1992) have not reported blinking during aiming tasks. Finally, during the flight phase the shooters fixated the hoop and the NE shooters tracked the ball on 35% trials as compared to 16% for the E shooters. Tracking the ball after release of the shot is a practice discouraged by experts of the game.

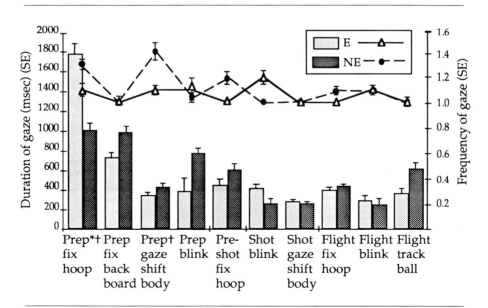

*Figure 4. Mean duration (bars) and frequency of gaze (points) of E and NE shooters during the preparation, preshot, shot and flight phases of the foul shot (hits and misses combined) * sig duration; † sig frequency.*

Mean frequency and duration of gaze. For each of the gaze behaviors shown in Figure 3, a 2 x 2 x 10 (skill x accuracy x trials) repeated measures ANOVA was carried out in order to determine the duration and frequency of gaze.[5] Figure 4 presents the mean duration (msec) of each gaze behavior on the left Y axes while on the right Y axis is the mean frequency of gaze (with SE). In the preparation phase, a significant difference was found due to skill level for fixation on the hoop for both duration (F (1,145) = 8.33, p <.008) and frequency of fixation (F (1,145) = 12.28, p <.002). The E shooters held their gaze fixated on the hoop for a significantly longer duration, E mean 1779 msec as compared to NE mean 1018 msec; frequency of fixation on the hoop also differed significantly, the E shooters averaged 1.1 gaze per fixation on the hoop as compared to 1.3 for the NE. A second difference found also in the preparation phase, was the E shooters moving their gaze less as a result of body movement than the NE (F (1, 174) = 9.60, p <.004). The E shooters averaged 1.1 shifts in gaze as compared to 1.4 for the NE.

Individual differences in fixation on the target

The analysis to this point has concentrated on group differences as opposed to individual differences. Analysis of each subject was therefore carried out in order to determine if there were any subjects who's gaze behavior were not consistent with the E or NE group in which they had been classified. The ANOVA was restricted to analysis of fixation frequency and duration on the hoop and on the backboard. It was found that a shooter rarely fixated both the hoop and the backboard in a trial or across trials but instead showed a clear preference for one location or the other, with the hoop being the preferred location. Six E and six NE subjects fixated the hoop, while two E and two NE fixated the backboard. These subjects are reflected in the percentages shown in Figure 3 and mean fixation durations found in Figure 4.

A second feature of gaze control determined for each individual was the time of fixation onset on the hoop or backboard (fixation 1 and fixation 2). Of interest was whether a shooter held the gaze fixated through to and including the shot phase. Mean onset of fixation on the hoop for the six E shooters was found to be 427 msec into the preparation phase, as shown in Figure 5 (with SE). Mean duration of fixation on the hoop for these subjects was earlier determined to be 1779 msec (Figures 4). Therefore mean fixation on the hoop extended through the preparation, preshot and into the initial 103 msec. of the shot phase. Fixation on the hoop was not held during the final 385 msec of the shooting action.

Onset of fixation for the six NE subjects who fixated the hoop occurred 513 msec into the preparation phase. Mean duration was 1081 msec (Figure 4),

---

[5] Analysis was also carried out using non-parametric techniques (Mann-Whitney U). In addition to the two significant results shown in Figure 4, the E and NE differed significantly in fixation duration on the backboard and duration of blink.

therefore a second fixation (f2) was usually needed, which was initiated either in the latter part of the preparation or during the preshot (onset of f2 is estimated in Figure 5). The NE then maintained fixation on the hoop longer into the shot phase than the E subjects.

Onset of fixation for the two E shooters who fixated the backboard was later in the preparation phase at 1450 msec. Mean duration of this fixation was 738 msec. and extended into the initial part of the shot phase. Mean fixation onset for the two NE subjects who fixated the backboard was 1214 msec with mean duration being 991 msec., thus extending to the midpoint of the shot phase.

In summary, foul shooters have a preference for fixating either the hoop or the backboard, with 12 of the 16 subjects preferring the hoop. The six E shooters

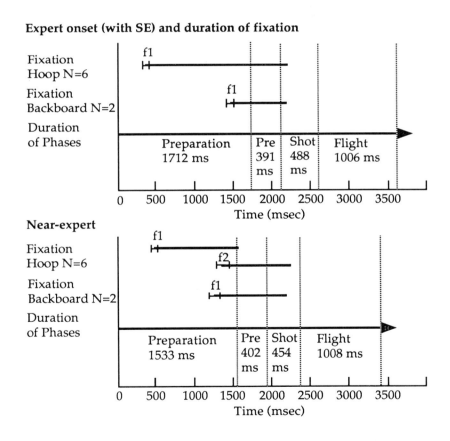

Figure 5. Mean onset (with SE) and duration of fixation (msec) on the hoop or backboard of E and NE shooters across phases of the foul shot. N indicates the number of subjects who fixated the hoop or backboard.

who fixated the hoop were significantly different from all others in that they were able to maintain a long fixation on the hoop from early in the preparation, through the preshot and into the initial 103 msec of the shot phase. From this point forward (for 385 msec) they appeared to suppress the intake of any new visual information by not initiating a fixation at this time, by letting their gaze shift with their body movement and/or by blinking. It would appear that maintaining visual control on the target was not a priority of the E shooters as the shot was executed. The six NE shooters who fixated the hoop were distinct in not only recording a higher frequency and lower duration of fixation on the hoop but also of maintaining fixation longer into the shot phase. The four subjects (2 E and 2 NE) who fixated the backboard initiated fixation on the backboard later in the preparation phase and for a shorter duration (under 1 sec).

Discussion

Eye movement and gaze control research in sport and physical activity has evolved to where a debate exists over two issues. There is disagreement about whether the visual search of performers differs due to expertise. There is also a debate whether the visual behavior of a performer is a critical factor in superior performance. Researchers using observer protocols (Abernethy, 1987; Goulet et al, 1989; Abernethy, 1990) have concluded that the visual search of athletes does not differ as a function of expertise. Using methods where expert and novice athletes view sport problems they have found that the intake of information is similar for all skill levels. Differences in performance are therefore not attributable to visual pick-up but in how information is processed internally. Alternatively, researchers using protocols where the gaze of the subjects is recorded while they perform skills have found gaze control differences between skilled and lesser skilled athletes (Ripoll et al, 1985; Helsen & Pauwels, 1992; Vickers, 1992). The research reported here falls into the latter category. Significant differences were found between E and NE foul shooters in mean frequency, duration and location of gaze.

Three distinctly different types of gaze control were found in foul shooting based on the location and duration of fixation on the target. The six E shooters who fixated the hoop were significantly different from all others in that they held their gaze for a long duration on the hoop (mean of 1779 msec) thus this fixation extended through the preparation, preshot and to the initial 103 msec of the shot phase. During the time when the hands executed the upward shooting action on the ball (for 385 msec) there appeared to be a suppression of vision accomplished through not fixating at this time, letting the gaze move with the action of the body and/or blinking. This style contrasted with the six NE subjects who fixated the hoop for shorter durations and higher frequencies but maintained fixation on the hoop longer into the shot phase. The remaining four subjects fixated the backboard and were all very late in the onset of the fixation which was of low duration (under one second). Interestingly, three of these subjects played the centre position, therefore their late initiation of fixation may have been due to the conditions under which they must shoot in the field, usually under extreme pressure and most often using the backboard. Also all of these subjects had competitive game percentages that were very similar, range 59 - 65% therefore they were right at the margin used to distinguish E from NE shooters in this study.

It was interesting what the subjects did not do. In all phases of the shot (except the flight), they did not fixate or track their hands or the ball. This was not due to the shooting action being so quick as to prevent fixating or tracking behaviors on the hands as the aiming movement was performed. Mean duration of the preshot and shot phases (combined) was 879 msec for the E and 855 msec for the NE (Figure 2), therefore there was enough time to monitor and make corrections to the hands as the shot was delivered (Zelaznik et al, 1983). Visual monitoring of the hands and ball was not a characteristic of these subjects.

A surprising result was the high percentage of trials in which the shooters blinked - the E in 77% of shots and the NE in 54% (all phases combined). Two reasons may account for this. During the preparation phase, the gaze was fixated for a mean duration of one to two seconds. Since we normally blink three to four times per second (Millodot, 1986) a blink may have been used to rest the eyes. However, in a number of other aiming studies blinking has not been reported (Ripoll et al, 1986; Vickers, 1992; Adolph & Vickers, in progress). A second, more intriguing reason may be unique to skills where potentially interfering objects or limbs enter the visual field of the performer. During the foul shot, as the hands and ball enter the visual field, the shooters may have learned to use the blink to prevent interference as the shot is executed. Volkmann, Riggs & Moore (1981) have shown that during the blink vision is suppressed and McCormick & Thompson (1984) have shown that the blinking reflex can be trained. Therefore it is possible that in basketball, shooters have developed a blink in order to preserve the original aiming commands to the hands.

Overall the results show that although the E and NE shooters temporally sequenced the movement in similar ways, the six E athletes used a different type of gaze control than all others. In complex spatial aiming skills to far targets, the sequence of gaze control is such that the target is fixated early and for a long duration (almost 2 seconds). The action is ultimately performed without maintaining fixation on the target. These results are supportive of a motor program control of the action, an open loop mode (Schmidt, 1975, 1991) where the action is programmed in advance and run off without interference. The results for the E and NE shooters were also very similar to those found for E and NE golfers where a fixation held on a the ball during the backswing, foreswing and contact was found to contribute to greater accuracy (Vickers. 1992). The ability to hold the gaze fixated on a critical location for a long duration early in the move may be a possible perceptual invariant underlying greater accuracy. Keeping the gaze fixated may act as a stabilizer in visual space as the movement is organized, an invisible line linked to the target and essential in ultimate accuracy. An open loop motor control perspective would argue that during this time the shooter sets the parameters for the shot such as the location of the hoop, distance to the hoop, trajectory on the ball, timing and coordination of the limbs and so on. (Schmidt, 1991). Is there any additional evidence that a fixated gaze may be a factor in setting and maintaining the aiming commands?

Gauthier, Nommay and Vercher (1990) have shown that pointing accuracy to spatial targets is affected by afferent (or inflow) signals arising from the movement of the eyes in orbit. A suction device was used to create an artificial strabismus in one eye (i.e. a pulling on the muscles of the eye). The deviated eye was then covered and subjects were asked to manually point to targets in space. All misjudged the location of the target, with pointing localization being consistently off in the direction of the deviated eye, an average of 6 degrees. Subjects were unaware of their error. Gauthier et al (1990) concluded that proprioceptive afferent inflow arising from the deviated eye was a factor in target localization by the hand. We can relate these results to the current study as follows. In foul shooting, if the eyes are active during the preparation and preshot phases as we see with the NE shooters, there may be a greater chance of inflow, thus increasing the potential of the targeting commands either not being set optimally or maintained as the action is prepared and initiated. Alternatively, if the gaze is held steady as we see with the six E shooters during the preparation and preshot and initial part of the shot phase, then this may permit a more optimal programming and maintenance of the aiming commands to the hands.

In conclusion, three gaze control characteristics appeared to underlie skill development and accuracy in foul shooting. A fixation of long duration was established on the hoop early in the preparation of the shot. The duration of this fixation for six of eight E shooters was exceptionally long (almost twice that of all others) and so extended through the preshot and to the initial part of the shot phase. Secondly, the E shooters exhibited gaze independence or the ability to perform the aiming movements while the gaze was fixated on the target. Third, as the shot was executed there appeared to be a suppression of the intake of any new visual information. This appeared to be accomplished in three ways: by not initiating a fixation during the preshot and shot phases, by letting the gaze move freely with the shooting action and/or by blinking. As the final aiming movements were executed, the highly skilled foul shooter appeared to be functioning with two independent systems, an aiming system that was shooting the ball and a gaze system that was functionally at rest.

References

Abernethy, B. (1990). Expertise, visual search, and information pick-up in squash. Perception. 19, 63-77.

Abernethy, B. & D. Russell (1987). Expert - novice differences in an applied selective attention task. Journal of Sport Psychology, 9, 326-345.

Adolph, R. & Vickers, J. N. (under review). Gaze control in the volleyball serve pass.

Astrand, F-O (1992). Endurance in other sports. In Ed. R. Shepard & P-O Astrand. Endurance in Sport. 616-617. Oxford. Blackwell Scientific.

Carl, J. & Gellman, R. (1987). Human smooth pursuit: Stimulus-dependent responses. Journal of Neurophysiology 57:1446-1463.

Fischer, B. & H. Weber (1993). Express saccades and visual attention. Behavioral & Brain Sciences. 16, 3, 553-610.

Gauthier, M., D. Nommay, & J. L. Vercher. (1990). The role of ocular muscle proprioception in visual localization of targets. Science. 249, 58-61.

Gauthier, G. M., J. L. Semmlow, J. L. Vercher, C. Pedrono, & G. Obrecht (1991). Adaptation of eye and head movements to reduced peripheral vision. In Eds. R. Schmid & D. Zambarbieri. Oculomotor Control and Cognitive Processes. North Holland, Elsevier.

Goulet, C., M. Bard & M. Fleury (1989). Expertise differences in preparing to return a tennis serve: A visual information processing approach. J. Sport Psychology. 11, 382-398.

Guitton D. & M. Volle (1987). Gaze control in humans: eye-head coordination during orienting movements to targets within and beyond the oculomotor range. Journal of Neurophysiology. 58, 427-459.

Hartley, H. (1992). Cardiac function and endurance. In Ed. R. Shepard & P-O Astrand. Endurance in Sport. 72-80. Oxford. Blackwell Scientific.

Helsen W. & J. M. Pauwels (1990). Analysis of visual search in solving tactical game problems. In Ed. D. Brogan. Visual Search. London. Taylor Francis.

Helsen W. & J. M. Pauwels (1993). A cognitive approach to visual search in sport. In Ed. D. Brogan, A. Gale & K. Carr. Visual Search 2. London. Taylor Francis.

Komi, P. V. (1992). Strength and Power in Sport. Oxford: Blackwell Scientific. Knudson, D. (1993). Biomechanics of the basketball jump shot. Journal of Physical Education, Recreation & Dance. February, 67-73. AAHIPERD.

McArdle, W. D., F. Katch, & V. Katch (1991). Exercise Physiology: Energy, Nutrition and Human Performance. Philadelphia. Lea & Febiger.

McCormick, D. A. & Thompson, R. F. (1984). Response of the rabbit cerebellum during acquisition and performance of a classically conditioned nicitating membrane/eyelid response. Journal of Neuroscience 4, 2811- 2822.

Millodot, M. (1986). Dictionary of Optometry. London. Butterworths.

Optican, L. M. (1985). Adaptive properties of the saccadic system. In Eds. A. Berthoz & M. Melville Jones. Adaptive Mechanisms in Gaze Control: Facts and Theories. New York: Elsevier Science.

Ron S. & A. Berthoz (1991). Coupled and disassociated modes of eye-head coordination in humans to flashed visual target. In (Eds). R. Schmid & D. Zambarbieri. Oculomotor Control and Cognitive Processes. Elsevier Science Publishers, North Holland.

Ripoll, H., C. Bard & J. Paillard (1986). Stabilization of the head and eyes on target as a factor in successful basketball shooting. Human Movement Science. 5, 47-58.

Ripoll, H., J. Papin, J. Guezennec & J. Verdy (1985). Analysis of visual scanning patterns of pistol shooters. Journal of Sport Sciences. 3, 99-101.

Schmid, R. & D. Zambarbieri (1991). Strategies of eye-head coordination. In Eds. Schmid, R. & D. Zambarbieri. Oculomotor Control and Cognitive Processes. North Holland, Elsevier.

Titel, K. & Wutscherk, H. (1992). Anatomical and anthropometric fundamentals of endurance. In Eds. Shepard, R & P-O Astrand. 35-46. Endurance in Sport. Oxford: Blackwell .

Schmidt, R. A (1975). The schema theory of discrete motor skill learning. Psychological Review. 82, 225-260.

Schmidt, R. A. (1991). Human Motor Learning & Performance. Champaign, IL: Vickers, J. N. (1992). Gaze control in putting. Perception. 21. 117-132. Vickers, J. N. (1988). Knowledge structures of expert-novice gymnasts. Human Movement Science 7, 47-72.

Volkmann, F., L. Riggs & R. Moore (1981). Eyeblinks and visual suppression. Science. 207, 900-902.

Wilmore, J. H. & D. L. Costill (1988). Training for Sport and Activity: The Physiological Basis of the Conditioning Process. 3rd ed. Dubuque. Iowa, Wm. C. Brown.

Wooden, J. R. (1988). Practical Modern Basketball. New York. Wiley & Sons. Zangemeister, W. & Stark, L. (1982). Gaze latency: Variable interaction of head and eye latency. Experimental Neurology. 75, 389-406.

Zelaznik, H., B. Hawkins & L. Kisselburgh (1983). Rapid visual feedback processing in single-aiming movements. Journal of Motor Behavior, 15, 217-236.

Acknowledgements

I would like to thank the athletes who volunteered to be the subjects in this study and their coach Donna Rudakis of the University of Calgary. My thanks also to Sandee Greatrex, Sue Forbes, Brian Gaines, Claire Michaels, John Findlay and two anonymous reviewers for their assistance and helpful advice in completing this study.

AUTHOR INDEX

SUBJECT INDEX